WILLIAM F. MAAG LIBRARY
YOUNGSTOWN STATE UNIVERSITY

SEMICONDUCTORS AND SEMIMETALS
VOLUME 35
Nanostructured Systems

SEMICONDUCTORS AND SEMIMETALS

Volume 35

Nanostructured Systems

Volume Editor
MARK REED
YALE UNIVERSITY
NEW HAVEN, CONNECTICUT

ACADEMIC PRESS, INC.
Harcourt Brace Jovanovich, Publishers

Boston   San Diego   New York
London   Sydney   Tokyo   Toronto

This book is printed on acid-free paper. ∞

COPYRIGHT © 1992 BY ACADEMIC PRESS, INC.
ALL RIGHTS RESERVED.
NO PART OF THIS PUBLICATION MAY BE REPRODUCED OR
TRANSMITTED IN ANY FORM OR BY ANY MEANS, ELECTRONIC
OR MECHANICAL, INCLUDING PHOTOCOPY, RECORDING, OR
ANY INFORMATION STORAGE AND RETRIEVAL SYSTEM, WITHOUT
PERMISSION IN WRITING FROM THE PUBLISHER.

ACADEMIC PRESS, INC.
1250 Sixth Avenue, San Diego, CA 92101

United Kingdom Edition published by
ACADEMIC PRESS LIMITED
24-28 Oval Road, London NW1 7DX

The Library of Congress has catalogued this serial title as follows:

Semiconductors and semimetals.—Vol. 1—New York: Academic Press, 1966–

   v.: ill.; 24 cm.

Irregular.
Each vol. has also a distinctive title.
Edited by R. K. Willardson, Albert C. Beer, and Eicke R. Weber.
ISSN 0080-8784 = Semiconductors and semimetals

1. Semiconductors—Collected works. 2. Semimetals—Collected works.
I. Willardson, Robert K. II. Beer, Albert C. III. Weber, Eicke R.
QC610.9.S48   621.3815'2—dc19   85-642319
                                             AACR2   MARC-S

Library of Congress   [8709]
ISBN 0-12-752135-6 (v 35)

PRINTED IN THE UNITED STATES OF AMERICA
92 93 94 95   9 8 7 6 5 4 3 2 1

Contents

CONTRIBUTORS	vii
PREFACE	ix

Chapter 1 Introduction   1
Mark Reed

References	6

Chapter 2 Quantum Point Contacts   9
H. van Houten, C. W. J. Beenakker, and B. J. van Wees

I.	Introduction	9
II.	Split-Gate Quantum Point Contacts	13
III.	Ballistic Quantum Transport	17
IV.	Adiabatic Transport in the Quantum Hall Effect Regime	81
	Acknowledgments	105
	References	106

Chapter 3 When Does a Wire Become an Electron Waveguide?   113
G. Timp

I.	Introduction	113
II.	Two-Terminal Resistance of an Electron Waveguide	117
III.	Three-Terminal Resistance	136
IV.	Four-Terminal Resistance	143
V.	Summary and Outlook	182
	Acknowledgments	185
	References	185

Chapter 4 The Quantum Hall Effect in Open Conductors 191

M. Büttiker

I.	Introduction	192
II.	Basic Elements of Electrical Conduction	195
III.	Quantization and Interference in the Absence of a Field	208
IV.	Motion in High Magnetic Fields: The Two-Terminal Resistance	221
V.	The Quantum Hall Effect in Open Conductors	238
VI.	Resonant Departures from the Hall Resistance	257
VII.	Discussion	271
	Acknowledgment	272
	References	273

Chapter 5 Electrons in Laterally Periodic Nanostructures 279

W. Hansen, J. P. Kotthaus, and U. Merkt

I.	Introduction	279
II.	Realization of Laterally Periodic Nanostructures	286
III.	Transport in Periodic Nanostructures	295
IV.	Far-Infrared Spectroscopy of Electronic Excitations	334
V.	Conclusions and Perspectives	372
	Acknowledgments	373
	References	374

INDEX . . . 381
CONTENTS OF VOLUMES IN THIS SERIES . . . 389

Contributors

Numbers in parentheses indicate the pages on which the authors' contributions begin.

C. W. J. BEENAKKER (9) *Philips Research Laboratories, 5600 JA Eindhoven, The Netherlands*

M. BÜTTIKER (191) *IBM Research Division, Thomas J. Watson Research Center, Yorktown Heights, New York 10598*

W. HANSEN (279), *Sektion Physik, Universität München, Geschwister-Scholl-Platz 1, 8000 München 22, Federal Republic of Germany*

J. P. KOTTHAUS (279), *Sektion Physik, Universität München, Geschwister-Scholl-Platz 1, 8000 München 22, Federal Republic of Germany*

U. MERKT (279) *Institut für Angewandte Physik, Universität Hamburg, Jungiusstr. 11, 2000 Hamburg 36, Federal Republic of Germany*

MARK A. REED (1) *Yale University, Department of Electrical Engineering, P.O. Box 2157, Yale Station, New Haven, Connecticut 06520-2157*

G. TIMP (113) *AT&T Bell Laboratories, Holmdel, New Jersey 07733*

H. VAN HOUTEN (9) *Philips Research Laboratories, 5600 JA Eindhoven, The Netherlands*

B. J. VAN WEES (9), *Department of Applied Physics, Delft University of Technology, 2600 GA Delft, The Netherlands*

Preface

This volume presents a selection of timely reviews in the emergent field of nanostructured systems. This collection is aimed at consolidating some of the more prominent subjects in this rapidly growing area for the expert as well as the novice. Each of these chapters is written by the pioneering researchers in their fields, who generously gave their time and effort for outstanding contributions.

The advent of ultra-thin epitaxial film growth techniques has ushered in the era of reduced-dimensionality condensed matter physics. Within the last decade, it has become routine to realize, investigate, and utilize two-dimensional electron, or hole, systems (2DEGs/2DHGs). This is an exceedingly rich field, since the ability to vary composition, band offset, periodicity, and other variables results in nearly limitless possibilities in creating structures for physical exploration as well as for electron and optical device applications. We have reached a stage where it is possible to observe and control large-scale quantum effects in a variety of materials and states of condensed matter; yet because quantum wells (or, more properly, *quantum planes*) and heterojunction interfaces inherently are two-dimensional, investigations generally have concentrated on quantum confinement in the epitaxial direction.

Within the last few years, advances in microfabrication technology have allowed laboratories around the world to impose additional lateral dimensions of quantum confinement on two-dimensional systems with length scales approaching those of epitaxial lengths in the growth direction. The achievements of quantum wires and quantum dots have demonstrated that electronic systems with different dimensionalities are now available to the experimentalist. The turning point in the understanding of nanometer-scale electronic transport was the development of reliable semiconductor fabrication techniques on the nanometer scale. With these techniques now common, the quantum limit of electronic transport can be explored.

Ballistic electronic structures had been done in the epitaxial dimension, but not in the lateral dimensions due to the small dimensional scales involved. The use of electron beam lithography and the incorporation of this dimensional scale onto high-mobility 2DEGs allowed one, for the first time, to create structures that are smaller than other relevant length scales. The recent discovery of quantized conductance in ballistic *quantum point contacts* has opened up a new era of nanostructure physics where one can fabricate nanostructures whose behavior is dominated by countably few electronic conducting channels. Chapter 2 by H. van Houten, C. W. J. Beenakker, and B. J. van Wees presents the first comprehensive review of this work to date.

In such systems, the *contacts* to the system are no longer ideal—they are now by definition part of the entire (interacting and interfering) electron wave system. Thus, electrical leads become intractably invasive. This topic has spawned an active area of current research, especially with regard to measurements in the quantum limit. The relation of measured global electronic properties and the relation to transmission coefficients, especially as applied to small electron systems, has recently been used to address a wide range of problems in nanostructures. A further extension of this formalism to observations of the quantum Hall effect in small systems is presented by M. Buttiker in Chapter 4. Transport along edge states provides a simple and elegant picture with which to explain the various magnetotransport phenomena observed in these nanostructures.

When the confinement dimensions become comparable to the Fermi wavelength in structures shorter than an electronic mean free path, electron *waveguides* surprisingly akin to their microwave counterparts (that are many orders of magnitude larger) can be created. Here, quantum transport becomes dominant and the wave nature of the electron becomes apparent. Elegant examples of the behavior of electrons on this length scale are discussed in Chapter 3 by G. Timp. The invasiveness of contacts to the system and the effects of geometry are apparent in these studies. The author also utilizes quantum point contacts as injection or detection filters, to examine the role of specific edge states in the transport.

The advantage of the gated-2DEG configuration over other confinement schemes is the ability to tune the confinement potential. An extension of the previous works is to impose not just one structure, but a periodic structure on a 2DEG, allowing one to examine the transition from a 2DEG to a 1DEG (in the case of parallel lines) or to a 0DEG (in the case of a periodic grid). Chapter 5, by W. Hansen, J. P. Kotthaus, and U. Merkt, presents a detailed study of the electronic transport and optical properties of these fascinating systems.

The field will continue to evolve from the point presented by this volume, though it is an impossible task to predict in which direction. However, the trend toward reduced-length scales compels us to understand condensed matter physics on the quantum size scale. It is hoped that this volume serves as a solid reference and building block upon which future insight into the realm of nanostructures will be gleaned.

Mark Reed
New Haven, Connecticut

CHAPTER 1

Introduction

Mark A. Reed

DEPARTMENT OF ELECTRICAL ENGINEERING
YALE UNIVERSITY
NEW HAVEN, CONNECTICUT

The ability to observe and control the solid state in dimensional extremes has been a compelling attraction to condensed matter scientists for over three decades. The demonstration in 1966 of a two-dimensional electron gas (2DEG) in a silicon inversion layer[1,2] initiated research in 2D systems that has continued unabated to date. The progress in this field increased dramatically with the arrival of atomically precise heterojunction epitaxial technology,[3] allowing the realization of heretofore *gedankenexperiment* structures (such as quantum wells) that exhibit manifest quantum effects. Today, we have reached a level of solid-state structural sophistication where we commonly design and fabricate devices based primarily on quantum mechanical phenomena.

Ultra-thin epitaxial semiconductor film growth became a reality to solid-state and device physicists in 1974 by the realization of the first quantum well[4] by Raymond Dingle and co-workers at Bell Labs, and the first resonant tunneling diode[5] by Leroy Chang and co-workers at IBM. The ability to vary composition, band offset, periodicity, and other variables in nearly limitless possibilities for physical exploration as well as for electron and optical device applications was irresistible. These structures were the physical realization of well-known examples of energy quantization and tunneling found in elementary quantum mechanics textbooks.

The explosion of interest in the following decade was tremendous; by the 1984 International Conference on the Physics of Semiconductors in San Fransisco, papers involving quantum wells, heteroepitaxy, or related subjects comprised nearly 40% of all the papers. The attraction was not just due to the pure excitement of the physics; quantum well lasers, HEMTs, and other related devices began to find their way into the commercial marketplace.

In the late 1970s, the field of artificially structured material and device research took a dramatic turn that dominated the thrust of research for nearly a decade. A technique to create an exceptionally pure 2D electron system was demonstrated by a simple trick known as *modulation doping*,[6] which spatially separates the charge carriers in a conducting channel from the impurity atoms from which they originate. This event opened up an exciting new area of research into electronic transport, parallel to the heterointerface, of ultrahigh mobility carrier systems.

By 1990, electron mobilities in GaAs/AlGaAs of $> 10^7$ cm^2/V-s had been achieved. This field was highlighted by a Nobel Prize for the *quantum Hall effect*,[7] the quantizing of sheet resistance values of 2DEGs in the high magnetic field quantum limit. Epitaxial control allowed physicists an unprecedented degree of control on experimental structure parameters, which opened up the possibilities for new, exciting discoveries (such as the fractional quantum Hall effect). In retrospect, this was just the beginning of a fascinating venture into artificially structured condensed matter systems.

During this time, a quiet renaissance was occurring in vertical electronic transport through multilayer heterostructures. Dominant nonclassical electron tunneling was demonstrated in resonant tunneling structures[8] nearly a decade after the initial seminal investigations by Chang et al.,[5] primarily due to improvements in heterojunction materials capability. The importance of this renaissance was not the specific realization of the tunneling structure, but a realization that artificially structured heterojunctions indeed could demonstrate predominantly quantum transport. It should be noted that there is a fundamental (in addition to the physically conceptual) distinction between parallel and perpendicular transport. Parallel transport utilizes only the quantum confinement of a heterointerface, whereas perpendicular transport also utilizes carrier tunneling (and perhaps artificial superlattice band formation), and thus is a direct spectroscopic probe of the electronic states. In addition, the related phenomena of ballistic transport (i.e., electronic transport in which carriers (electrons) traverse without undergoing a scattering event) in vertical hot-electron structures was reported about the same time.[9,10] These studies also demonstrated electronic transport through states that exhibited quantum reflections.

Yet as late as 1986, *low-dimensional structures* implicitly meant 2DEGs or quantum wells. At the 1986 International Conference on the Physics of Semiconductors in Stockholm, Sweden, there were less than 10 papers (out of a total of 405 papers) on fabricated nanostructures. However, these few papers framed the beginnings of a new era in semiconductor transport. One of the major motivational reasons for examining small electronic systems was the observation of *universal conductance fluctuations* (found later to be anything but universal) and weak localization effects in silicon[11] and GaAs[12] nanostructures. These works demonstrated that coherent and random quantum interference, respectively, are observable in few-electron systems, and perhaps even could be dominant in appropriate structures. However, fabrication processes and tolerances were not sufficiently well-developed to create structures that would exhibit large quantum interference or quantum-size effects.

A conceptually more pleasing investigation at the time, at least to transport physicists, was the observation of Aharonov–Bohm oscillations in small metallic ring structures.[13] These observations appeared to be the first clear observation of quantum interference of the electron wave function in a nanofabricated structure, the electronic analog of an *interferometer*. At the time, there appeared to be little connection between this esoteric effect and conductance fluctuation work, except for their embodiment in small electronic systems. This did not detract from the fact that these were rich microscopic laboratories for fun physics.

The turning point in the understanding of nanometer-scale electronic transport was the development of reliable semiconductor fabrication techniques on the nanometer scale, analogous to the renaissance of the resonant tunneling diode. An example was the demonstration of semiconductor quantum wires,[14] which allowed the fabrication of electronic systems in a totally new regime. The realization of ballistic structures had been done in the epitaxial dimension, but not in the lateral dimensions due to the small dimensional scales (tens of nanometers) involved. However, the advancement of electron-beam lithography and the incorporation of this dimensional scale onto high-mobility 2DEGs allowed one, for the first time, to create structures that dimensionally approached (and eventually surpassed) the relevant electronic-length scales (elastic, inelastic, phase-breaking, etc.). In this regime, quantum transport becomes dominant and the wave nature of the electron becomes apparent. An elegant example of the behavior of electrons on this length scale was the creation of electron *waveguides*.[15]

Microfabrication technology since has progressed tremendously in the last decade. Laboratories around the world now can controllably impose additional lateral dimensions of quantum confinement on two-dimensional systems with length scales approaching those of epitaxial lengths in the growth direction. The achievements of quantum wires[14] and quantum dots[16] have demonstrated that electronic systems with different dimensionalities now are available to the experimentalist.

How does one fabricate these structures of low dimensionality? The obvious approach is to utilize the existing technology of MBE or MOCVD to define a 2D system, and impose additional lateral confinement with nanometer lithographic techniques. State-of-the-art electron-beam lithography can define dimensions in the 15 nm regime, with pattern transfer techniques in the same dimensional regime, clearly sufficient to observe large quantum-size effects. However, the limitation comes in making the lithographic dimension the same as the confining potential dimension of the electron system.

The first technique that comes to mind to create the confining potential is anisotropic dry etching, which some call the *neolithic* approach. The basic principle is to use energetic ions either to erode or chemically react with the epitaxial material structure. By using a reactive gas species that forms volatile compounds with the material, semiconductor structures as small as 20nm have been demonstrated. Thus, hard wall potentials can be formed by etching either partially or completely through the epitaxial structure. However, there are two serious drawbacks to this approach. First, a serious side effect of dry etching is damage to the semiconductor by the energetic ions. The extent of the damage is poorly understood, and has shown large process variability. Second, the resulting free surfaces (in the compound semiconductor systems) have both Fermi level pinning and a large concentration of non-radiative recombination sites. Thus, both optical and electronic transport studies are problematic. Only in special arrangements has this technique yielded conclusive results. Thus, other avenues of fabrication have been explored.

A technique that bypasses the need for exposing critical surfaces is to define gates on the epitaxial structure that confine the underlying 2D system, often a two-dimensional electron gas (2DEG). By applying a negative potential, the underlying 2DEG is depleted underneath the gates and confined to the region between the gates. This approach results in a smooth electrostatic confinement that has the advantage of tunability, unlike the cutting technique described previously. It also has the advantage that nanometer-size imperfections in the definition of these structures are not transferred to the confining potential, since the screening length is significantly larger than these imperfections. However, this also means that the size of the potential is large (>100nm) and relatively shallow (compared to heterojunction confinement), resulting in quantum states spaced by ~ 1 meV; yet even with these limitations, this technique has proven to be sufficiently successful to demonstrate quantum confinement.

Exploration of the simplest configuration utilizing this technique, a pointlike construction of a 2DEG, has shown fascinating results.[17,18] Simply, it is found that the conductance of the point contacts will be quantized, in units of e^2/h per spin, in the limit of a few ballistic electron channels. This is experimental verification of the work of Rolf Landauer nearly 20 years ago, and is the subject of the chapters by van Houten *et al.* and Timp, pioneers in the field of quantum transport of mesoscopic systems. Such systems provide a fascinating laboratory for exploring the limit of electronic transport, exploring the ballistic transport regime, and demonstrating non-local quantum interference. Here, one can analyze transport in a countable number of conducting channels, leading to a fundamental understanding of electron transport in condensed matter.

Additionally, it is found in these systems that the contacts to such a system are non-trivial, intimately affecting the measurement.[19] If structures are created where the interference of electron waves is a dominant effect, how is the system measured? In such systems, the *contacts* to the system no longer are ideal—they now are by definition part of the entire (interacting and interfering) electron wave system. Thus, electrical leads become intractably invasive. This is addressed in the chapter by Büttiker, a recognized expert in the theoretical description of mesoscopic systems.

Generalizing the technique of electrostatic confinement to 1D quantum wire structures, or 0D quantum dot structures, is natural. The limit of this extension is to create periodic structures, with the hope of creating artificial bandstructure imposed by the lateral potentials. An extensive review is given by Hansen, Kotthaus, and Merkt, leaders in the field of periodic nanostructures. These structures eventually may lead to the creation of lateral superlattices of essentially arbitrary structure determined by the experimentalist.

These reviews only give a snapshot into the well-understood phenomena of quantum transport at this time. However, there are many unexplored paths that have yet to be examined, primarily due to an insufficient ability to structure the electronic systems to the required degree of control.

On the horizon are a number of enticing physical systems to be explored. At present, the degree of quantum confinement is relatively weak and uncontrolled. Confinement presently is achieved by *soft* potentials that give rise to subband splittings of only a few meV. The field awaits the technological achievement of a confining heterojunction technology that allows low-dimensional confinement equal to that presently done only in the epitaxial dimension. Once a lateral heterojunction technology is achieved, subband spacings of tens, or perhaps hundreds, of meV will be realized, with resulting quantum transport that is dominant and perhaps technologically useful. This degree of control, which has been achieved in the epitaxial dimension, may give rise to a host of new, promising electron devices that could find use in high-speed and novel transfer characteristics applications. Additionally, the viewpoint of this author and this series of reviews concentrates on electron device technology, though one should be cognizant of similar advances in low-dimensional optoelectronic devices that may benefit from the same advances.

An exciting contender for advances in this regime is the achievement of low-dimensional structures by *in situ* epitaxial growth. There exist a number of proposals for the realization of such structures, though none have been convincingly achieved to date. These proposals promise the realization of 3D heterojunction confinement, an important prerequisite for useful low-dimensional electron devices. Likewise, the spectrum of material systems that have been explored is relatively sparse, and is driven by those systems in which there are technological applications. Other promising material systems, too numerous to list here, exist on the horizon. Materials technology has continued to be the enabler for the exciting physics one sees today; it seems reasonable that this perspective will not change.

Within the last few years, it has been realized that charge quantization as well as energy quantization may well be an important consideration in low-dimensional structures. In fact, structures have been fabricated where the presence or absence of a single electron dominantly affects the transport.[20,21]

This effect has been termed *Coulomb blockade*, and manifests itself when the capacitance of the structure is so small that the charging energy of a single electron is the dominant energy in the system. This effect has led to the suggestion that single-electron devices can be realized; indeed, an electron *turnstile* has been realized.[22] It is clear that such effects will become dominant for electron devices on the nanoscale.

The last decade has been an exciting one for semiconductor physicists. The technological advancements have opened up a new era of mesoscopic physics where one can fabricate nanostructures whose behavior is dominated by quantum interference effects. This new capability has enthused the experimentalist and theorist alike, an excitement akin to the advent of quantum well technology, with limitless possibilities for physical exploration and device technology on the nanoscale. This book presents some of those key areas that have changed the way we look at physics in the nanoscale regime.

REFERENCES

1. A. B. Fowler, F. F. Fang, W. E. Howard, and P. J. Stiles, *Phys. Rev. Lett.* **16**, 901 (1966); A. B. Fowler, F. F. Fang, W. E. Howard, and P. J. Stiles, *J. Phys. Soc. Japan* **21**, Suppl. 331 (1966).
2. An exhaustive review of 2DEGs can be found by T. Ando, A. B. Fowler, and F. Stern, *Rev. Mod. Phys.* **54**, 437 (1982).
3. A. Y. Cho and J. R. Arthur, *Prog. Solid State Chem.* **10**, 157 (1975).
4. R. Dingle, A. C. Gossard, and W. Wiegmann, *Phys. Rev. Lett.* **33**, 827 (1974).
5. L. L. Chang, L. Esaki, and R. Tsu, *Appl. Phys. Lett.* **24**, 593 (1974).
6. R. Dingle, H. L. Stormer, A. C. Gossard, and W. Weigmann, *Appl. Phys. Lett.* **33**, 665 (1978).
7. K. von Klitzing, G. Dorda, and M. Pepper, *Phys. Rev. Lett.* **45**, 494 (1980).
8. T. C. L. G. Sollner, W. D. Goodhue, P. E. Tannenwald, C. D. Parker, and D. D. Peck, *Appl. Phys. Lett.* **43**, 588 (1983).
9. M. Heiblum, M. I. Nathan, D. C. Thomas, and C. M. Knoedler, *Phys. Rev. Lett.* **55**, 2200 (1985).
10. A. F. J. Levi, J. R. Hayes, P. M. Platzman, and W. Weigmann, *Phys. Rev. Lett.* **55**, 2071 (1985).
11. W. J. Skocpol, P. M. Mankiewich, R. E. Howard, L. D. Jackel, and D. M. Tennant, in *Proc. 18th International Conference on the Physics of Semiconductors* (O. Engstrom, ed.), World Scientific Publishing Co., Singapore p. 1491 (1987).
12. T. J. Thornton, M. Pepper, G. J. Davies, and D. Andrews, in *Proc. 18th International Conference on the Physics of Semiconductors* (O. Engstrom, ed.), World Scientific Publishing Co., Singapore p. 1503 (1987).
13. R. A. Webb, S. Washburn, C. P. Umbach, and R. B. Laibowitz, *Phys. Rev. Lett.* **54**, 2696 (1985).
14. H. van Houten, B. J. van Wees, M. G. J. Heijman, and J. P. Andre, *Appl. Phys. Lett.* **49**, 1781 (1986); and T. J. Thornton, M. Pepper, H. Ahmed, D. Andrews, and G. J. Davies, *Phys. Rev. Lett.* **56**, 1198 (1986).
15. G. Timp, A. M. Chang, P. Mankiewich, R. Behringer, J. E. Cunningham, T. Y. Chang, and R. E. Howard, *Phys. Rev. Lett.* **59**, 732 (1987).
16. M. A. Reed, J. N. Randall, R. J. Aggarwal, R. J. Matyi, T. M. Moore, and A. E. Wetsel, *Phys. Rev. Lett.* **60**, 535 (1988).
17. B. J. van Wees, H. van Houten, C. W. J. Beenakker, J. G. Williamson, L. P. Kouwenhoven, D. van der Marel, and C. T. Foxon, *Phys. Rev. Lett.* **60**, 848 (1988).

18. D. A. Wharam, T. J. Thornton, R. Newbury, M. Pepper, H. Ahmed, J. E. F. Frost, D. G. Hasko, D. C. Peacock, D. A. Ritchie, and G. A. C. Jones, *J. Phys. C* **21**, L209 (1988).
19. M. Büttiker, *Phys. Rev. Lett.* **57**, 1761 (1986).
20. T. A. Fulton and G. J. Dolan, *Phys. Rev. Lett.* **59**, 807 (1987).
21. L. S. Kuzmin, P. Delsing, T. Claeson, and K. K. Likharev, *Phys. Rev. Lett.* **62**, 2539 (1989).
22. L. J. Geerligs, V. F. Anderegg, P. A. M. Holweg, J. E. Mooij, H. Pothier, D. Esteve, C. Urbina, and M. H. Devoret, *Phys. Rev. Lett.* **64**, 2691 (1990).

CHAPTER 2

Quantum Point Contacts

H. van Houten and C. W. J. Beenakker

PHILIPS RESEARCH LABORATORIES
EINDHOVEN, THE NETHERLANDS

B. J. van Wees

DEPARTMENT OF APPLIED PHYSICS
DELFT UNIVERSITY OF TECHNOLOGY
DELFT, THE NETHERLANDS

I. INTRODUCTION 9
II. SPLIT-GATE QUANTUM POINT CONTACTS 13
III. BALLISTIC QUANTUM TRANSPORT 17
   1. *Introduction* 17
   2. *Conductance Quantization of a Quantum Point Contact* 19
   3. *Magnetic Depopulation of Subbands* 33
   4. *Magnetic Suppression of Backscattering at a Point Contact* 39
   5. *Electron Beam Collimation and Point Contacts in Series* 45
   6. *Coherent Electron Focusing* 56
   7. *Breakdown of the Conductance Quantization and Hot Electron Focusing* . . . 64
IV. ADIABATIC TRANSPORT IN THE QUANTUM HALL EFFECT REGIME 81
   8. *Introduction* 81
   9. *Anomalous Quantum Hall Effect* 84
  10. *Anomalous Shubnikov–de Haas Effect* 94
  11. *Aharonov–Bohm Oscillations and Inter-Edge Channel Tunneling* 99
   ACKNOWLEDGMENTS 105
   REFERENCES 106

I. Introduction

The subject of this chapter is quasi-one-dimensional quantum transport. Only a few years ago, a prevalent feeling was that there is a "limited purpose in elaborating on playful one-dimensional models" for quantum transport.[1] This situation has changed drastically since the realization of the quantum point contact, which now offers ample opportunity to study transport problems of textbook simplicity in the solid state. Interestingly, many of the phenomena treated in this chapter were not anticipated theoretically, even though they were understood rapidly after their experimental discovery.

In this chapter, we review the experimental and theoretical work by the Philips–Delft collaboration on electrical transport through quantum point contacts. These are short and narrow constrictions in a two-dimensional electron gas (2DEG), with a width of the order of the Fermi wave length λ_F. Throughout our presentation, we distinguish between ballistic and adiabatic transport. *Ballistic quantum transport* takes place in low magnetic fields, for which Landau level quantization is unimportant and the Fermi wavelength ($\lambda_F \approx 40$ nm) governs the quantization. In stronger fields in the quantum Hall effect regime, the Landau-level quantization dominates, characterized by the magnetic length ($l_m \equiv (\hbar/eB)^{1/2} \approx 10$ nm at $B = 5$ T). In the latter regime, inter-Landau-level scattering can be suppressed and *adiabatic quantum transport* may be realized. Because of the high mobility, elastic impurity scattering and inelastic scattering are of secondary importance in the ballistic and adiabatic transport regimes. Scattering is determined instead by the geometry of the sample boundary. The concept of a mean free path thus loses much of its meaning, and serves only as an indication of the length scale on which ballistic transport can be realized. (The transport mean free path in weak magnetic fields is about 10 μm in wide 2DEG regions.) Fully adiabatic transport in strong magnetic fields has been demonstrated over a short distance of the order of a μm, but may be possible on longer length scales. Separate and more detailed introductions to these two transport regimes are given in Part III (which is concerned with ballistic quantum transport) and Part IV (where adiabatic quantum transport is discussed). The following is intended only to convey the flavor of the subject, and to give an elementary introduction to some of the essential characteristics.

The common starting point for the structures investigated is the degenerate two-dimensional electron gas (2DEG), present at the interface between GaAs and $Al_xGa_{1-x}As$ layers in a heterostructure. (Experimental details are given in Part II; for general reviews of the 2DEG, see Refs. 2 and 3.) The electrons are confined in the GaAs by a potential well at the interface with the AlGaAs, which results from the repulsive barrier due to the conduction band offset between the two semiconductors (about 0.3 eV), and from the attractive electrostatic potential due to the positively charged ionized donors in the AlGaAs layer. The electrons thus are confined in a direction normal to the interface, but free to move along the interface. This implies that a two-dimensional subband is associated with each discrete confinement level in the well. Usually, the potential well is sufficiently narrow (about 10 nm) that only a single two-dimensional subband is occupied, and the density of states is strictly two-dimensional. At low temperatures, these states are occupied up to the Fermi energy, $E_F \approx 10$ meV. Additional confinement occurs in a lateral direction if a narrow channel is defined electrostatically in the 2DEG. This leads to the formation of one-dimensional subbands, characterized by free motion in a single direction.

Throughout this chapter we will use a magnetic field perpendicular to the 2DEG as a tool to modify the nature of the quantum states. In a wide 2DEG, a perpendicular magnetic field eliminates the two degrees of freedom, and forms dispersionless Landau levels (which correspond classically to the motion of electrons in cyclotron orbits). One thus has a purely discrete density of states. Near the boundary of the 2DEG, the Landau levels transform into magnetic edge channels, which are free to move along the boundary, and correspond classically to skipping orbits. These edge channels have a one-dimensional dispersion (i.e., the energy depends on the momentum along the boundary). In this respect, they are similar to the one-dimensional subbands resulting from a purely electrostatic lateral confinement in a channel. Because of the one-dimensional dispersion law, both edge channels and one-dimensional subbands can be viewed as propagating modes in an electron waveguide. This similarity allows a unified description of the quantum Hall effect and of quantum-size effects in narrow conductors in the ballistic transport regime.

A really unequivocal and striking manifestation of a quantum-size effect on the conductance of a single narrow conductor came, paradoxically, with the experimental realization by the Delft–Philips collaboration[4] and by the Cambridge group[5] of the quantum point contact—a constriction that one would have expected to be too short for one-dimensional subbands to be well-developed. A major surprise was the nature of the quantum-size effect: The conductance of quantum point contacts is quantized in units of $2e^2/h$. This is reminiscent of the quantum Hall effect, but measured in the absence of a magnetic field. The basic reason for the *conductance quantization* (a fundamental cancellation of group velocity and density of states for quantum states with a one-dimensional dispersion law) already was appreciated in the original publications. More complete explanations came quickly thereafter, in which the mode-coupling with the wide 2DEG at the entrance and exit of the narrow constriction was treated explicitly. Rapid progress in the theoretical understanding of the conductance quantization, and of its subsequent ramifications, was facilitated by the availability of a formalism,[6,7] which turned out to be ideally suited for quasi-one-dimensional transport problems in the ballistic and adiabatic transport regimes. The Landauer–Büttiker formalism treats transport as a transmission problem for electrons at the Fermi level. The ohmic contacts are modeled as current injecting and collecting reservoirs, in which all inelastic scattering is thought to take place exclusively. As described by Büttiker in Chapter 4, the measured conductances then can be expressed as rational functions of the transmission probabilities at the Fermi level between the reservoirs. The zero-field conductance quantization of an ideal one-dimensional conductor, and the smooth transition to the quantum Hall effect on applying a magnetic field, are seen to follow directly from the fact that a reservoir in equilibrium injects a current that is shared equally by all propagating modes (which can be one-dimensional subbands or magnetic edge channels).

Novel phenomena arise if a selective, non-equal distribution of current among the modes is realized instead. In the ballistic transport regime, directional selectivity can be effected by a quantum point contact, as a result of its horn-like shape and of the potential barrier present in the constriction.[8] The collimation of the electron beam injected by the point contact explains the strong non-additivity of the series resistance of two opposite point contacts observed in Ref. 9. A most striking manifestation of a non-equal distribution of current among the modes is realized in the adiabatic transport regime, where the selective population and detection of magnetic edge channels is the mechanism for the anomalous quantum Hall and Shubnikov–de Haas effects.[10-12]

Mode interference is another basic phenomenon. Its first unequivocal manifestation in quantum transport is formed by the large (nearly 100%) conductance oscillations found in the coherent electron focusing experiment.[13-15] They may be considered as the ballistic counterpart of the conductance fluctuations characteristic of the diffusive transport regime. In the adiabatic transport regime, mode interference is less important, because of the weakness in general of inter-edge channel coupling. Quantum interference phenomena still can be observed if a weak coupling exists between the edge channels at opposite edges of the conductor. Such a coupling can result naturally from the presence of an impurity in a narrow channel, or artificially at quantum point contacts. In this way, Aharonov–Bohm magnetoresistance oscillations can occur even in a singly connected geometry.[16,17]

In summary, transport phenomena in the ballistic and adiabatic regimes can be viewed as scattering or transmission experiments with modes in an electron waveguide. Quantization—i.e., the discreteness of the mode index—is essential for some phenomena (which necessarily require a description in terms of modes), but not for others (which could have been described semiclassically equally well in terms of the trajectories of electrons at the Fermi level). In this chapter, we consider the semiclassical limit along with a quantum mechanical formulation wherever this is appropriate. This serves to distinguish those aspects of the new phenomena that are intrinsically quantum mechanical from those that are not.

Most of the work described in the following has been published previously. The present chapter is the first comprehensive review. Many new details are added, and a critical overview of experimental as well as theoretical aspects is provided. This is not intended to be a comprehensive review of the whole field of quasi-one-dimensional quantum transport. Because of the limited amount of space and time available, we have not included a detailed discussion of related work by other groups (some of which is described extensively in other chapters in this volume). For the same reason, we have excluded work by ourselves and others on the quasi-ballistic transport regime, and on ballistic transport in narrow-channel geometries. For a broader perspective, we refer readers to a review[18] and to recent conference proceedings.[19-22]

II. Split-Gate Quantum Point Contacts

The study of ballistic transport through point contacts in metals has a long history. Point contacts in metals act like small conducting orifices in a thin insulating layer, separating bulk metallic conductors (with a mean free path l much larger than the size of the orifice). Actual point contacts usually are fabricated by pressing a metal needle on a metallic single crystal, followed by spot-welding. Ballistic transport has been studied successfully in this way in a variety of metals.[23-26] Point contacts in bulk doped semiconductors have been fabricated by pressing two wedge-shaped specimens close together.[27] One limitation of these techniques is that the size of a point contact is not continuously variable.

Point contacts in a 2DEG cannot be fabricated by the same method, since the electron gas is confined at the GaAs-Al$_x$Ga$_{1-x}$As interface in the sample interior. The point contacts used in our studies are defined electrostatically[28,29] by means of a split gate on top of the heterostructure. (See Fig. 1a.) In this way, one can define short and narrow constrictions in the 2DEG, of variable width comparable to the Fermi wavelength (a quantum point contact). Other techniques can be used to define constrictions of fixed width, such as a deep[30] or shallow[31] mesa etch, or ion implantation using focused ion beams,[32] but a variable constriction width is crucial for our purpose. (An alternative technique for the fabrication of variable width constrictions employing a gate in the plane—rather than on top—of the 2DEG recently has been demonstrated.)[33] Starting point for the fabrication of our quantum point contact structures is a GaAs-Al$_x$Ga$_{1-x}$As heterostructure ($x = 0.3$) grown by molecular beam epitaxy. The layer structure is drawn schematically in Fig. 1b. The width of the opening in the gate is approximately 250 nm, its length being much shorter (50 nm). The 2DEG sheet carrier density n_s, obtained from the periodicity of the Shubnikov–de Haas oscillations in the magnetoresistance,[2] has a typical value of 3.5×10^{15} m^{-2}.

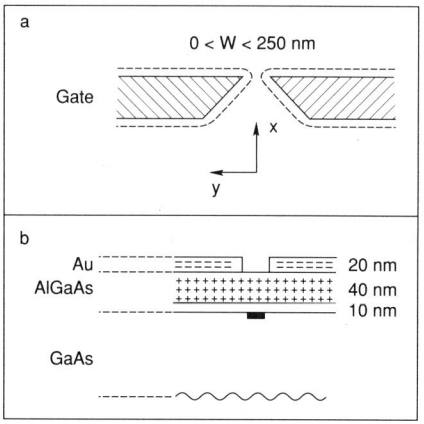

FIG. 1. (a) Top view of a quantum point contact, defined using a split gate (shaded) on top of a GaAs-Al$_x$Ga$_{1-x}$As heterostructure. The depletion boundary is indicated by the dashed curve. The width W of the constriction can be reduced by increasing the negative voltage on the gate. (b) Cross section of the quantum point contact. The narrow quasi-one-dimensional electron gas channel in the constriction is indicated in black. The positive ionized donors (+) in the AlGaAs layer are indicated, as well as the negative charge (−) on the gate.

The electrons at the Fermi level then have a wave vector $k_F \equiv (2\pi n_s)^{1/2} \approx 0.15 \times 10^9$ m^{-1}, a wavelength $\lambda_F \equiv 2\pi/k_F \approx 40$ nm, and a velocity $v_F \equiv \hbar k_F/m \approx 2.7 \times 10^5$ m/s. The transport mean free path $l \approx 10$ μm follows from the zero-field resistivity $\rho \equiv h/e^2 k_F l \approx 16$ Ω. Note that $m = 0.067 m_e$ is the effective mass in GaAs. Most of the experimental work presented in this chapter has been done on samples made by the Philips–Delft collaboration. An exception is formed by the experiments described in Sections 4 and 9.a.ii, which were done on a sample fabricated in a collaboration between Philips and the Cavendish group.[34]

The fabrication procedure essentially is the same for all the samples. A standard mesa-etched Hall bar geometry is defined by wet etching. The split gate is fabricated using a combination of electron beam and optical lithography. The gate pattern of many of our samples has been designed specifically for the electron focusing experiments, which require two point contacts positioned next to each other on the 2DEG boundary. (See Fig. 2a.) Note that the actual 2DEG boundary between the two point contacts is a depletion potential wall below the gate (which extends laterally beyond the gate pattern, up to about 150 nm for large negative gate voltages). The effect of a negative gate voltage is to deplete gradually the electron gas under the gate structure. Beyond the depletion threshold (typically -0.6 V), no mobile carriers are present under the gate, and two conducting constrictions are formed with a width of about 250 nm. Two high-mobility 2DEG regions thus

2. QUANTUM POINT CONTACTS

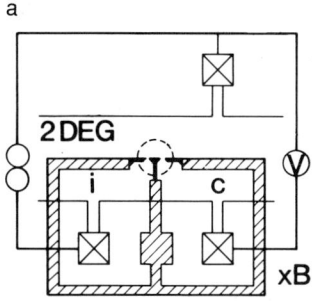

FIG. 2. Schematic layout (*a*) of a double point contact device, in a three-terminal measurement configuration used in some of the electron focusing experiments. The crossed squares are ohmic contacts to the 2DEG. The split gate (shaded) separates injector (i) and collector (c) areas from the bulk 2DEG. The fine details of the gate structure inside the dashed circle are shown in a scanning electron micrograph (*b*). The bar denotes a length of 1 μm. (From Ref. 15).

are isolated electrically from the rest of the 2DEG in the Hall bar, apart from the narrow constrictions, or point contacts, under the openings of the gate. A further increase of the negative gate voltage forces the constrictions to become progressively narrower, until they are fully pinched off. By this technique, it is possible to define point contacts of variable width W. To create constrictions with minimal length, the gates are tapered into a wedge. (See the scanning electron micrograph in Fig. 2b.) The precise functional dependence of width and carrier concentration on the gate voltage is dependent on the previous history of the sample. Thermal cycling and large positive gate voltages lead to a shift in the depletion threshold, although all transport measurements are quite reproducible if the sample is kept cold and the gate voltage is not varied strongly.

A low-frequency ac lock-in technique is used to measure the resistances. Several ohmic contacts (alloyed Au-Ge-Ni) are positioned at the sides of the Hall bar (Fig. 2a), to serve as current and voltage terminals. The resistance $R_{ij,kl} \equiv (V_k - V_l)/I$ is defined as the voltage difference between terminal k and l divided by the current I, which flows from terminal i to j. One distinguishes between two- and four-terminal resistance measurements, depending on whether or not the voltage difference is measured between the current source and drain ($i, j \equiv k, l$), or between two separate ohmic contacts. Section 2 deals with two-terminal measurements of the point contact resistance in zero magnetic field. This resistance contains a spurious contribution of several $k\Omega$ from the rather large contact resistance of the current-carrying ohmic contacts. This correction can be estimated from a measurement of the two-terminal resistance at zero gate voltage, or can be eliminated entirely by performing a four-terminal measurement. Apart from the presence of this contact resistance, there is no significant difference between two- and four-terminal measurements in the absence of a magnetic field, provided the voltage probes do not introduce additional scattering in the vicinity of the point contact. In an external magnetic field, the behavior of two- and four-terminal resistances is quite different, however, as we will discuss in Sections 3 and 4.

In addition to the series resistance of the ohmic contacts, there are two additional small corrections to the quantized point contact resistance that are gate voltage-independent *beyond* the depletion threshold of the gate (-0.6 V), as we now discuss briefly. At the depletion threshold, the two-terminal resistance increases abruptly for three reasons:

1. The formation of the ballistic point contact, which is the quantity of interest.
2. The increase of the diffusive resistance of the wide 2DEG lead on one side of the constriction, because of a change in the lead geometry. (See the gate layout in Fig. 2a.) This term is $\rho \approx 16\ \Omega$ multiplied by the extra number of squares in the lead.
3. The appearance of the two-dimensional Maxwell spreading resistance,[35] associated with the spreading from a region of radius l (the mean free path) surrounding the point contact to one of radius W_{wide} (the width of the wide 2DEG leads). This term is approximately[1] $\pi^{-1}\rho \ln(W_{\text{wide}}/l)$.

The contributions 2 and 3 are independent of gate voltage beyond the depletion threshold to a very good approximation, but they are difficult to determine very accurately from the device geometry. For this reason, we have opted to treat the total background resistance R_b in the two-terminal resistance R_{2t} of the point contact as a single adjustable parameter, chosen such that for one constant value of R_b, a uniform step height (between quantized plateaus) is obtained in the conductance $G = [R_{2t}(V_g) - R_b]^{-1}$ as a function of gate voltage V_g. This procedure always has yielded a uniform step height in G over the whole gate voltage range for a single value of R_b. Moreover, the resulting value of R_b is close to the value that one would have estimated from the preceding considerations.

III. Ballistic Quantum Transport

1. Introduction

In this section, we present a comprehensive review of the results of the study by the Philips–Delft collaboration of ballistic transport in geometries involving quantum point contacts in weak magnetic fields. To put this work in proper perspective, we first briefly discuss the two fields of research from which it has grown.

The first is that of point contacts in metals. Maxwell, in his *Treatise on Electricity and Magnetism*, investigated the spreading resistance of a small contact in the diffusive transport regime.[35] His results have been applied extensively in the technology of dirty metallic contacts.[36] The interest in point contacts gained new impetus with the pioneering work of Sharvin,[23] who proposed and subsequently realized[37] the injection and detection of a beam of electrons in a metal by means of point contacts much smaller than the mean free path. Sharvin's longitudinal electron focusing experiment was the analogue in the solid state of an experiment performed earlier in vacuum by Tricker[38] at the suggestion of Kapitza.[39] This technique since has been refined, especially with the introduction of the transverse electron focusing geometry by Tsoi.[24] (See Section 6.) Point contacts also can be used to inject electrons in a metal with an energy above the Fermi energy. This idea has been exploited in the field of point contact spectroscopy, and it has yielded a wealth of information on inelastic electron–phonon scattering.[25,26,40] Magnetotransport through ballistic point contacts and micro-bridges has been studied recently.[41,42] With the possible exception of the scanning tunneling microscope, which can be seen as a point contact on an atomic scale,[43–48] these studies in metals essentially are restricted to the *classical* ballistic transport regime, because of the extremely small Fermi wavelength ($\lambda_F \approx 0.5$ nm, of the same magnitude as the lattice spacing).

The second field is that of quasi-one-dimensional quantum transport in semiconductor microstructures, which started in the diffusive transport regime in narrow silicon MOSFETs. That work has focused on the study of reproducible (universal) conductance fluctuations, as discussed in this volume. Clear manifestations of the quasi-one-dimensional density of states of a single narrow wire proved to be elusive, mainly because the irregular conductance fluctuations mask the structure due to the one-dimensional subbands in the wire. Devices containing many wires in parallel were required to average out these fluctuations and resolve the subband structure in a transport experiment.[49] This situation changed with the realization by various techniques of narrow channels in the two-dimensional electron gas (2DEG) of a GaAs-Al_xGa_{1-x}As heterostructure.[28,29,31,50] This is an ideal model system because of the simple Fermi surface (a circle), the relatively long mean free path ($l \approx 10$ μm at low temperatures for material grown by molecular beam epitaxy), and the large Fermi wavelength ($\lambda_F \approx 40$ nm) resulting from the low electron density. Another essential advantage of this system is that its two-dimensionality allows the use of planar semiconductor technology to fabricate a rich variety of device structures. Finally, in contrast to metals, the low electron density in these semiconductor structures can be varied by means of a gate voltage. Thornton et al.[28] and Zheng et al.[29] have demonstrated that it is possible to realize structures of variable width and density by employing a split-gate lateral depletion technique. Other groups[51-54] have used the shallow mesa etch technique,[31] or other etch techniques,[55] to fabricate narrow channels of fixed width with many side probes for the study of quantum ballistic transport, as discussed by Timp in Chapter 3. An important result of these studies was the demonstration that in the ballistic transport regime, side probes are the dominant source of scattering.[56]

Our work on quantum ballistic transport builds on both fields summarized in the preceding. As discussed in Part I, the central vehicle for this investigation is the quantum point contact, a short and narrow constriction of variable width in the 2DEG, of dimensions comparable to λ_F and much smaller than l. This device yielded the first unequivocal demonstration of a quantum-size effect in a single narrow conductor,[4,5] in the form of the zero-field conductance quantization. We discuss the experiment and its theoretical explanation in Section 2. The quantization of the conductance provides us with an extremely straightforward way to determine the number of occupied subbands in the point contact. It is shown in Section 3 that a study of the magnetic depopulation of subbands directly yields the width and carrier density in the point contact.[57]

Two- and four-terminal resistance measurements are qualitatively different in the presence of an external magnetic field. The negative four-terminal magnetoresistance[34] arising from the suppression of backscattering at the point contact is the subject of Section 4. We then proceed to discuss in Section 5 the collimation of the beam injected by a point contact,[8] and its effect on transport measurements in geometries involving two opposite point contacts.[9,58] The variety of magnetoresistance effects[59] in such geometries is even richer than for single-point contacts, as we will discuss. An important application of point contacts is as point-like electron sources and detectors with a large degree of spatial coherence. The first such application was the coherent electron focusing experiment[13] described in Section 6. This experiment exhibits the characteristic features of the quantum ballistic transport regime in a most extreme way. The results are interpreted in terms of mode interference of magnetic edge channels.[14] Ballistic transport far from equilibrium is the subject of Section 7, where we discuss the breakdown of the conductance quantization in the nonlinear transport regime,[60] and hot electron focusing.[61] In the latter experiment, the kinetic energy of the injected electrons in the 2DEG is measured in a similar way as in a β-spectrometer in vacuum.

2. Conductance Quantization of a Quantum Point Contact

a. Experimental Observation of the Conductance Quantization

The first results on the resistance of a quantum point contact obtained by the Delft–Philips collaboration[4] are reproduced in Fig. 3. Equivalent results were obtained independently by the Cavendish group.[5] The resistance of the point contact is measured as a function of the voltage V_g on the split gate, at a temperature $T = 0.6$ K. The resistance measured at $V_g = 0$ V has been

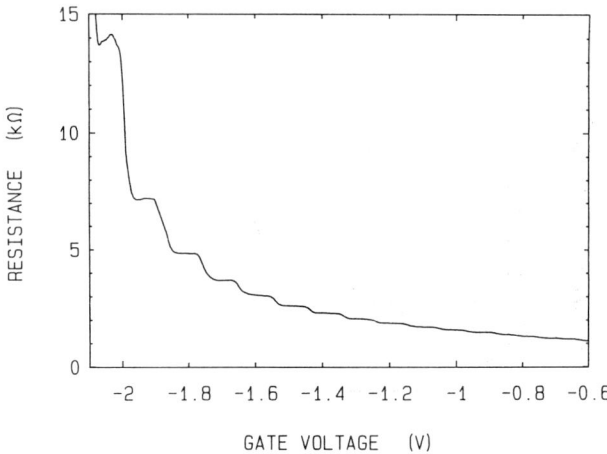

Fig. 3. Two-terminal point contact resistance measured as a function of gate voltage at 0.6 K. The resistance measured at zero gate voltage has been subtracted. (From Ref. 4.)

subtracted in Fig. 3, to eliminate the large ohmic contact resistance of about 4 kΩ in series with the point contact. At $V_g \approx -0.6$ V, the electron gas directly below the gate is depleted, and the constriction is formed. At this gate voltage, the constriction has its maximum width, roughly equal to the width of the opening in the gate (250 nm). On increasing the negative gate voltage, the width W gradually is reduced, and simultaneously the bottom of the conduction band is raised in the point contact region. The resulting bottleneck in real and energy space causes the point contact to have a nonzero resistance. This resistance increases without bound as the pinch-off voltage ($V_g \approx -2.2$ V) is approached. Classically, one expects this increase to be monotonic.

The unexpected characteristic of Fig. 3 is the sequence of plateaus and steps seen in the resistance versus gate voltage curve. The plateaus represent the conductance quantization of a quantum point contact in units of $2e^2/h$. This is seen most easily in Fig. 4, where the conductance is plotted (obtained by inverting the resistance of Fig. 3 after subtraction of an additional background resistance of 400 Ω, which accounts for the increase in lead resistance at the depletion threshold discussed in Part II). The conductance quantization is reminiscent of the quantum Hall effect,[62] but is observed in the absence of a magnetic field and thus can not have the same origin. The zero-field quantization is not as accurate as the quantum Hall effect. The deviations from exact quantization in the present experiments are estimated at 1%,[63] while in the quantum Hall effect, an accuracy of one part in 10^7 is obtained routinely.[64] It is very unlikely that in the case of the zero-field quantization, a similar accuracy can be achieved—if only because of the presence in series

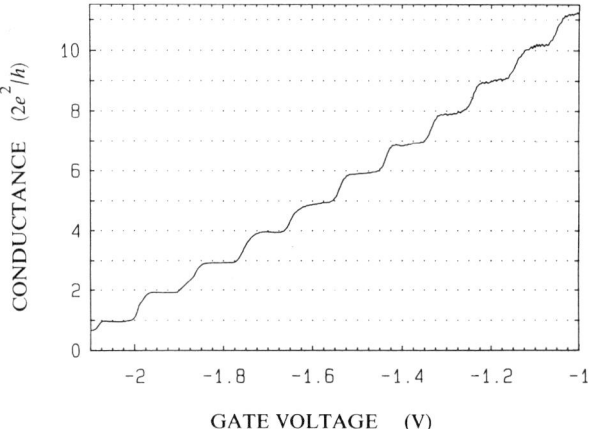

FIG. 4. Point contact conductance, obtained from the resistance in Fig. 3 after subtraction of an additional background resistance of 400 Ω. (From Ref. 4.)

with the point contact resistance of a background resistance whose magnitude can not be determined precisely. Both the degree of flatness of the plateaus and the sharpness of the transition between the plateaus vary among devices of identical design, indicating that the detailed shape of the electrostatic potential defining the constriction is important. There are many uncontrolled factors affecting this shape, such as small changes in the gate geometry, variations in the pinning of the Fermi level at the free GaAs surfaceor at the interface with the gate metal, not fully homogeneous doping of the heterostructure, and trapping of charges in the AlGaAs.

As will be discussed in Section 2.b, the sequence of plateaus is caused by the stepwise decrease of the number N of occupied one-dimensional subbands as the point contact gradually is pinched off, each subband contributing $2e^2/h$ to the conductance. In a simple approximation, the constriction is modeled as a straight channel of length L and width W, with a square-well lateral confining potential. The bottom of the well is at a height E_c above the conduction band bottom in the wide 2DEG. The density n_c in the constriction thus is reduced from the bulk density n_s by approximately a factor $(E_F - E_c)/E_F$. (This factor assumes a constant two-dimensional density of states in the constriction.) The stepwise reduction of N is due both to the decrease in W and the increase in E_c (or, equivalently, the reduction of n_c). If the latter effect is ignored, then the number of occupied subbands in the square well is $2W/\lambda_F$, with $\lambda_F = 40$ nm the Fermi wavelength in the wide 2DEG. The sequence of steps in Fig. 4 then would correspond to a gradual decrease in width from 320 nm (at $V_g \approx -1.0$ V) to 20 nm (at $V_g \approx -2.0$ V). This simple argument certainly overestimates the reduction in width, however, because of the unjustified neglect of the reduction in carrier density. By applying a perpendicular magnetic field, W and n_c can be determined independently, as discussed in Section 3.b. The length of the constriction is harder to assess, but the electrostatic depletion technique used is expected to create a constriction of length L, which increases with increasing negative gate voltage. Typically, $L \gtrsim W$. The actual two-dimensional shape of the confining potential certainly is smoother than a straight channel with hard walls. Nonetheless, for many applications, this simple model is adequate, and we will make use of it unless a more realistic potential is essential.

b. Theory of the Conductance Quantization

i. Classical point contact. It is instructive first to consider *classical* ballistic transport in a degenerate electron gas in some detail. The ballistic electron flow through a point contact is illustrated in Fig. 5a in real space, and in Fig. 5b in k-space, for a small excess electron density δn at one side of the point contact.[65,66] At low temperatures, this excess charge moves with

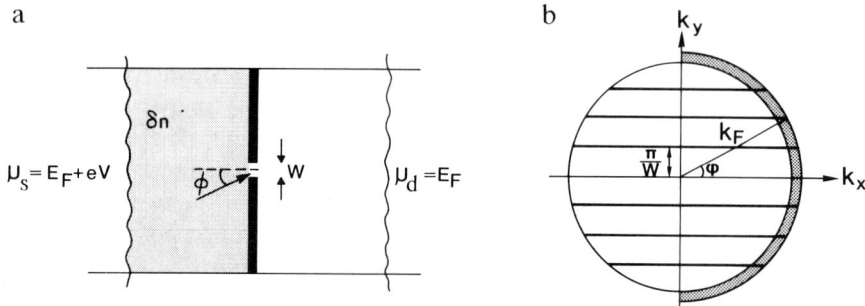

FIG. 5. (a) Classical ballistic transport through a point contact. The net concentration difference δn corresponds to a chemical potential difference eV between source (s) and drain (d). In reality, this concentration difference is eliminated by screening charges, but without changing the chemical potential difference or the current. (b) The net current through a quantum point contact is carried by the shaded region in k-space. In an ideal quasi-one-dimensional channel, the allowed states lie on the horizontal lines, which correspond to quantized values for $k_y = \pm n\pi/W$, and continuous values for k_x. The formation of these one-dimensional subbands gives rise to a quantized conductance. (From Refs. 65 and 66.)

the Fermi velocity v_F. The flux normally incident on the point contact is $\delta n\, v_F \langle \cos \phi\, \theta(\cos \phi) \rangle$, where $\theta(x)$ is the unit step function and the brackets denote an isotropic angular average. (The angle ϕ is defined in Fig. 5a.) In the ballistic limit $l \gg W$, the incident flux is fully transmitted, so that the total current I through the point contact is given by

$$I = eW\, \delta n\, v_F \int_{-\pi/2}^{\pi/2} \cos \phi\, \frac{d\phi}{2\pi} = \frac{e}{\pi} W v_F\, \delta n. \tag{1}$$

The transport problem in this formulation resembles the problem of Knudsen effusion in a non-degenerate gas. A fundamental difference is that in the former problem, I is linear in δn only for small $\delta n \ll n_s$, whereas this restriction is not necessary in the Knudsen problem. This distinction is a consequence of the interdependence of the velocity and the density in a degenerate gas described by Fermi–Dirac statistics.[66] We will return to this point in Section 7, when we discuss the nonlinear transport regime. To determine the conductance, we note that an electron density difference can not be maintained, because of the cost in electrostatic energy. Screening charges reduce δn without changing the chemical potential difference $\delta \mu \equiv eV$, which is assumed to be fixed by an external electron reservoir. The ratio $\delta n/\delta \mu$ is just the density of states at the Fermi level, $\delta n/\delta \mu = m/\pi \hbar^2$ for a two-dimensional electron gas with spin degeneracy. One thus finds from Eq. (1) for the conductance $G \equiv I/V$, the result,[4]

$$G = \frac{2e^2}{h} \frac{k_F W}{\pi}, \text{ in 2D.} \tag{2}$$

Equation (2) is the two-dimensional analogue of Sharvin's well-known expression[23] for the point contact conductance in three dimensions,

$$G = \frac{2e^2}{h} \frac{k_F^2 S}{4\pi}, \text{ in 3D,} \tag{3}$$

where S now is the area of the point contact.

The experimental constriction geometry differs from the hole in a screen of Fig. 5a, in having a finite length with a smoothly varying width W, and an electron gas density that decreases on entering the constriction. The reduced density leads to a smaller value for k_F in the constriction than in the wide 2DEG. Equation (2) still can be applied to this situation, if the product $k_F W$ is evaluated at the *bottleneck* (such that all electrons that reach the bottleneck are transmitted through the constriction). This typically is halfway into the constriction, where k_F and W take on their minimal values.

ii. Conductance quantization of an ideal quasi-one-dimensional conductor. The basic mechanism for the quantization of the conductance given classically by Eq. (2) can be understood in quite simple terms.[4] The argument, which we present here in a somewhat modified form, refers to an *ideal* quasi-one-dimensional conductor that behaves as an electron waveguide connecting two reservoirs in thermal equilibrium at chemical potentials E_F and $E_F + \delta\mu$. All inelastic scattering is thought to take place in the reservoirs, not in the conductor itself. This is the viewpoint introduced by Landauer.[6] The Landauer formula relates the conductance to the transmission probability through the conductor from one reservoir to the other. The net current is injected into the conductor within a narrow range $\delta\mu$ above E_F into the N one-dimensional subbands or waveguide modes that can propagate at these energies. The dispersion relation $E_n(k)$ of the subbands is

$$E_n(k) = E_n + \frac{\hbar^2 k^2}{2m}, \tag{4}$$

where k is the wave number for propagation along the conductor and E_n is the energy associated with the lateral confinement of the nth subband.

(Equivalently, $E_n/\hbar$ is the cutoff frequency of the nth mode.) The number N of occupied subbands (or propagating modes) at the Fermi energy is the largest integer such that $E_N < E_F$. The current per unit energy interval injected into a subband is the product of the group velocity and the one-dimensional density of states. The group velocity is $v_n = dE_n(k)/\hbar\, dk$, and the density of states for one velocity direction and including spin degeneracy is $\rho_n = (\pi\, dE_n(k)/dk)^{-1}$. The product of v_n and ρ_n is seen to be independent of both energy and subband index n. The injected current thus is *equipartitioned* among the subbands, each subband carrying the same amount of current $e\, v_n \rho_n\, \delta\mu = (2e/h)\,\delta\mu$. The equipartitioning of current, which is the basic mechanism for the conductance quantization, is illustrated in Fig. 5b for a square-well lateral confining potential of width W. The one-dimensional subbands then correspond to the pairs of horizontal lines at $k_y = \pm n\pi/W$, with $n = 1, 2, \ldots N$ and $N = \text{Int}[k_F W/\pi]$. The group velocity $v_n = \hbar k_x/m$ is proportional to $\cos\phi$, and thus decreases with increasing n. However, the decrease in v_n is compensated by an increase in the one-dimensional density of states. Since ρ_n is proportional to the length of the horizontal lines within the dashed area in Fig. 5b, ρ_n is proportional to $1/\cos\phi$ so that the product $v_n \rho_n$ does not depend on the subband index.

The total current $I = (2e/h)N\,\delta\mu$ yields a conductance $G = eI/\delta\mu$ given by

$$G = \frac{2e^2}{h} N. \tag{5}$$

This equation can be seen as a special limit of the Landauer formula for two-terminal conductances,[7,67–69]

$$G = \frac{2e^2}{h} \text{Tr}\, \mathbf{tt}^\dagger \equiv \frac{2e^2}{h} \sum_{n,m=1}^{N} |t_{nm}|^2, \tag{6}$$

where $\mathbf{t}$ is the matrix (with elements t_{nm}) of transmission probability amplitudes at the Fermi energy (from subband m at one reservoir to subband n at the other). The result, Eq. (5), follows from Eq. (6) if $\text{Tr}\,\mathbf{tt}^\dagger = N$. A sufficient condition for this is the absence of intersubband scattering, $|t_{nm}|^2 = \delta_{nm}$, a property that may be taken to define the ideal conductor. More generally, scattering among the subbands is allowed as long as it does not lead to backscattering (i.e., for zero reflection coefficients $r_{nm} = 0$ for all $n, m = 1, 2, \ldots N$). Equation (5) describes a stepwise increase in the conductance of an ideal quasi-one-dimensional conductor as the number of occupied subbands is increased. The conductance increases by $2e^2/h$ each time N increases by 1. For a square-well lateral confining potential, $N = \text{Int}[k_F W/\pi]$, so that in the classical limit, the result (2) for a two-dimensional point contact is recovered. Note that Eq. (5) also holds for three-dimensional point contacts, although in that case, no experimental system showing the conductance quantization as yet has been realized.

We emphasize that, although the classical formula, Eq. (2), holds only for a square-well lateral confining potential, the quantization, Eq. (5), is a general result for any shape of the confining potential. The reason simply is that the fundamental cancellation of the group velocity, $v_n = dE_n(k)/\hbar\, dk$, and the one-dimensional density of states, $\rho_n = (\pi\, dE_n(k)/dk)^{-1}$, holds *regardless* of the form of the dispersion relation $E_n(k)$. For the same reason, Eq. (5) is applicable equally in the presence of a magnetic field, when magnetic edge channels at the Fermi level take over the role of one-dimensional subbands. Equation (5) thus implies a continuous transition from the zero-field quantization to the quantum Hall effect, as we will discuss in Section 3.

The fact that the Landauer formula, Eq. (6), yields a *finite* conductance for a perfect (ballistic) conductor was a source of confusion in the early literature,[70-72] but now is understood as a consequence of the unavoidable contact resistances at the connection of the conductor to the reservoirs. The relation between ballistic point contacts and contact resistances of order h/e^2 for a one-dimensional subband first was pointed out by Imry.[68] This was believed to be only an *order of magnitude* estimate of the point contact resistance. One reason for this was that the Landauer formula follows from an idealized model of a resistance measurement; another one was that several multi-subband generalizations of the original Landauer formula[6] had been proposed,[67,73-76] which led to conflicting results. We refer to a paper by Stone and Szafer[69] for a discussion of this controversy, which now has been settled[7,77-79] in a way supported by the present experiments. This brief excursion into history may serve as a partial explanation of the fact that no prediction of the conductance quantization of a point contact was made, and why its experimental discovery came as a surprise.

iii. Conductance quantization of a quantum point contact. There are several reasons why the ideal quasi-one-dimensional conductor model given earlier, and related models,[80-82] are not fully satisfactory as an explanation of the experimentally observed conductance quantization. Firstly, to treat a point contact as a waveguide would seem to require a constriction much longer than wide, which is not the case in the experiments,[4,5] where $W \approx L$. (See Section 2.d. for a discussion of experiments in longer constrictions.) In a very short constriction, transmission through evanescent modes (with cutoff frequency above $E_F/\hbar$) becomes important. Secondly, the coupling of the modes in the wide 2DEG regions to those in the constriction has not been considered explicitly. In particular, diffraction and quantum mechanical reflection at the entrance and exit of the constriction have been ignored. Finally, alloyed ohmic contacts, in combination with those parts of the wide 2DEG leads that are more than an inelastic scattering length away from the constriction, only are approximate realizations of the reservoirs in thermal equilibrium of the idealized problem.[83] As an example, electrons may be scattered back into the constriction by an impurity without intervening equilibration. This modifies the point contact conductance, as has been studied extensively in the classical case.[26]

To resolve these issues, it is necessary to solve the Schrödinger equation for the wave functions in the narrow point contact and the adjacent wide regions, and match the wave functions at the entrance and exit of the constriction. Following the experimental discovery of the quantized conductance, this mode coupling problem has been solved numerically for point contacts of a variety of shapes,[84-91] and analytically in special geometries.[92-95] As described in detail in Ref. 96, the problem has a direct and obvious analogue in the field of classical electromagnetism. Although it can be solved by standard methods (which we will not discuss here), the resulting transmission steps appear not to have been noted before in the optical or microwave literature. When considering the mode coupling at the entrance and exit of the constriction, it is important to distinguish between the cases of a gradual (*adiabatic*) and an *abrupt* transition from wide to narrow regions.

The case of an *adiabatic* constriction has been studied by Glazman et al.[97] and is the easiest case to solve analytically (cf. also a paper by Imry in Ref. 20). If the constriction width W changes gradually on the scale of a wavelength, the transport within the constriction is adiabatic; i.e., there is no intersubband scattering from entrance to exit. At the exit, where connection is made to the wide 2DEG regions, intersubband scattering becomes unavoidable and the adiabaticity breaks down. However, if the constriction width at the exit W_{max} is much larger than its minimal width W_{min}, the probability for reflection back through the constriction becomes small. In the language of waveguide transmission, one has impedance-matched the constriction to the wide 2DEG regions.[98] Since each of the N propagating modes in the narrowest section of the constriction is transmitted without reflection, one has $\text{Tr}\, tt^\dagger = N$, provided evanescent modes can be neglected. The conductance quantization, Eq. (5), then follows immediately from the Landauer formula, Eq. (6). Glazman et al.[97] have calculated that the contributions from evanescent modes through an adiabatic constriction are small even for rather short constriction length L (comparable to W_{min}). The accuracy of the conductance quantization for an adiabatic constriction in principle can be made arbitrarily high, by widening the constriction more and more slowly, such that $(W_{max} - W_{min})/L \ll 1$. In practice, of course, the finite mean free path still poses a limitation. We note in this connection that the gradual widening of the constriction has an interesting effect on the angular distribution of the electrons injected into the wide 2DEF.[8] This *horn collimation* effect is discussed in detail in Section 5.

An adiabatic constriction is not necessary to observe the quantization of the conductance. The calculations[84–95] show that well-defined conductance plateaus persist for *abrupt* constrictions—even if they are rather short compared to the width. In fact an optimum length for the observation of the plateaus is found to exist, given by[85] $L_{opt} \approx 0.4 \, (W\lambda_F)^{1/2}$. In shorter constrictions, the plateaus acquire a finite slope, although they do not disappear completely even at zero length. For $L > L_{opt}$, the calculations exhibit regular oscillations that depress the conductance periodically below its quantized value. The oscillations are damped and usually have vanished before the next plateau is reached. A thermal average rapidly smears the oscillations and leads to smooth but non-flat plateaus. The plateaus disappear completely at elevated temperatures, when the thermal energy becomes comparable to the subband splitting. (See Section 2.c.) The plateaus also do not survive impurity scattering, either inside or near the constriction.[85,99,100]

Physical insight in these results can be obtained by treating the conduction through the constriction as a transmission problem, on the basis of the Landauer formula, Eq. (6). In the case of adiabatic transport discussed before, we had the simple situation that $|t_{nn'}|^2 = \delta_{n,n'}$ for $n, n' \leq N$, and zero otherwise. For an abrupt constriction, this is no longer true, and we have to consider the partial transmission of all the modes occupied in the wide regions. Semiclassically, the transverse momentum $\hbar n\pi/W$ of mode n is conserved at the abrupt transition from wide to narrow region. We thus can expect the coupling between modes n and n' in the narrow and wide regions (of width W_{min} and W_{wide}), respectively, to be strongest if $n/n' \sim W_{min}/W_{wide}$. This leads to a large increase in mode index at the exit of an abrupt constriction. Szafer and Stone[87] have formulated a *mean-field* approximation that exploits such ideas by assuming that a particular propagating or evanescent mode n in the constriction couples exclusively and uniformly to all modes n' in the wide region for which the energy $E_{n'} = (\hbar k_{n'})^2/2m$ of transverse motion is within a level splitting of E_n. Figure 6 contrasts the mode coupling for the abrupt constriction with the case of fully adiabatic transport from W_{min} to W_{wide}. Whereas in the adiabatic case, there is a one-to-one correspondence between the modes in the narrow and in the wide regions, in the abrupt case a mode in the constriction couples to a larger number (of order W_{wide}/W_{min}) of modes in the wide region.

Because of the abrupt widening of the constriction, there is a significant probability for backscattering at the exit of the constriction, in contrast to the adiabatic case considered previously. The conductance as a function of width, or Fermi energy, therefore is not a simple step function. On the nth

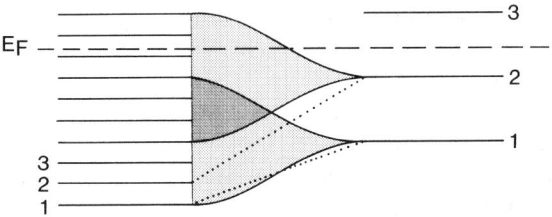

FIG. 6. Mode coupling between a constriction and a wide 2DEG region. The subband energies E_n are spaced closely in the wide region at the left. For an abrupt constriction, off-diagonal mode coupling is important (indicated by the shaded areas in the mean-field approximation of Ref. 87. The coupling is restricted between modes of the same parity.), while for an adiabatic constriction, this does not occur (dotted lines).

conductance plateau, backscattering occurs predominantly for the nth mode, since it has the largest longitudinal wavelength, $\lambda_n = h[2m(E_F - E_n)]^{-1/2}$. Resonant transmission of this mode occurs if the constriction length L is approximately an integer multiple of $\lambda_n/2$, and leads to the oscillations on the conductance plateaus found in the calculations referred to earlier. These transmission resonances are damped, because the probability for backscattering decreases with decreasing λ_n. The shortest value of λ_n on the nth conductance plateau is $h[2m(E_{n+1} - E_n)]^{-1/2} \approx (W\lambda_F)^{1/2}$ (for a square-well lateral confining potential). The transmission resonances thus are suppressed if $L \lesssim (W\lambda_F)^{1/2}$ (disregarding numerical coefficients of order unity). Transmission through evanescent modes, on the other hand, is predominant for the $(n+1)$th mode, since it has the largest decay length $\Lambda_{n+1} = h[2m(E_{n+1} - E_F)]^{-1/2}$. The observation of a clear plateau requires that the constriction length exceed this decay length at the population threshold of the nth mode, or $L \gtrsim h[2m(E_{n+1} - E_n)]^{-1/2} \approx (W\lambda_F)^{1/2}$. The optimum length,[85] $L_{opt} \approx 0.4\,(W\lambda_F)^{1/2}$, thus separates a short constriction regime, in which transmission via evanescent modes cannot be ignored, from a long constriction regime, in which transmission resonances obscure the plateaus.

c. Temperature Dependence of the Conductance

i. Thermal averaging of the point contact conductance. In Fig. 7, we show[12] the conductance of a quantum point contact in zero magnetic field as a function of gate voltage, for various temperatures between 0.3 K and 4.2 K. On increasing the temperature, the plateaus acquire a finite slope until they no longer are resolved. This is a consequence of the thermal smearing of the Fermi–Dirac distribution,

$$f(E - E_F) = \left(1 + \exp\frac{E - E_F}{k_B T}\right)^{-1}.$$

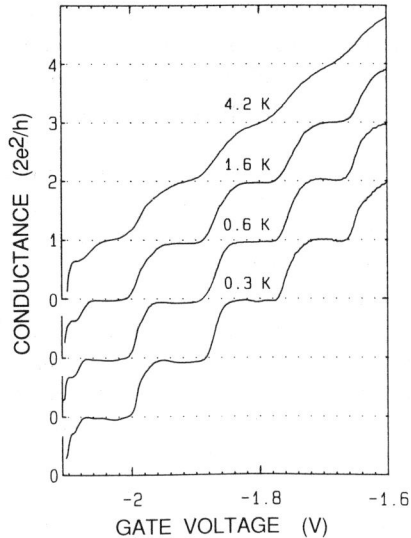

FIG. 7. Experimental temperature dependence of the conductance quantization in zero magnetic field. (From Ref. 12.)

If at $T = 0$ the conductance $G(E_F, T)$ has a step function dependence on the Fermi energy E_F, at finite temperatures it has the form,[80,101]

$$G(E_F, T) = -\int G(E, 0) \frac{df}{dE} dE = \frac{2e^2}{h} \sum_{n=1}^{\infty} f(E_n - E_F). \quad (7)$$

Here, as before, E_n denotes the energy of the bottom of the nth subband (cf. Eq. (4)). The width of the thermal smearing function df/dE is about $4k_B T$, so that the conductance steps should disappear above a characteristic temperature $T_{\text{char}} \approx \Delta E/4k_B$, with ΔE the subband splitting at the Fermi level. For the square-well confining potential, $\Delta E \approx 2(E_F - E_c)/N$. In Section 3.b, we estimate that ΔE increases from about 2 meV at $V_g = -1.0$ V (where $N = 11$) to 4 meV at $V_g = -1.8$ V (where $N = 3$). The increase in subband splitting thus qualitatively explains the experimental observation in Fig. 7 that the smearing of the plateaus is less pronounced for larger negative gate voltages. The temperature at which smearing becomes appreciable (≈ 4 K) implies $\Delta E \approx 2$ meV, which is of the correct order of magnitude.

It has been noted that a small but finite voltage drop across the constriction should have an effect that is qualitatively similar to that of a finite temperature.[101] This indeed is borne out by experiments.[12] Conduction at larger applied voltages in the nonlinear transport regime is discussed extensively in Section 7.

ii. Quantum interference effects at low temperatures. Interestingly, it was found experimentally[4,5] that, in general, a finite temperature yielded the best well-defined and flat plateaus as a function of gate voltage in the zero-field conductance. If the temperature is increased beyond this optimum (which is about 0.5 K), the plateaus disappear because of the thermal averaging discussed earlier. Below this temperature, oscillatory structure may be superimposed on the conductance plateaus, as demonstrated in Fig. 8, which shows[12] conductance traces at 40 mK (both in the absence and presence of a weak magnetic field). The strength and shape of the oscillations varies from device to device, probably due to the uncontrolled variations in the confining potential discussed in Part II. However, the data is quite reproducible if the sample is kept below 10 K. We believe that these oscillations are due at least in part to resonances in the transmission probability associated with reflections at the entrance and exit of the constriction. Indeed, similar oscillations were found in the numerical studies referred to earlier of the conductance of an abrupt constriction with $L \approx W$. Other groups[102,103] have measured comparable fine structure in the quantum point contact conductance.

In addition to these resonances, some of the structure may be a quantum interference effect associated with backscattering of electrons by impurities

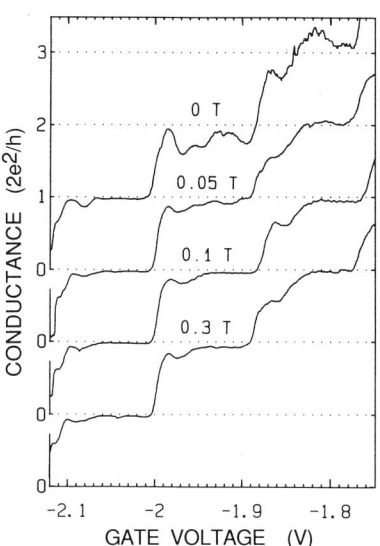

FIG. 8. Oscillatory structure observed in the zero-field conductance of a point contact at 40 mK (top curve). Some of the oscillatory structure is suppressed by a weak magnetic field (lower three curves). (From Ref. 12.)

near the opening of the constriction. The possibility that impurity scattering plays a role is supported by the fact that a very weak perpendicular magnetic field of 0.05 T leads to a suppression of some of the finest structure, leaving the more regular oscillations unchanged (Fig. 8). Increasing the magnetic field further has little effect on the flatness of the plateaus. The cyclotron radius for these fields is as long as 2 μm, so that the magnetic field hardly has any effect on the electron states *inside* the constriction. Such a field would be strong enough, however, to suppress the backscattering caused by one or a few impurities located within a few μm of the constriction. In contrast to the case of the conductance fluctuations in the diffusive transport regime,[104,105] the specific impurity configuration would be very important.

We thus believe that this data shows evidence of both impurity-related quantum interference oscillations and transmission resonances determined by the geometry. Only the latter survive in a weak perpendicular magnetic field. Provided this interpretation is correct, one in principle can estimate the length of the constriction from the periodicity of the relevant oscillations as a function of gate voltage. For a realistic modeling, one has to account for the complication that the gate voltage simultaneously affects the carrier density in the constriction, its width, and its length. Such calculations are not available, unfortunately. The effect of an increase in temperature on these quantum interference effects can be two-fold. Firstly, it leads to a suppression of the oscillations because of thermal averaging. Secondly, it reduces the phase coherence length as a result of inelastic scattering. The coherent electron focusing experiment discussed in Section 6 indicates that the latter effect is relatively unimportant for quantum ballistic transport at temperatures up to about 10 K. At higher temperatures, inelastic scattering induces a gradual transition to incoherent diffusive transport.

d. Length Dependence of the Conductance

Theoretically, one expects that the conductance quantization is preserved in longer channels than those used in the original publications[4,5] (in which, typically, $L \sim W \sim 100$ nm). Experiments on longer channels, however, did not show the quantization.[34,63,106] This is demonstrated in Fig. 9, where the resistance versus gate voltage is plotted[34] for a constriction with $L = 3.4$ μm. This is well below the transport mean free path in the bulk, which is about 10 μm in this material. The curve in Fig. 9 was taken at 50 mK, but the resistance is temperature-independent below 4 K. The sudden increase in the resistance at $V_g = -0.5$ V indicates the formation of the constriction. The lack of clear plateaus in Fig. 9 (compared with Fig. 3 for a short constriction) most likely is due to enhanced backscattering inside the constriction. Impurity

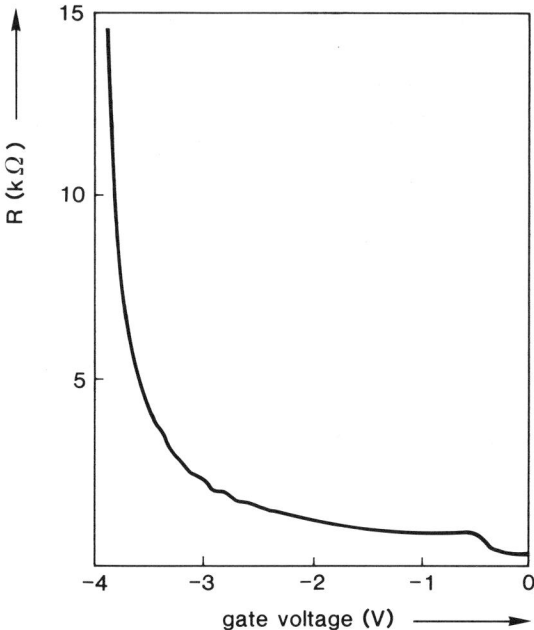

FIG. 9. Resistance of a long constriction ($L = 3.4$ μm) as a function of gate voltage, at T = 50 mK, showing the near absence of quantized plateaus. The shoulder at $V_g \approx -0.5$ V is a consequence of the formation of the constriction at the depletion threshold. (From Ref. 34.)

scattering may be one source of backscattering,[63,106] which is expected to be more severe in narrow channels due to the reduced screening in a quasi-one-dimensional electron gas.[107] Perhaps more importantly, backscattering can occur at channel wall irregularities. Thornton et al.[108] have found evidence of a small (5%) fraction of diffuse, rather than specular, reflections at boundaries defined electrostatically by a gate. In a 200 nm-wide constriction, this leads to an effective mean free path of about 200 nm/0.05 ≈ 4 μm, comparable to the constriction length in this device.

Long constrictions have been studied more extensively in the *quasi-ballistic* transport regime, where the mean free path is much larger than the channel width, but shorter than its length (*cf.* Refs. 50 and 109). Low-temperature transport in this regime is characterized by weak localization and electron–electron interaction effects, and by universal conductance fluctuations. It would be of interest to study the transition from the ballistic to the quasi-ballistic transport regime, by performing systematic studies on the length and width dependence of the quantum transport through smooth constrictions fabricated on material with different values for the mobility. Some recent work by Timp et al.[63,106] and Brown et al.[102] is in this direction.

3. Magnetic Depopulation of Subbands

a. Magneto-Electric Subbands

If a magnetic field B is applied perpendicular to a wide 2DEG, the kinetic energy of the electrons is quantized[110] at energies $E_n = (n - \tfrac{1}{2})\hbar\omega_c$, with $\omega_c = eB/m$ the cyclotron frequency. The quantum number $n = 1, 2, \ldots$ labels the Landau levels. The number of Landau levels below the Fermi energy $N \sim E_F/\hbar\omega_c$ decreases as the magnetic field is increased. This *magnetic depopulation* of Landau levels is observed in the quantum Hall effect, where each occupied Landau level contributes e^2/h (per spin direction) to the Hall conductance. The Landau level quantization is the result of the periodicity of the circular motion in a magnetic field. In a narrow channel or constriction, the cyclotron orbit is perturbed by the electrostatic lateral confinement, and this modifies the energy spectrum. Instead of Landau levels, one now speaks of *magneto-electric subbands*. The effect of the lateral confinement on the number N of occupied subbands becomes important when the cyclotron orbit at the Fermi energy (of radius $l_{\text{cycl}} = \hbar k_F/eB$) no longer fits fully into the channel. If $l_{\text{cycl}} \gg W$, the effect of the magnetic field on the trajectories (and thus on the energy spectrum) can be neglected, and N becomes approximately B-independent. Simple analytic expressions for the B-dependence of N can be obtained for a parabolic confining potential,[111] or for a square-well potential.[15] For the square well, one finds in a semiclassical approximation (with an accuracy of ± 1), and neglecting the spin-splitting of the energy levels,

$$N \approx \text{Int}\left[\frac{2}{\pi}\frac{E_F}{\hbar\omega_c}\left(\arcsin\frac{W}{2l_{\text{cycl}}} + \frac{W}{2l_{\text{cycl}}}\left[1 - \left(\frac{W}{2l_{\text{cycl}}}\right)^2\right]^{1/2}\right)\right], \tag{8a}$$

$$\text{if } l_{\text{cycl}} > \frac{W}{2},$$

$$N \approx \text{Int}\left[\frac{1}{2} + \frac{E_F}{\hbar\omega_c}\right], \text{ if } l_{\text{cycl}} < \frac{W}{2}. \tag{8b}$$

One easily verifies that for zero magnetic field, Eq. (8) yields $N = \text{Int}[k_F W/\pi]$, as it should. If Eq. (8) is applied to a constriction containing a potential barrier of height E_c, then one should replace $E_F \to E_F - E_c$ and, consequently, $l_{\text{cycl}} \to l_{\text{cycl}}(1 - E_c/E_F)^{1/2}$. In Fig. 10, we show the depopulation of Landau levels with its characteristic $1/B$ dependence of N (dashed curve), and the much slower depopulation of magneto-electric subbands for $W/2l_{\text{cycl}} < 1$ (solid curve). These results are calculated from Eq. (8) for a square-well potential with $k_F W/\pi = 10$. Smoother confining potentials (e.g., parabolic) give similar results.[111]

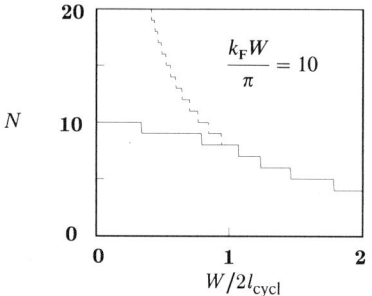

FIG. 10. Magnetic field dependence of the number of occupied subbands in a narrow channel, according to Eq. (8) (solid curve). The dashed curve gives the magnetic depopulation of Landau levels in a wide 2DEG, which has a $1/B$ dependence.

We note that in Fig. 10, a possible oscillatory B-dependence of E_F has been ignored, which would result from pinning of the Fermi level to the Landau levels—either in the narrow channel itself or in the adjacent wide 2DEG regions. To determine this B-dependence for a short constriction (where both pinning mechanisms compete) would require a self-consistent solution of the Schrödinger and Poisson equation, which has not been done yet in a quantizing magnetic field for such a geometry. In the application of Eq. (8) to the experiments in Section 3.b on a constriction containing a barrier, we similarly will neglect a possible oscillatory B-dependence of $E_F - E_c$.

In the Landau gauge for the vector potential $A = (0, Bx, 0)$ (for a channel along the y-axis), the translational invariance along the channel is not broken by the magnetic field, so that the propagating modes can still be described by a wave number k for propagation along the channel—just as in zero magnetic field (cf. Section 2.b). However, the dispersion relation $E_n(k)$ does not have the form of Eq. (4), and consequently, the group velocity $v_n \equiv dE_n/\hbar dk$ no longer is given by $\hbar k/m$ (as it is for $B = 0$). In a strong magnetic field ($l_{cycl} \lesssim W/2$), the propagating modes are extended along a boundary of the sample, and are referred to as magnetic *edge channels*. Classically, these states correspond to skipping orbits along a channel boundary (cf. Fig. 11a). In weaker fields ($l_{cycl} \gtrsim W/2$), the propagating modes extend throughout the bulk, and correspond to *traversing trajectories* that interact with both opposite channel boundaries (Fig. 11b). The wave functions and energy spectra for these various quantum states are very different, yet experimentally a gradual transition is observed from the zero-field conductance quantization of a quantum point contact to the strong-field quantum Hall effect. (See next section). The fundamental cancellation between group velocity and density of states for one-dimensional waveguide modes, which does not depend explicitly on the nature of the dispersion law $E_n(k)$, provides the theoretical explanation of the remarkable connection between these two quantum phenomena, which at first sight seem unrelated.

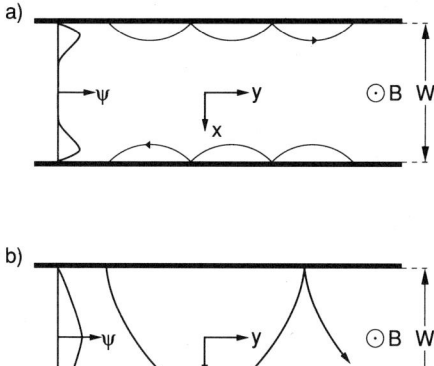

FIG. 11. Trajectories in a narrow channel in a perpendicular magnetic field (right) and the corresponding transverse profile of the wave function Ψ (left). Skipping orbits on both opposite edges and the corresponding edge states are shown in a, a traversing trajectory and the corresponding bulk state in b. Note that the wave functions shown correspond to the nodeless $n = 1$ mode.

b. *Conductance Quantization in an External Magnetic Field*

In Fig. 12, measurements[57] are shown of the conductance versus gate voltage for various values of the magnetic field (at $T = 0.6$ K). The point contact conductance has been obtained from the measured resistance after subtraction of a gate voltage-independent background resistance (*cf.* Part II). The measurements have been performed for values of the magnetic field where the 2DEG resistivity has a Shubnikov–de Haas minimum. The background resistance then is due mainly to the non-ideal ohmic contacts, and increases from about 4 kΩ to 8 kΩ between zero and 2.5 T.[57] Fig. 12 demonstrates that the conductance quantization is conserved in the presence of a magnetic field, and shows a smooth transition from zero-field quantization to quantum Hall effect. The main effect of the magnetic field is to reduce the number of plateaus in a given gate voltage interval. This provides a direct demonstration of depopulation of 1D subbands, as analyzed later. In addition, one observes that the flatness of the plateaus improves in the presence of the field. This is due to the spatial separation at opposite edges of the constriction of the left- and right-moving electrons (illustrated in Fig. 11a), which reduces the probability for backscattering in a magnetic field.[34,77] We return to the magnetic suppression of backscattering in Section 4. Finally, in strong magnetic fields, the spin degeneracy of the energy levels is removed, and additional plateaus appear at *odd* multiples of e^2/h. They are much less well-resolved than the even-numbered plateaus, presumably because the Zeeman

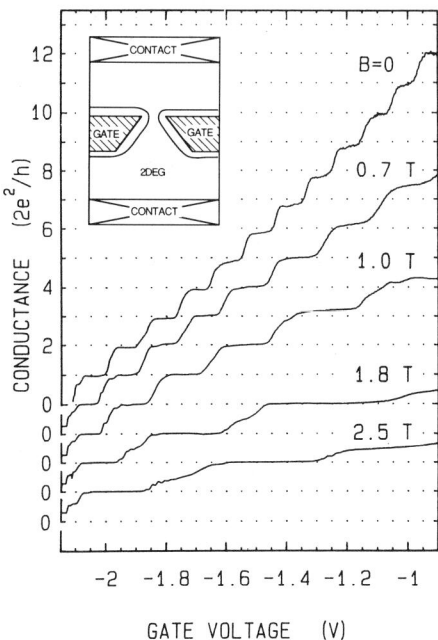

FIG. 12. Point contact conductance (corrected for a series lead resistance) as a function of gate voltage for several magnetic field values, illustrating the transition from zero-field quantization to quantum Hall effect. The curves have been offset for clarity. The inset shows the device geometry. (From Ref. 57.)

spin-splitting energy $|g\mu_B B|$ is considerably smaller than the subband splitting ΔE. (If one uses the low-field value $g = -0.44$ for the Landé g-factor in GaAs, and the definition $\mu_B = e\hbar/2m_e$ for the Bohr magneton, one finds a splitting as small as 0.025 meV per T, while ΔE in general is more than 1 meV, as discussed later.) We note that the spin degeneracy of the quantized plateaus also can be removed by a strong parallel (rather than perpendicular) magnetic field, as shown by Wharam et al.[5]

Because the arguments leading to Eq. (5) are valid regardless of the nature of the subbands involved, we can conclude that in the presence of a magnetic field, the conductance remains quantized according to $G = (2e^2/h)N$ (ignoring spin-splitting, for simplicity). Calculations[95,112-114] done for specific point contact geometries confirm this general conclusion. The number of occupied (spin degenerate) subbands N is given approximately by Eq. (8), for a square-well confining potential. In the high-magnetic field regime $W \gtrsim 2l_{\text{cycl}}$, the quantization of G with N given by Eq. (8b) is just the quantum Hall effect in a two-terminal configuration (which has been shown[115-117] to be equivalent to the quantization of the Hall resistance in the more usual four-terminal configuration, *cf.* Section 9.a; the ohmic contact resistance is exceptionally large in our sample, but usually is much smaller so that accurate quantization becomes possible in a two-terminal measurement in high magnetic fields). At lower magnetic fields, the quantization of the point contact conductance provides a direct and extremely straightforward method to measure, via $N = G(2e^2/h)^{-1}$, the depopulation of magneto-electric subbands in the constriction. Previously, this effect in a narrow channel had been studied indirectly by measuring the deviations from the $1/B$ periodicity of the Shubnikov–de Haas oscillations[118–120] (the observation of which is made difficult by the irregular conductance fluctuations that result from quantum interference in a disordered system).

Figure 13 shows the number N of occupied subbands obtained from the measured G (Fig. 12), as a function of reciprocal magnetic field for various gate voltages.[57] Also shown are the theoretical curves according to Eq. (8), with the potential barrier in the constriction taken into account. The barrier

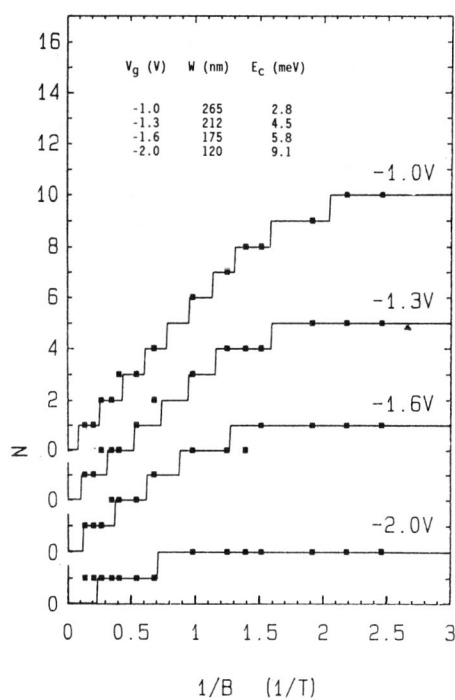

FIG. 13. Number of occupied subbands as a function of reciprocal magnetic field for several values of the gate voltage. Data points have been obtained directly from the quantized conductance (Fig. 12); solid curves are calculated from Eq. (8), with the parameters tabulated in the inset. (From Ref. 57.)

height E_c is obtained from the high-field conductance plateaus (where $N \approx (E_F - E_c)/\hbar\omega_c$), and the constriction width W then follows from the zero-field conductance (where $N \approx [2m(E_F - E_c)/\hbar^2]^{1/2} W/\pi$). The good agreement found over the entire field range confirms our expectation that the quantized conductance is determined exclusively by the number of occupied subbands—irrespective of their electric or magnetic origin. The present analysis is for a square-well confining potential. For the narrowest constrictions, a parabolic potential should be more appropriate; it has been used to analyze the data of Fig. 12 in Refs. 12 and 121. The most realistic potential shape is a parabola with a flat section inserted in the middle,[122,123] but this potential contains an additional undetermined parameter (the width of the flat section). Wharam et al.[124] have analyzed their depopulation data using such a model (cf. also Ref. 121). Because of the uncertainties in the actual shape of the potential, the parameter values tabulated in Fig. 13 only are rough estimates, but we believe that the observed trends in the dependence of W and E_c on V_g are significant.

In Fig. 14, we have plotted this trend (assuming a square-well confining potential) for the point contact discussed before (curves labeled 2) and for another (nominally identical) point contact (curves 3). For comparison, we also show the results obtained in Section 4 for a longer and wider constriction [34] (curves 1). The electron density n_c in the constriction has been calculated approximately by $n_c \approx (E_F - E_c)m/\pi\hbar^2$ (i.e., using the two-dimensional density of states, with neglect of the subband quantization). The dependence of the width and electron density on the gate voltage is qualitatively similar for the three devices. The quantitative differences between the two nominally identical quantum point contacts (curves 2 and 3) serve to emphasize the

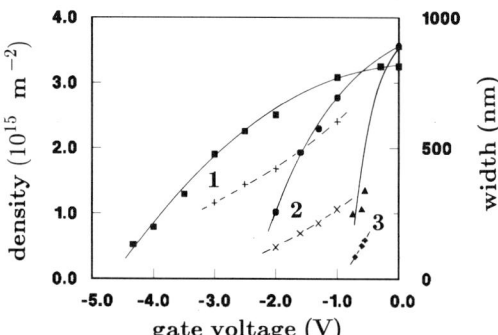

FIG. 14. Electron gas density (solid curves) and width (dashed curves) of the constrictions defined by three different split gate devices. The curves labeled 1 are for the wide split gate of Ref. 34 (discussed in Section 4). Curves 2 are for the point contact of Ref. 57 (cf. Fig. 13), and curves 3 for another point contact of identical design.

importance of the uncontrolled variations in the device electrostatics discussed in Part II. (It should be noted, though, that curve 2 is representative for several other samples studied.) The larger constriction (curve 1) needs a much higher gate voltage for pinch-off simply because of its different dimensions. It would be of interest to compare these results with a self-consistent solution of the three-dimensional Poisson and Schrödinger equation, which now are starting to become available.[122,123]

A significant reduction of the electron density n_c in the constriction with increasing negative gate voltage occurs in all the samples (*cf.* Fig. 14). The potential barrier in the constriction thus cannot be neglected (except at low gate voltages). As an example, one finds for a typical quantum point contact (Fig. 13 or curve 2 in Fig. 14) that E_c/E_F varies from 0 to 0.7 (with $E_F = 12.7$ meV) as the gate voltage is varied from 0 to -2.0 V. This corresponds to a reduction of n_c by a factor of 3.5. Because of the relatively large potential barrier, the N-dependence of the zero-field subband splitting at the Fermi energy $\Delta E \approx 2(E_F - E_c)/N$ for a small number of occupied subbands in the square well is found to be substantially reduced from the $1/N$ dependence that would follow on ignoring the barrier. For the typical sample mentioned previously, one finds at $V_g = -1.8$ V, where $N = 3$, a subband splitting $\Delta E \approx 3.5$ meV. This is only a factor of 2 larger than the splitting $\Delta E \approx 1.8$ meV that one finds at $V_g = -1.0$ V, although $N = 11$ has increased by almost a factor of 4.

4. MAGNETIC SUPPRESSION OF BACKSCATTERING AT A POINT CONTACT

Only a small fraction of the electrons injected by the current source into the 2DEG is transmitted through the point contact. The remaining electrons are scattered back into the source contact. This is the origin of the nonzero resistance of a ballistic point contact. In this section, we shall discuss how a relatively weak magnetic field leads to a suppression of the *geometrical backscattering* caused by the finite width of the point contact, while the amount of backscattering caused by the potential barrier in the point contact remains essentially unaffected.

The reduction of backscattering by a magnetic field is observed as a *negative* magnetoresistance (i.e., $R(B) - R(0) < 0$) in a *four-terminal* measurement of the point contact resistance.[34] The distinction between two- and four-terminal resistance measurements already has been mentioned in Part II. In Sections 2 and 3, we considered the two-terminal resistance R_{2t} of a point contact. This resistance is the total voltage drop between source and drain divided by the current, and has a particular significance as the quantity that determines the dissipated power I^2R_{2t}. Two-terminal resistance measurements, however, do not address the issue of the *distribution* of the voltage drop along the sample. In the ballistic (or adiabatic) transport regime, the measurement and analysis of the voltage distribution are non-trivial, because the concept of a local resistivity tensor (associated with that of local equilibrium) breaks down. (We will discuss non-local transport measurements in ballistic and adiabatic transport in Section 6 and Part IV, respectively.) In this section, we are concerned with the four-terminal *longitudinal* resistance R_L, measured with two adjacent (not opposite) voltage probes, one at each side of the constriction (*cf.* the inset in Fig. 15). We speak of a (generalized) longitudinal resistance, by analogy with the longitudinal resistance measured in a Hall bar, because the line connecting the two voltage probes does not intersect the line connecting the current source and drain (located at the far left and right of the conductor shown in Fig. 15). The voltage probes are positioned on wide 2DEG regions, well away from the constriction. This allows the establishment of local equilibrium near the voltage probes, at least in weak magnetic fields (*cf.* Part IV), so that the measured four-terminal resistance does not depend on the properties of the probes.

The experimental results[34] for R_L in this geometry are plotted in Fig. 15. This quantity shows a negative magnetoresistance, which is temperature-independent (between 50 mK and 4 K), and is observed in weak magnetic fields once the narrow constriction is defined (for $V_g \lesssim 0.3$ V). (The very small

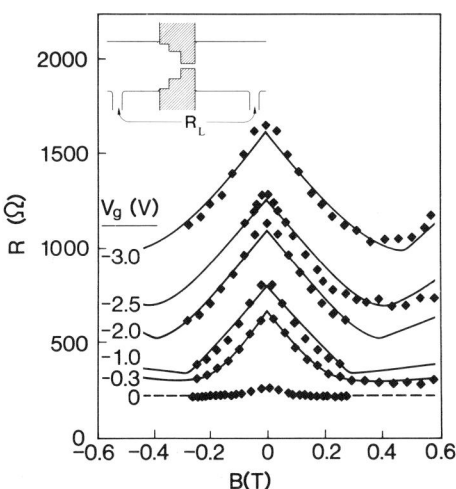

FIG. 15. Four-terminal longitudinal magnetoresistance R_L of a constriction for a series of gate voltages from 0 V (lowest curve) to -3 V. Solid lines are according to Eqs. (8) and (10), with the constriction width as adjustable parameter. The inset shows schematically the device geometry, with the two voltage proves used to measure R_L. (From Ref. 34.)

effect seen in the trace for $V_g = 0$ V probably is due to a density reduction by the Schottky barrier under the gate.) At stronger magnetic fields ($B > 0.4$ T), a crossover is observed to a positive magnetoresistance. The zero-field resistance, the magnitude of the negative magnetoresistance, the slope of the positive magnetoresistance, as well as the crossover field, all increase with increasing negative gate voltage.

The magnetic field dependence of the four-terminal resistance shown in Fig. 15 is qualitatively different from that of the two-terminal resistance R_{2t} considered in Section 3. In fact, R_{2t} is approximately B-independent in weak magnetic fields (below the crossover fields of Fig. 15). We recall that R_{2t} is given by (*cf.* Eq. (5))

$$R_{2t} = \frac{h}{2e^2} \frac{1}{N_{\min}}, \qquad (9)$$

with $N_{\min}$ the number of occupied subbands at the bottleneck of the constriction (where it has its minimum width and electron gas density). In weak magnetic fields such that $2l_{\text{cycl}} > W$, the number of occupied subbands remains approximately constant (*cf.* Fig. 10 or Eq. (8)), which is the reason for the weak dependence on B of the two-terminal resistance in this field regime. For stronger fields, Eq. (9) describes a *positive* magnetoresistance, because $N_{\min}$ decreases due to the magnetic depopulation of subbands discussed in Section 3. Why then do we find a *negative* magnetoresistance in the four-terminal measurements of Fig. 15? Qualitatively, the answer is shown in Fig. 16 for a constriction without a potential barrier. In a magnetic field the left- and right-moving electrons are separated spatially by the

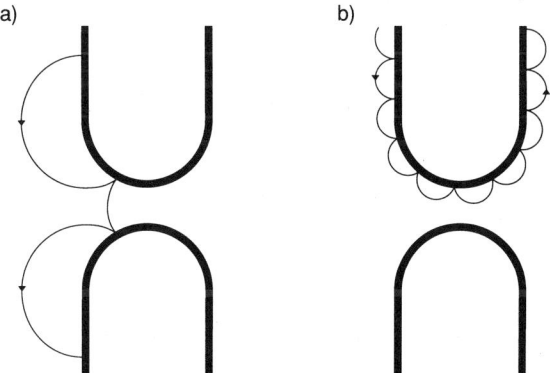

FIG. 16. Illustration of the reduction of backscattering by a magnetic field, which is responsible for the negative magnetoresistance of Fig. 15. Shown are trajectories approaching a constriction without a potential barrier, in a weak (*a*) and strong (*b*) magnetic field.

Lorentz force at opposite sides of the constriction. Quantum mechanically, the skipping orbits in Fig. 16 correspond to magnetic edge states (cf. Fig. 11). Backscattering thus requires scattering across the width of the constriction, which becomes increasingly improbable as l_{cycl} becomes smaller and smaller compared to the width. (Compare Figs. 16a, b.) For this reason, in a magnetic field the role of the constriction as the dominant bottleneck limiting the current is taken over increasingly by the contact resistance at the connection of the current contacts with the 2DEG. Quantitatively, this can be treated as follows.[34]

Consider the four-terminal geometry of Fig. 17. A current I flows through a constriction due to a chemical potential difference between the source (at chemical potential $\mu_s = E_F + \delta\mu$) and the drain (at chemical potential $\mu_d = E_F$). Unless the magnetic field is very weak, we can assume that the left- and right-moving electrons are separated spatially at the lower and upper boundary of the wide 2DEG, and that backscattering can occur only at the constriction. The four-terminal longitudinal resistance is defined as $R_L \equiv (\mu_l - \mu_r)/eI$, where μ_l and μ_r are the chemical potentials measured by the two voltage probes shown in Fig. 17, at the upper boundary to the left and right of the constriction. The left voltage probe, which is in equilibrium with the electrons coming from the source, has $\mu_l = E_F + \delta\mu$. A fraction $N_{\text{min}}/N_{\text{wide}}$ of these electrons is transmitted through the constriction, the remainder returning to the source contact via the opposite edge. Here, N_{min} and N_{wide} are the number of propagating modes in the narrow and wide regions, respectively. If we assume a local equilibrium near the voltage probes, the excess chemical potential, $\mu_r - E_F$, is reduced by the same factor, $\mu_r - E_F = (N_{\text{min}}/N_{\text{wide}})(\mu_l - E_F)$. (In the absence of local equilibrium, the measured chemical potential depends on how the voltage probe couples to the 2DEG.) The transmitted current itself is determined by the two-terminal resistance, from Eq. (9), which gives $I = (2e/h)N_{\text{min}}\delta\mu$. Collecting results, we find the

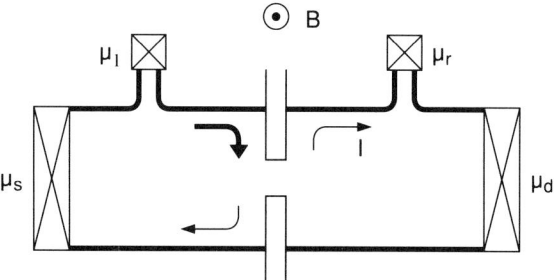

FIG. 17. Schematic arrangement of a four-terminal conductor containing a constriction, used in the text to derive Eq. (10). The spatial separation of the left- and right-moving electrons in the wide regions is indicated (cf. Fig. 11a).

simple formula,[34]

$$R_L = \frac{h}{2e^2}\left(\frac{1}{N_{min}} - \frac{1}{N_{wide}}\right), \tag{10}$$

obtained independently by Büttiker.[77] For a more formal derivation of this result using the Landauer–Büttiker formalism, see Section 9.

At small magnetic fields, N_{min} is approximately constant, while N_{wide} decreases linearly with B (cf. Eq. (8)). Equation (10) thus predicts a *negative* magnetoresistance. Physically, the resistance reduction is due to the fact that, as B is increased, a larger and larger fraction of the edge states is transmitted through the constriction (as is illustrated in Fig. 16). If the electron density in the wide and narrow regions is equal (i.e., the barrier height $E_c = 0$), then the resistance R_L vanishes for fields $B > B_{crit} \equiv 2\hbar k_F/eW$. This follows from Eq. (10), because in this case, N_{min} and N_{wide} are identical. If, on the other hand, the electron density in the constriction is less than its value in the wide region, then Eq. (10) predicts a crossover at B_{crit} to a strong-field regime of *positive* magnetoresistance described by

$$R_L \approx \frac{h}{2e^2}\left(\frac{\hbar\omega_c}{E_F - E_c} - \frac{\hbar\omega_c}{E_F}\right), \text{ if } B > B_{crit}. \tag{11}$$

The solid curves in Fig. 15 have been obtained from Eqs. (8) and (10) (after addition of the background resistance found at gate voltage zero), with the constriction width W_{min} as the single freeparameter. The barrier height E_c has been determined independently from the two-terminal resistance in high magnetic fields (cf. Section 3). The agreement found is quite good, confirming the validity of Eq. (10) in the weak-field regime, and providing a means to determine the constriction width (which is found to vary from 0.8 to 0.3 μm as V_g varies from -0.3 to -3.0 V; see the curves labeled 1 in Fig. 14). The constriction in the present experiment is relatively long ($L \approx 3.4$ μm), so that it does not exhibit clear quantized plateaus in the zero-field two-terminal conductance (cf. Section 2.d and Fig. 9, measured on this same sample). For this reason, the discreteness of N was ignored in the theoretical curves in Fig. 15. We emphasize, however, that the preceding analysis is equally applicable to the quantum case (as will be discussed in Section 9). For example, Eq. (10) describes the quantization in *fractions* of $2e^2/h$ of the longitudinal conductance R_L^{-1} of a point contact observed experimentally.[125,126] (See Section 9.a).

In high magnetic fields in the quantum Hall effect regime, the validity of the result, Eq. (10), is not restricted to point contacts, but holds also for a wide Hall bar (having N_{wide} occupied Landau levels), of which a segment has a reduced electron density (so that only N_{min} Landau levels are occupied in that region). Many such experiments have been performed.[127–132] In these papers, the simplicity of an analysis in terms of transmitted and reflected edge states had not been appreciated yet, in contrast to more recent experimental work[133–135] in which a narrow gate across the Hall bar induces a potential barrier in the 2DEG. Deviations from Eq. (10) can result from interedge channel scattering; cf. Refs. 136, 137, and 138.

The preceding argument predicts a Hall resistance, $R_H = R_{2t} - R_L$, in the wide regions given by

$$R_H = \frac{h}{2e^2} \frac{1}{N_{\text{wide}}}, \tag{12}$$

unaffected by the presence of the constriction. This is a direct consequence of our assumption of local equilibrium near the voltage probes. In the weak-field regime of Fig. 15, this result, Eq. (12), has been confirmed experimentally, but deviations were found for higher magnetic fields.[34] These are discussed further in Section 9. Anticipating that discussion, we note that in a strong magnetic field, the assumption of local equilibrium near the voltage probes is a sufficient but not necessary condition for Eqs. (10) and (12) to hold. Even in the absence of local equilibrium, these equations remain valid if the voltage probes are much wider than l_{cycl}, so that all edge states on one edge of the wide 2DEG region are fully absorbed by the voltage contact. Such an *ideal* contact induces itself a local equilibrium among the edge states[77] (cf. Section 9.a).

For completeness, we mention that one also can measure the two four-terminal *diagonal* resistances R_{D^+} and R_{D^-} across the constriction, in such a way that the two voltage probes are on opposite (not adjacent) edges of the 2DEG, on either side of the constriction. (See Fig. 18). Additivity of voltages on contacts tells us that $R_{D^\pm} = R_L \pm R_H$ (for the magnetic field direction of Fig. 18), so that

$$R_{D^+} = \frac{h}{2e^2} \frac{1}{N_{\text{min}}}; \quad R_{D^-} = \frac{h}{2e^2} \left(\frac{2}{N_{\text{wide}}} - \frac{1}{N_{\text{min}}} \right). \tag{13}$$

On field reversal, R_{D^+} and R_{D^-} are interchanged. Thus, a four-terminal resistance (R_{D^+} in Eq. (13)) can be equal in principle to the two-terminal resistance (R_{2t} in Eq. (9)). The main difference between these two quantities is that the additive contribution of the ohmic contact resistance, and of a part of the diffusive background resistance, is eliminated in the four-terminal resistance measurement (cf. Part II).

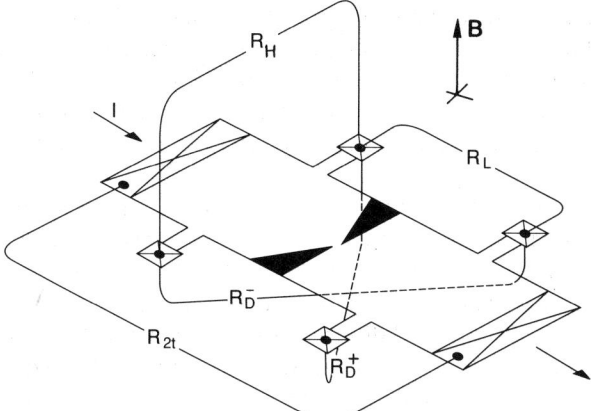

FIG. 18. Perspective view of a six-terminal Hall bar containing a point contact, showing the various two- and four-terminal resistances mentioned in the text.

In conclusion, we have demonstrated that a magnetic field suppresses the backscattering at the point contact that results from its finite width. In Ref. 34, it was suggested that a magnetic field also may suppress the backscattering caused by impurities in a narrow channel, thereby explaining a negative magnetoresistance effect in the quasi-ballistic transport regime first observed by Choi et al.[50] (See also Ref. 109.) A theory of the quantum Hall effect based on the suppression of backscattering by a magnetic field has been developed by Büttiker;[77] see Part IV.

5. Electron Beam Collimation and Point Contacts in Series

a. Introduction

The first experimental study of ballistic transport through two opposite point contacts was carried out by Wharam et al.,[9] who discovered that the series resistance is considerably less than the sum of the two individual resistances. Subsequent experiments confirmed this result.[139,151] To explain this observation theoretically, two of us proposed[8,66] that *collimation* of the electron beam injected by a point contact enhances the direct transmission probability from one point contact to the other, thereby significantly reducing the series resistance below its ohmic value. An alternative measurement configuration was suggested, in which the deflection of the beam by a magnetic field can be sensitively detected, to provide direct experimental proof of collimation. Such an experiment now has been performed,[58] and will be discussed shortly in some detail. We will not consider here the obvious alternative geometry of two adjacent point contacts in parallel (studied in Refs. 140, 141, and 142). In that geometry, the collimation effect can not enhance the coupling of the two point contacts, so only small deviations from Ohm's law are to be expected.

The collimation effect has an importance in ballistic transport that goes beyond the point contact geometry for which it originally was proposed. Recent theoretical[143,144] and experimental[145,146] work has shown that collimation is at the origin of the phenomenon of the quenching of the Hall effect (a suppression of the Hall resistance at low magnetic fields).[52] In fact, collimation is one of a set of semiclassical mechanisms[144] that together can explain a whole variety of magnetoresistance anomalies found experimentally in ballistic narrow-channel geometries, including quenched and negative Hall resistances, the last Hall plateau, bend resistances, and geometrical resonances. These recent developments emphasize the general importance of the collimation effect, but will not be discussed here any further. In this chapter, we restrict ourselves to point contact geometry, which happens to be an ideal geometry for the study of collimated electron beams.

b. *Collimation*

Collimation follows from the constraints on the electron momentum imposed by the potential energy barrier in the point contact (barrier collimation), and by the gradual flaring of the confining potential at the entrance and exit of the point contact (horn collimation).[8,66] Semiclassically, collimation results from the adiabatic invariance of the product of channel width W and absolute value of the transverse momentum $\hbar k_y$. (This product is proportional to the action for motion transverse to the channel.)[147] Therefore, if the electrostatic potential in the point contact region is sufficiently smooth, the quantity $S = |k_y|W$ is approximately constant from point contact entrance to exit. Note that S/π corresponds to the quantum mechanical one-dimensional subband index n. The quantum mechanical criterion for adiabatic transport thus is that the potential in the point contact region does not cause intersubband transitions. To this end, it should be smooth on the scale of a wavelength, and the width should change gradually on the same length scale. As we discussed in Section 2.b.iii, adiabatic transport breaks down at the exit of the point contact, where it widens abruptly into a 2DEG of essentially infinite width. Barrier- and horn-collimation reduce the *injection/acceptance cone* of the point contact from its original value of π to a value of $2\alpha_{max}$. This effect is illustrated in Fig. 19. Electrons incident at an angle $|\alpha| > \alpha_{max}$ from normal incidence are reflected. (The geometry

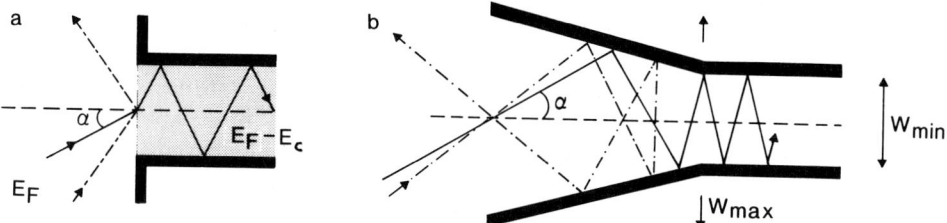

FIG. 19. Illustration of the collimation effect for an abrupt constriction (*a*) containing a potential barrier of height E_c, and for a horn-shaped constriction (*b*) that is flared from a width W_{min} to W_{max}. The dash-dotted trajectories approaching at an angle α outside the injection/acceptance cone are reflected. (From Ref. 66.)

of Fig. 19b is known in optics as a *conical reflector*.)[148] On the other hand, all electrons leave the constriction at an angle $|\alpha| < \alpha_{max}$; i.e., the injected electrons form a collimated beam of angular opening $2\alpha_{max}$.

To obtain an analytic expression for the collimation effect, we describe the shape of the potential in the point contact region by three parameters: W_{min}, W_{max}, and E_c. (See Fig. 19.) We consider the case that the point contact has its minimal width W_{min} at the point where the barrier has its maximal height E_c above the bottom of the conduction band in the broad regions. At that point, the largest possible value of S is

$$S_1 \equiv (2m/\hbar^2)^{1/2}(E_F - E_c)^{1/2} W_{min}.$$

We assume that adiabatic transport (i.e., S = constant) holds up to a point of zero barrier height and maximal width W_{max}. The abrupt separation of adiabatic and non-adiabatic regions is a simplification that can be—and has been—tested by numerical calculations. (See the following.) At the point contact exit, the largest possible value of S is

$$S_2 \equiv (2m/\hbar^2)^{1/2}(E_F)^{1/2} \sin\alpha_{max} W_{max}.$$

The invariance of S implies that $S_1 = S_2$, so that

$$\alpha_{max} = \arcsin\left(\frac{1}{f}\right); \quad f \equiv \left(\frac{E_F}{E_F - E_c}\right)^{1/2} \frac{W_{max}}{W_{min}}. \qquad (14)$$

The *collimation factor* $f \geq 1$ is the product of a term describing the collimating effect of a barrier of height E_c, and a term describing collimation due to a gradual widening of the point contact width from W_{min} to W_{max}. In the adiabatic approximation, the angular injection distribution $P(\alpha)$ is proportional to $\cos\alpha$ with an abrupt truncation at $\pm\alpha_{max}$. The cosine angular dependence follows from the cosine distribution of the incident flux in combination with time reversal symmetry, and thus is not affected by the reduction of the *injection/acceptance* cone. (This also can be seen from the quantum mechanical correspondence, by noting that in the absence of intersubband transitions, the relative magnitude of the contributions of the transmitted one-dimensional subbands to the injected current cannot change.) We conclude, therefore, that in the adiabatic approximation, $P(\alpha)$ (normalized to unity) is given by

$$P(\alpha) = \frac{1}{2} f \cos \alpha, \text{ if } |\alpha| < \alpha_{\max} \equiv \arcsin(1/f), \tag{15}$$

$$P(\alpha) = 0, \text{ otherwise.}$$

We defer to Section 5.d a comparison of the analytical result, Eq. (15), with a numerical calculation.

The injection distribution, Eq. (15), can be used to obtain (in the semiclassical limit) the direct transmission probability T_d through two identical opposite point contacts separated by a large distance L. To this end, first note that T_d/N is the fraction of the current injected through the first point contact, which is transmitted through the second point contact (since the transmission probability through the first point contact is N, for N occupied subbands in the point contact). Electrons injected within a cone of opening angle $W_{\max}/L$ centered at $\alpha = 0$ reach the opposite point contact, and are transmitted. If this opening angle is much smaller than the total opening angle $2\alpha_{\max}$ of the beam, then the distribution function $P(\alpha)$ can be approximated by $P(0)$ within this cone. This approximation requires $W_{\max}/L \ll 1/f$, which is satisfied experimentally in devices with a sufficiently large point contact separation. We thus obtain $T_d/N = P(0) W_{\max}/L$, which, using Eq. (15), can be written as[8]

$$T_d = f \frac{W_{\max}}{2L} N. \tag{16}$$

This simple analytical formula can be used to describe the experiments on transport through identical opposite point contacts in terms of one empirical parameter f, as discussed in the following two subsections.

c. Series Resistance

The expression for the series resistance of two identical opposite point contacts in terms of the preceding transmission probability can be obtained directly from the Landauer–Büttiker formalism,[7] as was done in Ref. 8.

We give here an equivalent, somewhat more intuitive derivation. Consider the geometry shown in Fig. 20a. A fraction T_d/N of the current GV injected through the first point contact by the current source is transmitted directly through the second point contact (and then drained to ground). Here, $G = (2e^2/h)N$ is the conductance of the individual point contacts, and V is the source-drain voltage. The remaining fraction, $1 - T_d/N$, equilibrates in the region between the point contacts, as a result of inelastic scattering. (Elastic scattering is sufficient if phase coherence does not play a role.) Since that region cannot drain charge (as contacts 2 and 4 are not connected to ground in Fig. 20a), these electrons eventually will leave via one of the two point contacts. For a symmetric structure, we may assume that the fraction $\frac{1}{2}(1 - T_d/N)$ of the injected current GV is transmitted through the second point contact after equilibration. The total source-drain current I is the sum of the direct and indirect contributions,

$$I = \frac{1}{2}\left(1 + \frac{T_d}{N}\right)GV.$$

The series conductance $G_{\text{series}} = I/V$ becomes

$$G_{\text{series}} = \frac{1}{2}G\left(1 + \frac{T_d}{N}\right). \tag{17}$$

In the absence of direct transmission ($T_d = 0$), one recovers the ohmic addition law for the resistance, as expected for the case of complete intervening

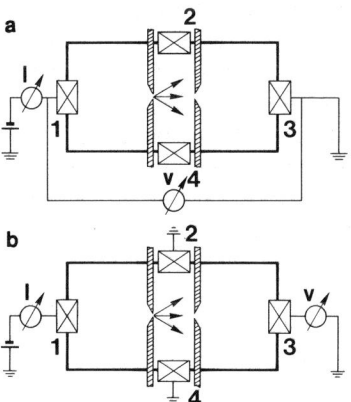

FIG. 20. Configuration for a series resistance measurement (*a*) and for a measurement of the collimation peak (*b*). The gates defining the two opposite point contacts are shaded; the squares indicate ohmic contacts to the 2DEG. (From Ref. 66.)

equilibration (*cf.* the related analysis by Büttiker of tunneling in series barriers).[149,150] At the opposite extreme, if all transmission is direct ($T_d = N$), the series conductance is identical to that of the single-point contact. Substituting the expression, Eq. (16), into Eq. (17), we obtain the result[8] for small but nonzero direct transmission,

$$G_{\text{series}} = \frac{1}{2}G\left(1 + f\frac{W_{\text{max}}}{2L}\right). \tag{18}$$

The plateaus in the series resistance as a function of gate voltage, observed experimentally,[9] of course are not obtained in the semiclassical calculation leading to Eq. (18). However, since the non-additivity essentially is a semiclassical collimation effect, the present analysis should give a reasonably reliable estimate of deviations from additivity for not too narrow point contacts. Once the point contact width becomes less than a wavelength, diffraction inhibits collimation of the electron beam. In the limit $k_F W \ll 1$, the injection distribution becomes proportional to $\cos^2 \alpha$ for all α, independent of the shape of the potential in the point contact region.[15] Here, we will compare Eq. (18) only with experiments on rather wide point contacts. A fully quantum mechanical calculation of the series conductance, with which the experiments and the semiclassical result could be compared, unfortunately is not available. (The calculation of Ref. 89 can not be used for this purpose because equilibration in the region between the point contacts due to inelastic scattering is ignored.)

Wharam *et al.*[9] find for relatively large point contact widths ($G = 5 \times (2e^2/h)$ for both point contacts) a ratio $G/G_{\text{series}} = 1.4$, considerably less than the ratio of 2 expected from ohmic addition. From Eq. (17), we infer that a fraction $T_d/N \approx 0.4$ of the injected current is transmitted directly. As discussed in Ref. 8, this number is consistent with Eq. (16) using an estimate of $f \approx 2.4$ and $W_{\text{max}}/L \approx 0.4$ in their geometry.

In a recent study, Beton *et al.*[151] have applied the preceding results to their experiment on transport through point contacts in series. For the two widest identical point contacts considered ($G = 4 \times (2e^2/h)$), they infer a fraction $T_d/N = 0.6$ of directly transmitted current. From their scale diagram, we estimate $L \approx 440$ nm; From the relation,

$$G = \left(\frac{2e^2}{h}\right)\left(\frac{2m}{\hbar^2}\right)^{1/2}(E_F - E_c)^{1/2}\frac{W_{\text{min}}}{\pi}$$

(with $E_c \approx 0$ and the experimentally given value of E_F), we estimate $W_{\text{min}} \approx 90$ nm. The maximal width for adiabatic transport W_{max} is difficult to estimate reliably, and the lithographic opening of 240 nm in the gates defining the point contact presumably is an overestimate. Using $E_c \approx 0$ and $W_{max} \lesssim 240$ nm, we find from Eq. (14) a collimation factor, $f \lesssim 2.7$, which from Eq. (16) gives the theoretical value, $T_d/N \lesssim 0.7$—consistent with the experimental value[151] of 0.6. (In Ref. 151, a much larger theoretical value is stated without derivation.) At smaller point contact widths, the agreement between experiment and theory becomes worse, possibly as a result of the diffraction effects mentioned earlier.

In both these experiments (as well as in a similar experiment of Hirayama and Saku)[139] L is not much larger than W_{max}, so that the requirement for the validity of Eqs. (16) and (18) of small fW_{max}/L (or, equivalently, small T_d/N) is not well satisfied. A more significant comparison between the present analytical theory and experiment would require a larger point contact separation. Unfortunately, the non-additivity of the resistance then is only a small correction to the series resistance (since $T_d \ll N$). For a more sensitive study of the collimated electron beam, one needs to eliminate the background signal from electrons transmitted after equilibration, which obscures the direct transmission in a series resistance measurement at large point contact separation. As proposed in Refs. 8 and 66 and discussed in Sec. 5.d, this uninteresting background can be largely eliminated by maintaining the region between the point contacts at ground potential and operating one of the point contacts as a voltage probe drawing no net current. (See Fig. 20b.)

So far, we have considered only the case of zero magnetic field. In a weak magnetic field ($2l_{cycl} > L$), the situation is rather complicated. As discussed in detail in Ref. 8, there are two competing effects in weak fields: On the one hand, the deflection of the electron beam by the Lorentz force reduces the direct transmission probability, with the effect of decreasing the series conductance; on the other hand, the magnetic field enhances the indirect transmission, with the opposite effect. The result is an initial *decrease* in the series conductance for small magnetic fields in the case of strong collimation, and an *increase* in the case of weak collimation. This is expected to be a relatively small effect compared to the effects at stronger fields discussed next (*cf*. Ref. 8).

In stronger fields ($2l_{cycl} < L$), the direct transmission probability vanishes, which greatly simplifies the situation. If we assume that all transmission between the opposite point contacts is with intervening equilibration, then the result is[8]

$$G_{series} = \frac{2e^2}{h} \left(\frac{2}{N} - \frac{1}{N_{wide}} \right)^{-1/2}. \quad (19)$$

Here, N is the (B-independent) number of occupied subbands in the point contacts, and N_{wide} is the number of occupied Landau levels in the 2DEG between the point contacts. The physical origin of the simple addition rule, Eq. (19), is additivity of the four-terminal longitudinal resistance as in Eq. (10). From this additivity, it follows that for n different point contacts in series, Eq. (19) generalizes to

$$\frac{1}{G_{\text{series}}} - \frac{h}{2e^2}\frac{1}{N_{\text{wide}}} = \sum_{i=1}^{n} R_{\text{L}}(i), \quad (20)$$

where

$$R_{\text{L}}(i) = \left(\frac{h}{2e^2}\right)\left(\frac{1}{N_i} - \frac{1}{N_{\text{wide}}}\right)$$

is the four-terminal longitudinal resistance of point contact i. Equation (19) predicts a non-monotonic B-dependence for G_{series}. This can be seen most easily by disregarding the discreteness of N and N_{wide}. We then have $N_{\text{L}} \approx E_{\text{F}}/\hbar\omega_{\text{c}}$, while the magnetic field dependence of N (for a square-well confining potential in the point contacts) is given by Eq. (8). The resulting B-dependence of G_{series} is shown in Fig. 21 (dotted curves). The non-monotonic behavior is due to the delayed depopulation of subbands in the point contacts, compared to the broad 2DEG. While the number of occupied Landau levels N_{wide} in the region between the point contacts decreases steadily with B for $2l_{\text{cycl}} < L$, the number N of occupied subbands in the point contacts remains approximately constant until $2l_{\text{c,min}} \approx W_{\text{min}}$ (with $l_{\text{c,min}} \equiv l_{\text{cycl}}(1 - E_{\text{c}}/E_{\text{F}})^{1/2}$ denot-

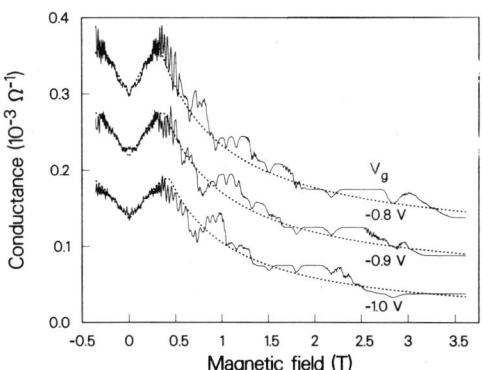

FIG. 21. Magnetic field dependence of the series conductance of two opposite point contacts ($L = 1.0 \ \mu$m) for three different values of the gate voltage (solid curves) at $T = 100$ mK. For clarity, subsequent curves from bottom to top are offset by $0.5 \times 10^{-4} \ \Omega^{-1}$, with the lowest curve shown at its actual value. The dotted curves are calculated from Eqs. (8) and (19), with the point contact width as adjustable parameter. (From Ref. 59.)

ing the cyclotron radius in the point contact region). In this field interval, G_{series} increases with B, according to Eq. (19). For stronger fields, depopulation in the point contacts begins to dominate G_{series}, leading finally to a decreasing conductance (as is the rule for single point contacts; see Section 3). The peak in G_{series} thus occurs at $2l_{c,\text{min}} \approx W_{\text{min}}$.

The remarkable camelback shape of G_{series} versus B predicted by Eq. (19) now has been observed experimentally.[59] The data is shown in Fig. 21 (solid curves) for three values of the gate voltage V_g, at $T = 100$ mK. The measurement configuration is as shown in Fig. 20a, with a point contact separation $L = 1.0$ μm. The voltage is the same on all the gates defining the point contacts, so that we expect the two opposite point contacts to be similar. Moreover, since $|V_g|$ is rather small, we can neglect the barrier in the point contact and assume that the electron density in the point contacts is the same as in the 2DEG channel that separates the point contacts. (The latter density can be obtained independently from the Hall resistance of the channel, and is found to decrease from 1.1 to 0.8×10^{15} m^{-2} as V_g varies from -0.8 to -1.0 V.) The point contact width W_{min} then remains the only free parameter, which is determined by a fit using Eqs. (8) and (19). (A square-well confining potential is assumed in the point contacts; W_{min} is found to decrease from 320 to 200 nm over the gate voltage range of Fig. 21). The dotted curves in Fig. 21 show the result of such a fit[59] (after correction for a constant background resistance of 2.0 kΩ, estimated from the two-terminal measurement of the quantized resistance of the individual point contacts). It is seen that Eq. (19) provides a good description of the overall magnetoresistance behavior from low magnetic fields up to the quantum Hall effect regime. The additional structure in the experimental curves has several different origins, for which we refer to Ref. 59. Similar structure in the two-terminal resistance of a single point contact will be discussed in detail in Section 11.

We emphasize that Eq. (19) is based on the assumption of complete equilibration of the current-carrying edge states in the region between the point contacts. In a quantizing magnetic field, local equilibrium is reached by inter-Landau-level scattering. If the potential landscape (both in the point contacts themselves and in the 2DEG region in between) varies by less than the Landau-level separation $\hbar\omega_c$ on the length scale of the magnetic length $(\hbar/eB)^{1/2}$, then inter-Landau-level scattering is suppressed in the absence of other scattering mechanisms. (See Part IV.) This means that the transport from one point contact to the other is adiabatic. The series conductance then simply is $G_{\text{series}} = (2e^2/h)N$ for two identical point contacts, where $N \equiv \min(N_1, N_2)$ for two different point contacts in series. This expression differs from Eq. (19) if a barrier is present in the point contacts, since that causes the number N of occupied Landau levels in the point contact to be less than the number N_{wide} of occupied levels in the wide 2DEG. (In a strong magnetic field, $N \approx (E_F - E_c)/\hbar\omega_c$, while $N_{\text{wide}} \approx E_F/\hbar\omega_c$.) Adiabatic transport in a magnetic field through two point contacts in series has been studied experimentally in Ref. 152.

d. Magnetic Deflection of a Collimated Electron Beam

Consider the geometry of Fig. 20b. The current I_i through the injecting point contact is drained to ground at the two ends of the 2DEG channel separating the point contacts. The opposite point contact, the collector, serves as a voltage probe (with the voltage V_c being measured relative to ground). Since the terminals 2 and 4 are grounded, the indirect transmission probability from injector to collector is suppressed.[8,66] The collector voltage divided by the injected current is given by

$$\frac{V_c}{I_i} = \frac{1}{G}\frac{T_d}{N}, \quad T_d \ll N, \tag{21}$$

with $G = (2e^2/h)N$ the two-terminal conductance of the individual point contacts (which are assumed to be identical), and T_d the direct transmission probability through the two point contacts, as calculated in Section 5b. Equation (21) can be obtained either from the Landauer–Büttiker formalism (as done in Ref. 66), or simply by noting that the incoming current $I_i T_d/N$ through the collector has to be counterbalanced by an equal outgoing current GV_c (since the collector draws no net current). Using Eq. (16), we find

$$\frac{V_c}{I_i} = \frac{h}{2e^2} f^2 \frac{\pi}{2k_F L}, \tag{22}$$

where we have used the relation

$$G = \left(\frac{2e^2}{h}\right)\left(\frac{k_F W_{\text{max}}}{\pi}\right)\left(\frac{1}{f}\right)$$

($k_F \equiv (2mE_F/\hbar^2)^{1/2}$ being the Fermi wave vector in the region between the point contacts). In an experimental situation, L and k_F are known, so that the collimation factor f can be determined directly from the collector voltage by means of Eq. (22).

The result Eq. (22) holds in the absence of a magnetic field. A small magnetic field B will deflect the collimated electron beam past the collector. Simple geometry leads to the criterion, $L/2l_{\text{cycl}} = \alpha_{\text{max}}$, for the cyclotron radius at which T_d is reduced to zero by the Lorentz force. One thus would expect to see in V_c/I_i a peak around zero field, of height given by Eq. (22)

and of width,

$$\Delta B = \frac{4\hbar k_F}{eL} \arcsin \frac{1}{f} \quad (23)$$

according to Eq. (14).

In Fig. 22, this collimation peak is shown[58] (solid curve), at $T = 1.8$ K in a device with an $L = 4.0$ μm separation between injector and collector. The actual measurement configuration differed from Fig. 20b in an inessential way, in that only one end of the region between the point contacts was grounded. The current I_i thus flows from contact 1 to 2, and the voltage V_c is measured between contacts 3 and 4. In the notation of Part II, $V_c/I_i \equiv R_{12,34}$. This four-terminal resistance is referred to in narrow Hall bar geometries as a *bend resistance* measurement.[54,56] One can show,[58] using the Landauer–Büttiker formalism,[7] that the height of the collimation peak still is given by Eq. (22) if one replaces f^2 by $f^2 - 1/2$. The expression (23) for the width is not modified. The experimental result in Fig. 22 shows a peak height of 150 Ω (measured relative to the background resistance at large magnetic fields). Using $L = 4.0$ μm and the value $k_F = 1.1 \times 10^8$ m^{-1} obtained from Hall resistance measurements in the channel between the point contacts, one deduces a collimation factor, $f \approx 1.85$. The corresponding opening angle of the injection/acceptance cone is $2\alpha_{max} \approx 65°$. The calculated value of f would imply a width, $\Delta B \approx 0.04$ T, which is not far from the measured full width at half maximum of 0.03 T.

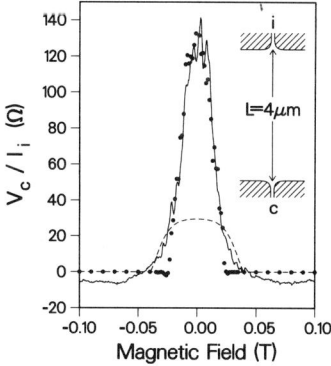

FIG. 22. Detection of a collimated electron beam over a distance of 4 μm. In this four-terminal measurement, two ohmic contacts to the 2DEG region between the point contacts are used: One of these acts as a drain for the current I_i through the injector, and the other is used as a zero-reference for the voltage V_c on the collector. The drawn curve is the experimental data, at $T = 1.8$ K. The black dots are the result of a semiclassical simulation, using a hard-wall potential with contours as shown in the inset. The dashed curve results from a simulation without collimation (corresponding to rectangular corners in the potential contour). (From Ref. 58.)

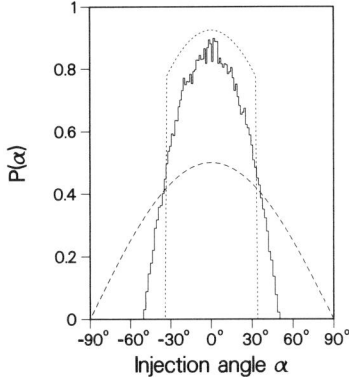

Fig. 23. Calculated angular injection distributions in zero magnetic field. The solid histogram is the result of a simulation of the classical trajectories at the Fermi energy in the geometry shown in the inset of Fig. 22; The dotted curve follows from the adiabatic approximation, Eq. (15), with the experimental collimation factor $f = 1.85$. The dashed curve is the cosine distribution in the absence of any collimation. (From Ref. 58.)

The experimental data in Fig. 22 are compared with the result[58] from a numerical simulation of classical trajectories of the electrons at the Fermi level (following the method of Ref. 144). This semiclassical calculation was performed to relax the assumption of adiabatic transport in the point contact region, and of small T_d/N, on which Eqs. (16) and (21) are based. The dashed curve is for point contacts defined by hard-wall contours with straight corners (no collimation); the dots are for the smooth hard-wall contours shown in the inset, which lead to collimation via the horn effect (cf. Fig. 19b; the barrier collimation of Fig. 19a presumably is unimportant at the small gate voltage used in the experiment, and is not taken into account in the numerical simulation). The angular injection distributions $P(\alpha)$ that follow from these numerical simulations are compared in Fig. 23 (solid histogram) with the result, Eq. (15), from the adiabatic approximation for $f = 1.85$ (dotted curve). The uncollimated distribution $P(\alpha) = (\cos \alpha)/2$ also is shown for comparison (dashed curve). Taken together, Figs. 22 and 23 unequivocally demonstrate the importance of collimation for the transport properties, as well as the adequateness of the adiabatic approximation as an estimator of the collimation cone.

6. Coherent Electron Focusing

a. Introduction

Electron focusing in metals was pioneered by Sharvin[23] and Tsoi[24] as a powerful tool to investigate the shape of the Fermi surface, surface scattering, and the electron–phonon interaction.[40] The experiment is the analogue in the solid state of magnetic focusing in vacuum. Required is a large mean free path for the carriers at the Fermi surface, to ensure *ballistic* motion as in vacuum. The mean free path (which can be as large as 1 cm in pure metallic single crystals) should be much larger than the length L on which the focusing takes place. Experimentally, $L = 10^{-2} - 10^{-1}$ cm is the separation of two metallic needles or point contacts (of typical width $W \sim 1$ μm) pressed on the crystal surface, which serve to inject a divergent electron beam and detect its focusing by the magnetic field. In metals, electron focusing essentially is a *classical* phenomenon because of the small Fermi wavelength λ_F (typically 0.5 nm, on the order of the inter-atomic separation). Both the ratios λ_F/L and λ_F/W are much larger in a 2DEG than in a metal, typically by factors of 10^4 and 10^2, respectively. *Coherent* electron focusing[13-15] is possible in a 2DEG because of this relatively large value of the Fermi wavelength, and turns out to be strikingly different from classical electron focusing in metals.

The geometry of the experiment[13] in a 2DEG is the transverse focusing geometry of Tsoi,[24] and consists of two point contacts on the same boundary in a perpendicular magnetic field. (In metals, one also can use the geometry of Sharvin[23] with opposite point contacts in a longitudinal field. This is not possible in two dimensions.) Two split-gate quantum point contacts and the intermediate 2DEG boundary are created electrostatically by means of split gates, as described in Part II. On applying a negative voltage to the split-gate electrode shown in Fig. 2a, the electron gas underneath the gate structure is depleted, creating two 2DEG regions (i and c) electrically isolated from the rest of the 2DEG—apart from the two quantum point contacts under the 250 nm wide openings in the split gate. The devices studied had point contact separations L of 1.5 and 3.0 μm, both values being below the mean free path of 9 μm estimated from the mobility.

Electron focusing can be seen as a transmission experiment in electron optics; *cf.* Ref. 96 for a discussion from this point of view. An alternative point of view (emphasized in Ref. 15) is that coherent electron focusing is a prototype of a non-local resistance[153] measurement in the quantum ballistic transport regime, such as that studied extensively in narrow-channel geometries (cf. Chapter 3 by Timp). Longitudinal resistances that are negative, not $\pm B$ symmetric, and dependent on the properties of the current and voltage contacts as well as on their separation; periodic and aperiodic magneto-resistance oscillations; absence of local equilibrium—these all are characteristic features of this transport regime that appear in a most extreme and bare form in the electron focusing geometry. One reason for the simplification offered by this geometry is that the current and voltage contacts, being point contacts, are not nearly as invasive as the wide leads in a Hall bar geometry. Another reason is that the electrons interact with only one boundary (instead of two in a narrow channel). Apart from the intrinsic interest of electron focusing in a 2DEG, the experiment also can be seen as a method to study electron scattering—as in metals. For two such applications, see Refs. 154

and 155. A search for inelastic scattering far from equilibrium by means of hot electron focusing[61] is the subject of Section 7c.

The outline of this section is as follows. In Section 6.b, the experimental results on electron focusing[13,15] are presented. A theoretical description[14,15] is given in Section 6.c, in terms of mode interference in the waveguide formed by the magnetic field at the 2DEG boundary. A discussion of the anomalous quantum Hall effect in the electron focusing geometry[10,12,15] is deferred to Section 9, where we consider adiabatic quantum transport. The present section is based on our earlier review[156] of this subject.

b. *Experiment*

Figure 24 illustrates electron focusing in two dimensions as it follows from the classical mechanics of electrons at the Fermi level. The injector (i) injects electrons ballistically into the 2DEG. The injected electrons all have the same Fermi velocity, but in different directions. Electrons are detected if they reach the adjacent collector (c), after one or more specular reflections at the boundary connecting i and c. These *skipping orbits* are composed of translated circular arcs of cylotron radius, $l_{\text{cycl}} \equiv \hbar k_F/eB$. The focusing action of the magnetic field is evident in Fig. 24 (top) from the black lines of high density of trajectories. These lines are known in optics as *caustics*, and are plotted separately in Fig. 24 (bottom). The caustics intersect the 2DEG boundary at multiples of the cyclotron diameter from the injector. As the magnetic field is

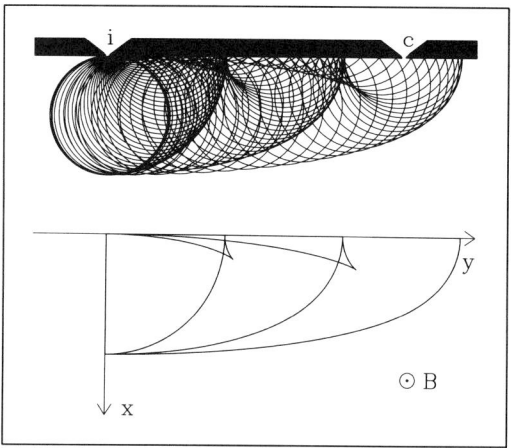

FIG. 24. (Top) Skipping orbits along the 2DEG boundary. The trajectories are drawn up to the third specular reflection. (Bottom) Plot of the caustics, which are the collection of focal points of the trajectories. (From Ref. 15.)

increased, a series of these focal points shifts past the collector. The electron flux incident on the collector thus reaches a maximum whenever its separation L from the injector is an integer multiple of $2l_{\text{cycl}}$. This occurs when

$$B = pB_{\text{focus}}, p = 1, 2, \ldots, \text{with}$$

$$B_{\text{focus}} = \frac{2\hbar k_F}{eL}. \tag{24}$$

For a given injected current I_i, the voltage V_c on the collector is proportional to the incident flux. The classical picture thus predicts a series of equidistant peaks in the collector voltage as a function of magnetic field.

In Fig. 25 (top), we show such a classical focusing spectrum, calculated for parameters corresponding to the experiment discussed in this section ($L = 3.0\ \mu\text{m}, k_F = 1.5 \times 10^8\ \text{m}^{-1}$). The spectrum consists of equidistant focusing peaks of approximately equal magnitude superimposed on the Hall resistance (dashed line). The pth peak is due to electrons injected perpendicularly to the boundary that have made $p - 1$ specular reflections between injector and collector. Such a classical focusing spectrum commonly is observed in metals,[157] albeit with a decreasing height of subsequent peaks

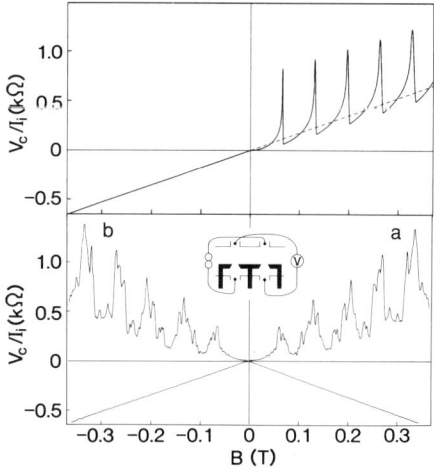

FIG. 25. (Bottom) Experimental electron focusing spectrum ($T = 50\ \text{mK}, L = 3.0\ \mu\text{m}$) in the generalized Hall resistance configuration depicted in the inset. The two traces a and b are measured with interchanged current and voltage leads, and demonstrate the injector–collector reciprocity as well as the reproducibility of the fine structure. (Top) Calculated classical focusing spectrum corresponding to the experimental trace a. (50 nm-wide point contacts were assumed.) The dashed line is the extrapolation of the classical Hall resistance seen in reverse fields. (From Ref. 15.)

because of partially diffuse scattering at the metal surface. Note that the peaks occur in one field direction only; in reverse fields, the focal points are at the wrong side of the injector for detection; and the normal Hall resistance is obtained. The experimental result for a 2DEG is shown in the bottom half of Fig. 25 (trace a; trace b is discussed later.) A series of five focusing peaks is evident at the expected positions. Note that the observation of multiple focusing peaks immediately implies that the electrostatically defined 2DEG boundary scatters predominantly *specularly*. This conclusion is supported by magnetotransport experiments in a narrow channel defined by a split gate.[108] In contrast, it has been found that a 2DEG boundary defined by ion beam exposure induces a large amount of *diffuse* scattering.[108,155]

Fig. 25 is obtained in a measuring configuration (inset) in which an imaginary line connecting the voltage probes crosses that between the current source and drain. This is the configuration for a generalized Hall resistance measurement. Alternatively, one can measure a generalized longitudinal resistance (*cf.* Section 4) in the configuration shown in the inset of Fig. 26. One then measures the focusing peaks without a superimposed Hall slope. Note that the experimental longitudinal resistance (Fig. 26, bottom) becomes *negative*. This is a classical result of magnetic focusing, as demonstrated by the calculation shown in the top half of Fig. 26. Büttiker[158,159] has studied negative longitudinal resistances in a different (Hall bar) geometry.

On the experimental focusing peaks, a fine structure is evident. The fine structure is quite reproducible (as is evident when comparing Figs. 25 and 26), but sample-dependent. It is resolved only at low temperatures (below 1 K) and small injection voltages. (The measurements shown are taken at 50 mK and a few μV ac voltage over the injector.) A nice demonstration of the reproducibility of the fine structure is obtained upon interchanging current and voltage leads, so that the injector becomes the collector and vice versa. The resulting focusing spectrum shown in Fig. 25 (trace b) is almost

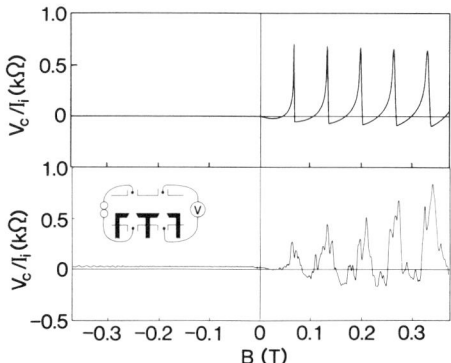

FIG. 26. Same as Fig. 25, but in the longitudinal resistance configuration. (From Ref. 15.)

the precise mirror image of the original one (trace *a*)—although this particular device had a strong asymmetry in the widths of injector and collector. The symmetry in the focusing spectra is a consequence of the fundamental reciprocity relation derived by Büttiker,[7,158] which generalizes the familiar Onsager–Casimir symmetry relation for the resistivity tensor to resistances.

The fine structure on the focusing peaks in Figs. 25 and 26 is the first indication that electron focusing in a 2DEG is qualitatively different from the corresponding experiment in metals. At higher magnetic fields, the resemblance to the classical focusing spectrum is lost. (See Fig. 27.) A Fourier transform of the spectrum for $B \geq 0.8$ T (inset in Fig. 27) shows that the large-amplitude high-field oscillations have a dominant periodicity of 0.1 T, which is approximately the same as the periodicity B_{focus} of the much smaller focusing peaks at low magnetic fields. (B_{focus} in Fig. 27 differs from Fig. 25 because of a smaller $L = 1.5$ μm.) This dominant periodicity is the result of quantum interference between the different trajectories in Fig. 24 that take an electron from injector to collector. In Section 6.c, we demonstrate this in a mode picture, which in the WKB approximation is equivalent to calculating the interferences of the (complex) probability amplitude along classical trajectories. The latter ray picture is treated extensively in Ref. 15. The theoretical analysis implies for the experiment that the injector acts as a *coherent point source* with the coherence maintained over a distance of several microns to the collector.

c. Edge Channels and Mode Interference

To explain the characteristics features of coherent electron focusing mentioned previously, it is necessary to go beyond the classical description.[14,15]

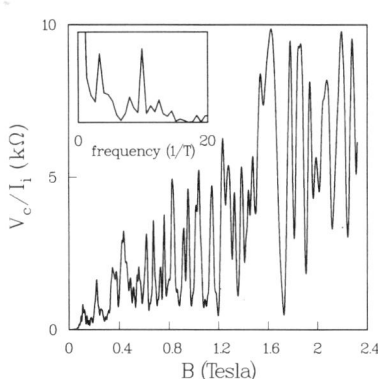

FIG. 27. Experimental electron focusing spectrum over a larger field range and for very narrow point contacts (estimated width 20–40 nm; $T = 50$ mK, $L = 1.5$ μm). The inset gives the Fourier transform for $B \geq 0.8$ T. The high-field oscillations have the same dominant periodicity as low-field focusing peaks—but with a much larger amplitude. (From Ref. 15.)

As discussed briefly in Section 3 (cf. Fig. 11a), quantum ballistic transport along the 2DEG boundary in a magnetic field takes place via magnetic edge states, which are the propagating modes of this problem.[160] The modes at the Fermi level are labeled by a quantum number $n = 1, 2, \ldots N$. Since the injector has a width below λ_F, it excites these modes coherently. For $k_F L \gg 1$, the interference of modes at the collector is dominated by their rapidly varying phase factors $\exp(ik_n L)$. The wave number k_n in the y direction (along the 2DEG boundary; see Fig. 24 for the choice of axes) corresponds classically to the x coordinate of the center of the cyclotron orbit, which is a conserved quantity upon specular reflection at the boundary.[161] In the Landau gauge $A = (0, Bx, 0)$, this correspondence may be written as $k_n = k_F \sin \alpha_n$, where α is the angle with the x axis under which the cyclotron orbit is reflected from the boundary ($|\alpha| < \pi/2$). The quantized values α_n follow in this semiclassical description from the Bohr–Sommerfeld quantization rule[160,161] that the flux enclosed by the cyclotron orbit and the boundary equals $(n - \tfrac{1}{4}) h/e$ (for an infinite barrier potential). Simple geometry shows that this requires that

$$\frac{\pi}{2} - \alpha_n - \frac{1}{2}\sin 2\alpha_n = \frac{2\pi}{k_F l_{\text{cycl}}}\left(n - \frac{1}{4}\right), \quad n = 1, 2, \ldots N, \tag{25}$$

with N the largest integer smaller than $\tfrac{1}{2} k_F l_{\text{cycl}} + \tfrac{1}{4}$. As plotted in Fig. 28, the dependence on n of the phase $k_n L$ is close to linear in a broad interval. This also follows from expansion of Eq. (25) around $\alpha_n = 0$, which gives

$$k_n L = \text{constant} - 2\pi n \frac{B}{B_{\text{focus}}} + k_F L \times \text{order}\left(\frac{N - 2n}{N}\right)^3. \tag{26}$$

If B/B_{focus} is an integer, a fraction of order $(1/k_F L)^{1/3}$ of the N edge states interferes constructively at the collector. (The edge states outside the domain of linear n-dependence of the phase give rise to additional interference struc-

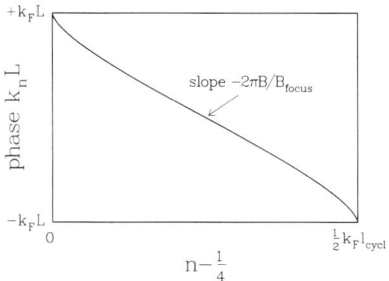

FIG. 28. Phase $k_n L$ of the edge channels at the collector, calculated from Eq. (25). Note the domain of approximately linear n-dependence of the phase, responsible for the oscillations with B_{focus}-periodicity. (From Ref. 15.)

ture which, however, does not have a simple periodicity.) Because of the 1/3 power, this is a substantial fraction even for the large $k_F L \sim 10^2$ of the experiment. The resulting mode interference oscillations with B_{focus}-periodicity can become much larger than the classical focusing peaks. This has been demonstrated in Refs. 14 and 15, where the collector voltage has been determined in WKB approximation with neglect of the finite width of the injector and detector. The result obtained there can be written in the form,

$$\frac{V_c}{I_i} = \frac{h}{2e^2} \left| \frac{1}{N} \sum_{n=1}^{N} e^{ik_n L} \right|^2. \tag{27}$$

Note that this equation implies that in the absence of interference among the modes, the normal quantum Hall resistance $h/2Ne^2$ is obtained. This is *not* a general result, but depends specifically on the properties of the injector and collector point contacts—as we will discuss in Section 9.

Figure 29 gives the focusing spectrum from Eq. (27), with parameter values corresponding to the experimental Fig. 27. The inset shows the Fourier transform for $B \geq 0.8$ T. There is no detailed one-to-one correspondence between the experimental and theoretical spectra. No such correspondence was to be expected in view of the sensitivity of the experimental spectrum to small variations in the voltage on the gate defining the point contacts and the 2DEG boundary.[13,15] Those features of the experimental spectrum that are insensitive to the precise measurement conditions are well-produced however, by the calculation: We recognize in Fig. 29 the low-field focusing peaks and the large-amplitude high-field oscillations with the same periodicity. (The reason that the periodicity B_{focus} in Fig. 29 is somewhat larger than in Fig. 27 most likely is the experimental uncertainty in the effective point contact separation of the order of the split-gate opening of 250 nm.) The high-field oscillations

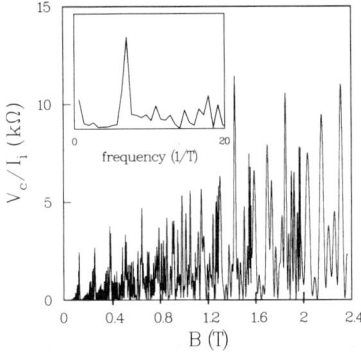

FIG. 29. Focusing spectrum calculated from Eq. (27) for parameters corresponding to the experimental Fig. 27. The inset shows the Fourier transform for $B \geq 0.8$ T. Infinitesimally small point contact widths are assumed in the calculation. (From Ref. 156.)

range from about 0 to 10 kΩ in both theory and experiment. This maximum amplitude is not far below the theoretical upper bound of $h/2e^2 \approx 13$ kΩ, which follows from Eq. (27) if we assume that *all* the modes interfere constructively. This indicates that a *maximal phase coherence* is realized in the experiment, and implies that:

1. The experimental injector and collector point contacts resemble the idealized point source/detector in the calculation.
2. Scattering events other than specular scattering on the boundary can be largely ignored (since any other inelastic as well as elastic scattering events would scramble the phases and reduce the oscillations with B_{focus}-periodicity).

The theory can be improved in several ways. This will affect the detailed form of the spectra, but probably not the fundamental periodicity. Since the exact wave functions of the edge states are known (Weber functions), one could go beyond the WKB approximation. This will become important at large magnetic fields, when the relevant edge states have small quantum numbers. In this regime, one also would have to take into account a possible B-dependence of E_F relative to the conduction band bottom (due to pinning of the Fermi energy at the Landau levels; *cf.* the related discussion in Section 3.a). It would be interesting to find out to what extent this bulk effect is reduced at the 2DEG boundary by the presence of edge states to fill the gap between the Landau levels. Another direction of improvement is towards a more realistic modeling of the injector and collector point contacts. Since the maximum amplitude of the theoretical and experimental oscillations is about the same (as is evident when comparing Figs. 27 and 29), the loss of spatial coherence due to the finite point contact size does not seem to be particularly important in this experiment. (Infinitesimal point contact width was assumed in the calculation.) On the other hand, the experimental focusing spectrum does not contain as many rapid oscillations as the calculation would predict. This may be due to the collimating properties of the point contacts. (See Ref. 156).

7. Breakdown of the Conductance Quantization and Hot Electron Focusing

a. *Mechanisms for Nonlinear Ballistic Transport*

Nonlinear transport in semiconductor devices is the rule, rather than an exception, but its analysis can be quite complicated.[162–164] In this section, we are concerned with several experiments that extend the study of ballistic transport to the regime of a nonlinear dependence of current on voltage. Although our theoretical understanding is far from being complete, we shall demonstrate that the overall behavior is readily understood on the basis of simple considerations. This simplicity is due to the ballistic nature of the transport on the length scales probed by the experiments. This allows us to ignore the energy dependence of the transport mean free path, which underlies much of the complexity of nonlinear transport in the diffusive transport regime.

Nonlinear transport through metallic point contacts has been investigated widely because of the possibility to observe phonon-related structure in the second derivative of the current-voltage characteristics. This is known as *point contact spectroscopy*.[25,26] Since typical optical phonon energies are of the order of 30 meV, while the Fermi energy in a metal is several eV, classical nonlinearities governed by the parameter eV/E_F do not play a significant role in metals. In a 2DEG, where E_F typically is only 10 meV, the situation is reversed, and the latter effects so far have obscured possible structure due to inelastic scattering processes in single quantum point contacts. The typical energy scale for nonlinearities in the quantum ballistic transport regime is even smaller than E_F. In experiments on the conductance quantization of a quantum point contact, the maximum breakdown voltage is found to be the energy separation between consecutive subbands.[60] This is discussed in Section 7.b. A complication is formed by the presence of a potential barrier in the point contact. The barrier height E_c depends on the applied bias voltage, thereby forming an additional mechanism for nonlinear transport. This complication may be turned into an advantage because it allows the injection of hot electrons over the barrier for sufficiently large bias voltage. Hot electron transport has been studied widely in devices inspired by the hot electron transistors pioneered by Shannon,[165] and this field of research has matured since the advent of vertically layered structures. A high sensitivity to inelastic scattering processes may be achieved in geometries involving two barriers, if one is used as a hot electron injector and the other as an energy-selective collector. This technique, known as *hot-electron spectroscopy*,[166-168] was adapted recently to transport in the plane of a 2DEG.[169-171] Among the results, we quote the demonstration by Palevski *et al.*[169] of emission of single longitudinal optical phonons (which in GaAs have an energy of 36 meV) and the discovery by Sivan *et al.*[171] of a long inelastic mean free path (exceeding 2 μm) for hot electrons with an excess energy up to the phonon energy. This finding contradicts theoretical predictions that the inelastic mean free path for electron–electron interactions should be one or two orders of magnitude smaller.[172,173] An entirely new way of detecting ballistic hot electrons is the *electron focusing* technique, which is the solid state analog of a β-spectrometer. As discussed in Section 7.c, the experimental results[61] corroborate the finding of Sivan *et al.* of ballistic hot-electron transport on long-length scales.

The hot-electron focusing technique is special for another reason: It can be used to determine the magnitude of the voltage drop in the immediate vicinity of a quantum point contact.[61] This may be the first realization of a really non-invasive voltage probe.

Two additional sources of nonlinearity are characteristic for quantum transport, but are not discussed beyond this introduction. The first is (resonant) tunneling through potential barriers, which has been investigated widely in vertical layered semiconductor structures following original work by Chang, Esaki, and Tsu.[174-176] Only recently has tunneling been studied in double-barrier structures defined in a 2DEG.[169,177] Clear signatures of tunneling currents have not been found yet in the current-voltage characteristics of single quantum point contacts, but the physics should be very similar to that of Ref. 169. An interesting difference between point contacts and wide tunnel barriers is the spatial resolution, which in a geometry with two opposite point contacts may be exploited to impose constraints on the lateral momentum of the tunneling electrons.[171]

A second source of quantum mechanical nonlinearity[178] arises in experiments that, like coherent electron focusing, probe the coherence of the injected electrons. Energy averaging due to a small nonzero bias voltage V is very similar to that due to a finite temperature $T \approx eV/k_B$, as discussed in Section 2.c (Ref. 101). This similarity is supported by experimental observations. (For the coherent electron focusing experiment, one finds,[15] for example, that energy averaging becomes important if the temperature is raised above 1 K, or for injection voltages beyond 100 μV.)

b. Breakdown of the Conductance Quantization

i. Experiments. The breakdown of the conductance quantization has been studied[60] by measuring the dc current-voltage $(I-V)$ characteristics of a quantum point contact device at a temperature of 0.6 K. The $I-V$ traces were obtained for a set of gate voltages V_g in an interval from -2.0 to -2.1 V, corresponding to the $N = 1$ conductance plateau at small bias voltages. (See inset of Fig. 30a.) These measurements are representative for the case that the point contact at zero bias is not yet pinched off, which implies that in equilibrium the Fermi energy E_F exceeds the energy E_1 of the bottom of the lowest subband in the constriction. We will refer to E_1 as the height of the effective barrier in the constriction. The case $E_F < E_1$ of a pinched-off point contact is discussed in the next paragraph. The experimental $I-V$ traces reproduced in Figs. 30a,b have been obtained from a two-terminal resistance measurement, after a correction for the background resistance originating in the ohmic contacts and the wide 2DEG regions. (See Part II and Ref. 60.) The dotted lines in Figs. 30a,b represent the quantized conduc-

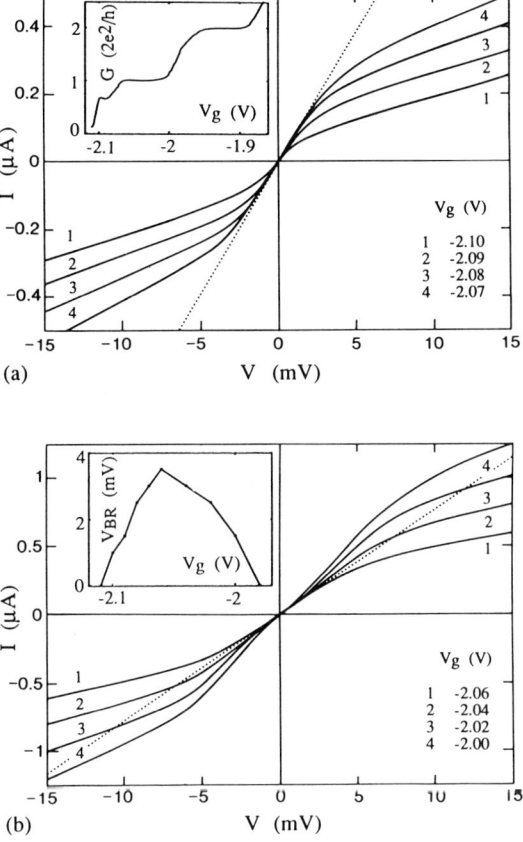

FIG. 30. Current-voltage characteristics of a quantum point contact at 0.6 K for different values of the gate voltage V_g at which the conductance G for small bias voltages is quantized at $2e^2/h$. The insets show the gate voltage dependence of the small-bias conductance G, and of the voltage V_{BR} beyond which the quantization $I = (2e^2/h)V$ (indicated by the dotted line) breaks down. The I–V characteristics shown have been corrected for a background resistance. (From Ref. 60.)

tance $I/V = 2e^2/h$. The conductance quantization breaks down beyond a critical voltage V_{BR}, which depends sensitively on the gate voltage (inset of Fig. 30b). The dependence of V_{BR} on V_g has a characteristic triangular shape, with a maximum of about 3 mV. Note that the maximum of V_{BR} is comparable to the subband separation at the Fermi level (calculated in Sec. 3b). On increasing the bias voltage V beyond V_{BR}, the differential conductance $\partial I/\partial V$ is seen initially to be either smaller (Fig. 30a) or larger (Fig. 30b) than the quantized value. As discussed shortly, this can be understood qualitatively as a consequence of the unequal number of populated subbands for the two opposite velocity directions in the constriction (whereby the bias voltage-induced lowering of the effective barrier height E_1 in the constriction plays a central role). Eventually, $\partial I/\partial V$ is found to drop below $2e^2/h$, regardless of the value of the gate voltage (in the interval considered in Fig. 30). We note that no evidence for a saturation of the current has been found for voltages up to 200 mV.

On increasing the negative gate voltage beyond -2.1 V, the point contact is pinched off, and hardly any current flows for small voltages. The residual current in the regime in which the point contact is just pinched off may be caused by a combination of tunneling[179] and thermionic emission over the barrier[180] (depending on the temperature and the value of the gate voltage). Measurements in this regime will be meaningful only if the leakage current through the gate is much less than this current. We have not investigated this regime systematically yet. (See in this connection Ref. 169.) When the point contact is pinched off, the effective barrier height in the constriction exceeds the Fermi level in equilibrium ($E_F < E_1$). An appreciable current starts to flow only if the bias voltage is sufficiently large to tilt the barrier such that electrons injected by one reservoir can pass over it freely. The experimental results are shown in Fig. 31. It is seen that a clear conduction threshold exists, which depends sensitively on the gate voltage. Beyond this threshold, the differential conductance $\partial I/\partial V$ following from these $I-V$ curves is roughly independent of the gate voltage (with a value of $(80 \text{ k}\Omega)^{-1}$), while it also is seen to be of comparable magnitude as the *trans*-conductance $\partial I/\partial V_g$. The latter result is significant in view of the symmetry relation discussed next.

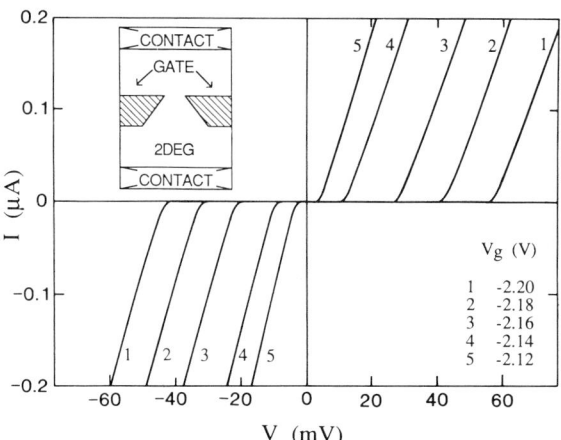

FIG. 31. $I-V$ characteristics at different gate voltages V_g for which the constriction is pinched off at small bias voltage. The inset shows the sample layout. (From Ref. 60.)

ii. Symmetry of the current-voltage characteristics. Before we turn to an account of some models that explain the basic features of the experimental results, it is of interest to examine the symmetry of the $I-V$ curves under a polarity change of the voltage V between source and drain. None of the experimental $I-V$ curves shown in Figs. 30 and 31 is anti-symmetric in V. This does not necessarily imply an intrinsic asymmetry in the device geometry. The reason is that point contacts are three- rather than two-terminal devices, and one of the current-carrying contacts also serves as the zero-reference of the gate voltage V_g, for either polarity of V. The presented current-voltage characteristics have been obtained without changing the choice of zero reference or the value of the gate voltage. The proper symmetry relation for a three-terminal device with mirror symmetry between source and drain is

$$I(V, V_g) = -I(-V, V_g - V), \quad (28)$$

rather than simply $I(V) = -I(-V)$. Here, we have assumed that V as well as V_g are defined relative to the same current contact, as is the case in Figs. 30 and 31. The difference between the two relations is significant, even though V is smaller than V_g by up to two orders of magnitude, because of the comparable magnitude of the differential conductance and the differential trans-conductance noted earlier. Equation (28) can be applied to Figs. 30 and 31 by comparing selected data points on curves measured for different gate voltages. It becomes readily apparent from such a procedure that gross deviations from Eq. (28) are found for the point contact under investigation, indicating that it did not have the mirror symmetry between source and drain. Deviations from Eq. (28) in other devices were found to be much smaller, however.

It will be clear from the preceding considerations that a precise modeling of the observed results would require a knowledge of the complex interdependence of the shape and height of the effective barrier in the constriction on the bias voltage and the gate voltage. In the following, we employ a highly simplified model, which suffices nevertheless for a qualitative description of the experiments.

iii. I–V characteristic of a semiclassical point contact. Consider semiclassical transport at large voltages through a point contact in a 2DEG defined by a hard-wall confining potential of width W. In contrast to our earlier treatment in the linear transport regime (Section 2.b.i), the net current now can be considered no longer to be carried exclusively by electrons at the Fermi energy. Instead, electrons in a finite energy interval contribute to the current. The fundamental origin of the resulting nonlinear $I-V$ characteristic is the interdependence of the Fermi velocity v_F and the electron gas density n_s in a degenerate electron gas, given by $v_F = (2\pi n_s)^{1/2} \hbar / m$ in two dimensions. We illustrate this by a simple model,[66] related to those commonly used in the literature on vertical transport in hot-electron spectroscopy devices,[166–168] and also to models for tunneling through a potential barrier under finite bias.[179,181,182]

The reservoirs on either side of the constriction are assumed to maintain a Fermi–Dirac distribution in the 2DEG for the electrons that move *towards* the constriction, with a Fermi energy difference eV. The electric field thus is assumed to be nonzero only in the constriction itself. The distribution of electrons moving *away* from the constriction can be far from equilibrium, however. According to these assumptions, the current is carried by conduction electrons in an energy interval from max $(E_F - e|V|, 0)$ to E_F, and can be written in the form (compared with Section 2.b.i),

$$I = eW \int_{\max(E_F - e|V|, 0)}^{E_F} \rho(E) \left(\frac{2E}{m}\right)^{1/2} dE \int_{-\pi/2}^{\pi/2} \cos\phi \frac{d\phi}{2\pi}, \qquad (29)$$

with $\rho(E) = m/\pi\hbar^2$ the density of states, which is energy-independent in two dimensions. The resulting current no longer is linear in the applied voltage,

$$I = I_{\max}\left[1 - \left(1 - \frac{e|V|}{E_F}\right)^{3/2}\right], \text{ for } e|V| < E_F, \qquad (30a)$$

$$I = I_{\max}, \text{ for } e|V| > E_F, \qquad (30b)$$

with

$$I_{\max} = e\frac{m}{\pi\hbar^2}\frac{2W}{3\pi}\left(\frac{2}{m}\right)^{1/2} E_F^{3/2} = en_s v_F \frac{2W}{3\pi}. \qquad (31)$$

For $e|V| \ll E_F$, one recovers Eqs. (1) and (2). For $e|V| > E_F$, we find that the current is limited by a saturation value $I_{\max}$. This saturation primarily is a consequence of our assumption that the entire voltage drop is localized at the point contact, with neglect of any accelerating fields outside the point contact region. We return to this point shortly.

The effect of a potential barrier of height E_c in the constriction can be taken into account by replacing E_F with $E_F - E_c$ in Eqs. (30) and (31). A complete description also has to determine the dependence of the barrier height on the applied bias voltage. (See next section).

iv. Breakdown of the conductance quantization. A model for the nonlinear I–V characteristics in the quantum ballistic case has been given in Ref. 60, on which our present discussion is based. This problem also has been considered in Refs. 183 and 101. To be specific, we consider an idealized model of a trapezoidal effective potential barrier in the constriction,[125] as illustrated in Fig. 32. The applied source-drain voltage V is assumed to result in a constant electric field between entrance and exit of the point contact. Due to this electric field, the potential barrier in the point contact is tilted, and thereby lowered, as illustrated in Figs. 32b,c. Due to the lateral confinement, one-dimensional subbands are formed in the constriction. On entering the constriction, the bottom of the nth subband rises relative to the bulk 2DEG, as a combined result of the increased lateral confinement and the electrostatic barrier (responsible for the reduced density in the constriction). The nth subband bottom has a maximal energy E_n, constituting a bottleneck for the current. We calculate the net current I_n through the constriction carried by the nth subband by considering the occupation of the right- and left-moving states at the bottleneck. The right-moving states are filled from E_n up to μ_s, the electrochemical potential of the source at the left of the constriction (provided $\mu_s > E_n$). Analogously, provided that $\mu_d > E_n$, the left-moving states

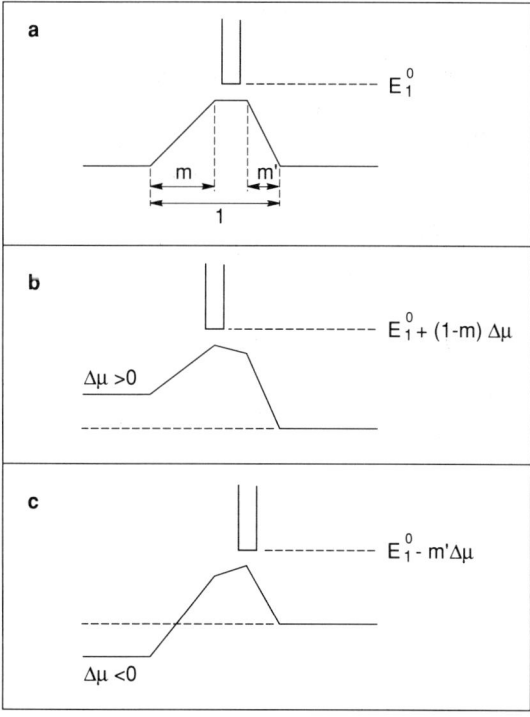

FIG. 32. Illustration of the voltage-induced tilting of the barrier in the constriction, for the case of an asymmetric trapezoidal shape of the barrier, with aspect ratios m and m'. The lowest one-dimensional subband is indicated. The energy E_1^0 of the subband bottom at the maximum of the trapezium is the effective barrier height at zero bias. (From Ref. 125.)

are filled from E_n up to μ_d, the electrochemical potential of the drain to the right of the constriction. We assume that the electrons in the nth subband with energy $\mu > E_n$ are transmitted fully through the constriction, and neglect intersubband scattering. In the following, we take $\mu_s > \mu_d$ (as in Fig. 32b), and define $\Delta\mu \equiv \mu_s - \mu_d \equiv eV$. The difference in occupation of right- and left-moving states for nonzero $\Delta\mu$ gives rise to a net current in the nth subband, which, according to the cancellation of group velocity and one-dimensional density of states (discussed in Section 2.b.ii), is given by

$$I_n = \frac{2e}{h}[\mu_s - \max(\mu_d, E_n)]\theta(\mu_s - E_n), \tag{32}$$

with $\theta(x)$ the unit step function. The bottom of the subband E_n at the bottleneck differs from its equilibrium position (which we denote by E_n^o in this section) because of the tilting of the potential barrier, and is given by $E_n = E_n^o + (1-m)\Delta\mu$. (See Fig. 32b.) Here, m is a phenomenological parameter between 0 and 1, which gives the fraction of $\Delta\mu$ that drops to the left of the bottleneck. (The value of m is determined by the shape of the effective potential barrier, as shown in Fig. 32a.) At a characteristic voltage V_c or V_c', either μ_s or μ_d crosses the subband bottom E_n, thereby changing the contribution $\partial I_n/\partial V$ from the nth subband to the differential conductance, $g \equiv \partial I/\partial V$. From Eq. (32), one finds[60] for an initially unpopulated subband,

$$\frac{\partial I_n}{\partial V} = \frac{2e^2}{h} m\theta(V - V_c), \text{ if } E_F < E_n^o, \tag{33}$$

while for an initially populated subband, one has

$$\frac{\partial I_n}{\partial V} = \frac{2e^2}{h}[1 - (1-m)\theta(V - V_c')], \text{ if } E_F > E_n^o. \tag{34}$$

The breakdown voltages are given by

$$V_c = \frac{(E_n^o - E_F)}{me}, \tag{35}$$

$$V_c' = \frac{(E_F - E_n^o)}{(1-m)e}, \tag{35b}$$

and they depend on the subband index n. (Note that m itself depends on n as well, since the effective potential barrier is due in part to the lateral confinement.)

Eq. (35b) applies to a subband that is occupied in equilibrium ($E_F > E_n^o$). Beyond a critical voltage V_c', the differential conductance due to this subband *decreases* from its normal quantized value of $2e^2/h$. Equation (35a) applies to a subband that in equilibrium is not occupied at the bottleneck of the constriction ($E_F < E_n^o$). The differential conductance of this subband *increases* beyond a critical voltage V_c to a value that is smaller than the quantized value. Although the expressions for the critical voltage depend on the parameter m, these general conclusions presumably are model-independent. The smallest breakdown voltage (which corresponds either to the highest occupied subband or the lowest unoccupied one) determines the breakdown voltage observed experimentally. If V is smaller than the breakdown voltages V_c, V_c' for all n, then we recover the conductance quantization $G \equiv I/V = (2e^2/h)N$ (with N the number of subbands occupied in equilibrium). This result exemplifies an interesting difference between quantum ballistic and semiclassical ballistic transport, in that it predicts linear transport *exactly* in a finite voltage interval, in contrast to Eq. (30).

v. Discussion. To illustrate the consequences[60] of Eqs. (33) and (34) for the differential conductance $g \equiv \partial I/\partial V$, we have shown schematically in Fig. 33

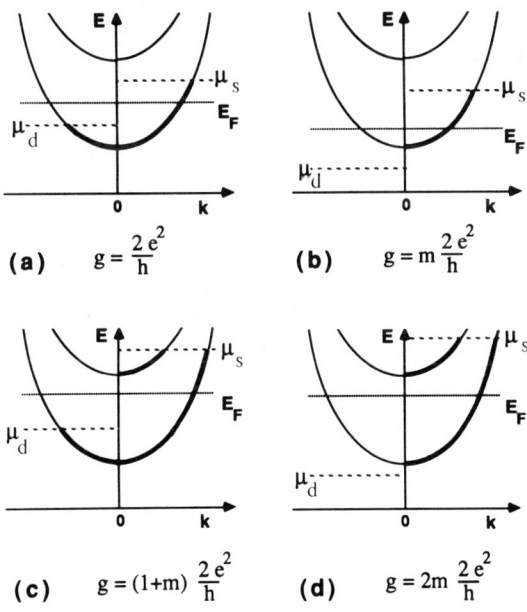

FIG. 33. Occupation of two one-dimensional subbands (represented by the parabolic dispersion curves $E_n(k)$) at the bottleneck in the constriction. Four situations are illustrated for different voltages $V = (\mu_s - \mu_d)/e$ across the point contact, and for different values of E_F (relative to the potential barrier height). (From Ref. 60.)

the energy of the two lowest one-dimensional subbands at the bottleneck as a function of longitudinal wave vector k. In equilibrium ($V = 0$), the subbands are occupied up to the Fermi energy E_F. A voltage V across the constriction gives a difference $\mu_s - \mu_d = eV$ in occupation between the two velocity directions (Fig. 33a), resulting in a net current. As long as the number of occupied subbands is the same for the two velocity directions, the conductance is quantized. However, at larger applied voltages, μ_d can fall below the bottom of a subband. Then g is reduced to a fraction m of $2e^2/h$ as shown in Fig. 33b, where E_F is near the bottom of the lowest subband. This is observed experimentally in Fig. 30a. The subband occupation of Fig. 33b also can be reached from the situation $E_F < E_1$, where there are no occupied states in equilibrium. For low voltages $g = 0$, but at a voltage V_c, the chemical potential μ_s crosses E_1, so that g increases to $m(2e^2/h)$ (cf. Eq. (33)). This is the case studied experimentally in Fig. 31.

Figures 33c,d correspond to the situation where E_F is close to the bottom of the second subband, as in the experimental Fig. 30b. On increasing V, first the second subband starts to be populated (Fig. 30c), leading to an increase in g to $(1 + m)2e^2/h$. A further increase of V causes μ_d to fall below the bottom of the first subband (Fig. 30d), which then reduces g to $2m(2e^2/h)$. This explains qualitatively the increasing and then decreasing slope in Fig. 30b. We note that the situation of Fig. 33d also can be reached directly from that in Fig. 33a, which actually is happening at $V_g = -2.06$ V in Fig. 30b.

So far, we have considered the case $\Delta\mu > 0$. If $\Delta\mu < 0$, then the parameter m is replaced by m', which will be different from m in the case of an assymmetric barrier in the constriction. (See Fig. 32c.) This may explain part of the asymmetry in the $I-V$ characteristics of Figs. 30 and 31, discussed earlier.

Both the quantum mechanical result, Eqs. (33)–(34), and its semiclassical counterpart Eq. (30), predict a saturation current of order I_{max} given by Eq. (31). The magnitude of the expected saturation current for $E_F = 10$ meV is $I_{max} \sim 1$ μA for a constriction with a few occupied one-dimensional subbands. Such a saturation has not been observed experimentally. The likely reason is that for large applied voltages, the electric field no longer is localized at the point contact. The electric field outside the point contact region will give rise to an acceleration of the electrons before they enter the point contact (i.e., in a wider region), thereby enhancing I_{max}. This point is discussed further in Section 7.c.

More recently, Brown et al.[102,184,185] have studied the differential resistance of a quantum point contact. In particular, they predict that the current saturation of Eq. (31) should cross over to a negative differential resistance regime at sufficiently large bias voltages as a consequence of quantum mechanical reflection. Further experiments will be required to test this prediction.

Here, we have considered only $I-V$ characteristics at fixed V_g. Glazman and Khaetskii[183] have predicted that the differential conductance as a function of gate voltage at finite V should exhibit additional plateaus between those at multiples of $2e^2/h$. Some evidence has been found for such plateaus (which also follow from Eqs. (33) and (34)), but these were not well resolved.[60]

The results Eqs. (33) and (34) predict a maximum breakdown voltage given by the one-dimensional subband separation at the Fermi level. This criterium is equivalent to that obtained by Jain and Kivelson[186] for the breakdown of the quantum Hall effect, where the Landau levels take over the role of the one-dimensional subbands. The triangular dependence of the breakdown voltage on the gate voltage shown in Fig. 30b also is reminiscent of experiments on the breakdown of the quantum Hall effect, where the breakdown voltage has a triangular dependence on the magnetic field. (Recent discussions of this topic are Refs. 30, 136, 187–191.) These similarities emphasize the correspondence between the zero-field quantization and the quantum Hall effect, discussed in Section 3.

c. Hot Electron Focusing

The voltage drop in a current-carrying sample containing a point contact in general is highly localized at or near the constriction. As mentioned in Section 4, two-terminal resistance measurements do not yield information on the voltage distribution. Four-terminal measurements could do so in principle, but such measurements in practice are hampered by the fact that conventional voltage probes located within a mean free path of the point contact are *invasive*. Indeed, in narrow multi-probe conductors, the voltage probes are the dominant source of scattering, as discussed by Timp in Chapter 3. In addition, conventional probes measure an electrochemical potential rather than an electrostatic voltage. In the linear response regime, a knowledge of the actual electric field distribution is not required to know the dissipation in the system (*cf*. Section 4). For many applications beyond linear transport, however, the electric field distribution does matter, as emphasized repeatedly by Landauer.[6,83,178,192–194] We thus are faced with a challenge to overcome the limitations of conventional resistance measurements in the ballistic regime. In this section, we discuss how an extension of the electron focusing technique of Section 6 to finite applied voltages meets this challenge at least in part.[61] (A promising alternative technique is to use a scanning tunneling microscope.)[194,195]

In Section 6, we emphasized the great difference in *length scales* between electron focusing in metals and in a 2DEG. The experiments discussed next demonstrate another qualitative difference, that of *energy scales*. In metals,[196] electrons are injected at energies above E_F, which generally are much less than $E_F \approx 5$ eV. In contrast, the Fermi energy in a 2DEG is only about 10 meV, so that dc-biasing the small ac injection voltage used in the electron focusing experiment leads to a noticeable shift in the focusing peaks. The magnitude of the shift allows a direct determination of the kinetic energy of the injected electrons, in direct analogy with a β-spectrometer. In metals, shifts in the peak position in electron focusing experiments also have been observed,[37] but are attributed to the magnetic field induced by the current at large dc bias voltages.[196] This field is totally negligible at the current levels used in a 2DEG.

i. Experiment. Hot electron focusing[61] spectra have been measured in a geometry identical to that used for the coherent electron focusing experiments of Section 6. (See inset of Fig. 36.) The electrons are injected from a 2DEG region i through the injector point contact, and are collected by a second point contact in region c. The 2DEG region where the focusing takes place is denoted by s. The point contact spacing was approximately 1.5 μm. A four-terminal generalized longitudinal resistance configuration was employed, with a small ac modulation voltage of 100 μV superimposed on a dc bias voltage V_{DC} on the order of a few mV between terminals 1 and 2. The differential focusing signal $\partial V_c/\partial I_i$ is obtained by measuring the ac collector volt-

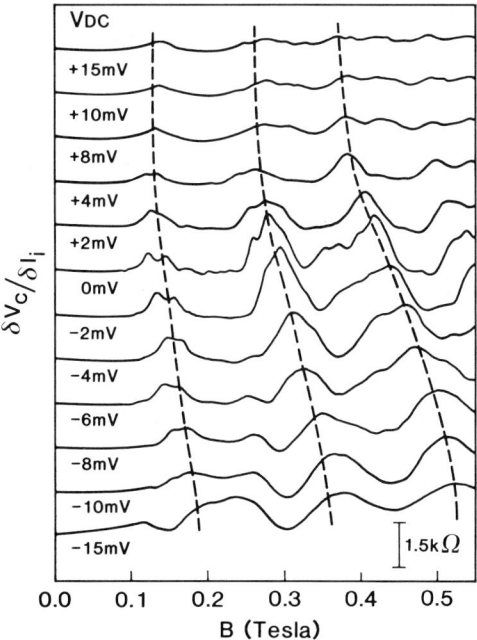

FIG. 34. Differential electron focusing spectra for various applied dc bias voltages, demonstrating the effects of electron acceleration and deceleration over the point contact region. The dashed lines connect peaks of the same index (From Ref. 61.)

age between terminals 3 and 4, and dividing by the ac current from terminal 1 to 2. In Fig. 34, the evolution of the focusing spectra with bias voltage V_{DC} is shown in the range $+15$ to -15 meV. Because of the rather large ac voltage, the fine structure due to quantum interference (studied in detail in Section 6) is smeared out, and only the classical focusing peaks remain—which are of primary interest in this experiment. A clear shift of these peaks is observed as the dc bias is increased, in particular for negative bias voltages. The peak positions are directly related to the energy of the injected electrons, as we now will demonstrate with a simple model,[61] which is based on the one used to explain the nonlinear conductance of a single-point contact,[60] discussed in Section 7.b.

ii. Model. Figure 35 illustrates the injection process at large bias voltages for four different positions of the chemical potentials μ_i and μ_s in regions i and s, relative to the subband bottom E_1 at the bottleneck of the injector

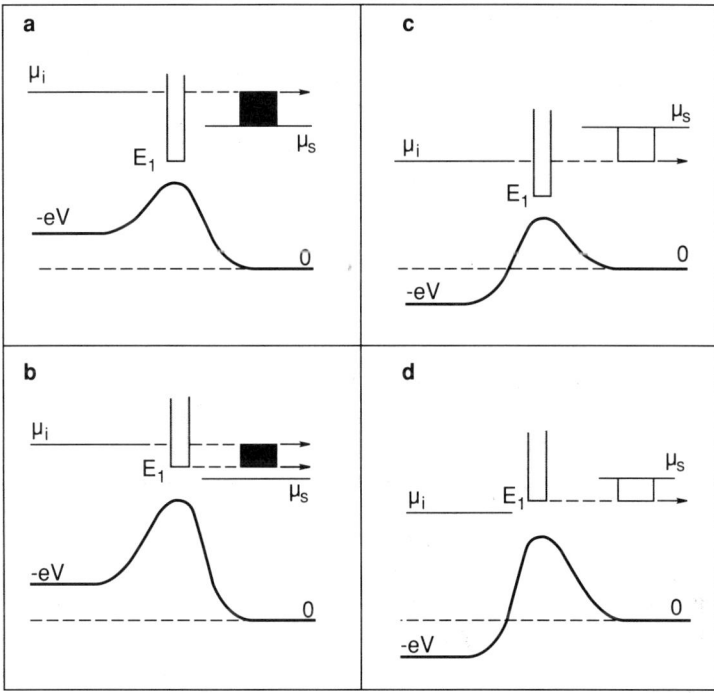

FIG. 35. Schematic representation of the injection of hot electrons (in an energy range indicated in black) or cold holes (indicated in white) from the injector at the left to the wide 2DEG region at the right. The lowest one-dimensional subband is indicated by the column. Depending on the positions of the Fermi levels μ_i and μ_s, the subband bottom E_1 can act as a low-energy cutoff for the injected carriers. The arrows denote the extremal energy of carriers detected in a differential focusing experiment. (From Ref. 61.)

point contact. The electric field caused by the applied voltage is assumed to be small outside the immediate vicinity of the point contact. In the point contact, adiabatic transport is assumed, and for simplicity only the lowest subband in the point contact is considered. (In fact, in the experiment of Fig. 34, only a single subband is occupied.) A negative voltage ($-eV > 0$) implies injection from i to s of electrons with energy in excess of E_F, as shown in panels *a* and *b* of Fig. 35. The electron flow at positive voltages is from region s into the injector, or equivalently, *unoccupied* electron states in region s move away from the injector (panels *c* and *d*). These states will be referred to as *holes* in the conduction band, below the local Fermi level μ_s. Peaks in the focusing spectrum also are observed for positive bias voltages (Fig. 34) resulting from focusing of ballistic holes rather than electrons.

The distribution of injected electrons or holes extends over a wide range of energies, corresponding to the height of the black and white boxes in Fig. 35. The differential measurement technique selects primarily only those electrons with maximal or minimal injection energy, as indicated by arrows in Fig. 35. One thus can study the transport of the injected *hot* electrons for $V < 0$ and of *cool* holes for $V > 0$, with an energy resolution determined by the magnitude of the ac voltage. The kinetic energy E_{focus} of the injected carriers is directly related to the position B_{focus} of the nth focusing peak if E_1 is below both μ_i and μ_s (the cases shown in Figs. 35a,c). By requiring that the point contact separation L is a multiple n of the cyclotron diameter at energy E_{focus} and magnetic field B_{focus}, one obtains the relation,

$$E_{\text{focus}} = \frac{(eLB_{\text{focus}})^2}{8mn^2}. \tag{36}$$

The two cases shown in Figs. 35c,d (where E_1 is below either μ_i or μ_s) are somewhat more complicated, because one has to take into account that the effective barrier height E_1 depends on the bias voltage. (See Sec. 7.b.) As analyzed in Ref. 61, for $V < 0$, Eq. (36) remains valid also when $E_1 < \mu_i$, but in that case, an additional small dip in the focusing spectrum is expected due to the appearance of a second extremal energy. (Note the two arrows in Fig. 35b.) This feature is not resolved clearly in Fig. 34. For $V > 0$, Eq. (36) no longer holds when $E_1 < \mu_s$ (the case of Fig. 35d). In addition, one expects a suppression of the focusing peaks for large positive bias voltages due to the cutoff imposed by the barrier in the collector. This does not play a role for negative bias voltages.

Figure 36 shows a plot of E_{focus} obtained via Eq. (36) from the position of the $n = 3$ focusing peak, as a function of the dc bias voltage.[61] For V_{DC} between -8 and $+3$ mV, E_{focus} varies linearly with V_{DC}. A linear least-squares

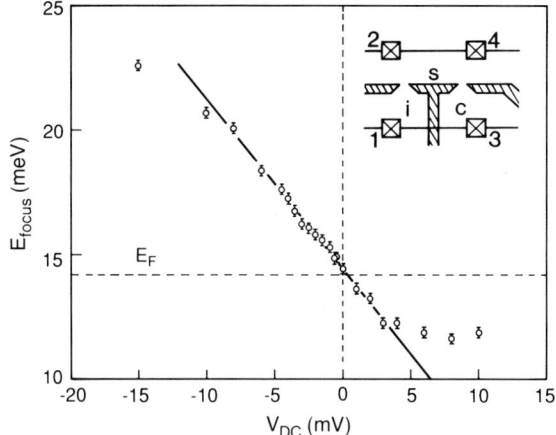

FIG. 36. Energy E_{focus} extracted from the focusing peak spacing as a function of applied dc bias voltage. The error bars shown reflect the estimated uncertainty in the measurement of the peak position, and the full straight line represents a least-squares fit. The Fermi energy E_F obtained from the Shubnikov–de Haas oscillations in the wide 2DEG region is indicated by the horizontal dashed line. The inset shows a schematic device diagram. The shaded parts indicate the gate used to define the point contacts and the 2DEG boundary, and the squares denote the ohmic contacts. (From Ref. 61.)

fit in this region yields

$$E_{\text{focus}} = -0.68 e V_{\text{DC}} + 14.4 \text{ meV}. \quad (37)$$

For $V_{\text{DC}} = 0$, the electron kinetic energy E_{focus} agrees very well with $E_F = 14.2$ meV obtained from the Shubnikov–de Haas oscillations, as it should. The deviations from linearity below -8 mV and above $+3$ mV are the subject of Section 7.c.iv. We first discuss the implication of Eq. (37) for the electric field at the injector point contact.

iii. The dipole field at the injector point contact. The slope of the straight line in Fig. 36 tells us what fraction of the total dc voltage drop V_{DC} across the sample is localized in the immediate vicinity of the injector point contact. The experimental results represented by Eq. (37) indicate that this fraction is 0.68.

We now will argue that the part of the voltage drop across the sample associated with the (quantized) contact resistance nearly is completely localized at the injector point contact. (In this subsection, the term *voltage drop* refers to that part of the electrochemical potential difference that is associated with the electric field.) The total sample resistance in the experiment[61] was 19.4 ± 0.3 kΩ. Applying the voltage division rule for series resistors, one finds that the voltage drop localized at the point contact corresponds to a fraction 0.68 of this resistance, i.e., to 13.2 ± 0.3 kΩ. This value agrees within the experimental uncertainty with the quantized resistance $h/2e^2 = 12.9$ kΩ of a quantum point contact with a single occupied one-dimensional subband. The remainder of the total sample resistance corresponds to the background discussed in Part II, which is due primarily to the ohmic contacts. It is important to realize that the observation of focusing peaks requires that the acceleration through the injector is completed on a length scale that is short compared to the cyclotron radius, which in this experiment is about 0.25 μm (for a point contact separation of 1.5 μm and a peak index $n = 3$). The fact that a correct value for the quantized point contact resistance is found using no information but the focusing peak spacing and the total sample resistance thus implies that the point contact resistance is associated with an electric field localized within this length scale. The significance of such dipole fields for transport measurements was first stressed by Landauer in the context of the residual resistivity of metals.[6] For a more recent discussion of the same topic, see Ref. 194. We stress that the spatial scale (of 0.25 μm or less) determined here for the dipole field at the point contact is about two orders of magnitude below the elastic or inelastic mean free path.

iv. Further observations. For hot electron injection at large negative bias voltages $V_{DC} < -8$ mV, the kinetic energy E_{focus} inferred from the focusing peaks (via Eq. (36)) increases more weakly with V_{DC} than at smaller biases. In addition, there is some evidence of new peaks in the focusing spectra, with positions corresponding roughly to injection of electrons with the Fermi energy. (See Fig. 34.) These two features may be indicative of a rapid energy relaxation process close to the injector point contact. Another possibility is that in this large bias regime, the voltage drop no longer is well localized at the point contact, in contrast to the case for smaller biases. We recall in this connection the absence of saturation in the experimental $I-V$ characteristics discussed in Section 7.b. It would seem that the observation of well-defined peaks in a focusing experiment precludes energy relaxation on length scales longer than the cyclotron radius as a possible explanation.

The deviation from linearity in Fig. 36 for positive injection voltages sets in for relatively small biases, $V_{DC} > +3$ mV. According to the model of Ref. 61, this deviation may arise when the subband bottom E_1 in the injector exceeds the Fermi energy μ_i in region i. In addition, the barrier in the collector point contact may impose an additional energy selection on the electron focusing signal. This will be important especially at high positive V_{DC} (but not for negative biases). Further experiments done in the regime where the injector point contact is close to pinch-off are discussed in Ref. 61.

In this device, hot electrons travel $\pi L/2 = 2.3$ μm between injector and collector. From theoretical work,[172,173] we estimate that the mean free path for electrons with an excess energy 50% of the Fermi energy of 14 meV (which still is considerably smaller than the optical phonon energy of 36 meV) should be limited to about 400 nm as a result of electron–electron interaction effects. Such a short mean free path seems irreconcilable with the present data,[61] since it would imply a two order of magnitude reduction of the focusing peak height. An even larger discrepancy was found by Sivan et al.[171] in a different experiment involving two opposite point contacts. It remains a theoretical challenge to explain these unanticipated long scattering lengths observed by two independent experiments.

IV. Adiabatic Transport in the Quantum Hall Effect Regime

8. INTRODUCTION

Both the quantum Hall effect (QHE) and the quantized conductance of a ballistic point contact are described by one and the same relation, $G = (2e^2/h)N$, between the conductance G and the number N of propagating modes at the Fermi level. The smooth transition from zero-field quantization to QHE that follows from this relation was the subject of Section 3. The resemblance between ballistic quantum transport and transport in the QHE regime becomes superficial, however, if one considers the entirely different role of scattering processes in weak and strong magnetic fields. First of all, the zero-field conductance quantization is destroyed by a small amount of elastic scattering (due to impurities or roughness of the channel boundaries; cf. Section 2.d), while the QHE is not. This difference is a manifestation of the suppression of backscattering by a magnetic field, discussed in Section 4. Absence of backscattering by itself does not imply adiabatic transport, which requires a total suppression of scattering among the modes. In weak magnetic fields, adiabaticity is of importance within a point contact, but not on longer length scales (cf. Sections 2.b.iii and 5). In the wide 2DEG region, scattering among the modes in weak fields establishes local equilibrium on a length scale given by the inelastic scattering length (which, in a high-mobility material, presumably is not much longer than the elastic scattering length $l \sim 10 \, \mu\text{m}$). The situation is strikingly different in a strong magnetic field, where the *selective* population and detection of the modes at the Fermi level demonstrated in Ref. 10 (to be discussed in Section 9.b) is made possible by the persistence of adiabaticity outside the point contact. In the words of Ref. 197, application of a magnetic field induces a transition from a *local* to a *global* adiabatic regime. Over some longer distance (which is not yet known precisely), adiabaticity breaks down, but surprisingly enough, local equilibrium remains absent even on macroscopic length scales (exceeding 0.25 mm).[11,12,34,198–200] Since local equilibrium is a prerequisite for the use of a local resistivity tensor, these findings imply a non-locality of the transport that had not been anticipated in theories of the QHE (which commonly are expressed in terms of a local resistivity; cf. Ref. 64 for a review). An important exception, to which we will return, is Büttiker's theory[77] on the role of contacts in the establishment of the quantized Hall resistance.

In the QHE regime, the propagating modes at the Fermi level are located at the edges of the sample,[201,202] under circumstances such that in the bulk, the Fermi level lies in a band of localized states.[62,203] These edge states originate from Landau levels that in the bulk lie below the Fermi level, but rise in energy on approaching the sample boundary. The point of intersection of the nth Landau level ($n = 1, 2, \ldots$) with the Fermi level forms the site of edge states belonging to the nth edge *channel*. The energy of each state can be separated into a part $(n - \tfrac{1}{2})\hbar\omega_c$ due to the quantized cyclotron motion, a part E_G due to the potential energy in an electrostatic potential $V(x, y)$, and the Zeeman energy $\pm\tfrac{1}{2}g\mu_B B$ depending on the spin direction. For a given total Fermi energy E_F, one has

$$E_G = E_F - (n - \tfrac{1}{2})\hbar\omega_c \pm \tfrac{1}{2}g\mu_B B. \tag{38}$$

Adiabatic transport is motion with constant n, implying that the cyclotron orbit center is guided along contours of constant $V(x, y) = E_G$. The energy E_G of this equipotential is referred to as the *guiding center energy*.

The simplicity of this guiding center drift along equipotentials has been exploited in the percolation theory[204,205] of the QHE, soon after its experimental discovery.[62] The physical requirements for the absence of inter-Landau-level scattering have received considerable attention[206,207] in that context, and more recently[197,208,209] in the context of adiabatic transport in edge channels. One requirement is that strong potential variations occur on a spatial scale that is large compared to the magnetic length $l_m \equiv (\hbar/eB)^{1/2}$, which corresponds to the cyclotron radius in the QHE ($l_{\text{cycl}} \equiv l_m(2n - 1)^{1/2} \approx l_m$ if the Landau level index $n \approx 1$). More rapid potential fluctuations may be present provided their amplitude is much less than $\hbar\omega_c$ (the energy separation of Landau levels).

Because edge channels at opposite edges of the sample move in opposite directions (as the drift velocity along an equipotential is given by $\mathbf{E} \times \mathbf{B}/B^2$, with $\mathbf{E}$ the electric field), backscattering requires scattering from one edge to the other. *Selective backscattering* for all edge channels with $n \geq n_0$ is imposed by a potential barrier[133–135] across the sample if its height is between the guiding center energies of edge channel n_0 and $n_0 - 1$. (Note that the edge channel with a larger index n has a smaller value of E_G.) The anomalous Shubnikov–de Haas effect,[11] which we will discuss in Section 10.a, has demonstrated that selective backscattering also can occur *naturally* in the absence of an imposed potential barrier. The edge channel with the highest index $n = N$ is selectively backscattered when the Fermi level approaches the energy $(N - \frac{1}{2})\hbar\omega_c$ of the Nth bulk Landau level (disregarding the Zeeman energy for simplicity of notation). The guiding center energy of the Nth edge channel then approaches zero, and backscattering either by tunneling or by thermally activated proccesses becomes effective—but only for that edge channel, which remains almost completely decoupled from the other $N - 1$ edge channels over distances as large as 250 μm. It was believed initially[11] that the transport might be fully adiabatic over this macroscopic length scale. However, Alphenaar et al.[200] now have demonstrated experimentally that on this length scale, the edge channels wth $n \leq N - 1$ have equilibrated to a large extent. The absence of scattering was found to persist only between this group of edge channels and the Nth edge channel, and then only if the Fermi level lies in (or near) the Nth bulk Landau level. As a qualitative explanation of these observations, it was proposed[200,210] that states from the highest-index edge channel hybridize with the localized states from the bulk Landau level of the same index when both types of states coexist at the Fermi level. Such a coexistence does not occur for the lower-index edge channels. A complete theoretical description is not available, and our present understanding of how fully adiabatic transport breaks down (as well as of the length scale on which this occurs) remains incomplete.

To avoid misunderstanding, it should be emphasized that the fact that the measured resistance can be expressed in terms of the transmission probabilities of edge states at the Fermi level does *not* imply that these few states carry a macroscopic current, *nor* does it imply that the current flows at the edges. A determination of the spatial current distribution, rather than just the total current, requires consideration of all the states below the Fermi level, which acquire a net drift velocity because of the Hall field. Within the range of validity of a linear response theory,[79] however, knowledge of the current distribution is not necessary to know the resistances. (See Ref. 18 for a more extensive discussion of this issue, which has caused some confusion in the literature.)

The outline of Part IV is as follows. Section 9 deals with anomalies in the quantum Hall effect due to the absence of local equilibrium at the current and voltage contacts.[10,34,198−200] The ideality of the contacts then affects the QHE in a fundamental way,[77,211] in contrast to the weak-field case where non-ideal contacts lead only to an uninteresting additive contact resistance (*cf.* Part II). An *ideal* contact in the QHE is one that establishes an equilibrium population among the outgoing edge channels by distributing the injected current equally among these propagating modes (*cf.* Section 2.b.ii). The selective population of edge channels by quantum point contacts[10] is discussed as an extreme example of a non-equilibrium population. Selective backscattering within a single edge channel, and the resulting anomalies in the Shubnikov–de Haas effect,[11,12] are the subjects of Section 10. In Section 11, inter-edge channel tunneling in a quantum point contact is discussed, which can explain the unusual observation of Aharonov–Bohm magnetoresistance oscillations in a *singly connected* geometry.[16] By combining two such point contacts to form a disc-shaped cavity, it has been possible to study this effect in a highly controlled way.[17]

In this chapter, we restrict ourselves to the *integer* QHE, where the edge channels can be described by single-electron states. Recent theoretical[212] and experimental[213–216] work on adiabatic transport in the *fractional* QHE (which fundamentally is a many-body effect)[64,217,218] indicates that many of the phenomena discussed here have analogues in that context as well.

9. Anomalous Quantum Hall Effect

a. *Ideal versus Disordered Contacts*

The quantization of the Hall resistance was discovered in a four-terminal measurement.[62] Under conditions in which the Hall resistance is quantized, the longitudinal resistance vanishes. Since the two-terminal resistance is the sum of the Hall and longitudinal resistances, a two-terminal measurement also shows a quantized resistance in a strong magnetic field[115–117] (to within an experimental uncertainty on the order of one part in 10^6). Nevertheless, investigators interested in high-precision determinations of the quantized Hall resistance generally have preferred a four-terminal measurement,[219] under the assumption that one thereby eliminates the contact resistances (*cf.* Part II). That is correct if local equilibrium is established near the contacts. A surprising conclusion of the work described in this section and the next is that local equilibrium is *not* the rule in the QHE regime, and that the effects of contacts can persist on macroscopic-length scales. Four-terminal measurements then in general will not yield a more accurate determination of the quantized Hall resistance than a two-terminal measurement does.

A necessary condition for the accurate quantization of the two-terminal Hall resistance is that the source and drain contacts are ideal,[77,211] in the sense that the edge states at the Fermi level have unit transmission probability through the contacts. This condition also plays a central role in four-terminal geometries. However, in the latter case, the requirement of ideal contacts is not necessary if the edge channels close to the contacts are equilibrated (as a result of inelastic scattering).

i. Ideal contacts. We return to the four-terminal measurements on a quantum point contact considered in Section 4, but now in the QHE regime, where the earlier assumption of local equilibrium near the contacts no longer is applicable in general. We assume strong magnetic fields, so that the four-terminal longitudinal resistance R_L of the quantum point contact is determined by the potential barrier in the constriction. Before we present the experimental results, we first describe briefly how the Landauer–Büttiker formalism[7] can be applied to multi-probe measurements in the QHE regime.[77] For a more detailed discussion of this formalism, we refer to Chapter 4 by Büttiker.

Consider the geometry of Fig. 37. Reservoirs at chemical potential μ_α are connected by leads to the conductor. The current I_α in the lead to reservoir α is related to these chemical potentials via the transmission probabilities $T_{\alpha \to \beta}$ (from reservoir α to β), and reflection probabilities R_α (from reservoir α back to the same reservoir). These equations have the form,[7]

$$\frac{h}{2e} I_\alpha = (N_\alpha - R_\alpha)\mu_\alpha - \sum_{\beta \neq \alpha} T_{\beta \to \alpha} \mu_\beta, \tag{39}$$

with N_α the number of propagating modes (or *quantum channels*) in lead α. In a strong magnetic field, N_α equals the number of edge channels at the Fermi energy E_F in lead α (which is the same as the number of bulk Landau levels below E_F in view of the one-to-one correspondence between edge channels and bulk Landau levels discussed in Section 8). As before, we denote the number of edge channels in the wide 2DEG and in the constriction by N_{wide} and N_{min}, respectively.

An *ideal* contact to the wide 2DEG has the property that all N_{wide} edge channels are fully transmitted into the contact, where they equilibrate.[77,211] Such a contact thus has $N_\alpha - R_\alpha = N_{\text{wide}}$. The constriction transmits only N_{min} channels, and the remaining $N_{\text{wide}} - N_{\text{min}}$ channels are reflected back

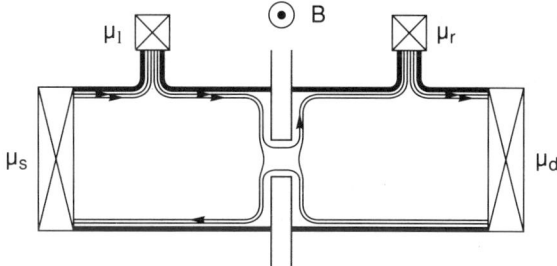

FIG. 37. Guiding center motion along equipotentials in the QHE regime, in a four-terminal geometry with a saddle-shaped potential formed by a split gate. Ideal contacts are assumed. The thin lines indicate the location of the edge channels at the Fermi level, with the arrows pointing in the direction of motion of edge channels that are populated by the contacts (crossed squares).

along the opposite 2DEG boundary (cf. Fig. 37). We denote by μ_l and μ_r the chemical potentials of adjacent voltage probes to the left and to the right of the constriction. The current source is at μ_s, and the drain at μ_d. Applying Eq. (39) to this case, using $I_s = -I_d \equiv I$, $I_r = I_l = 0$, one finds for the magnetic field direction indicated in Fig. 37,

$$\frac{h}{2e}I = N_{\text{wide}}\mu_s - (N_{\text{wide}} - N_{\text{min}})\mu_l, \tag{40a}$$

$$0 = N_{\text{wide}}\mu_l - N_{\text{wide}}\mu_s, \tag{40b}$$

$$0 = N_{\text{wide}}\mu_r - N_{\text{min}}\mu_l. \tag{40c}$$

We have used the freedom to choose the zero level of chemical potential by fixing $\mu_d = 0$, so that we have three independent (rather than four dependent) equations. The four-terminal longitudinal resistance $R_L \equiv (\mu_l - \mu_r)/eI$ that follows from Eq. (40) is

$$R_L = \frac{h}{2e^2}\left(\frac{1}{N_{\text{min}}} - \frac{1}{N_{\text{wide}}}\right). \tag{41}$$

In reversed-field direction, the same result is obtained. Equation (41), derived for ideal contacts without assuming local equilibrium near the contacts, is identical to Eq. (10). Similarly, applying Eq. (39) to a six-terminal geometry, one recovers the results from Eqs. (12) and (13) for the four-terminal Hall and diagonal resistances.

The fundamental reason that the local equilibrium approach of Section 4 (appropriate for weak magnetic fields) and the ideal contact approach of this section (for strong fields) yield identical answers is that an ideal contact attached to the wide 2DEG regions *induces* a local equilibrium by equipartitioning the outgoing current among the edge channels.[77] (This is illustrated in Fig. 37, where the current entering the voltage probe to the right of the constriction is carried by a *single* edge channel, while the equally large current flowing out of that probe is distributed equally over the *two* edge channels available for transport in the wide region.) In weaker magnetic fields, when the cyclotron radius exceeds the width of the narrow 2DEG region connecting the voltage probe to the Hall bar, not all edge channels in the wide 2DEG region are transmitted into the voltage probe, which therefore is not effective in redistributing the current. This is the reason that the weak-field analysis in Section 4 required the assumption of a local equilibrium in the wide 2DEG near the contacts.

We now discuss some experimental results, which confirm the behavior predicted by Eq. (41) in the QHE regime, to complement the weak-field experiments discussed in Section 4. In this field regime, the split-gate point contact geometry of Fig. 37 essentially is equivalent to a geometry studied recently by several authors,[133–135,198,199] in which a potential barrier across the Hall bar is created by means of a narrow continuous gate. The quantization of the longitudinal conductance R_L^{-1} in *fractions* of $2e^2/h$ (for unresolved spin degeneracy), predicted by Eq. (41), is shown in Fig. 38 for a quantum point contact sample[125] at $T = 0.6$ K. (Similar point contact data is reported in Ref. 126.) The magnetic field is kept fixed at 1.4 T (such that $N_{\text{wide}} = 5$) and the gate voltage is varied (such that N_{min} ranges from 1 to 4). Conductance plateaus close to 5/4, 10/3, 15/2, and $20 \times (2e^2/h)$ (solid horizontal lines) are observed, in accord with Eq. (41). Spin-split plateaus (dashed lines) are barely resolved at this rather low magnetic field. Observations of such a *fractional* quantization due to the integer QHE were made before on wide Hall bars with regions of different electron density in series,[127,129] but the theoretical explanation[130] given at that time was less straightforward than Eq. (41).

ii. Disordered contacts. The validity of Eq. (41) in the QHE regime breaks down for non-ideal contacts, if local equilibrium near the contacts is not established. As discussed in Section 4, Eq. (41) implies that the Hall voltage over the wide 2DEG regions adjacent to the constriction (cf. Fig. 18) is

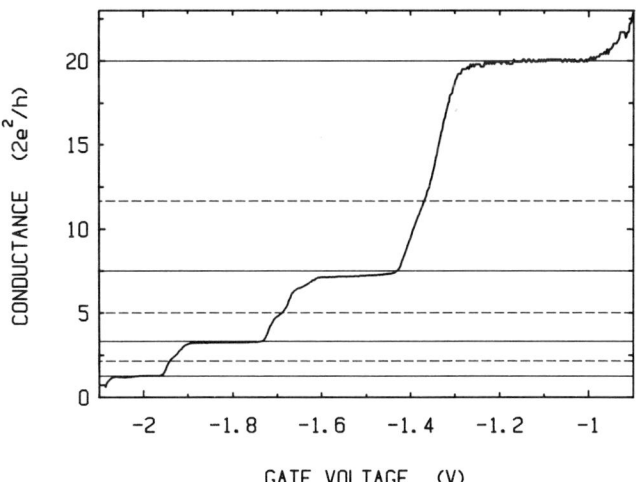

FIG. 38. *Fractional* quantization of the four-terminal longitudinal conductance R_L^{-1} of a point contact in a magnetic field of 1.4 T at $T = 0.6$ K. The solid horizontal lines indicate the quantized plateaus predicted by Eq. (41), with $N_{\text{wide}} = 5$ and $N_{\text{min}} = 1, 2, 3, 4$. The dashed lines give the location of the spin-split plateaus, which are not well-resolved at this magnetic field value. (From Ref. 125.)

unaffected by the presence of the constriction. In Fig. 39, we show the four-terminal longitudinal resistance R_L and Hall resistance R_H obtained on the sample considered in Section 4, but now over a wider field range, for both a small gate voltage (-0.3 V) and a large gate voltage (-2.5 V). In addition to the weak-field negative magnetoresistance discussed before in Section 4, a crossover to a positive magnetoresistance with superimposed Shubnikov–de Haas oscillations is seen.

The data for $V_g = -0.3$ V exhibits Shubnikov–de Haas oscillations with zero minima in the longitudinal resistance R_L, and the normal quantum Hall resistance $R_H = (h/2e^2)N_{\text{wide}}^{-1}$ is determined by the number of Landau levels occupied in the wide regions. (N_{wide} can be obtained from the quantum Hall effect measured in the absence of the constriction, or from the periodicity of the Shubnikov–de Haas oscillations.)

At the higher gate voltage $V_g = -2.5$ V, non-vanishing minima in R_L are seen in Fig. 39, as a result of the formation of a potential barrier in the constriction. At the minima, R_L has the fractional quantization predicted by Eq. (41) (See earlier.) For example, the plateau in R_L around 2.2 T for $V_g = -2.5$ V is observed to be at $R_L = 2.1$ k$\Omega \approx (h/2e^2) \times (\frac{1}{2} - \frac{1}{3})$, in agreement with the fact that the two-terminal resistance yields $N_{\min} = 2$, and the number of Landau levels in the wide regions $N_{\text{wide}} = 3$. In spite of this agreement, and in apparent conflict with the analysis leading to Eq. (41), it is seen in Fig. 39 that the Hall resistance R_H measured across a wide region for $V_g = -2.5$ V has *increased* over its value for small gate voltages. Indeed,

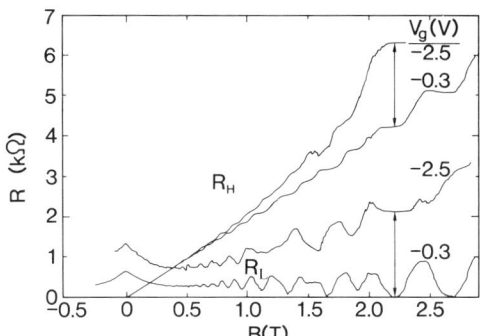

FIG. 39. Non-vanishing Shubnikov–de Haas minima in the longitudinal resistance R_L and anomalous quantum Hall resistance R_H, measured in the point contact geometry of Fig. 18 at 50 mK. These experimental results are extensions to higher fields of the weak-field traces shown in Fig. 15. The Hall resistance has been measured across the wide region, more than 100 μm away from the constriction, yet R_H is seen to increase if the gate voltage is raised from -0.3 V to -2.5 V. The magnitude at $B = 2.2$ T of the deviation in R_H and the Shubnikov–de Haas minimum in R_L are indicated by arrows, which for both R_H and R_L have a length of $(h/2e^2)(\frac{1}{2} - \frac{1}{3})$, in agreement with the analysis given in the text. (From Ref. 34.)

around 2.2 T, a Hall plateau at $R_H = 6.3\ \text{k}\Omega \approx (h/2e^2) \times \frac{1}{2}$ is found, as if the number of occupied Landau levels was given by $N_{min} = 2$ rather than by $N_{wide} = 3$. This unexpected deviation was noted in Ref. 34, but was not understood at the time. The temperature dependence of this effect has not been studied systematically, but it was found that the deviations in the Hall resistance persist at least up to 1.6 K. At higher magnetic fields (not shown in Fig. 39), the $N = 1$ plateau is reached, and the deviation in the Hall resistance vanishes. Following the similar experiment by Komiyama et al.,[198,199] and the demonstration of the anomalous QHE measured with quantum point contacts by the Delft–Philips collaboration,[10] the likely explanation of the data of Fig. 39 is that one or more of the ohmic contacts used as voltage probes are *disordered*, in the sense of Büttiker[77] that not all edge channels have unit transmission probability into the voltage probe. (Note that a point contact containing a potential barrier also is disordered in this sense.)

We now will demonstrate, following Refs. 77 and 211, how the anomalies in Fig. 39 can be accounted for nicely if one assumes that one of the probes used to measure the Hall voltage is disordered, because of a potential barrier in the probe with a height not below that of the barrier in the constriction. This is illustrated schematically in Fig. 40. A net current I flows through

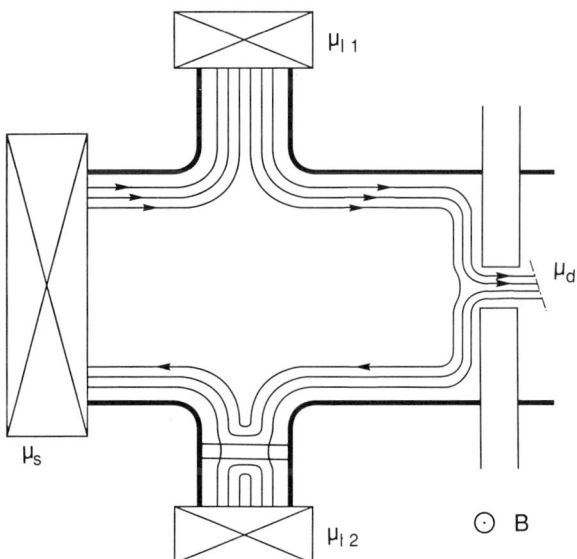

FIG. 40. Illustration of the flow of edge channels along equipotentials in a sample with a constriction (defined by the split gates) and a disordered voltage probe. (A potential barrier in the lower probe is indicated by the double bar.)

the constriction, determined by its two-terminal resistance according to $I = (2e/h)N_{\min}\mu_s$, with μ_s the chemical potential of the source reservoir. (The chemical potential of the drain reservoir μ_d is taken as a zero reference.) Equation (39), applied to the two opposite Hall probes l_1 and l_2 in Fig. 40, takes the form (using $I_{l_1} = I_{l_2} = 0$, $\mu_s = (h/2e)I/N_{\min}$, and $\mu_d = 0$),

$$0 = N_{\text{wide}}\mu_{l_1} - T_{s \to l_1}\frac{h}{2e}\frac{I}{N_{\min}} - T_{l_2 \to l_1}\mu_{l_2}, \tag{42a}$$

$$0 = N_{l_2}\mu_{l_2} - T_{s \to l_2}\frac{h}{2e}\frac{I}{N_{\min}} - T_{l_1 \to l_2}\mu_{l_1}, \tag{42b}$$

where we have assumed that the disordered Hall probe l_2 transmits only $N_{l_2} < N_{\text{wide}}$ edge channels because of some barrier in the probe. For the field direction shown in Fig. 40, one has, under the assumption of no inter-edge channel scattering from constriction to probe l_2, $T_{s \to l_1} = N_{\text{wide}}$, $T_{s \to l_2} = T_{l_2 \to l_1} = 0$, and $T_{l_1 \to l_2} = \max(0, N_{l_2} - N_{\min})$. Equation (42) then leads to a Hall resistance $R_H \equiv (\mu_{l_1} - \mu_{l_2})/eI$ given by

$$R_H = \frac{h}{2e^2}\frac{1}{\max(N_{l_2}, N_{\min})}. \tag{43}$$

In the opposite field direction, the normal Hall resistance $R_H = (h/2e^2)N_{\text{wide}}^{-1}$ is recovered.

The assumption of a single disordered probe, plus absence of inter-edge channel scattering from constriction to probe, thus explains the observation in Fig. 39 of an anomalously high quantum Hall resistance for large gate voltages, such that $N_{\min} < N_{\text{wide}}$. Indeed, the experimental Hall resistance for $V_g = -2.5$ V has a plateau around 2.2 T, close to the value $R_H = (h/2e^2)N_{\min}^{-1}$ (with $N_{\min} = 2$), in agreement with Eq. (43) if $N_{l_2} \leq N_{\min}$ at this gate voltage. This observation demonstrates the absence of inter-edge channel scattering over 100 μm (the separation of constriction and probe)—but only between the highest-index edge channel (with index $n = N_{\text{wide}} = 3$) and the two lower-index channels. Since the $n = 1$ and $n = 2$ edge channels either are both empty or both filled (cf. Fig. 40, where these two-edge channels lie closest to the sample boundary), any scattering between $n = 1$ and 2 would have no measurable effect on the resistances. As discussed in Section 8, we now know from the work of Alphenaar et al.[200] that (at least in the present samples) the edge channels with $n \leq N_{\text{wide}} - 1$, in fact, do equilibrate to a large extent on a length scale of 100 μm.

In the absence of a constriction, or at small gate voltages (where the constriction is just defined), one has $N_{\min} = N_{\text{wide}}$ so that the normal Hall effect is observed in both field directions. This is the situation realized in the experimental trace for $V_g = -0.3$ V in Fig. 39. In very strong fields such that $N_{min} = N_{l_2} = N_{wide} = 1$ (still assuming non-resolved spin-splitting), the normal result $R_H = h/2e^2$ would follow even if the contacts contain a potential barrier, in agreement with the experiment (not shown in Fig. 39). This is a more general result, which also holds for a barrier that only partially transmits the $n = 1$ edge channel.[12,77,220–222]

A similar analysis as the preceding predicts that the longitudinal resistance measured on the edge of the sample that contains ideal contacts retains its regular value, as in Eq. (41). The observation in the experiment of Fig. 39 for $V_g = -2.5$ V of a regular longitudinal resistance (in agreement with Eq. (41)), along with an anomalous quantum Hall resistance, thus is consistent with this analysis. On the opposite sample edge, the measurement would involve the disordered contact, and one finds instead

$$R_L = \frac{h}{2e^2}\left(\frac{1}{N_{min}} - \frac{1}{\max(N_{l_2}, N_{min})}\right) \quad (44)$$

The experiment[10] discussed in the following subsection is topologically equivalent to the geometry of Fig. 40, but involves quantum point contacts rather than ohmic contacts. This gives the possibility to populate and detect edge channels selectively, thereby enabling a study of the effects of a non-equilibrium population of edge channels in a controlled manner.

b. *Selective Population and Detection of Edge Channels*

In Section 6, we have seen in the coherent electron focusing experiment how a quantum point contact can inject a *coherent* superposition of edge channels at the 2DEG boundary. In that section, we restricted ourselves to weak magnetic fields. Here, we will show how in the QHE regime, the point contacts can be operated in a different way as *selective* injectors (and detectors) of edge channels.[10] We recall that electron focusing can be measured as a generalized Hall resistance, in which case the pronounced peaked structure due to mode interference is superimposed on the weak-field Hall resistance (*cf.* Fig. 25). If the weak-field electron focusing experiments are extended to stronger magnetic fields, a transition to the quantum Hall effect is observed, provided the injecting and detecting point contacts are not pinched off too strongly.[13] The oscillations characteristic of mode interference disappear in this field regime, suggesting that the coupling of the edge channels (which form the propagating modes from injector to collector) is suppressed and adiabatic transport is realized. It now is sufficient no longer to model the point contacts by a point source/detector of infinitesimal width (as was done in Section 6), but a somewhat more detailed description of the electrostatic potential $V(x, y)$ defining the point contacts and the 2DEG boundary between them is required. Schematically, $V(x, y)$ is represented in Fig. 41. Fringing fields from the split gate create a potential barrier in the point contacts, so that V has a saddle form as shown. The heights of the barriers E_i, E_c in the injector and collector are adjustable separately by means of the voltages on the split gates, and can be determined from the two-terminal conductances of the individual point contacts. The point contact separation in the experiment of Ref. 10 is small (1.5 μm), so that we can assume fully adiabatic transport from injector to collector in strong magnetic fields. (The experiment has been repeated for much larger point contact separations (60 and 130 μm) by Alphenaar et al.,[200] who find only partial absence of inter-edge channel scattering over these distances; See Section 8.) As discussed in Section 8, the adiabatic transport is along equipotentials at the guiding center energy E_G. Note that the edge channel with the smallest index n has the largest guiding center energy (according to Eq. (38)). In the absence of inter-edge channel scattering, edge channels only can be transmitted through a point contact if E_G exceeds the potential barrier height (disregarding tunneling through the barrier). The injector thus injects

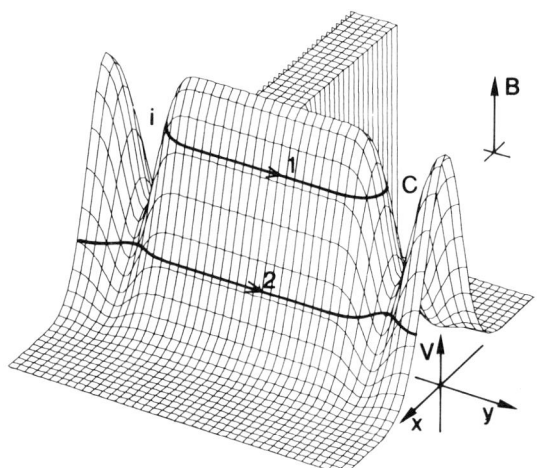

FIG. 41. Schematic potential landscape, showing the 2DEG boundary and the saddle-shaped injector and collector point contacts. In a strong magnetic field, the edge channels are extended along equipotentials at the guiding center energy, as indicated here for edge channels with index $n = 1, 2$. (The arrows point in the direction of motion.) In this case, a Hall conductance of $(2e^2/h)N$ with $N = 1$ would be measured by the point contacts—in spite of the presence of two occupied spin-degenerate Landau levels in the bulk 2DEG. (From Ref. 156.)

$N_i - R_i \approx (E_F - E_i)/\hbar\omega_c$ edge channels into the 2DEG, while the collector is capable of detecting $N_c - R_c \approx (E_F - E_c)/\hbar\omega_c$ channels. Along the boundary of the 2DEG, however, a larger number of $N_{\text{wide}} \approx E_F/\hbar\omega_c$ edge channels, equal to the number of occupied bulk Landau levels in the 2DEG, are available for transport at the Fermi level. The selective population and detection of Landau levels leads to deviations from the normal Hall resistance.

These considerations can be put on a theoretical basis by applying the Landauer–Büttiker formalism discussed in Section 9.a to the electron focusing geometry.[15] We consider a three-terminal conductor as shown in Fig. 2a, with point contacts in two of the probes (injector i and collector c), and a wide ideal drain contact d. The collector acts as a voltage probe, drawing no net current, so that $I_c = 0$ and $I_d = -I_i$. The zero of energy is chosen such that $\mu_d = 0$. One then finds from Eq. (39) the two equations,

$$0 = (N_c - R_c)\mu_c - T_{i \to c}\mu_i, \tag{45a}$$

$$\frac{h}{2e}I_i = (N_i - R_i)\mu_i - T_{c \to i}\mu_c, \tag{45b}$$

and obtains for the ratio of collector voltage $V_c = \mu_c/e$ (measured relative to the voltage of the current drain) to injected current I_i the result,

$$\frac{V_c}{I_i} = \frac{2e^2}{h} \frac{T_{i \to c}}{G_i G_c - \delta}. \tag{46}$$

Here, $\delta \equiv (2e^2/h)^2 T_{i \to c} T_{c \to i}$, and $G_i \equiv (2e^2/h)(N_i - R_i)$, $G_c \equiv (2e^2/h)(N_c - R_c)$ denote the conductances of injector and collector point contact. The injector–collector reciprocity demonstrated in Fig. 25 is present manifestly in Eq. (46), since G_i and G_c are even functions of B while[7] $T_{i \to c}(B) = T_{c \to i}(-B)$.

For the magnetic field direction indicated in Fig. 2a, the term δ in Eq. (46) can be neglected, since $T_{c \to i} \approx 0$. An additional simplification is possible in the adiabatic transport regime. We consider the case that the barrier in one of the two point contacts is sufficiently higher than in the other, to ensure that electrons are transmitted over the highest barrier will have a negligible probability of being reflected at the lowest barrier. Then $T_{i \to c}$ is dominated by the transmission probability over the highest barrier, $T_{i \to c} \approx \min(N_i - R_i, N_c - R_c)$. Substitution in Eq. (46) gives the remarkable result[10] that the *Hall conductance* $G_H \equiv I_i/V_c$ measured in the electron focusing geometry can be expressed entirely in terms of the *contact conductances* G_i and G_c,

$$G_H \approx \max(G_i, G_c). \tag{47}$$

Equation (47) tells us that quantized values of G_H occur not at $(2e^2/h)N_{wide}$, as one would expect from the N_{wide} populated Landau levels in the 2DEG — but at the smaller value of $(2e^2/h)\max(N_{min,i}, N_{min,c})$. As shown in Fig. 42, this indeed is observed experimentally.[10] Notice in particular how any deviation from quantization in $\max(G_i, G_c)$ is reproduced faithfully in G_H, in complete agreement with Eq. (47).

10. Anomalous Shubnikov–de Haas Effect

Shubnikov–de Haas oscillations periodic in $1/B$ occur in the longitudinal resistance of a 2DEG at low temperatures ($k_B T < \hbar\omega_c$), provided the mobility μ (or scattering time τ) is large enough to allow the formation of Landau levels ($\mu B \equiv \omega_c \tau > 1$, so that many cyclotron orbits occur on average between scattering events). In weak magnetic fields, where a theoretical description in

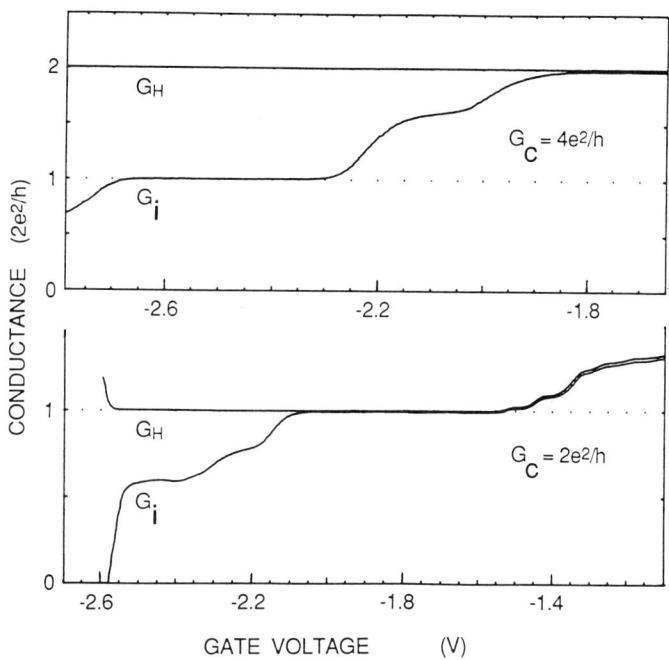

FIG. 42. Experimental correlation between the conductances G_i, G_c of injector and collector, and the Hall conductance $G_H \equiv I_i/V_c$, shown to demonstrate the validity of Eq. (47) ($T = 1.3$ K; point contact separation is 1.5 μm). The magnetic field was kept fixed (top: $B = 2.5$ T, bottom: $B = 3.8$ T, corresponding to a number of occupied bulk Landau levels $N = 3$ and 2, respectively). By increasing the gate voltage on one half of the split gate defining the injector, G_i was varied at constant G_c. (From Ref. 10.)

terms of a local resistivity tensor is meaningful, a satisfactory agreement between theory and experiment is obtained.[2] In the strong magnetic field regime of interest here, one has the complication that the concept of a local resistivity tensor may break down entirely because of the absence of local equilibrium. A theory of the Shubnikov–de Haas effect then has to take into account explicitly the properties of the contacts used for the measurement. The resulting anomalies seem not to have been anticipated in the theoretical literature. In Section 10.a, we discuss the anomalous Shubnikov–de Haas oscillations[11,12] in a geometry with a quantum point contact as a selective edge channel detector, which demonstrates the selective backscattering within a single Landau level and the absence of local equilibrium on a length scale of 250 μm. More general consequences of the absence of local equilibrium for the Shubnikov–de Haas effect are the subject of Section 10.b.

a. Selective Backscattering

To discuss the anomalous Shubnikov–de Haas effect, we consider the three-terminal geometry of Fig. 43, where a single voltage contact is present on the boundary between source and drain contacts. (An alternative two-terminal measurement configuration also is possible; see Ref. 11.) The voltage probe p is formed by a quantum point contact, while source s and drain d are normal ohmic contacts. (Note that *two* special contacts were required for the anomalous quantum Hall effect of Section 9.b.) One finds directly from Eq. (39) that the three-terminal resistance $R_{3t} \equiv (\mu_p - \mu_d)/eI$ measured between point

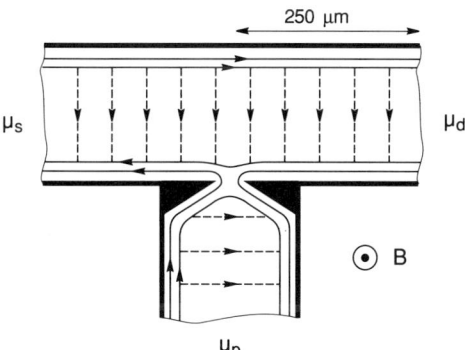

FIG. 43. Illustration of the mechanism for the suppression of Shubnikov–de Haas oscillations due to selective detection of edge channels. The black area denotes the split gate point contact in the voltage probe, which is at a distance of 250 μm from the drain reservoir. Dashed arrows indicate symbolically the selective backscattering in the highest-index edge channel, via states in the highest-bulk Landau level that coexist at the Fermi level.

contact probe and drain is given by

$$R_{3t} = \frac{h}{2e^2} \frac{T_{s \to p}}{(N_s - R_s)(N_p - R_p) - T_{p \to s} T_{s \to p}}. \tag{48}$$

This three-terminal resistance corresponds to a generalized *longitudinal* resistance if the magnetic field has the direction of Fig. 43. In the absence of backscattering in the 2DEG, one has $T_{s \to p} = 0$, so that R_{3t} vanishes, as it should for a longitudinal resistance in a strong magnetic field.

Shubnikov–de Haas oscillations in the longitudinal resistance arise when backscattering leads to $T_{s \to p} \neq 0$. The resistance reaches a maximum when the Fermi level lies in a bulk Landau level, corresponding to a maximum probability for backscattering (which requires scattering from one edge to the other across the bulk of the sample, as indicated by the dashed lines in Fig. 43). From the preceding discussion of the anomalous quantum Hall effect, we know that the point contact voltage probe in a high magnetic field functions as a selective detector of edge channels with index n less than some value determined by the barrier height in the point contact. If, as discussed in Section 8, backscattering itself occurs selectively for the channel with the highest index $n = N_{\text{wide}}$, and if the edge channels with $n \leq N_{\text{wide}} - 1$ do not scatter to that edge channel, then a suppression of the Shubnikov–de Haas oscillations is to be expected when R_{3t} is measured with a point contact containing a sufficiently high potential barrier. This indeed was observed experimentally,[11] as is shown in Fig. 44. The Shubnikov–de Haas maximum at 5.2 T, for example, is found to disappear at gate voltages such that the point contact conductance is equal to or smaller than $2e^2/h$, which means that the point contact only transmits two spin-split edge channels. The number of occupied spin-split Landau levels in the bulk at this magnetic field value is three. This experiment thus demonstrates that the Shubnikov–de Haas oscillations result from the highest-index edge channel only, and that this edge channel does not scatter to the lower-index edge channels over the distance of 250 μm from point contact probe to drain.

In Section 9.a, we discussed how an *ideal* contact at the 2DEG boundary *induces* a local equilibrium by equipartitioning the outgoing current equally among the edge channels. The anomalous Shubnikov–de Haas effect provides a direct way to study this contact-induced equilibration by means of a second point contact between the point contact voltage probe p and the current drain d in Fig. 43. This experiment is described in Ref. 12. Once again, use was made of the double-split-gate point contact device (Fig. 2), in this case with a 1.5 μm separation between point contact p and the second point contact. It is found that the Shubnikov–de Haas oscillations in R_{3t} are suppressed only if the second point contact has a conductance of

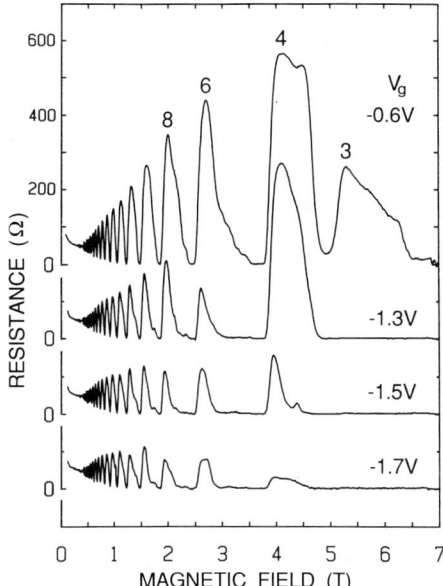

FIG. 44. Measurement of the anomalous Shubnikov–de Haas oscillations in the geometry of Fig. 43. The plotted longitudinal resistance is the voltage drop between contacts p and d divided by the current from s to d. At high magnetic fields, the oscillations increasingly are suppressed as the point contact in the voltage probe is pinched off by increasing the negative gate voltage. The number of occupied spin-split Landau levels in the bulk is indicated at several of the Shubnikov–de Haas maxima. (From Ref. 11.)

$(2e^2/h)(N_{\text{wide}} - 1)$ or smaller. At larger conductances, the oscillations in R_{3t} return, because this point contact now can couple to the highest-index edge channel and distribute the backscattered electrons over the lower-index edge channels. The point contact positioned between contacts p and d thus functions as a controllable *edge channel mixer*.

b. *Anomalous Scaling of the Shubnikov–de Haas Effect*

The conclusions of the previous subsection have interesting implications for the Shubnikov–de Haas oscillations in the strong-field regime even if measured with contacts that do *not* detect selectively certain edge channels only. Consider again the geometry of Fig. 43, in the low-gate voltage limit where the point contact voltage probe transmits all edge channels with unit probability. (This is the case of an *ideal* contact; *cf*. Section 9.a.) To simplify the expression, Eq. (48), for the three-terminal longitudinal resistance R_{3t}, we note that the transmission and reflection probabilities $T_{s \to p}$, R_s, and R_p

refer to the highest-index edge channel only (with index $n = N_{\text{wide}}$), under the assumptions of selective backscattering and absence of scattering to lower-index edge channels discussed before. As a consequence, $T_{\text{s}\to\text{p}}$, R_s, and R_p each are at most equal to 1, so that up to corrections smaller by a factor N_{wide}^{-1}, we may put these terms equal to zero in the denominator on the right-hand side of Eq. (48). In the numerator, the transmission probability $T_{\text{s}\to\text{p}}$ may be replaced by the backscattering probability $t_{\text{bs}} \leq 1$, which is the probability that the highest-index edge channel injected by the source contact reaches the point contact probe following scattering across the wide 2DEG (dashed lines in Fig. 43). With these simplifications, Eq. (48) takes the form (assuming spin degeneracy),

$$R_{3t} = \frac{h}{2e^2} \frac{t_{\text{bs}}}{N_{\text{wide}}^2} \times (1 + \text{order } N_{\text{wide}}^{-1}). \tag{49}$$

Only if $t_{\text{bs}} \ll 1$ may the backscattering probability be expected to scale linearly with the separation of the two contacts p and d (between which the voltage drop is measured). If t_{bs} is not small, then the upper limit $t_{\text{bs}} \leq 1$ leads to the novel prediction of a *maximum* possible amplitude,

$$R_{\max} = \frac{h}{2e^2} \frac{1}{N_{\text{wide}}^2} \times (1 + \text{order } N_{\text{wide}}^{-1}), \tag{50}$$

of the Shubnikov–de Haas resistance oscillations in a given large magnetic field, independently of the length of the segment over which the voltage drop is measured—provided equilibration does not occur on this segment. Equilibration might result, for example, from the presence of additional contacts between the voltage probes, as discussed in Sections 9.a and 10.a. One easily verifies that the high-field Shubnikov–de Haas oscillations in Fig. 44 at $V_\text{g} = -0.6$ V (when the point contact is just defined, so that the potential barrier is small) lie well below the upper limit, from Eq. (50). For example, the peak around 2 T corresponds to the case of four occupied spin-degenerate Landau levels, so that the theoretical upper limit is $(h/2e^2) \times \frac{1}{16} \approx 800\ \Omega$, well above the observed peak value of about $350\ \Omega$. The prediction of a maximum longitudinal resistance implies that the linear scaling of the amplitude of the Shubnikov–de Haas oscillations with the distance between voltage probes found in the weak-field regime, and expected on the basis of a description in terms of a local resistivity tensor,[2] breaks down in strong magnetic fields. Anomalous scaling of the Shubnikov–de Haas effect has been observed experimentally,[188,223,224] and also has been interpreted[225] recently in terms of a non-equilibrium between the edge channels.

Selective backscattering and the absence of local equilibrium have consequences as well for the two-terminal resistance in strong magnetic fields. In weak fields, one usually observes in two-terminal measurements a superposition of the Shubnikov–de Haas longitudinal resistance oscillations and the quantized Hall resistance (*cf.* Section 9.a). This superposition shows up as a characteristic *overshoot* of the two-terminal resistance as a function of magnetic field, as it increases from one quantized Hall plateau to the next. (The plateaus coincide with minima of the Shubnikov–de Haas oscillations.) In the strong-field regime (in the absence of equilibration between source and drain contacts), no such superposition is to be expected. Instead, the two-terminal resistance would increase monotonically from $(h/2e^2)N_{\text{wide}}^{-1}$ to $(h/2e^2)(N_{\text{wide}} - 1)^{-1}$ as the transmission probability from source to drain decreases from N_{wide} to $N_{\text{wide}} - 1$. We are not aware of an experimental test of this prediction. In Chapter 4, Büttiker discusses an experiment by Fang *et al.* on the Shubnikov–de Haas oscillations in the short-channel regime, to which our analysis does not apply.

11. Aharonov–Bohm Oscillations and Inter-Edge Channel Tunneling

a. Aharonov–Bohm Effect in a Singly Connected Geometry

The *Aharonov–Bohm effect* is a fundamental manifestation of the influence of a magnetic field on the phase of the electron wave function.[226,227] In the solid state, Aharonov–Bohm magnetoresistance oscillations have been studied extensively in metal rings and cylinders,[228,229] and more recently with a much larger amplitude in 2DEG rings.[230,231] (See Chapter 3 by Timp.) In such experiments, oscillations in the resistance of the ring are observed as a function of the applied perpendicular magnetic field B. The oscillations are periodic in B, with a fundamental period $\Delta B = h/eA$ determined by the area A of the ring. Their origin is the field-induced phase difference between the two paths (one clockwise, one counterclockwise), which take an electron from one side of the ring to the other. Normally, a *multiply*-connected geometry (a ring) is required to see the effect. It thus came as a surprise when a magnetoresistance oscillation periodic in B was discovered in a quantum point contact,[16] which constitutes a *singly* connected geometry. A similar effect more recently has been reported by Wharam *et al.*[232]

The magnetic field dependence of the two-terminal resistance in the experiment of Ref. 16 is shown in Fig. 45. The periodic oscillations occur predominantly between quantum Hall plateaus, in a range of gate voltages only, and only at low temperatures. (In Fig. 45, $T = 50$ mK; the effect has disappeared at 1 K.) The fine structure is very reproducible if the sample is kept in the cold, but changes after cycling to room temperature. As one can see

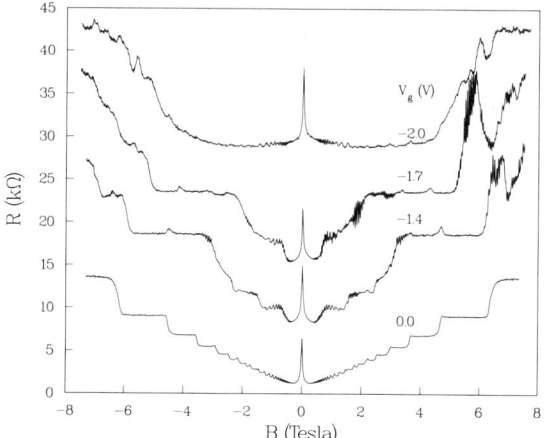

FIG. 45. Two-terminal magnetoresistance of a point contact for a series of gate voltages at $T = 50$ mK, showing oscillations that are periodic in B between the quantum Hall plateaus. The second, third and fourth curves from the bottom have offsets, respectively, of 5, 10, and 15 kΩ. The rapid oscillations below 1 T are Shubnikov–de Haas oscillations periodic in $1/B$, originating from the wide 2DEG regions. The sharp peak around $B = 0$ T originates from the ohmic contacts. (From Ref. 16.)

from the enlargements in Fig. 46, a splitting of the peaks occurs in a range of magnetic fields, presumably as spin-splitting becomes resolved. A curious aspect of the effect (which has remained unexplained) is that the oscillations have a much larger amplitude in one field direction than in the other (Fig. 45), in apparent conflict with the $\pm B$ symmetry of the two-terminal resistance required by the reciprocity relation[7,158] in the absence of magnetic impurities. Other devices of the same design did not show oscillations of well-defined periodicity, and had a two-terminal resistance that was approximately $\pm B$ symmetric.

Figure 47 illustrates the tunneling mechanism for the periodic magnetoresistance oscillations as it was originally proposed[16] to explain the observations. Because of the presence of a barrier in the point contact, the electrostatic potential has a saddle form. Equipotentials at the guiding center energy (cf. Section 8) are drawn schematically in Fig. 47. (Arrows indicate the direction of motion along the equipotential.) An electron that enters the constriction at a can be reflected back into the broad region by tunneling to the opposite edge, either at the potential step at the entrance of the constriction (from a to b) or at its exit (from d to c). These two tunneling paths acquire an Aharonov–Bohm phase difference[186,223] of $eBA/\hbar$ (where A is the enclosed area $abcd$), leading to periodic magnetoresistance oscillations. (Note that the periodicity ΔB may differ[17,208] somewhat from the usual expression $\Delta B = h/eA$, if A itself is B-dependent due to the B-dependence of the guiding

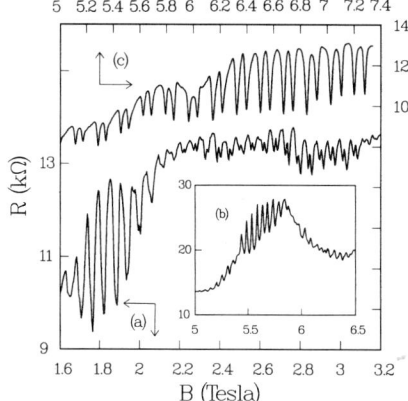

FIG. 46. Curves *a* and *b* are close-ups of the curve for $V_g = -1.7$ V in Fig. 45. Curve *c* was measured three months earlier on the same device. (Note the different field scale due to a change in electron density in the constriction.) (From Ref. 16.)

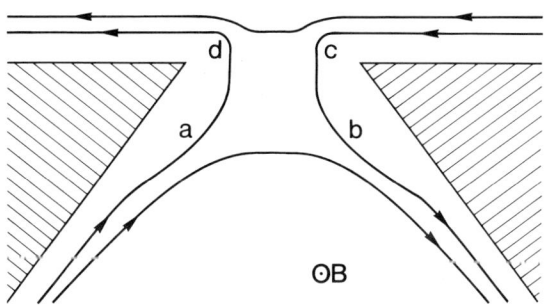

FIG. 47. Equipotentials at the guiding center energy in the saddle-shaped potential created by a split gate (shaded). Aharonov–Bohm oscillations in the point contact magnetoresistance result from the interference of tunneling paths *ab* and *adcb*. Tunneling from *a* to *b* may be assisted by an impurity at the entrance of the constriction. (From Ref. 16.)

center energy from Eq. (38).) This mechanism shows how an Aharonov–Bohm effect is possible in principle in a singly connected geometry: The point contact behaves as if it were multiply connected, by virtue of the spatial separation of edge channels moving in opposite directions. (Related mechanisms, based on circulating edge currents, have been considered for Aharonov–Bohm effects in small conductors.)[221,222,234–236] Unlike in the original Aharonov–Bohm effect[226,227,237] in vacuum, where only the effect of the magnetic field on the *phase* of the wave function matters, the *Lorentz force* plays an essential role here. This explains why the oscillations periodic in *B* are observed only at large magnetic fields in Fig. 45 (above about 1 T; the oscillations at lower fields are Shubnikov–de Haas oscillations periodic in $1/B$, due to the series resistance of the wide 2DEG regions). At low magnetic fields, the spatial separation of edge channels responsible for the Aharonov–Bohm effect is not effective yet. The spatial separation also can be destroyed by a large negative gate voltage (top curve in Fig. 45) when the width of the point contact becomes so small that the wave functions of edge states at opposite edges overlap.

Although the mechanism illustrated in Fig. 47 is attractive because it is an intrinsic consequence of the point contact geometry, the observed well-defined periodicity of the magnetoresistance oscillations requires that the potential induced by the split gate varies rapidly over a short distance (to have a well-defined area A). A smooth saddle potential seems more realistic. Moreover, one would expect the periodicity to vary more strongly with gate voltage than the small 10% variation observed experimentally as V_g is changed from -1.4 to -1.7 V. Glazman and Jonson[208] have proposed an alternative *impurity-assisted* tunneling mechanism as a more likely (but essentially equivalent) explanation of the experiment.[16] In their picture, one of the two tunneling processes (from a to b in Fig. 47) is mediated by an impurity outside but close to the constriction. The combination of impurity and point contact introduces a well-defined area even for a smooth saddle potential, which moreover will not be strongly gate voltage-dependent. We note that the single sample in which the effect was observed was special in having an anomalously small pinch-off voltage. (See the curves labeled 3 in Fig. 14.) This well may be due to the accidental presence of an impurity in the immediate vicinity of the constriction. To study the Aharonov–Bohm effect due to inter-edge channel tunneling under more controlled conditions, a different device geometry is necessary,[17] as discussed in the following subsection.

b. *Tunneling through an Edge State Bound in a Cavity*

i. *Experiment.* The geometry of Ref. 17 is shown schematically in Fig. 48. A cavity with two opposite point contact openings is defined in the 2DEG by split gates A and B. The diameter of the cavity is approximately 1.5 μm. The conductances G_A and G_B of the two point contacts can be measured independently (by grounding one set of gates), with the results plotted in Figs. 49a,b (for $V_g = -0.35$ V on either gate A or B). The conductance G_C of the cavity (for $V_g = -0.35$ V on both the split gates) is plotted in Fig. 49c. A long series of periodic oscillations is observed between two quantum Hall plateaus. Similar series of oscillations (but with a different periodicity) have been observed between other quantum Hall plateaus. The oscillations are suppressed on the plateaus themselves. The amplitude of the oscillations is comparable to that observed in the experiment on a single point contact[16] (discussed in Section 11.a), but the period is much smaller (consistent with a larger effective area in the double point contact device), and also no splitting of

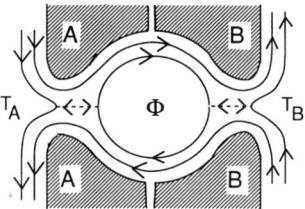

FIG. 48. Tunneling through an edge state bound in a cavity. The cavity (of 1.5 μm diameter) is defined by a double set of split gates A and B. For large negative gate voltages, the 2DEG region under the narrow gap between gates A and B is fully depleted, while transmission remains possible over the potential barrier in the wider openings at the left and right of the cavity. Tunneling through the left and right barrier (as indicated by dashed lines) occurs with transmission probabilities T_A and T_B, which are adjustable separately by means of the voltages on gates A and B. On increasing the magnetic field, resonant tunneling through the cavity occurs periodically each time the flux Φ enclosed by the circulating edge state increases by one flux quantum h/e. (From Ref. 17.)

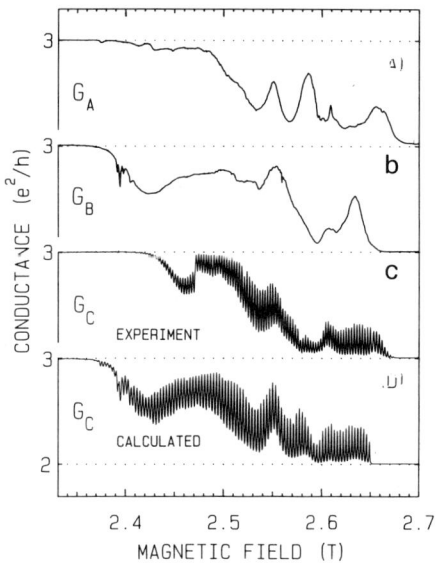

FIG. 49. Magnetoconductance experiments on the device of Fig. 48 at 6 mK, for a fixed gate voltage of -0.35 V. (a) Conductance of point contact A, measured with gate B grounded. (b) Conductance of point contact B (gate A grounded). (c) Measured conductance of the entire cavity. (d) Calculated conductance of the cavity, obtained from Eqs. (51) and (52) with the measured G_A and G_B as input. (From Ref. 17.)

the peaks is observed (presumably due to a fully resolved spin degeneracy). No gross $\pm B$ asymmetries were found in the present experiment, although an accurate test of the symmetry on field reversal was not possible because of difficulties with the reproducibility. The oscillations are quite fragile, disappearing when the temperature is raised above 200 mK or when the voltage across the device exceeds 40 μV. (The data in Fig. 49 was taken at 6 mK and 6 μV.) The experimental data is well described by resonant transmission through a circulating edge state in the cavity,[17] as illustrated in Fig. 48 and described in detail shortly. Aharanov–Bohm oscillations due to resonant transmission through a similar structure have been reported more recently by Brown et al.[238]

ii. Model. As discussed in Sections 9.b and 11.a, the electrostatic potential defining each point contact has a saddle shape, due to the combination of the lateral confinement and the potential barrier. The height of the barrier can be adjusted by means of the gate voltage. An edge state with a guiding center energy below the barrier height forms a bound state in the cavity formed by two opposite point contacts, as is illustrated in Fig. 48. Tunneling of edge channels through the cavity via this bound state occurs with transmission probability T_{AB}, which for a single edge channel is given by[221,239]

$$T_{AB} = \left| \frac{t_A t_B}{1 - r_A r_B \exp(i\Phi e/\hbar)} \right|^2 = \frac{T_A T_B}{1 + R_A R_B - 2(R_A R_B)^{1/2} \cos(\phi_0 + \Phi e/\hbar)}. \tag{51}$$

Here, t_A and r_A are the transmission and reflection probability amplitudes through point contact A, $T_A \equiv |t_A|^2$ and $R_A \equiv |r_A|^2 = 1 - T_A$ are the transmission and reflection probabilities, and t_B, r_B, T_B, R_B denote the corresponding quantities for point contact B. In Eq. (51), the phase acquired by the electron on one revolution around the cavity is the sum of the phase ϕ_0 from the reflection probability amplitudes (which can be assumed to be only weakly B-dependent) and the Aharanov–Bohm phase $\Phi \equiv BA$, which varies rapidly with B. (Φ is the flux through the area A enclosed by the equipotential along which the circulating edge state is extended.) Resonant tunneling occurs periodically with B, whenever $\phi_0 + \Phi e/\hbar$ is a multiple of 2π.

In the case that only a single (spin-split) edge channel is occupied in the 2DEG, the conductance $G_C = (e^2/h)T_{AB}$ of the cavity follows directly from Eq. (51). The transmission and reflection probabilities can be determined independently from the individual point contact conductances, $G_A = (e^2/h)T_A$ (and similarly for G_B)—at least if one may assume that the presence of the cavity has no effect itself on T_A and T_B (but only on the total transmission probability T_{AB}). If $N > 1$ spin-split edge channels are occupied, we assume that the $N - 1$ lowest-index edge channels are fully transmitted and do not modify the transmission probability of the Nth edge channel, so that we can write[17]

$$G_C = \frac{e^2}{h}(N - 1 + T_{AB}), \quad G_A = \frac{e^2}{h}(N - 1 + T_A), \quad G_B = \frac{e^2}{h}(N - 1 + T_B). \quad (52)$$

This simple model is compared with the experiment in Fig. 49. The trace in Fig. 49d has been calculated from Eqs. (51) and (52) using the individual point contact conductances in Figs. 49a,b as input for T_A and T_B. The flux Φ has been adjusted to the experimental periodicity of 3 mT, and the phase ϕ_0 in Eq. (51) has been ignored (since that only would amount to a phase shift of the oscillations). Energy averaging due to the finite temperature and voltage has been taken into account in the calculation.[17] The agreement with the experimental trace (Fig. 49c) is quite satisfactory.

The results shown in Fig. 49 demonstrate once again how some textbook quantum mechanics becomes reality when one studies transport through quantum point contacts. This seems an appropriate place to conclude our review of the subject. From initial work on the properties of single quantum point contacts, we have moved on to structures involving two adjacent point contacts (as in the electron focusing geometry), or two opposite point contacts (as in the cavity geometry considered before). The basic principles of the ballistic and adiabatic transport regime are now well-understood, and have been demonstrated experimentally. Future work in other transport regimes and in more complicated structures undoubtedly will reveal more interesting and potentially useful phenomena. It is our belief that the quantum point contact, because of its simplicity and fundamental nature, will continue to be used as a versatile tool and building block in such investigations.

Acknowledgments

We thank our collaborators from the Delft University of Technology, and the Philips Research Laboratories in Eindhoven and Redhill for their generous permission to reproduce results from our joint publications.[4,10-13,15-17,57-61] Reference 34 has been part of a collaboration between Philips and the University of Cambridge. The stimulating support of Professors J. A. Pals and M. F. H. Schuurmans from the Philips Research Laboratories is gratefully acknowledged. Technological support has been provided by the Delft Center for Submicron Technology and the Philips Mask Center. The work at Delft is supported financially by the Stichting voor Fundamenteel Onderzoek der Materie (F.O.M.).

References

1. R. Landauer, in *Localization, Interaction, and Transport Phenomena* (B. Kramer, G. Bergmann, and Y. Bruynseraede, eds.), Springer, Berlin (1985).
2. T. Ando, A. B. Fowler, and F. Stern, *Rev. Mod Phys.* **54**, 437 (1982).
3. J. J. Harris, J. A. Pals, and R. Woltjer, *Rep. Prog. Phys.* **52**, 1217 (1989).
4. B. J. van Wees, H. van Houten, C. W. J. Beenakker, J. G. Williamson, L. P. Kouwenhoven, D. van der Marel, and C. T. Foxon, *Phys. Rev. Lett.* **60**, 848 (1988).
5. D. A. Wharam, T. J. Thornton, R. Newbury, M. Pepper, H. Ahmed, J. E. F. Frost, D. G. Hasko, D. C. Peacock, D. A. Ritchie, and G. A. C. Jones, *J. Phys. C* **21**, L209 (1988).
6. R. Landauer, *IBM J. Res. Dev.* **1**, 223 (1957).
7. M. Büttiker, *Phys. Rev. Lett.* **57**, 1761 (1986).
8. C. W. J. Beenakker and H. van Houten, *Phys. Rev. B* **39**, 10445 (1989).
9. D. A. Wharam, M. Pepper, H. Ahmed, J. E. F. Frost, D. G. Hasko, D. C. Peacock, D. A. Ritchie, and G. A. C. Jones, *J. Phys. C* **21**, L887 (1988).
10. B. J. van Wees, E. M. M. Willems, C. J. P. M. Harmans, C. W. J. Beenakker, H. van Houten, J. G. Williamson, C. T. Foxon, and J. J. Harris, *Phys. Rev. Lett.* **62**, 1181 (1989).
11. B. J. van Wees, E. M. M. Willems, L. P. Kouwenhoven, C. J. P. M. Harmans, J. G. Williamson, C. T. Foxon, and J. J. Harris, *Phys. Rev. B* **39**, 8066 (1989).
12. B. J. van Wees, L. P. Kouwenhoven, E. M. M. Willems, C. J. P. M. Harmans, J. E. Mooij, H. van Houten, C. W. J. Beenakker, J. G. Williamson, and C. T. Foxon, *Phys. Rev. B* **43**, 12431 (1991).
13. H. van Houten, B. J. van Wees, J. E. Mooij, C. W. J. Beenakker, J. G. Williamson, and C. T. Foxon, *Europhys. Lett.* **5**, 721 (1988).
14. C. W. J. Beenakker, H. van Houten, and B. J. van Wees, *Europhys. Lett.* **7**, 359 (1988).
15. H. van Houten, C. W. J. Beenakker, J. G. Williamson, M. E. I. Broekkaart, P. H. M. van Loosdrecht, B. J. van Wees, J. E. Mooij, C. T. Foxon, and J. J. Harris, *Phys. Rev. B* **39**, 8556 (1989).
16. P. H. M. van Loosdrecht, C. W. J. Beenakker, H. van Houten, J. G. Williamson, B. J. van Wees, J. E. Mooij, C. T. Foxon, and J. J. Harris, *Phys. Rev. B* **38**, 10162 (1988).
17. B. J. van Wees, L. P. Kouwenhoven, C. J. P. M. Harmans, J. G. Williamson, C. E. T. Timmering, M. E. I. Broekaart, C. T. Foxon, and J. J. Harris, *Phys. Rev. Lett.* **62**, 2523 (1989).
18. C. W. J. Beenakker and H. van Houten, *Solid State Physics* **44**, 1 (1991).
19. H. Heinrich, G. Bauer, and F. Kuchar, eds., *Physics and Technology of Submicron Structures*, Springer, Berlin (1988).
20. M. Reed and W. P. Kirk, eds., *Nanostructure Physics and Fabrication*, Academic Press, New York (1989).
21. S. P. Beaumont and C. M. Sotomayor-Torres, eds., *Science and Engineering of 1- and 0-Dimensional Semiconductors*, Plenum, London (1990).
22. J. M. Chamberlain, L. Eaves, and J. C. Portal, eds., *Electronic Properties of Multilayers and Low-Dimensional Semiconductor Structures*, Plenum, New York (1990).
23. Yu. V. Sharvin, *Zh. Eksp. Teor. Fiz.* **48**, 984 (1965) [*Sov. Phys. JETP* **21**, 655 (1965)].
24. V. S. Tsoi, *Pis'ma Zh. Eksp. Teor. Fiz.* **19**, 114 (1974) [*JETP Lett.* **19**, 701 (1974)]; *Zh. Eksp. Teor. Fiz.* **68**, 1849 (1975) [*Sov. Phys. JETP* **41**, 927 (1975)].
25. I. K. Yanson, *Zh. Eksp. Teor. Fiz.* **66**, 1035 (1974) [*Sov. Phys. JETP* **39**, 506 (1974)].
26. A. G. M. Jansen, A. P. van Gelder, and P. Wyder, *J. Phys. C* **13**, 6073 (1980).
27. R. Trzcinski, E. Gmelin, and H. J. Queisser, *Phys. Rev. B* **35**, 6373 (1987).
28. T. J. Thornton, M. Pepper, H. Ahmed, D. Andrews, and G. J. Davies, *Phys. Rev. Lett.* **56**, 1198 (1986).
29. H. Z. Zheng, H. P. Wei, D. C. Tsui, and G. Weimann, *Phys. Rev. B* **34**, 5635 (1986).

30. J. P. Kirtley, Z. Schlesinger, T. N. Theis, F. P. Milliken, S. L. Wright, and L. F. Palmateer, *Phys. Rev. B* **34**, 5414 (1986).
31. H. van Houten, B. J. van Wees, M. G. J. Heijman, and J. P. André, *Appl. Phys. Lett.* **49**, 1781 (1986).
32. Y. Hirayama, T. Saku, and Y. Horikoshi, *Phys. Rev. B* **39**, 5535 (1989); Y. Hirayama and T. Saku, *Appl. Phys. Lett.* **54**, 2556 (1989).
33. A. D. Wieck and K. Ploog, *Surf. Sci.* **229**, 252 (1990); *Appl. Phys. Lett.* **56**, 928 (1990).
34. H. van Houten, C. W. J. Beenakker, P. H. M. van Loosdrecht, T. J. Thornton, H. Ahmed, M. Pepper, C. T. Foxon, and J. J. Harris, *Phys. Rev. B* **37**, 8534 (1988); and unpublished.
35. J. C. Maxwell, *A Treatise on Electricity and Magnetism,* Clarendon, Oxford (1904).
36. R. Holm, *Electrical Contacts Handbook*, Springer, Berlin (1958).
37. Yu. V. Sharvin and N. I. Bogatina, *Zh. Eksp. Teor. Fiz.* **56**, 772 (1969) [*Sov. Phys. JETP* **29**, 419 (1969)].
38. R. A. R. Tricker, *Proc. Camb. Phil. Soc.* **22**, 454 (1925).
39. A. B. Pippard, *Magnetoresistance in Metals*, Cambridge University Press, Cambridge (1989).
40. P. C. van Son, H. van Kempen, and P. Wyder, *Phys. Rev. Lett.* **58**, 1567 (1987).
41. E. E. Vdovin, A. Yu. Kasumov, Ch. V. Kopetskii, and I. B. Levinson, *Zh. Eksp. Teor. Fiz.* **92**, 1026 (1987) [*Sov. Phys. JETP* **65**, 582 (1987)].
42. H. M. Swartjes, A. P. van Gelder, A. G. M. Jansen, and P. Wyder, *Phys. Rev. B* **39**, 3086 (1989).
43. J. K. Gimzewski and R. Möller, *Phys. Rev. B* **36**, 1284 (1987).
44. N. D. Lang, *Phys. Rev. B* **36**, 8173 (1987).
45. J. Ferrer, A. Martin-Rodero, and F. Flores, *Phys. Rev. B* **38**, 10113 (1988).
46. N. D. Lang, *Comm. Cond. Matt. Phys.* **14**, 253 (1989).
47. N. D. Lang, A. Yacoby, and Y. Imry, *Phys. Rev. Lett.* **63**, 1499 (1989).
48. N. Garcia and H. Rohrer, *J. Phys. Condens. Matter* **1**, 3737 (1989).
49. A. C. Warren, D. A. Antoniadis, and H. I. Smith, *Phys. Rev. Lett.* **56**, 1858 (1986).
50. K. K. Choi, D. C. Tsui, and S. C. Palmateer, *Phys. Rev. B* **33**, 8216 (1986).
51. G. Timp, A. M. Chang, P. Mankiewich, R. Behringer, J. E. Cunningham, T. Y. Chang, and R. E. Howard, *Phys. Rev. Lett.* **59**, 732 (1987).
52. M. L. Roukes, A. Scherer, S. J. Allen, Jr., H. G. Craighead, R. M. Ruthen, E. D. Beebe, and J. P. Harbison, *Phys. Rev. Lett.* **59**, 3011 (1987).
53. A. M. Chang, G. Timp, T. Y. Chang, J. E. Cunningham, P. M. Mankiewich, R. E. Behringer, and R. E. Howard, *Solid State Comm.* **67**, 769 (1988).
54. Y. Takagaki, K. Gamo, S. Namba, S. Ishida, S. Takaoka, K. Murase, K. Ishibashi, and Y. Aoyagi, *Solid State Comm.* **68**, 1051 (1988).
55. M. L. Roukes, T. J. Thornton, A. Scherer, J. A. Simmons, B. P. van der Gaag, and E. D. Beebe, in *Science and Engineering of 1- and 0-Dimensional Semiconductors* (S. P. Beaumont and C. M. Sotomayer-Torres, eds.), Plenum, London (1990).
56. G. Timp, H. U. Baranger, P. de Vegvar, J. E. Cunningham, R. E. Howard, R. Behringer, and P. M. Mankiewich, *Phys. Rev. Lett.* **60**, 2081 (1988).
57. B. J. van Wees, L. P. Kouwenhoven, H. van Houten, C. W. J. Beenakker, J. E. Mooij, C. T. Foxon, and J. J. Harris, *Phys. Rev. B* **38**, 3625 (1988).
58. L. W. Molenkamp, A. A. M. Staring, C. W. J. Beenakker, R. Eppenga, C. E. Timmering, J. G. Williamson, C. J. P. M. Harmans, and C. T. Foxon, *Phys. Rev. B* (**41**, 1274 (1990).
59. A. A. M. Staring, L. W. Molenkamp, C. W. J. Beenakker, L. P. Kouwenhoven, and C. T. Foxon, *Phys. Rev. B* **41**, 8461 (1990).
60. L. P. Kouwenhoven, B. J. van Wees, C. J. P. M. Harmans, J. G. Williamson, H. van Houten, C. W. J. Beenakker, C. T. Foxon, and J. J. Harris, *Phys. Rev. B* **39**, 8040 (1989).
61. J. G. Williamson, H. van Houten, C. W. J. Beenakker, M. E. I. Broekaart, L. I. A. Spendeler, B. J. van Wees, and C. T. Foxon, *Phys. Rev. B* **41**, 1207 (1990); *Surf. Sci.* **229**, 303 (1990).

62. K. von Klitzing, G. Dorda, and M. Pepper, *Phys. Rev. Lett.* **45**, 494 (1980).
63. G. Timp. R. Behringer, S. Sampere, J. E. Cunningham, and R. E. Howard, in *Nanostructure Physics and Fabrication* (M. A. Reed and W. P. Kirk, eds.), Academic Press, New York (1989).
64. R. E. Prange and S. M. Girvin, eds., *The Quantum Hall Effect*, Springer, New York (1987).
65. H. van Houten, B. J. van Wees, and C. W. J. Beenakker, in *Physics and Technology of Submicron Structures* (H. Heinrich, G. Bauer, and F. Kuchar, eds.), Springer, Berlin (1988).
66. H. van Houten and C. W. J. Beenakker, in *Nanostructure Physics and Fabrication* (M. A. Reed and W. P. Kirk, eds.), Academic Press, New York (1989); see also M. Knudsen, *The Kinetic Theory of Gases*, Methuen, London (1934).
67. D. S. Fisher and P. Λ. Lee, *Phys. Rev. B* **23**, 6851 (1981); E. N. Economou and C. M. Soukoulis, *Phys. Rev. Lett.* **46**, 618 (1981)
68. Y. Imry, in *Directions in Condensed Matter Physics*, Vol. 1 (G. Grinstein and G. Mazenko, eds.), World Scientific Publishing Co. Singapore (1986).
69. A. D. Stone and A. Szafer, *IBM J. Res. Dev.* **32**, 384 (1988).
70. D. J. Thouless, *Phys. Rev. Lett.* **47**, 972 (1981).
71. R. Landauer, *Phys. Lett. A* **85**, 91 (1981).
72. H. L. Engquist and P. W. Anderson, *Phys. Rev. B* **24**, 1151 (1981).
73. D. C. Langreth and E. Abrahams, *Phys. Rev. B* **24**, 2978 (1981).
74. P. W. Anderson, D. J. Thouless, E. Abrahams, and D. S. Fisher, *Phys. Rev. B* **22**, 3519 (1980).
75. M. Büttiker, Y. Imry, R. Landauer, and S. Pinhas, *Phys. Rev. B* **31**, 6207 (1985).
76. M. Ya. Azbel, *J. Phys. C* **14**, L225 (1981).
77. M. Büttiker, *Phys. Rev. B* **38**, 9375 (1988).
78. J. Kucera and P. Streda. *J. Phys. C* **21**, 4357 (1988).
79. H. U. Baranger and A. D. Stone, *Phys. Rev. B* **40**, 8169 (1989).
80. Y. Isawa, *J. Phys. Soc. Japan* **57**, 3457 (1988).
81. B. Kramer and J. Masek, *J. Phys. C* **21**, L1147 (1988).
82. R. Johnston and L. Schweitzer, *J. Phys. C* **21**, L861 (1988).
83. R. Landauer, *J. Phys. Condens. Matter* **1**, 8099 (1989); also in *Nanostructure Physics and Fabrication* (M. A. Reed and W. P. Kirk, eds.), Academic Press, New York (1989).
84. L. Escapa and N. Garcia, *J. Phys. Condens. Matter* **1**, 2125 (1989).
85. E. G. Haanappel and D. van der Marel, *Phys. Rev. B* **39**, 5484 (1989); D. van der Marel and E. G. Haanappel, *Phys. Rev. B* **39**, 7811 (1989).
86. G. Kirczenow, *Solid State Comm.* **68**, 715 (1988); *J. Phys. Condens. Matter* **1**, 305 (1989).
87. A. Szafer and A. D. Stone, *Phys. Rev. Lett.* **62**, 300 (1989).
88. E. Tekman and S. Ciraci, *Phys. Rev. B* **39**, 8772 (1989); *Phys. Rev. B* **40**, 8559 (1989).
89. Song He and S. Das Sarma, *Phys. Rev. B* **40**, 3379 (1989).
90. N. Garcia and L. Escapa, *Appl. Phys. Lett.* **54**, 1418 (1989).
91. E. Castaño and G. Kirczenow, *Solid State Comm.* **70**, 801 (1989).
92. A. Kawabata, *J. Phys. Soc. Japan* **58**, 372 (1989).
93. I. B. Levinson, *Pis'ma Zh. Eksp. Teor. Fiz.* **48**, 273 (1988) [*JETP Lett.* **48**, 301 (1988)].
94. A. Matulis and D. Segzda, *J. Phys. Condens. Matter* **1**, 2289 (1989).
95. M. Büttiker, *Phys. Rev. B* **41**, 7906 (1990).
96. H. van Houten and C. W. J. Beenakker, in *Analogies in Optics and Microelectronics* (W. van Haeringen and D. Lenstra, eds.), Kluwer, Deventer, (1990).
97. L. I. Glazman, G. B. Lesovik, D. E. Khmel'nitskii, R. I. Shekhter, *Pis'ma Zh. Teor. Fiz.* **48**, 218 (1988) [*JETP Lett.* **48**, 238 (1988)].
98. R. Landauer, *Z. Phys. B* **68**, 217 (1987).
99. C. S. Chu and R. S. Sorbello, *Phys. Rev. B* **40**, 5941 (1989).
100. J. Masek, P. Lipavsky, and B. Kramer, *J. Phys. Condens. Matter* **1**, 6395 (1989).
101. P. F. Bagwell and T. P. Orlando, *Phys. Rev. B* **40**, 1456 (1989).

102. R. J. Brown, M. J. Kelly, R. Newbury, M. Pepper, B. Miller, H. Ahmed, D. G. Hasko, D. C. Peacock, D. A. Ritchie, J. E. F. Frost, and G. A. C. Jones, *Solid State Electron.* **32**, 1179 (1989).
103. Y. Hirayama, T. Saku, and Y. Horikoshi, *Jap. J. Appl. Phys.* **28**, L701 (1989).
104. B. L. Al'tshuler, *Pis'ma Zh. Eksp. Teor. Fiz.* **41**, 530 (1985) [*JETP Lett.* **41**, 648 (1985)].
105. P. A. Lee and A. D. Stone, *Phys. Rev. Lett.* **55**, 1622 (1985); P. A. Lee, A. D. Stone, and H. Fukuyama, *Phys. Rev. B* **35**, 1039 (1987).
106. G. Timp, in *Mesoscopic Phenomena in Solids* (P. A. Lee, R. A. Webb, and B. L. Al'tshuler, eds.), Elsevier Science Publ., Amsterdam (1989).
107. J. H. Davies and J. A. Nixon, *Phys. Rev. B* **39**, 3423 (1989); J. H. Davies in *Nanostructure Physics and Fabrication* (M. A. Reed and W. P. Kirk, eds.), Academic Press, New York (1989).
108. T. J. Thornton, M. L. Roukes, A. Scherer, and B. van der Gaag, *Phys. Rev. Lett.* **63**, 2128 (1989).
109. H. van Houten, C. W. J. Beenakker, M. E. I. Broekaart, M. G. H. J. Heijman, B. J. van Wees, H. E. Mooij, and J. P. André, *Acta Electronica*, **28**, 27 (1988).
110. L. D. Landau and E. M. Lifshitz, *Quantum Mechanics*, Pergamon, Oxford, (1977).
111. K.-F. Berggren, G. Roos, and H. van Houten, *Phys. Rev. B* **37**, 10118 (1988).
112. L. I. Glazman and A. V. Khaetskii, *J. Phys. Condens. Matter* **1**, 5005 (1989).
113. Y. Avishai and Y. B. Band, *Phys. Rev. B* **40**, 3429 (1989).
114. K. B. Efetov, *J. Phys. Condens. Matter* **1**, 5535 (1989).
115. F. F. Fang and P. J. Stiles, *Phys. Rev. B* **27**, 6487 (1983).
116. T. G. Powell, C. C. Dean, and M. Pepper, *J. Phys. C* **17**, L359 (1984).
117. G. L. J. A. Rikken, J. A. M. M. van Haaren, W. van der Wel, A. P. van Gelder, H. van Kempen, P. Wyder, J. P. André, K. Ploog, and G. Weimann, *Phys. Rev. B* **37**, 6181 (1988).
118. K. F. Berggren, T. J. Thornton, D. J. Newson, and M. Pepper, *Phys. Rev. Lett.* **57**, 1769 (1986).
119. S. B. Kaplan and A. C. Warren, *Phys. Rev. B* **39**, 1346 (1986).
120. H. van Houten, B. J. van Wees, J. E. Mooij, G. Roos, and K. -F. Berggren, *Superlattices and Microstructures*, **3**, 497 (1987).
121. J. F. Weisz and K. -F. Berggren, *Phys. Rev. B* **40**, 1325 (1989).
122. S. E. Laux, D. J. Frank, and F. Stern, *Surf. Sci.* **196**, 101 (1988).
123. A. Kumar, S. E. Laux, and F. Stern, *Appl. Phys. Lett.* **54**, 1270 (1989); *Bull. Am. Phys. Soc.* **34**, 589 (1989).
124. D. A. Wharam, U. Ekenberg, M. Pepper, D. G. Hasko, H. Ahmed, J. E. F. Frost, D. A. Ritchie, D. C. Peacock, and G. A. C. Jones, *Phys. Rev. B* **39**, 6283 (1989).
125. L. P. Kouwenhoven, Master Thesis, Delft University of Technology, Delft, The Netherlands (1988).
126. B. R. Snell, P. H. Beton, P. C. Main, A. Neves, J. R. Owers-Bradley, L. Eaves, M. Henini, O. H. Hughes, S. P. Beaumont, and C. D. W. Wilkinson, *J. Phys. Condens. Matter* **1**, 7499 (1989).
127. K. von Klitzing and G. Ebert, in *Two-Dimensional Systems, Heterostructures, and Superlattices* (G. Bauer, F. Kuchar, and H. Heinrich, eds.), Springer, Heidelberg, Germany (1984); **61**, 1001 (1988).
128. F. F. Fang and P. J. Stiles, *Phys. Rev. B* **29**, 3749 (1984).
129. D. A. Syphers, F. F. Fang, and P. J. Stiles, *Surf. Sci.* **142**, 208 (1984).
130. D. A. Syphers and P. J. Stiles, *Phys. Rev. B* **32**, 6620 (1985).
131. A. B. Berkut, Yu. V. Dubrovskii, M. S. Nunuparov, M. I. Reznikov, and V. I. Tal'yanski, *Pis'ma Zh. Teor. Fiz.* **44**, 252 (1986) [*JETP Lett.* **44**, 324 (1986)].
132. G. L. J. A. Rikken, J. A. M. M. van Haaren, A. P. van Gelder, H. van Kempen, P. Wyder, H. -U. Habermeier, and K. Ploog, *Phys. Rev. B* **37**, 10229 (1988).
133. R. J. Haug, A. H. MacDonald, P. Streda, and K. von Klitzing, *Phys. Rev. Lett.* **61**, 2797 (1988).
134. S. Washburn, A. B. Fowler, H. Schmid, and D. Kern, *Phys. Rev. Lett.* **61**, 2801 (1988).

135. H. Hirai, S. Komiyama, S. Hiyamizu, and S. Sasa, in *Proc. 19th International Conference on the Physics of Semiconductors* (W. Zawadski, ed.), Inst. of Phys., Polish Acad. Sci. (1988).
136. B. E. Kane, D. C. Tsui, and G. Weimann, *Phys. Rev. Lett.* **61**, 1123 (1988).
137. R. J. Haug, J. Kucera, P. Streda, and K. von Klitzing, *Phys. Rev. B* **39**, 10892 (1989).
138. Y. Zhu, J. Shi, and S. Feng, *Phys. Rev. B* **41**, 8509 (1990).
139. Y. Hirayama and T. Saku, *Solid State Commun.* **73**, 113 (1990); **41**, 2927 (1990).
140. E. Castaño and G. Kirczenow, *Phys. Rev. B* **41**, 5055 (1990).
141. Y. Avishai, M. Kaveh, S. Shatz, and Y. B. Band, *J. Phys. Condens. Matter* **1**, 6907 (1989).
142. C. G. Smith, M. Pepper, R. Newbury, H. Ahmed, D. G. Hasko, D. C. Peacock, J. E. F. Frost, D. A. Ritchie, G. A. C. Jones, and G. Hill, *J. Phys. Condens. Matter* **1**, 6763 (1989).
143. H. U. Baranger and A. D. Stone, *Phys. Rev. Lett.* **63**, 414 (1989).
144. C. W. J. Beenakker and H. van Houten, *Phys. Rev. Lett.* **63**, 1857 (1989); and in *Electronic Properties of Multilayers and Low-Dimensional Semiconductor Structures* (J. M. Chamberlain, L. Eaves, and J. C. Portal, eds.), Plenum, New York (1990).
145. C. J. B. Ford, S. Washburn, M. Büttiker, C. M. Knoedler, and J. M. Hong, *Phys. Rev. Lett.* **62**, 2724 (1989).
146. A. M. Chang, T. Y. Chang, and H. U. Baranger, *Phys. Rev. Lett.* **63**, 996 (1989).
147. L. D. Landau and E. M. Lifshitz, *Mechanics*, Pergamon, Oxford, (1976).
148. N. S. Kapany, in *Concepts of Classical Optics* (J. Strong, ed.), Freeman, San Francisco (1958).
149. M. Büttiker, *Phys. Rev. B* **33**, 3020 (1986).
150. M. Büttiker, *IBM J. Res. Dev.* **32**, 63 (1988).
151. P. H. Beton, B. R. Snell, P. C. Main, A. Neves, J. R. Owers-Bradley, L. Eaves, M. Henini, O. H. Hughes, S. P. Beaumont, and C. D. W. Wilkinson, *J. Phys. Condens. Matter* **1**, 7505 (1989).
152. L. P. Kouwenhoven, B. J. van Wees, W. Kool, C. J. P. M. Harmans, A. A. M. Staring, and C. T. Foxon, *Phys. Rev. B* **40**, 8083 (1989).
153. A. D. Benoit, C. P. Umbach, R. B. Laibowitz, and R. A. Webb, *Phys. Rev. Lett.* **58**, 2343 (1987).
154. J. Spector, H. L. Stormer, K. W. Baldwin, L. N. Pfeiffer, and K. W. West, *Surf. Sci.* **228**, 283 (1990).
155. K. Nakamura, D. C. Tsui, F. Nihey, H. Toyoshima, and T. Itoh, *Appl. Phys. Lett.* **56**, 385 (1990).
156. C. W. J. Beenakker, H. van Houten, and B. J. van Wees, *Festkörperprobleme/Advances in Solid State Physics* **29**, 299 (1989).
157. P. A. M. Benistant, G. F. A. van de Walle, H. van Kempen, and P. Wyder, *Phys. Rev. B* **33**, 690 (1986).
158. M. Büttiker, *IBM J. Res. Dev.* **32**, 317 (1988).
159. M. Büttiker, *Phys. Rev. B* **38**, 12724 (1988).
160. R. E. Prange and T.-W. Nee, *Phys. Rev.* **168**, 779 (1968); M. S. Khaikin, *Av. Phys.* **18**, 1 (1969).
161. A. M. Kosevich and I. M. Lifshitz, *Zh. Eksp. Teor. Fiz.* **29**, 743 (1955) [*Sov. Phys. JETP* **2**, 646 (1956)].
162. J. R. Barker and D. K. Ferry, *Solid State Electron.* **23**, 519 (1980); *ibid.*, **23**, 531 (1980); D. K. Ferry and J. R. Barker, *Solid State Electron.* **23**, 545 (1980).
163. W. Hänsch and M. Mima-Mattausch, *J. Appl. Phys.* **60**, 650 (1986).
164. E. M. Conwell and M. O. Vassell, *IEEE Trans. Electron Devices* **ED-13**, 22 (1966).
165. J. M. Shannon, *IEEE J. Solid State Electron Devices*, **3**, 142 (1979).
166. J. R. Hayes, A. F. J. Levi, and W. Wiegman, *Phys. Rev. Lett.* **54**, 1570 (1985).
167. J. R. Hayes and A. F. J. Levi, *IEEE J. Quantum Electr.* **22**, 1744 (1986).
168. M. Heiblum, M. I. Nathan, D. C. Thomas, and C. M. Knoedler, *Phys. Rev. Lett.* **55**, 2200 (1985).
169. A. Palevski, M. Heiblum, C. P. Umbach, C. M. Knoedler, A. N. Broers, and R. H. Koch, *Phys. Rev. Lett.* **62**, 1776 (1989).

170. A. A. Palevski, C. P. Umbach, and M. Heiblum, *Appl. Phys. Lett.* **55**, 1421 (1989).
171. U. Sivan, M. Heiblum, and C. P. Umbach, *Phys. Rev. Lett.* **63**, 992 (1989).
172. P. Hawrylak, G. Eliasson, and J. J. Quinn, *Phys. Rev. B* **37**, 10187 (1988).
173. R. Jalabert and S. Das Sarma, *Solid State Electron.* **32**, 1259 (1989); *Phys. Rev. B* **40**, 9723 (1989).
174. L. Esaki and R. Tsu, *IBM J. Res. Dev.* **14**, 61 (1970).
175. L. L. Chang, L. Esaki, and R. Tsu, *Appl. Phys. Lett.* **24**, 593 (1974).
176. L. Esaki, *Rev. Mod. Phys.* **46**, 237 (1974).
177. S. Y. Chou, D. R. Allee, R. F. W. Pease, and J. S. Harris, Jr., *Appl. Phys. Lett.* **55**, 176 (1989).
178. R. Landauer, in *Nonlinearity in Condensed Matter*, Springer, Berlin (1987).
179. C. B. Duke, *Tunneling in Solids*, Academic Press, New York (1969).
180. R. F. Kazarinov and S. Luryi, *Appl. Phys. Lett.* **38**, 810 (1981).
181. P. Hu, *Phys. Rev. B* **35**, 4078 (1987).
182. D. Lenstra and R. T. M. Smokers, *Phys. Rev. B* **38**, 6452 (1988).
183. L. I. Glazman and A. V. Khaetskii, *Pis'ma Zh. Teor. Fiz.* **48**, 546 (1988) [*JETP Lett.* **48**, 591 (1988)]; *Europhys. Lett.* **9**, 263 (1989).
184. R. J. Brown, M. J. Kelly, M. Pepper, H. Ahmed, D. G. Hasko, D. C. Peacock, J. E. F. Frost, D. A. Ritchie, and G. A. C. Jones, *J. Phys. Condens. Matter* **1**, 6285 (1989).
185. M. J. Kelly, *J. Phys. Condens. Matter* **1**, 7643 (1989).
186. J. K. Jain and S. A. Kivelson, *Phys. Rev. B* **37**, 4276 (1988).
187. L. Bliek, E. Braun, G. Hein, V. Kose, J. Niemeyer, G. Weimann, and W. Schlapp, *Semiconductor Sci. Techn.* **1**, 1101 (1986).
188. B. E. Kane, D. C. Tsui, and G. Weimann, *Phys. Rev. Lett.* **59**, 1353 (1987).
189. P. G. N. de Vegvar, A. M. Chang, G. Timp, P. M. Mankiewich, J. E. Cunningham, R. Behringer, and R. E. Howard, *Phys. Rev. B* **36**, 9366 (1987).
190. P. M. Mensz and D. C. Tsui, preprint.
191. P. C. van Son, G. H. Kruithof, and T. M. Klapwijk, *Surf. Sci.* **229**, 57 (1990); P. C. van Son and T. M. Klapwijk, *Eur Phys. Lett.* **12**, 429 (1990).
192. R. Landauer, *Phil. Mag.* **21**, 863 (1970).
193. R. Landauer, *Z. Phys. B* **21**, 247 (1975).
194. R. Landauer, *IBM J. Res. Dev.* **32**, 306 (1988).
195. J. R. Kirtley, S. Washburn, and M. J. Brady, *Phys. Rev. Lett.* **60**, 1546 (1988).
196. P. C. van Son, H. van Kempen, and P. Wyder, *J. Phys. F* **17**, 1471 (1987).
197. L. I. Glazman and M. Jonson, *J. Phys. Condens. Matter* **1**, 5547 (1989).
198. S. Komiyama, H. Hirai, S. Sasa, and T. Fuji, *Solid State Comm.* **73**, 91 (1990).
199. S. Komiyama, H. Hirai, S. Sasa, and S. Hiyamizu, *Phys. Rev. B* **40**, 12566 (1989).
200. B. W. Alphenaar, P. L. McEuen, R. G. Wheeler, and R. N. Sacks, *Phys. Rev. Lett.* **64**, 677 (1990).
201. B. I. Halperin, *Phys. Rev. B* **25**, 2185 (1982).
202. A. H. MacDonald and P. Streda, *Phys. Rev. B* **29**, 1616 (1984); P. Streda, J. Kucera, and A. H. MacDonald, *Phys. Rev. Lett.* **59**, 1973 (1987).
203. R. B. Laughlin, *Phys. Rev. B* **23**, 5632 (1981).
204. R. F. Kazarinov and S. Luryi, *Phys. Rev. B* **25**, 7626 (1982); S. Luryi, in *High Magnetic Fields in Semiconductor Physics* (G. Landwehr, ed.), Springer, Berlin (1987).
205. S. V. Iordansky, *Solid State Comm.* **43**, 1 (1982).
206. R. Joynt and R. E. Prange, *Phys. Rev. B* **29**, 3303 (1984).
207. R. E. Prange, in *The Quantum Hall Effect* (R. E. Prange and S. M. Girvin, eds.), Springer, New York (1987).
208. L. I. Glazman and M. Jonson, *Phys. Rev. B* **41**, 10686 (1990).
209. T. Martin and S. Feng, *Phys. Rev. Lett.* **64**, 1971 (1990).
210. J. K. Jain, unpublished.
211. S. Komiyama and H. Hirai, *Phys. Rev. B* **40**, 7767 (1989).
212. C. W. J. Beenakker, *Phys. Rev. Lett.* **64**, 216 (1990).

213. A. M. Chang and J. E. Cunningham, *Solid State Comm.* **72**, 651 (1989).
214. L. P. Kouwenhoven, B. J. van Wees, N. C. van der Vaart, C. J. P. M. Harmans, C. E. Timmering, and C. T. Foxon, *Phys. Rev. Lett.* **64**, 685 (1990).
215. J. A. Simmons, H. P. Wei, L. W. Engel, D. C. Tsui, and M. Shayegan, *Phys. Rev. Lett.* **63**, 1731 (1989).
216. G. Timp, R. Behringer, J. E. Cunningham, and R. E. Howard, *Phys. Rev. Lett.* **63**, 2268 (1989).
217. D. C. Tsui, H. L. Störmer, and A. C. Gossard, *Phys. Rev. Lett.* **48**, 1559 (1982); D. C. Tsui and H. L. Störmer, *IEEE J. Quantum Electr.* **22**, 1711 (1986); T. Chakraborty and P. Pietiläinen, *The Fractional Quantum Hall Effect*, Springer, Berlin (1988).
218. R. B. Laughlin, *Phys. Rev. Lett.* **50**, 1395 (1983).
219. M. E. Cage, in *The Quantum Hall Effect* (R. E. Prange and S. M. Girvin, eds.), Springer, New York (1987).
220. M. Büttiker, *Surf. Sci.* **229**, 201 (1990).
221. U. Sivan, Y. Imry, and C. Hartzstein, *Phys. Rev. B* **39**, 1242 (1989).
222. U. Sivan, and Y. Imry, *Phys. Rev. Lett.* **61**, 1001 (1988).
223. H. Z. Zheng, K. K. Choi, D. C. Tsui, and G. Weimann, *Phys. Rev. Lett.* **55**, 1144 (1985).
224. K. von Klitzing, G. Ebert, N. Kleinmichel, H. Obloh, G. Dorda, and G. Weimann, *Proc. 17th International Conference on the Physics of Semiconductors* (J. D. Chadi and W. A. Harrison, eds.), Springer, New York (1985).
225. R. J. Haug and K. von Klitzing, *Europhys. Lett.* **10**, 489 (1989).
226. Y. Aharonov and D. Bohm, *Phys. Rev.* **115**, 485 (1959).
227. Y. Imry and R. A. Webb, *Sci. Am.*, **259**, 56 (1989).
228. S. Washburn and R. A. Webb, *Adv. Phys.* **35**, 375 (1986).
229. A. G. Aronov and Yu. V. Sharvin, *Rev. Mod. Phys.* **59**, 755 (1987).
230. G. Timp, A. M. Chang, J. E. Cunningham, T. Y. Chang, P. Mankiewich, R. Behringer, and R. E. Howard, *Phys. Rev. Lett.* **58**, 2814 (1987); A. M. Chang, K. Owusu-Sekyere, and T. Y. Chang, *Solid State Comm.* **67**, 1027 (1988); A. M. Chang, G. Timp, J. E. Cunningham, P. M. Mankiewich, R. E. Behringer, R. E. Howard, and H. U. Baranger, *Phys. Rev. B* **37**, 2745 (1988); G. Timp, P. M. Mankiewich, P. DeVegvar, R. Behringer, J. E. Cunningham, R. E. Howard, H. U. Baranger, and J. K. Jain, *Phys. Rev. B* **39**, 6227 (1989).
231. C. J. B. Ford, T. J. Thornton, R. Newbury, M. Pepper, H. Ahmed, C. T. Foxon, J. J. Harris, and C. Roberts, *J. Phys. C* **21**, L325 (1988).
232. D. A. Wharam, M. Pepper, R. Newbury, H. Ahmed, D. G. Hasko, D. C. Peacock, J. E. F. Frost, D. A. Ritchie, and G. A. C. Jones, *J. Phys. Condens. Matter* **1**, 3369 (1989).
233. J. K. Jain and S. Kivelson, *Phys. Rev. B* **37**, 4111 (1988).
234. E. N. Bogachek and G. A. Gogadze, *Zh. Eksp. Teor. Fiz.* **63**, 1839 (1972) [*Sov. Phys. JETP* **36**, 973 (1973)].
235. N. B. Brandt, D. V. Gitsu, A. A. Nikolaevna, and Ya. G. Ponomarev, *Zh. Eksp. Teor. Fiz.* **72**, 2332 (1977) [*Sov. Phys. JETP* **45**, 1226 (1977)]; N. B. Brandt, D. B. Gitsu, V. A. Dolma and Ya. G. Ponomarev, *Zh. Eksp. Teor. Fiz.* **92** 913 (1987) [*Sov. Phys. JETP* **65**, 515 (1987)].
236. Y. Isawa, *Surf. Sci.* **170**, 38 (1986).
237. A. Tonomura, *Rev. Mod. Phys.* **59**, 639 (1987).
238. R. J. Brown, C. G. Smith, M. Pepper, M. J. Kelly, R. Newbury, H. Ahmed, D. G. Hasko, J. E. F. Frost, D. C. Peacock, D. A. Ritchie, and G. A. C. Jones, *J. Phys. Condens. Matter* **1**, 6291 (1989).
239. J. K. Jain, *Phys. Rev. Lett.* **60**, 2074 (1988).

CHAPTER 3

When Does a Wire Become an Electron Waveguide?

G. Timp

AT&T BELL LABORATORIES
HOLMDEL, NEW JERSEY

I. INTRODUCTION		113
II. TWO-TERMINAL RESISTANCE OF AN ELECTRON WAVEGUIDE		117
1. How to Make an Electron Waveguide		117
2. Measurements of the Two-Terminal Resistance		123
3. Theoretical Estimate of the Two-Terminal Resistance		127
4. Discussion of the Two-Terminal Results		132
III. THREE-TERMINAL RESISTANCE.		136
5. Theoretical Estimates of the Three-Terminal Resistance		136
6. Measurements of the Three-Terminal Resistance		137
7. Discussion of the Three-Terminal Results		142
IV. FOUR-TERMINAL RESISTANCE		143
8. Theoretical Estimates of the Four-Terminal Magnetoresistance		143
9. Four-Terminal Resistance Measurements for $HeW^2/hc < 1$		147
10. Four-Terminal Resistance Measurements for $1 < HeW^2/hc < 10$—the Integer Quantized Hall Effect		164
11. The Magnetoresistance of an Annulus in the Quantized Hall Regime		167
12. Four-Terminal Resistance Measurements for $HeW^2/hc > 10$—the Fractional Quantized Hall Effect		174
V. SUMMARY AND OUTLOOK		182
ACKNOWLEDGMENTS		185
REFERENCES		185

I. Introduction

The low-temperature electrical resistance of a high-mobility, semiconducting wire 100 nm wide, 10 nm thick, and 200 nm long is nonlinear; it does not scale with length; it can be quantized as a function of the width; it depends upon the leads used to measure it; and it is nonlocal; i.e., the current at one point in the wire depends not only on the electric field at that point, but on electric fields micrometers away. Classical models for the resistance cannot account for these peculiarities entirely.[1,2] Classical models for the resistance treat charge carriers like billiard balls, and average over the velocities to predict transport coefficients; but if a wire is so narrow that the width and thickness are comparable to, or less than, the wavelength of an electron at the Fermi energy, and the length is less than the elastic and inelastic mean free paths, then the electronic motion is *coherent*, the energy and momenta are quantized, and so the average velocity is not an appropriate basis for a description of the resistance. Instead, the resistance has a quantum mechanical aspect, and the wire resembles an *electron waveguide*.

An electron waveguide is a wire that is so clean and so small that electron waves can propagate in guided modes, which are characteristic of the geometry, without loss of phase coherence. It is supposed to be reminiscent of an optical or microwave waveguide, but unlike an electromagnetic wave, an electron wave is sensitive to an applied electric or magnetic field because it possesses a charge. Moreover, an electron is a fermion. Consequently, the number of electrons in a specific mode of the waveguide is limited by the Pauli exclusion principle.

In response to an applied electric field (or to an applied current), an electron waveguide has a resistance that is related to the quantum mechanical transmission through the wire.[3] The transmission probability is affected by both elastic and inelastic scattering. *Elastic scattering*, such as might occur at an impurity, for example, changes the distribution of the electrons between the modes of the guide, but it is phase-deterministic; i.e., information associated with the phase of the electronic wave is not ruined by an elastic scattering event. In contrast, *inelastic scattering* destroys the phase memory of the wave. Coherent electronic transport is possible whenever the device is smaller than the inelastic scattering length, whether or not there is elastic scattering.[4] Only when the device is smaller than both the inelastic and the elastic scattering lengths and comparable to the wavelength, like it is in an electron waveguide, is the modal distribution of the current important.

Now it is possible to make an electron waveguide.[5] Using molecular beam epitaxy (MBE) in conjunction with high-resolution electron beam lithography, wires shorter than the mean free path (≈ 1 μm), with a width and thickness comparable to the Fermi wavelength of the electron (≈ 50 nm) already have been made by constricting a high-mobility, two-dimensional electron gas (2DEG) in a GaAs–AlGaAs heterostructure to submicron dimensions. With a few exceptions,[6,7] low-temperature resistance measurements have been used to characterize them. An especially conspicuous advertisement of the quantum mechanical nature of the resistance of an electron waveguide has been found—quantized resistances.

Measurements have revealed that the resistance can be quantized in three different circumstances:

1. Two-terminal measurements of a variable-width, submicron constriction in a high-mobility 2DEG *in the absence of a magnetic field* have revealed a resistance that is quantized as a function of the width W at $h/2e^2N$, where N is an integer and h is Planck's constant.[8-10] The index N counts the number of discrete transverse energy levels occupied in the constriction.
2. It has been shown that the four-terminal Hall resistance R_{xy} of a submicron constriction, with N discrete energy levels occupied in the absence of a magnetic field, is quantized as a function of magnetic field H at $R_{xy} = h/e^2 i$ for integers $i \leq 2N$, where i counts the number of discrete spin-polarized energy levels occupied.[11-13]
3. Finally, it has been shown that for $HeW^2/hc > 10$, when only one energy level is occupied, the four terminal Hall resistance is quantized as a function of magnetic field at $R_{xy} = h/e^2 i$, where i is a simple rational fraction.[14,15]

The quantization of the resistance is due to the discrete transverse electronic energies that develop from the confinement of the electron on the scale of its wavelength, and the quantization of the electronic charge. According to this interpretation, the quantization of the two-terminal resistance in the absence of a magnetic field, and the integer quantization of the Hall effect, measure the number of energy levels occupied by electrons with an elementary charge e, while the fractional quantization of the Hall effect may represent fractional quantization of a quasi-particle charge.[16-18]

A quantized resistance is a paradigm of quantum mechanics in transport; but the wave nature of the electron can be recognized in other features of the resistance, too. In this chapter, we explore the quantum mechanical aspects of transport in an electron waveguide using two-, three-, and four-terminal resistance measurements. This chapter is not a comprehensive review, but rather an incomplete summary of work done at AT&T Bell Laboratories, Holmdel, New Jersey. We show many features of the resistance—such as the Aharonov–Bohm effect[19] and the four-terminal resistance of a bend in an electron waveguide—that must have a quantum mechanical interpretation to be explained entirely, but we especially emphasize the observations of a quantized resistance as unambiguous manifestations of quantum mechanics. Specifically, we examine how robust the quantization is, and investigate the influence of impurities, contacts, and electron–electron scattering on the values and the precision of the quantized resistance found as a function of width, and as a function of magnetic field.

In the second section of this chapter, we examine the two-terminal resistance of a submicron constriction in a high-mobility 2DEG. Like van Wees et al.[8] and Wharam et al.,[9] we show that the two-terminal conductance of constrictions with a lithographic length of 200 nm can be quantized in steps of $2e^2/h$ with about 95–99% accuracy. We examine the effect of impurity scattering on the two-terminal resistance of a constriction through an investigation of the length dependence, the mobility dependence, and carrier density dependence of the quantization, and by using *telegraph noise* to modulate the scattering. We demonstrate that an impurity in the constriction generally is detrimental to the quantization of the resistance, while the effect of impurity scattering outside the constriction is suppressed.

Scattering from a lead or contact also can be detrimental to the quantization of the resistance. In Section III, we examine the effect of a contact on the quantization by using a lead juxtaposed between two constrictions in series. While the three-terminal resistance, measured using the lead between the two constrictions, can be quantized, the quantization is associated only with electron waves propagating in the lowest transverse energy levels in the narrowest constriction. Electrons occupying the lowest energy levels do not propagate effectively into the voltage lead, but instead are collimated or focused in the forward direction. As a result of the collimation, we show that the resistance of a constriction even can vanish under certain conditions.

The distinctive way an electron wave propagates around a bend has important implications for four-terminal measurements, too. In Section IV, we consider the four-terminal magnetoresistance of a constriction in a high-mobility 2DEG in three different field regimes measured relative to the width of the constriction:

1. The low-field regime, where the magnetic length is less than the width of the wire, $HeW^2/hc < 1$.
2. The regime in which Hall conductance is quantized in steps of e^2/h, $1 < HeW^2/hc < 10$.
3. The regime where the Hall conductance is quantized at fractions of e^2/h, $HeW^2/hc > 10$.

In the absence of a magnetic field, and for low fields, we show that the resistance can be negative,[20,21] and that the Hall resistance can be suppressed for low N from the conventional 2D value[11] due to the collimation of the electron wave. For magnetic fields beyond $HeW^2/hc \approx 1$, we show that the longitudinal resistance still can be negative, and that the Hall resistance is quantized at $h/e^2 i$ for integers $i \leq 2N$, but not precisely. This is in contrast with the quantized Hall effect in a 2DEG where the longitudinal resistance vanishes, and the Hall resistance is quantized at $h/e^2 i$ for integers $i \leq 40$ with 99.9999% accuracy.

We explicitly show that in the quantized Hall regime, the current is carried by states of spatial extent $r_0 = \sqrt{hc/eH}$ near the edges of a wire by investigating the suppression of the Aharonov–Bohm effect found in the magnetoresistance of an annulus made from two submicron constrictions in parallel.[22–26] We then attribute the deterioration of the quantization of the Hall resistance found in an electron waveguide to backscattering between edge states when $W \approx r_0$.

Although *noninteracting* quantum mechanical models for transport in an electron waveguide give an adequate explanation for most of the measurements discussed in this chapter, this correspondence may be coincidental. The deficiencies of a noninteracting model are acutely apparent in an intense magnetic field, when only one edge state is occupied. Under these conditions, the Coulomb interaction between electrons in a constriction can be comparable to the intersubband spacing. When only one edge state is occupied, we show that the four-terminal Hall resistance of a junction comprised of constrictions, each 200 nm long and about 900 nm wide, is quantized with an accuracy of about 96% at $(h/e^2)/i$, where i is a rational fraction. In a 2DEG, the fractional quantum Hall effect is attributed to the Coulomb interaction between electrons and the formation of an incompressible quantum fluid, which is described by the Laughlin wave function. Because the ground state wave function is required to be *antisymmetric* under particle exchange, the resistance is expected to be quantized only if i is an *odd-denominator* rational fraction. What is especially provocative is that the four-terminal Hall resistance of a junction between electron waveguides also is quantized for $(h/e^2)/0.55 < R_{xy} < (h/e^2)/0.45$. This is surprising because the Hall resistance of a macroscopic 2DEG is featureless[27] near $(h/e^2)/(1/2)$.

In Section V, we summarize our results.

II. Two-Terminal Resistance of an Electron Waveguide

1. How to Make an Electron Waveguide

Recently, van Wees et al.[8] and Wharam et al.[9] discovered that the two-terminal conductance of a one-dimensional (1D) constriction in a high-mobility 2DEG is quantized in steps of approximately $2e^2/h$, *in the absence of a magnetic field*, as the width of the constriction is varied. The high-mobility 2DEG is achieved in an AlGaAs–GaAs heterostructure. The electron gas is not strictly two-dimensional; it is dynamically two-dimensional because it is confined within an electron wavelength of the heterointerface between an insulator (AlGaAs) and a semiconductor (GaAs), but is free to move in the two spatial dimensions along the interface. Using molecular beam epitaxy, it is possible to grow AlGaAs–GaAs heterojunctions routinely with two-dimensional electron mobilities $\mu \geq 100$ m^2/Vs for a carrier density of $n \approx 3 \times 10^{15}$ m^{-2}. Using high-mobility heterojunctions in conjunction with electron beam[28] or focused ion beam lithographies,[29] it is possible to make devices in which the 2DEG is laterally constrained to a submicron width over a distance of less than one micrometer. Provided it is short compared to the elastic mean free path, which typically is 10 μm, and comparable in

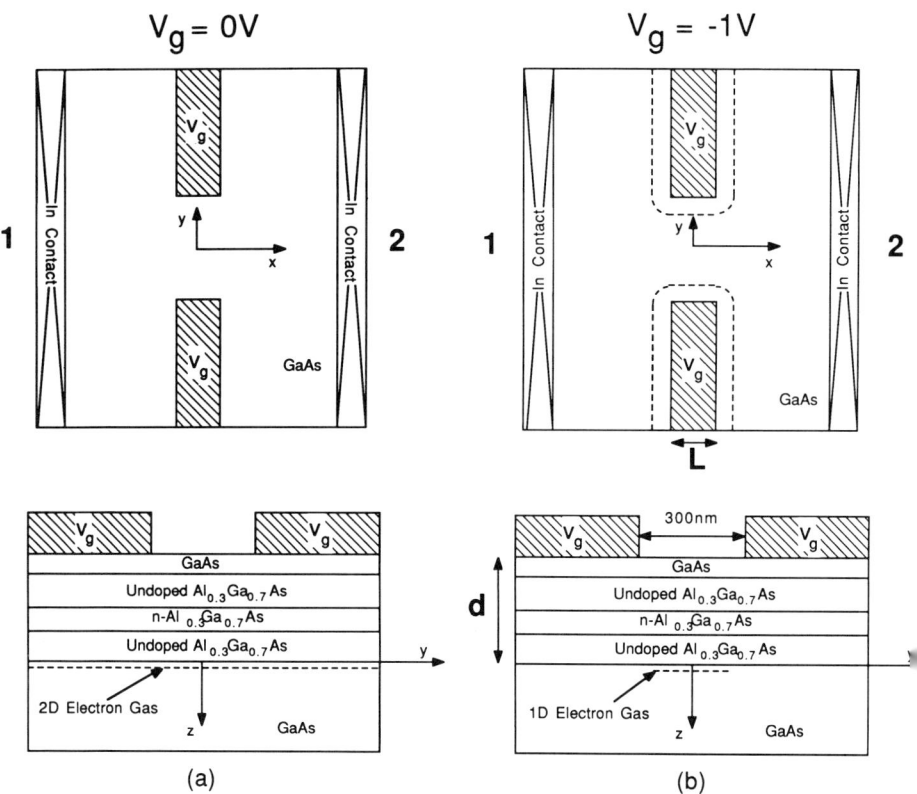

FIG. 1. (a) and (b), respectively, show the geometry of the split gate used to investigate the two-terminal resistance of a constriction in a 2DEG for two different bias voltages: $V_g = 0$ V and $V_g = -1$ V. The top portions of the figures show schematically the top plan of the device and the location of indium contacts. Actually, the contacts are approximately 300 μm from the gate electrodes and there are intervening voltage terminals. The depletion around the gate electrodes is indicated by the dashed lines in (b). The lower portions of (a) and (b) represent a cross section through the device. The 2DEG is localized with 10 nm of the lower AlGaAs/GaAs heterointerface. For gate voltages $V_g < -0.5$ V, the 2D electron gas is constrained within the gap to the region indicated by the dashed lines in (b). (From Ref.10.)

3. When Does A Wire Become An Electron Waveguide?

width to the electron wavelength, which is about 50 nm, a constriction in a 2DEG can be ballistic and dynamically one-dimensional (1D).

Following Pepper,[30-32] to make a 1D constriction in a 2DEG, we have used the electrostatic potential provided by a split gate geometry to constrain laterally the electron gas to the region within the gap between the gates. Figures 1a and b schematically represent a typical device in which both gate electrodes are biased at $V_g = 0V$ and $-1V$, respectively. The device is made on top of wide leads (100 μm wide) etched into a heterostructure in the shape of a Hall bar. (Only two of the eight leads are indicated in the figure.) On top of the Hall bar geometry, a split gate is fabricated using electron beam lithography. Electron beam lithography is used to prepare a mask for lift-off; then a Ti/Au or Ti/AuPd film approximately 7.5/50 nm thick is evaporated and the mask is removed to give the split gate electrodes illustrated schematically in Fig. 1. Typically, the split gate electrodes are 200 nm wide with an intervening gap of 300 nm. Table I gives the dimensions of the same heterostructure. The heterostructure consists of a GaAs cap layer, an AlGaAs spacer layer, a Si δ-doped layer, and an AlGaAs buffer layer on top of a 1 μm GaAs buffer layer grown on a semi-insulating GaAs substrate. Electron micrographs of a top view of actual split gate electrodes are shown in the insets to Fig. 3 and in Fig. 8.

By applying a negative voltage to the split gates (as shown in Fig. 1b), the 2DEG gas at the AlGaAs/GaAs interface immediately beneath the gate

TABLE I

A list of n (the 2D carrier density), μ (the 2D mobility), d_s (the $Al_{0.3}Ga_{0.7}As$ spacer layer thickness), d_b (the $Al_{0.3}Ga_{0.7}As$ buffer layer thickness), d_c the thickness of the GaAs cap layer, the low-temperature mean free path L_e estimated from the mobility, and the electron wavelength λ_F for the six different materials examined in this work. The concentration of Si in the delta-doped layer is approximately 4×10^{16} m^{-2} except for heterostructure 6, which has a doping concentration of 5×10^{16} m^{-2}.

	Heterostructure					
	#1	#2	#3	#4	#5	#6
$n/10^{15}$ m^{-2}	1.0 ± 0.07	1.6 ± 0.05	2.0 ± 0.08	2.75 ± 0.05	4.4 ± 0.2	7.1 ± 0.2
μ (m^2/Vs)	85 ± 4	93 ± 2	107 ± 5	115 ± 5	87 ± 2	16 ± 2
d_s (nm)	78	50	47	42	27	7.5
d_b (nm)	52	40	28	24	53	70
d_c (nm)	6	9	9	6	7	6
L_e (μm)	3.0	4.3	5.6	6.7	6.6	1.6
λ_F	84	63	56	48	39	30

electrodes is depleted, and so the 2DEG is laterally constrained (along the y axis in Fig. 1) within the gap between the electrodes. The electrostatic potential ϕ, due to split gate electrodes, not accounting for the δ-doped impurity layer or the charge density associated with the 2DEG or the dielectric constant, is [33]

$$\phi(x, y, z) = V_g \left\{ f\left[\left(\frac{-l}{2} + \frac{x}{z}\right), \left(\frac{w}{2} + \frac{y}{z}\right)\right] - f\left[\left(\frac{l}{2} + \frac{x}{z}\right), \left(\frac{w}{2} + \frac{y}{z}\right)\right] \right.$$
$$\left. + f\left[\left(\frac{-l}{2} + \frac{x}{z}\right), \left(\frac{w}{2} - \frac{y}{z}\right)\right] - f\left[\left(\frac{l}{2} + \frac{x}{z}\right), \left(\frac{w}{2} - \frac{y}{z}\right)\right] \right\}, \quad (1)$$

where

$$f(u, v) = \frac{1}{2\pi}\left[\frac{\pi}{2} - \tan^{-1}(u) - \tan^{-1}(v) + \tan^{-1}\left(\frac{uv}{\sqrt{1 + u^2 + v^2}}\right)\right],$$

and where l and w denote, respectively, the lithographic length of and gap between the gate electrodes. We assume that the gate electrodes are a distance $z = d$ above the 2DEG, extending along y from $+w/2$ to ∞ and from $-w/2$ to ∞, and along x from $-l/2$ to $l/2$ with the origin in the center of the gap.

Figures 2a and b represent this potential for constrictions with $l = 200$ nm and 600 nm long, and $w = 300$ nm. We expect that the actual form of the potential near the edges of the constriction in the 2DEG and for $y > w/2$ to be well-represented by Fig. 2 because it is appropriate to ignore the effect of the two-dimensional charge density there, but the potential within the constriction requires a self-consistent solution to Poisson's equation as well as Schrödinger's equation,[34] so that the form for the potential shown in Fig. 2 generally is an inadequate approximation for $|y| < W/2$, where W is the width of the narrowest part of the constriction at the Fermi energy. The potential represented in Fig. 2 may be an adequate approximation to the actual potential for $|y| < W/2$ if the carrier density is low, however.[35]

The two salient features of the electrostatic confinement potential are represented in Fig. 2. The first is the smooth taper from the 2DEG to the 1D constriction. The contact is not abrupt because the constriction is formed by depletion using gate electrodes that are at least d away from the 2DEG and typically $w - W/2$ away. A self-consistent calculation of the potential produces a smooth taper even if there are imperfections in the gate electrodes for this reason.[34] The potential is relatively flat near the center of the

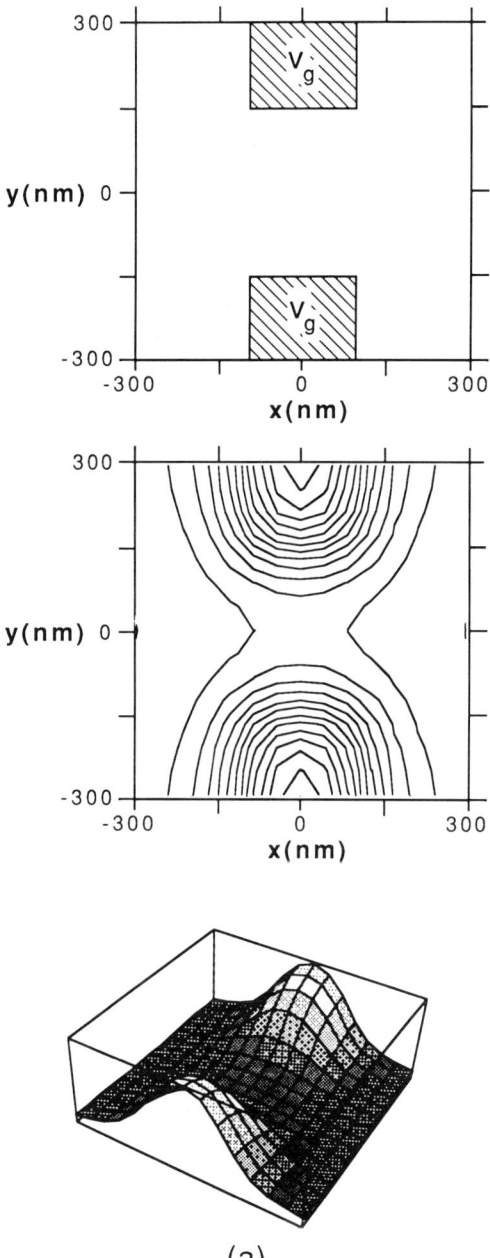

FIG. 2. A graphical representation of Eq. (1) that gives both a contour plot and a three-dimensional plot of the potential due to the split gate electrodes (a) 200 nm long spaced 300 nm apart and (b) 600 nm long spaced 300 nm apart. The gate electrode configuration is indicated in the top portion of each figure. It is assumed that −1.0 V is applied to each electrode. The middle plots show 15 equipotential contours at intervals of 0.066 V. At the bottom of each figure is a corresponding three-dimensional representation of the electronic potential where the height represents the size of the potential.

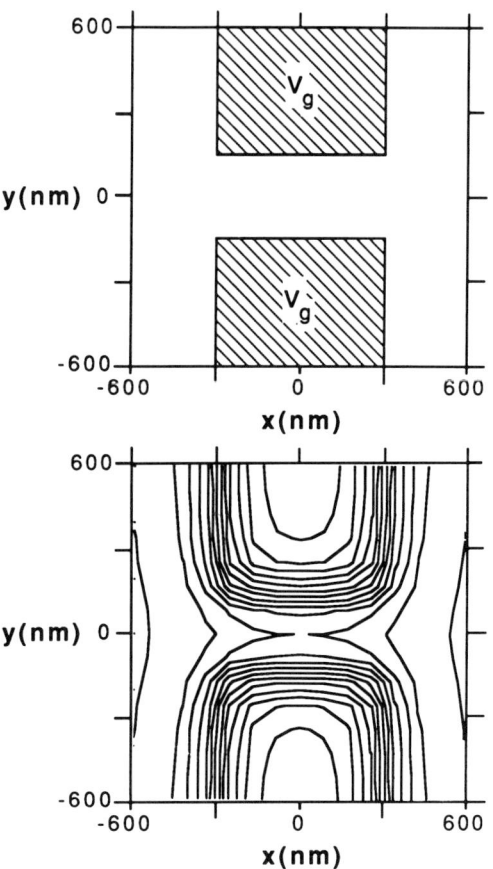

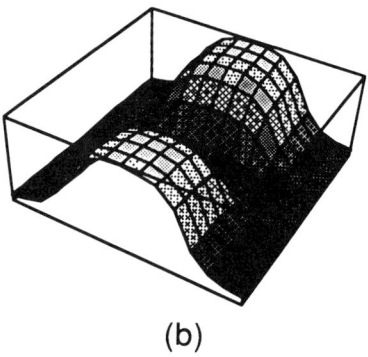

FIG. 2. (*Continued*)

constriction, but rises abruptly near the edges. When the gate voltage is large and negative, the constriction gradually widens to the 2D contact over approximately 250 nm, which is about 5–10 λ_F. A second feature of this solution is the potential barrier between the constriction and the 2DEG along the z axis; the constriction is at a higher potential than the 2D contact. Since we assume that the Fermi energy is constant throughout the device, the higher potential within the constriction means that the carrier density is lower there.

2. MEASUREMENTS OF THE TWO-TERMINAL RESISTANCE

Figure 3 shows the two-terminal resistances (conductances), $R_{12,12}$ ($G_{12,12}$), as a function of the applied gate voltage that we found in devices like that shown schematically in Fig. 1. These results are representative of the data obtained from between four and eight different devices in two different heterostructures; the top figures represent data from heterostructure 4 in Table I, while the bottom figures represent data from heterostructure 5. Following the convention of Büttiker,[36] $R_{lm,jk}$($G_{lm,jk}$) denotes a resistance (conductance) measurement in which positive current flows into lead l and out lead m with a positive voltage measured between leads j and k, respectively. The convention for numbering the leads is given in Fig. 3. The series resistance, found at $V_g = 0$ V, was subtracted from the measured resistance obtained as a function of gate voltage to give the data plotted in Fig. 3.[37]

The depletion of the 2DEG immediately beneath the gate electrodes occurs near −0.325 V and −0.5 V for the devices represented, respectively, by the top and bottom of Fig. 3a, and is not shown. The depletion of the 2D electron gas beneath the electrodes depends on the heterostructure—the lower the carrier density, the lower the negative voltage needed to deplete the 2D electron gas—but it is approximately the same from device to device for a particular heterostructure (within about 25 mV). As the gate voltage decreases further, the constriction within the gap between the electrodes narrows, the carrier density within the constriction decreases, and plateaus are observed in the resistance. The two-terminal conductance obtained by inverting the resistance-versus-gate voltage also is shown in Fig. 3. The *average* conductance (minus the series resistance) is approximately $2e^2N/h$ with N an integer ranging from 1 to about 10, and evidently is quantized in steps of $2e^2/h$ with about 1–5% accuracy as an increasingly negative gate voltage makes the constriction narrower. The quantized conductance represents a dramatic departure from the Sharvin resistance associated with a point contact.[38–40]

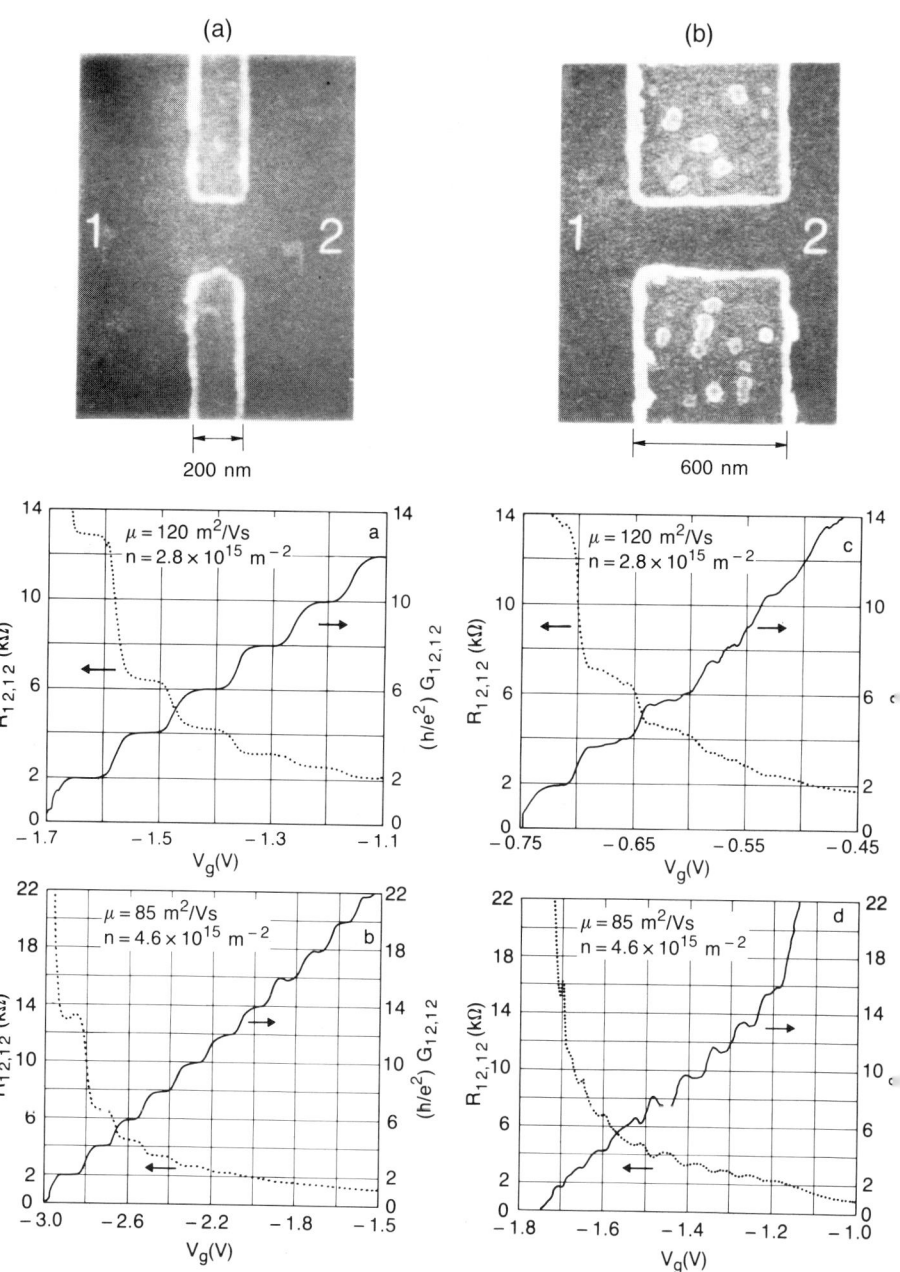

FIG. 3. The two-terminal resistances (conductance), $R_{12,12}(G_{12,12})$, of a constriction in a 2DEG as a function of gate voltage V_g, obtained at 280 mK for two different heterostructures with a mean free path, estimated from the mobility, of about 7 μm. The data shown in the top of (a) and (b) were obtained from devices in heterostructure 4, while the data of the bottom figures were obtained from devices in heterostructure 5. (See Table I.) In (a), the split gate is 200 nm long with a 300 nm gap between the electrodes. (a) shows that the conductance, $G_{12,12}$, is quantized in steps of $2e^2/h$ as a function of V_g. In (b), the split-gate electrodes are 600 nm long with a 300 nm gap between electrodes. Although the low-temperature mean free path in the 2DEG is much longer than the length of the constriction, (b) shows deterioration of the quantization of the two-terminal resistance. (From Ref. 10.)

TABLE II

A LIST OF THE CARRIER DENSITY n, FERMI WAVEVECTOR k_F, AND WIDTH W VERSUS 1D SUBBAND INDEX N, FOUND IN A TYPICAL DEVICE 200 NM LONG WITH A GAP 300 NM WIDE MADE ON HETEROSTRUCTURE 5 IN TABLE I. W IS INFERRED FROM A HARD-WALL POTENTIAL WELL.

N	$n(\text{m}^{-2})$	$k_F(\text{nm}^{-1})$	$W(\text{nm})$
	$4.6 \pm 0.1 \times 10^{15}$	0.172	≈ 3000
12	$3.5 \pm 0.1 \times 10^{15}$	0.148	≈ 110
7	$2.3 \pm 0.2 \times 10^{15}$	0.120	≈ 80
5	$1.9 \pm 0.2 \times 10^{15}$	0.109	≈ 60
3	$1.8 \pm 0.2 \times 10^{15}$	0.106	≈ 45
2	$1.3 \pm 0.2 \times 10^{15}$	0.090	≈ 25

The gate voltage does not correspond directly to either the carrier density, the width of the constriction, or the chemical potential.[34] Table II lists the carrier density n and the Fermi wavevector k_F, deduced from Shubnikov–de Haas (SdH) measurements taken at constant gate voltage versus the number of 1D subbands found in the absence of a magnetic field for a device in heterostructure 5. It is apparent from Table II that the constriction has a lower carrier density than the 2D contact. As indicated in Fig. 2, the difference in density is consistent with an electrostatic potential difference along the longitudinal direction (the x direction in Fig. 1).

The precision of the quantization can be accurate to about 1%. Figure 4 compares the conductance obtained in a two-terminal measurement in a device made in heterostructure 4 at zero magnetic field with the four-terminal Hall conductance of a 2DEG, the prototypical quantized resistance, measured using a bridge of 100 μm-wide wires. The accuracy of the quantization of the two-terminal conductance as a function of gate voltage to $2e^2/h$ does not approach the accuracy or precision that can be achieved in a routine measurement of the quantized Hall effect in a 2DEG, however. The inset to Fig. 4 shows the deviation from perfect quantization that we find in the two-terminal resistance as a function of the step N. The deviation can be represented qualitatively as an additional resistance, $r_s \approx 100\ \Omega$, in series with the ballistic constriction—a resistance we have no justification for subtracting.[41]

The gate length (the length L along the x axis in Fig. 1b) for the device of Fig. 3a is about 200 nm. The mobilities of the 2DEG used to fabricate the

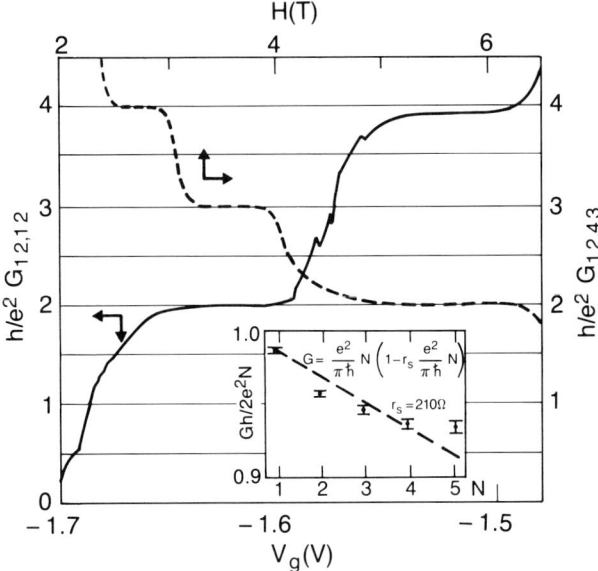

FIG. 4. The two-terminal conductance $G_{12,12}$ of a ballistic constriction as a function of the applied gate voltage at zero magnetic field, and the four-terminal magnetoconductance $G_{12,43}$, measured in the same device but with $V_g = 0$ V. The Hall conductance is quantized precisely in steps of e^2/h with 99.99% accuracy, while the two-terminal conductance of a ballistic constriction is quantized in steps of $2e^2/h$ with only 99% accuracy. The inset shows the deviations from exact quantization found in the two-terminal conductance of a constriction. The deviations are consistent with a series resistance, $r_s = 210\ \Omega$. (From Ref. 10.)

devices of Fig. 3 exceed 85 m^2/Vs (at $T = 280$ mK), which corresponds to a low-temperature mean free path of at least $L_e = \hbar k_F \mu / e \approx 7$ μm, where $\hbar = h/2\pi$. Since the 300 nm gap in each device pinches off for large negative gate voltages, the depletion around the gate must be less than 150 nm under conditions where the constriction conducts, and so we estimate the actual length of the constriction to be less than 500 nm in Fig. 3a—much less than the mean free path deduced from the mobility of the 2DEG.

We find that as the length of the constriction increases, the quantization deteriorates. Figure 3b shows the resistance and conductance determined as before for a constriction with gate electrodes of length $L = 600$ nm, spaced with a 300 nm gap in heterostructures 4 (top) and 5 (bottom). For this geometry, the length of the constriction is estimated to be less than 900 nm, which still is a factor of eight less than the mean free path estimated from the 2D mobility and carrier density, yet the quantization has deteriorated dramatically from that shown in Fig. 3a. While the quantization of the resistance still is apparent in the higher-mobility device represented in the top of Fig. 3b, the accuracy of the quantization is only about 90% for the first three steps, with no quantization observed for higher N. The bottom of Fig. 3b shows the complete deterioration of the quantized resistance found in a device 600 nm long in heterostructure 5, which has lower mobility than heterostructure 4, but higher carrier density.

The quantization of the two-terminal resistance also deteriorates as L_e, estimated from the mobility of the 2DEG, becomes smaller. Figure 5 (a–f) shows the two-terminal resistances (conductances) measured at 280 mK in devices made in the six different heterostructures identified in Table I. For large negative voltages, the 200 nm-long, 300 nm-wide gap pinches off in every case, but the threshold voltage generally is sensitive to the heterostructure. Generally, the lighter the carrier density, the less negative the threshold gate voltage. The threshold voltage can vary appreciably (as much as 600 mV) from device to device for the same heterostructure. The variation of the threshold from device to device may be indicative of the effect of the unscreened impurity potentials in the neighborhood of the constriction on the potential profile within the constriction. (See the following).

3. THEORETICAL ESTIMATE OF THE TWO-TERMINAL RESISTANCE

The measured resistance of a 1D constriction is due to the redistribution of the current among the electronic states in the wide 2D contact and the 1D constriction. The quantization of the two-terminal resistance of a 1D ballistic constriction was not anticipated theoretically because it was held widely that the nature of the contact, used to measure the resistance of a ballistic constriction, was not ideal.[42] While Imry[42] observed that theoretically, the two-terminal conductance is a quantized function of the number of occupied 1D subbands, he concluded that fluctuations of magnitude e^2/h in the measured potential due to scattering at the contacts would obscure the effect. Subsequently, Glazman et al.[43] demonstrated that the conductance of a constriction measured between two semi-infinite contacts could be quantized provided that: (1) the contacts are tapered adiabatically to match to the constriction, and (2) the constriction is both short enough to be ballistic and long enough to prevent evanescent modes from carrying appreciable current; these are in qualitative agreement with the observations by van Wees et al.[8]

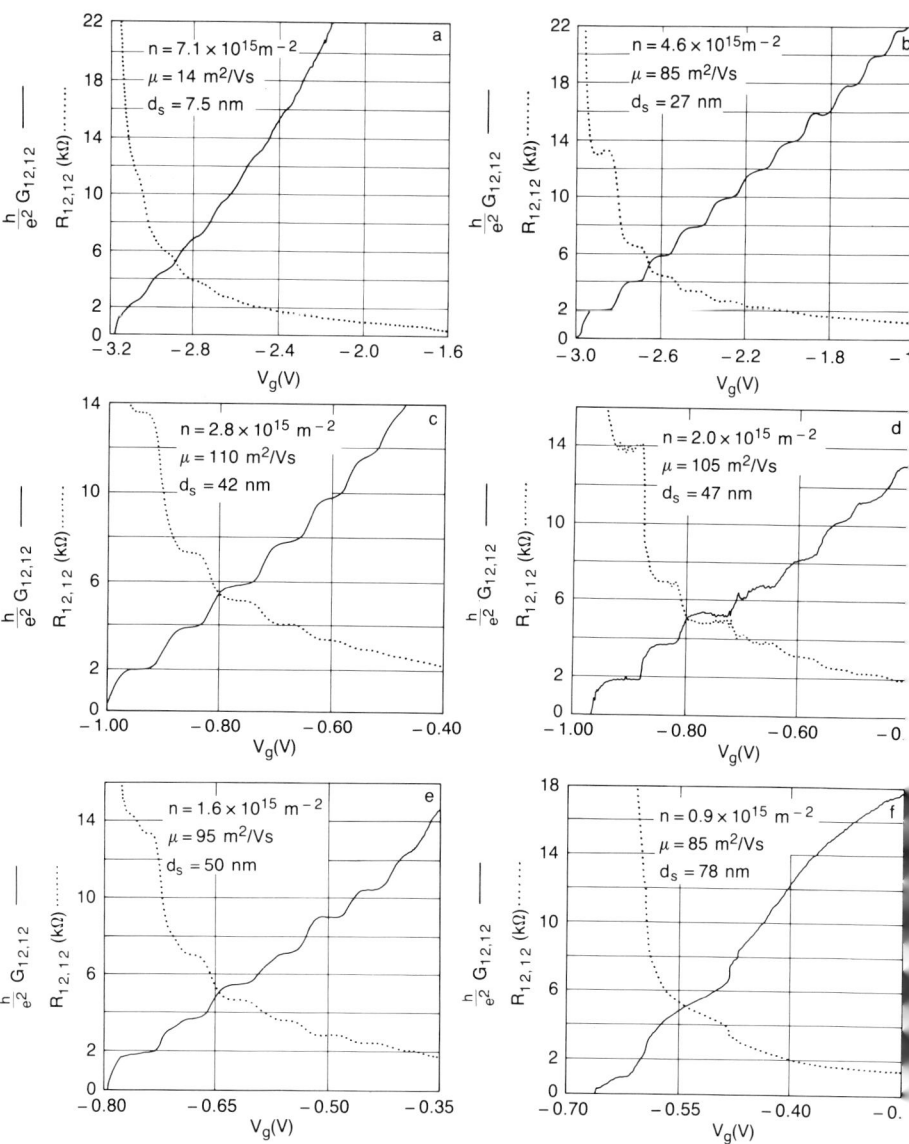

FIG. 5. The two-terminal resistance (conductance) $R_{12,12}(G_{12,12})$, measured in devices made in six different heterostructures, is shown. The data of (a), (b), (c), (d), (e), and (f) were obtained from devices made in heterostructures 6, 5, 4, 3, 2, and 1 of Table I, respectively, using the same split-gate electrode geometry. The split gate electrodes were 200 nm long with a 300 nm gap. The resistance (conductance) of (b)–(f) was measured at 280 mK, while the resistance (conductance) of (a) was obtained at 4 K.

3. When Does A Wire Become An Electron Waveguide?

Following the analysis of Imry,[42] Büttiker,[36] and Glazman et al.,[43] the Hamiltonian of an electron waveguide can be written as

$$-\frac{\hbar^2}{2m^*}\left[\frac{\partial^2 \Psi}{\partial x^2} + \frac{\partial^2 \Psi}{\partial y^2}\right] + V(y)\Psi = E\Psi, \qquad (2)$$

where the x coordinate is along the length of the wire, y is in the transverse direction, and m^* is the effective mass. $V(y)$ is the confining potential. If we assume that $V(y)$ is a harmonic potential, i.e., $V(y) = \frac{1}{2}m^*\omega_0^2 y^2$, where ω_0 is the classical oscillation frequency, then the wave function is given by $\Psi = \exp(ik_x x)\exp(-\xi^2)\chi_N(\xi)$, where χ_N is the Hermite polynomial of order $(N-1)$ and $\xi = y\sqrt{m^*\omega_0/\hbar}$. In the x direction, we find running waves characterized by a wavevector k_x, while in the y direction, we find the solutions characteristic of a harmonic oscillator. Because of the confinement, the transverse energy, $E_N = \hbar\omega_0(N - \frac{1}{2})$, is quantized with integer $N = 1, 2, \ldots$, The total energy E is the sum of the transverse energy E_N and the energy for longitudinal motion along the constriction, $\hbar^2 k_x^2/2m^*$. The energy levels E_N constitute a series of minima in the total energy called a *subband ladder*, where the index N is used to denote the Nth subband in the series. At the Fermi energy, E_F, there are $2N$ states where N is the total number of (spin-degenerate) transverse 1D subbands below E_F. The size of the confinement potential measured at the Fermi energy is $W = \sqrt{N - \frac{1}{2}}\sqrt{\hbar/m^*\omega_0}$. This estimate of the width systematically is a factor of 2–3 smaller than that obtained using a hard wall confinement potential for $N < 10$.

In a conventional two-terminal measurement, two wider conductors contact each end of the electron waveguide. If the contacts are wide enough, they become reservoirs in which the electronic motion approaches thermal equilibrium. Each contact then can be characterized by chemical potentials μ_1 and μ_2. When $\mu_1 > \mu_2$, there is a net current through the constriction. Generally, only a fraction of the flux incident in a particular subband (j) is transmitted through the constriction; i.e., $I_j = ev_j \delta n = ev_j(dg/dE)_j \sum_{k=1}^{N} t_{jk} \Delta\mu$, where v_j and $(dg/dE)_j$ are, respectively, the Fermi velocity and the density of states at the Fermi energy for subband j; t_{jk} is the probability intensity for transmission from subband k into subband j; N is the number of subbands in the constriction; and $\Delta\mu$ is the chemical potential difference between the two wide contacts to either side of the constriction; i.e., $\Delta\mu = \mu_1 - \mu_2$. (We ignore the energy dependence of the t_{jk} that arises from the difference in chemical potential between the two contacts.) In 1D, the density of states is inversely proportional to the velocity, and so $I_j = (2e/h)\sum_{k=1}^{N} t_{jk}\Delta\mu$. Summing the contributions from each of the subbands gives the total current, and since the voltage difference between the wide reservoirs is $V = \Delta\mu/e$, the two-terminal conductance is

$$G_{12,12} = (2e^2/h) \sum_{j,k=1}^{N} t_{jk}. \tag{3}$$

This is the Landauer formula for the two-terminal conductance. If we assume that: (1) the temperature is much lower than the separation between transverse energy levels, (2) the length of the perfect wire is much longer than the decay length for evanescent modes, and (3) there is no scattering (i.e., $t_{ij} = \delta_{ij}$), then the current injected into subband j is $I_j = (2e/h)\,\Delta\mu$, independent of the specific 1D subband under consideration, and so the total current is $I = N(2e/h)\,\Delta\mu$. The two-terminal conductance of constriction with N occupied subbands then is

$$G_{12,12} = (2e^2/h)N \tag{4}$$

exactly. Thus, the ideal two-terminal conductance is a quantized function that measures the number of occupied 1D subbands.

The correspondence between Eq. (4) and the data of Fig. 3 is surprising because to obtain Eq. (4), it is assumed that: (1) $T = 0$, (2) the transmission through the contacts into the consideration and through the constriction is perfect, and (3) the contacts to the constriction are thermal reservoirs. Thus, the observation of a well-quantized two-terminal conductance implies that:

1. The energy separation between subbands is much larger than 280 mK.
2. The constriction is ballistic for $L = 200$ nm when $L_e \gg L$ and the reflections associated with contacts to the constriction are negligible.
3. The finite width of the contacts can be ignored.

While the 2D contacts to the ballistic constriction appear ideal phenomenologically because the resistance is quantized, it is not necessarily so that there is no scattering at the contacts. As Szafer and Stone[44] have shown, the resistance may be well-quantized even if there are substantial reflections in the neighborhood of the contacts. Because of depletion, the 1D constriction gradually widens to embrace the 2D contact as shown in Fig. 2. Glazman et al.[43] first appreciated this and developed an adiabatic model for conductance a constriction that relies upon a gradual taper from the 1D constriction to the 2D contact.

Following Glazman et al.,[43] we imagine the electron waveguide of Eq. (2) to have a classical frequency that is a function of the longitudinal position, i.e., $\omega_0(x)$, yielding an adiabatically smooth constriction of width W_i tapering from a width W_f. If the variation along x is slow, the Hamiltonian separates into terms in the longitudinal variable x and the variable y. The y-dependent part of the Hamiltonian is the harmonic well problem with solution: $E_N(x) = (N - \frac{1}{2})\hbar\omega_0(x)$, where $N = 1, 2,\ldots$. The x-dependent part of the Hamiltonian represents scattering from a barrier. For a given Fermi energy, E_F, only a finite number of modes, N, will be above the barrier. If tunneling below and reflections above the barrier are negligible, then N modes, with $T_{ij} = \delta_{ij}$, carry current while the transmission of all other modes vanishes exponentially, and so Eq. (4) is recovered from Eq. (3).

It is implicit in the model by Glazman that the electron wave is focused. A wave propagating adiabatically along a variable-width guide conserves the mode number N. As the width of the guide narrows, the transverse energy increases while the longitudinal energy decreases continuously. When the wave abruptly encounters the end of the adiabatic taper, the transverse wavevector now is conserved (at least in the interval $\pm\pi/W_f$), and the wave is focused to a width $\pi k W_f$. It is unlikely that the actual constrictions are entirely adiabatic; Fig. 2 shows that they are not. However, even if the taper is not adiabatic, the reflections associated with the mismatch between the 2D contact and the constriction may not be large because the taper is *gradual* occurring on the scale of 250 nm or 5–10 λ_F. It is sufficient that the adiabaticity holds only up to a certain width beyond which reflections occur or that the reflections are small.[45] Glazman evaluated corrections to the transmission in each mode due to reflections at the contacts (disregarding energies near the threshold for occupation of another mode), assuming a constriction formed between two circles of radius R, and found only exponentially small deviations from exact quantization: i.e., $\delta G = 2e^2/h[1 + \exp(-\eta\pi^2\sqrt{2R/W})]$,[43] where $\eta = k_F W/\pi - N$.

While the adiabatic approximation for the taper between the 2DEG and the 1D constriction is sufficient to produce the quantization of the two-terminal conductance in the absence of a magnetic field, in particular, and to reproduce the results of multi-terminal resistance measurements generally (as seen shortly), it may not be necessary. Other nonadiabatic models for the contact to the constriction also have produced the quantization of the two-terminal conductance of a constriction.[44,46–49] However, the collimation of the electron wave and the ballistic nature of the transport are necessary elements for the explanation of these phenomena. The collimation can be achieved either with tapered constriction or by the potential step between the 2D contact and the 1D constriction.[50]

4. Discussion of the Two-Terminal Results

If the constriction is ballistic, we expect the precision of quantization of the conductance plateaus to improve as L increases because the current carried by evanescent modes diminishes.[43] However, oscillations in the conductance below the quantized value may occur when the condition $k_x L = b\pi$ is satisfied, where $b = 1, 2, \ldots$, corresponding to resonant scattering from the ends of the constriction.[44,46] We find the quantization deteriorates in the longer constrictions with no conclusive evidence of regular oscillations in the conductance below the quantized values associated with resonant scattering from the ends of the constriction.

The top trace in Fig. 6a shows the two-terminal conductance quantized in steps of $2e^2/h$ with an accuracy of 99%. The same figure also shows the two-terminal conductance of the same device after cycling it to room temperature and cooling it to 280 mK again. Warming the device to room temperature apparently gives rise to a different configuration of impurities or disorder. The lower trace in Fig. 6a shows reproducible fluctuations observed in the conductance near a plateau found in a device made from heterostructure 4 that are similar phenomenologically to the oscillations in the conductance found experimentally by van Wees et al. Oscillations like these have been attributed to resonant scattering from the ends by Szafer and Stone[44] and Kirczenow.[46] When the device was refrigerated for the first time, oscillations were not observed on the $N = 1$ conductance plateau where $G_{12,12} = 2e^2/h$. However, after the device was cycled to room temperature and cooled again, it exhibited oscillations in the conductance, and the accuracy of the quantization deteriorated to about 94%.

Van Wees et al.[51] have interpreted the oscillations found in the conductance, which appear similar to those shown here, as resonant scattering from the ends of the constriction, but this interpretation is dubious. While the temperature dependence of the fluctuations found in the conductance is qualitatively consistent with their interpretation, the systematic variation with the occupation number N predicted by Kirczenow[46] has not been observed, and the oscillations were found in only two devices. The oscillations we found in the resistance are reproducible, but these features are not observed frequently and probably are not associated with reflections from edges near the contacts. These oscillations also have been attributed to scattering from disorder; specifically, to resonant tunneling through a single impurity in the constriction.[52] The deterioration of the quantization is consistent with this interpretation.

We also attribute the deterioration of the quantization in the long constrictions to disorder or impurity scattering within the constriction.[47,53] Depending on the position, impurity scattering may reduce or enhance the

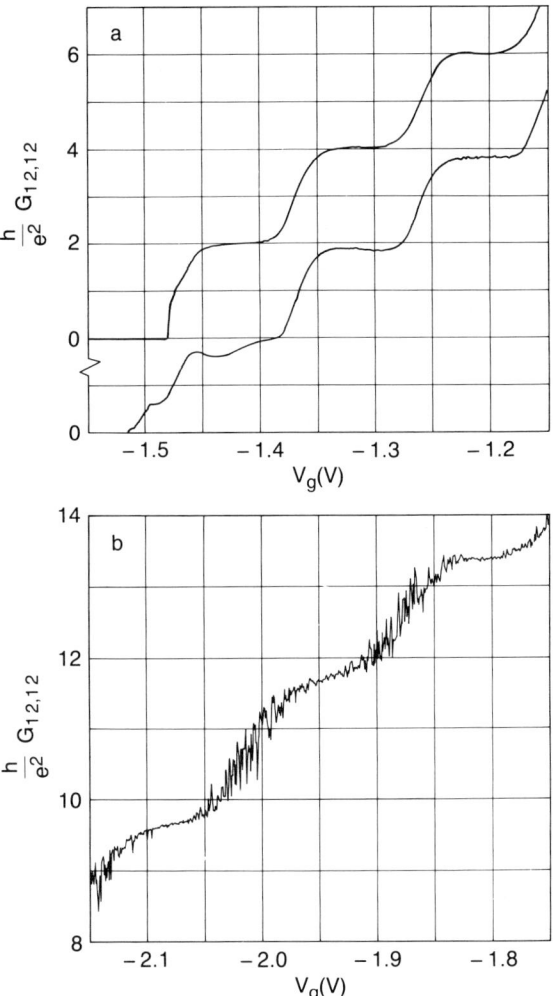

FIG. 6. (a) shows the two-terminal conductance $G_{12,12}$ as a function of gate voltage of a constriction at 280 mK (top) and the same measurement after it had been warmed to room temperature and refrigerated to 280 mK again (bottom). The conductance in the top trace of (a) is quantized as a function of V_g with an accuracy of 99%, while the same device later showed an accuracy of only 94% with reproducible fluctuation near the plateau at $V_g = -1.4$ V. We attribute this effect to disorder scattering within the constriction. The effect of an impurity outside the constriction on the quantization of the conductance is illustrated in (b). Here, time-dependent fluctuations (the rapid changes in the conductance that occur as a function of V_g) due to *telegraph noise* are found between the plateaus, but the noise is suppressed in the middle of the plateaus.

conductance. If an impurity is in the constriction, the conductance is reduced from the quantized value. Alternatively, if the impurity is outside the constriction, the conductance either may be enhanced or reduced.[47] In either case, an impurity destroys the quantization of the conductance.

While it is used widely as an estimate for the mean free path and a measure of the extent of impurity scattering, L_e, inferred from the mobility of the wide wire, is not necessarily an appropriate estimate for the elastic scattering length in the constriction, especially for gate voltages near pinch-off. (Parenthetically, the estimate of L_e obtained from electron focusing may not be relevant either. Electron focusing data obtained by using ballistic constrictions as both a monochromator and analyzer do not provide an indication of L_e within the constriction, but measure L_e within the 2DEG where the focusing occurs.)[54] A 2DEG effectively screens potential fluctuations due to Coulombic impurities, but as the constriction in the 2DEG becomes narrower with more negative gate voltages and the carrier density is reduced, the Fermi wavevector k_F becomes shorter, and the screening of the impurities in the doped layer by the electron gas becomes less effective.[55] Impurity potentials beneath the gate will be screened by the metal electrodes, but the impurities in the gap between the electrodes are not. Consequently, the elastic scattering length may be reduced to approximately 1 μm or less because the 1D electron gas ineffectively screens potential fluctuations.[25]

There also is a large disparity between the mean free path estimated from the mobility and the electron lifetime estimated from Shubnikov–de Haas.[56,57] The mean free path estimated from the mobility principally measures the large-angle backscattering rate (because the screening length in GaAs/AlGaAs is comparable to the electron wavelength), but there is at least 10 times more small-angle forward scattering than backscattering.[58] However, as illustrated in Fig. 7, small-angle forward scattering probably is irrelevant to the deterioration quantization of the resistance for small values of N, because of the restricted possibilities for scattering. Even if small-angle scattering through θ_2, for example, at the Fermi surface is more likely than backscattering through θ_1, the quantization that results from the narrow width of the constriction restricts the possibilities for both forward and backscattering severely.[59] Moreover, the quantization is not affected by a single forward scattering event in the constriction because the sum in $\Sigma_{i,j=1}^{N} t_{ij}$ is not affected. Although scattering through θ_2 may change the subband index, the sum does not change. An electron must be backscattered or reflected from the constriction to affect the quantization of the conductance. The cumulative effect of many small-angle scattering events, or a backscattering event, has this result. According to this interpretation, the deterioration of the quantization as L_e of the 2DEG becomes smaller, as shown in Fig. 5, illustrates the effect of increased backscattering on the quantization. Furthermore, as

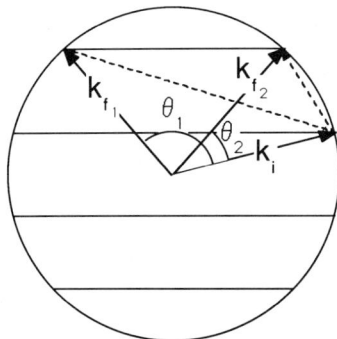

FIG. 7. The Fermi surface of a narrow constriction in a 2DEG is shown. The circle represents the Fermi surface of a 2DEG; the horizontal lines reflect the allowed transverse momenta states in the 1D constriction. The intersection between the circle and the lines represents the Fermi surface of the constriction. Two scattering events, occurring at the Fermi surface of the constriction, are depicted in this Figure. In the first event, k_i is backscattered through θ_1 to k_{f_1}. Alternatively, in the second event, k_i is forward scattered through a small angle θ_2 to k_{f_2}. Small-angle scattering becomes less likely as the number of allowed states is reduced by the quantization condition.

shown in Fig. 3b, we expect that the quantization of the high-index subbands to be more fragile than the low-index subbands because, as the number of subbands increases, the possibilities for backscattering increase.[60]

Another demonstration of the resiliency of the quantization with respect to impurity scattering is given by Fig. 6b. The rapid fluctuations shown in the conductance of Fig. 6b as a function of gate voltage are *time-dependent* and may be related to the *telegraph noise* or *switching noise* examined in experiments by Rogers and Buhrman[61] and Skocpol *et al.*[62] We have found switching noise in the resistance of about one-fifth of the devices we have fabricated. Presumably, the rapid fluctuations found in the conductivity as a function of time are due to the occupation or scattering cross section of an impurity site(or sites) changing on a time scale shorter than that used to acquire the data. We notice that the switching noise is suppressed near the center of the plateau in the conductance, but increases in amplitude away from the center in either direction. The site of the imperfection associated with the switching phenomenon of Fig. 6b is likely to be near the entrance or exit of the constriction in the 2DEG, since the switching affects the distribution of the current carried by the subbands, as is evident from the fluctuations observed between plateaus, but not the quantization of the resistance. The site of the imperfection associated with the switching phenomena shown in Fig. 6b cannot be in the 2D contact because it would contribute an additive series resistance.

III. Three-Terminal Resistance

5. Theoretical Estimates of the Three-Terminal Resistance

Two-terminal measurements of a ballistic constriction do not measure the potential of the constriction directly because of the contact resistance between the 2DEG and the constriction.[42] Three- and four-terminal measurements generally do not measure the local potential of the conductor either, even if the leads are from the same material and have the same dimensions as the conductor and there is no net current in the voltage leads. Although the net current in the voltage leads vanishes, a carrier still can propagate coherently from the conductor into a lead and back into the conductor. The lead then becomes part of the measured resistance and must be included in the computation of the resistance. As illustrated next by three-terminal measurements, the leads or contacts completely determine the resistance of a 1D constriction in a high-mobility 2DEG.

Following the developments of the Landauer formula made by Büttiker,[36] we can generalize the computation of the resistance to a multi-terminal measurement. The current in lead i is given by

$$I_j = -\frac{2e}{h} \sum_{k=1}^{N} T_{jk}\mu_k, \tag{5}$$

where T_{jk} is the trace over the subbands, i.e., $T_{jk} = \Sigma_{m,n} t_{jk,mn}$, with $t_{jk,mn}$ representing the probability for a carrier in subband n and lead k to be transmitted to subband m in lead j, and $T_{jj} = \Sigma_{m,n} r_{jj,mn} - N_j$, with N_j the number of subbands in lead j and $r_{jj,mn}$ the probability for a carrier in subband n and lead j to be reflected into subband j and lead m. The potential V_k associated with a reservoir of chemical potential μ_k is given by $V_k = \mu_k/e$.

From Eq. (5), we can obtain formulae for the three-terminal or four-terminal resistance by imposing a current I between two leads, requiring the net current in the other leads to vanish, and solving for the voltage difference. For the purpose of illustration, we consider the mathematically simpler problem associated with the determination of the three-terminal resistance first. If we impose a current between leads 1 and 2 so that a positive current enters through lead 1 and exits through lead 2, then calculate the potential differences $\mu_1 - \mu_3$ and $\mu_3 - \mu_2$, we find the conductances,

$$G_{12,13} \equiv \frac{eI}{\mu_1 - \mu_3} = \frac{e^2}{h} \frac{D}{T_{32}}, \tag{6a}$$

$$G_{12,32} \equiv \frac{eI}{\mu_3 - \mu_2} = \frac{e^2}{h} \frac{D}{T_{31}}, \tag{6b}$$

with $D = T_{21}T_{31} + T_{21}T_{32} + T_{31}T_{23}$, whereas the two-terminal conductances associated with leads 1, and 2, and 1 and 3, are, respectively,

$$G_{12,12} = \frac{e^2}{h} \frac{D}{(T_{31} + T_{32})}, \quad (7a)$$

$$G_{13,13} = \frac{e^2}{h} \frac{D}{(T_{21} + T_{23})}. \quad (7b)$$

The two-terminal conductance in the presence of one additional lead generally differs from Eq. (3). The additional lead scatters carriers according to the probabilities T_{13} and T_{32}.

6. Measurements of the Three-Terminal Resistance

To demonstrate the effect of a lead on the conductance, we examined two constrictions in series and in close proximity to one another. Figure 8 shows an electron micrograph of such a device. The device is composed of two sets of split-gate electrodes in series; each set of electrodes is 200 nm long with a gap of 300 nm between the electrodes. The two sets are separated by 300 nm to make the device symmetric. Associated with each port of the device, there are contacts to the 2DEG, which are not shown. The convention for numbering the leads is given in the figure.

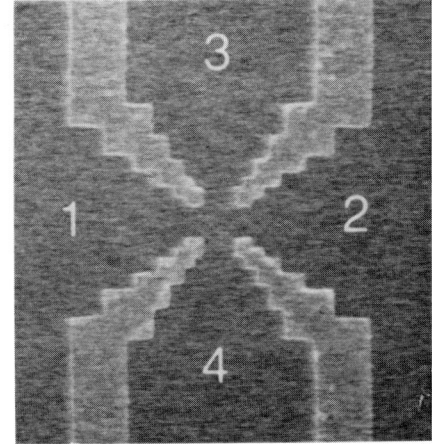

FIG. 8. An electron micrograph of two series split-gate electrodes in close proximity fabricated on a high-mobility heterostructure. Associated with each port in the device, numbered 1 through 4, there is a contact to the 2DEG so that multi-terminal resistance measurements can be performed on a junction in which the geometry can be manipulated. (From Ref. 10.)

Figure 9 illustrates the potential contours calculated for such a device made on heterostructure 4 with the gate electrodes biased at $V_g = -1$ V. The potential is defined by

$$\varphi(x,y,z) = V_g \left\{ f\left[\left(\frac{w}{2} - \frac{x}{z}\right), \left(\frac{w}{2} - \frac{y}{z}\right)\right] + f\left[\left(\frac{w}{2} + \frac{x}{z}\right), \left(\frac{w}{2} - \frac{y}{z}\right)\right] \right.$$

$$\left. + f\left[\left(\frac{w}{2} - \frac{x}{z}\right), \left(\frac{w}{2} + \frac{y}{z}\right)\right] + f\left[\left(\frac{w}{2} + \frac{x}{z}\right), \left(\frac{w}{2} + \frac{y}{z}\right)\right] \right\}, \quad (8)$$

where $f(u,v)$ is defined in Eq. (1). In Fig. 9, the potential due to the δ-doped layer and the 2DEG is neglected, and the dielectric permittivity of the heterostructure is ignored.[33] With the gate electrodes biased, the device forms a symmetric junction composed of four constrictions in a cross geometry. Each constriction can be used as a current or voltage lead. Although the disposition of the split-gate electrodes in a typical device is symmetric, each electrode can be biased independently so the symmetry of the device can be manipulated. Similar junctions have been made by reactive ion etching[12,63] or by ion damage around a mask defined by electron beam lithography,[64] but the properties of the individual constrictions comprising the junction cannot be examined or varied so easily.

Figure 10 shows the conductances $G_{12,12}$, $G_{12,13}$, and $G_{13,13}$ obtained in such a device. The top trace in Fig. 10 shows the dependence of $G_{12,12}$ on the voltage applied to the electrodes that define lead 1, V_{g1}, when the gate voltage on the constriction that defines lead 2, V_{g2}, is zero. Under these circumstances, $G_{12,12} = G_{13,13} = G_{12,13}$ because contacts 2 and 3 are identical. The traces at the bottom of Fig. 10 were obtained by biasing the gate electrodes defining lead 2 in Fig. 8 at $V_{g2} = 0.90$ V and varying the potential V_{g1} from 0.0 V to -1.7 V. In this particular device, the thresholds for the constrictions comprising the cross were such that lead 4 did not conduct if $V_{g1} < -0.8$ V for $V_{g2} < -0.8$ V. Using Eqs. (6) and (7), we can estimate the transmission coefficients directly from the data of Fig. 10. For example, near $V_{g1} = -1.6$ V, we find that $T_{13} = 0.00 \pm 0.02$, $T_{21} = 1.98 \pm 0.02$, and $T_{23} = 2.94 \pm 0.04$; while near $V_g = -1.4$ V, we find that $T_{13} = 0.6627 \pm 0.05$, $T_{21} = 5.0 \pm 0.05$, and $T_{23} = 1.82 \pm 0.05$. Under these experimental conditions, we deduce that $T_{31} \ll T_{21}$ and $T_{32} \approx 1.5$ for low N, although T_{31} increases and becomes comparable with T_{21} as N increases.

In Fig. 11, we compare the dependence of the two- and three-terminal conductances $G_{12,12}$ and $G_{12,13}$ as a function of gate voltage applied to the second constriction, V_{g2}. Traces a, b, c, and d were obtained by biasing V_{g2} at voltages of -0.55 V, -0.67 V, -0.9 V, and -1.15 V, respectively, and varying V_{g1}. In the inset to Fig. 11, the two-terminal conductance $G_{12,12}$

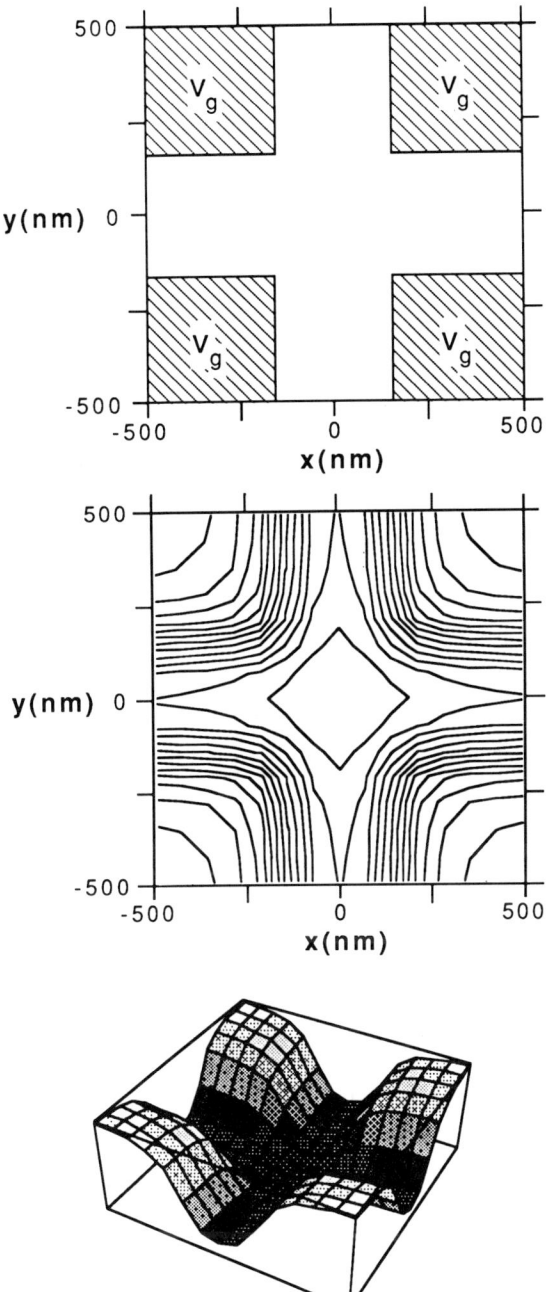

FIG. 9. A graphical representation of Eq. (8) that gives both a contour plot and a three-dimensional plot of the potential due to two split-gate electrodes 200 nm long spaced 300 nm apart in series. The gate electrode configuration is indicated in the top portion of each figure. It is assumed that −1.0 V is applied to each electrode. The middle plot shows 15 equipotential contours at an interval of 0.066 V. At the bottom of each figure is a corresponding three-dimensional representation of the electron potential in which the height represents the size of the potential.

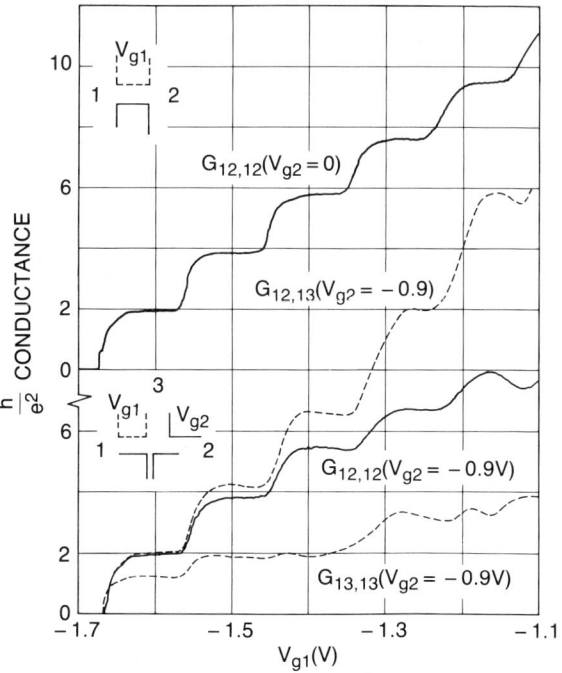

FIG. 10. The two-terminal resistances, $G_{12,12}$ and $G_{13,13}$, and the three-terminal resistance $G_{12,13}$ as a function of gate voltage for two series constrictions. In the top trace, all three conductances coincide because there is no second constriction. When the second constriction is biased at $V_{g2} = -0.90$ V so that only six subbands are occupied, the two-and three-terminal resistances no longer coincide because of the discrepancy between the transmission coefficients they measure.

through the second constriction as a function of gate voltage is shown when $V_{g1} = 0$. The voltages corresponding to traces a–d also are indicated in the inset. The shift in the threshold for conduction as a function of V_{g1} found in Fig. 11 for different values of V_{g2} is only about 50 mV, but becomes greater as V_{g2} becomes more negative. The shift occurs because near threshold, the electrostatic potential beneath a particular electrode depends upon the voltage applied to the other electrode as well.

The solid lines in traces a, b, c, and d of Fig. 11 indicate that the conductance $G_{12,12}$ is determined by the value of the narrowest constriction.[65] For example, for trace d, $G_{12,12}$ is suppressed to values below $6e^2/h$ at a negative voltage of about -1.1 V applied to V_{g1}, corresponding to the conductance of the second constriction where only three 1D subbands are occupied, i.e., $N = 3$. Figures 10 and 11 show that a second constriction in series with and in close proximity (300 nm) to the first can be detrimental to the quantization for $N > 1$. Although regular steps are found in the conductance for

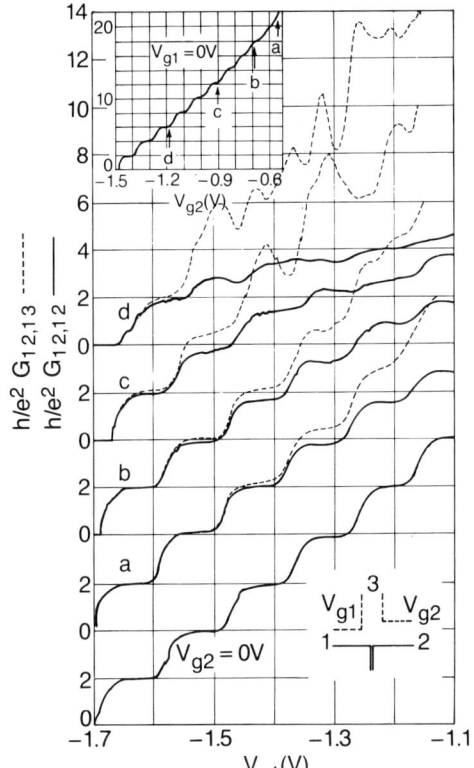

FIG. 11. The three-terminal conductance $G_{12,13}$ and the two-terminal conductance $G_{12,12}$ of two series constrictions as a function of gate voltage applied to the forward constriction V_{g1}. The gate voltage applied to the second constriction, V_{g2}, is the parameter varied between traces (a–d). For the biases shown, lead 4 in Fig. 8 is pinched off and does not conduct. $G_{12,13}$ and $G_{12,12}$ coincide only for low gate voltages, $V_{g1} < -1.6$ V, where only one subband is occupied in the forward constriction. The inset depicts the quantization of the two-terminal conductance $G_{12,12}$ versus V_{g2} for the second series constriction when there is no forward constriction, i.e., $V_{g1} = 0$ V. (From Ref. 10.)

$V_{g1} > -1.6$ V, the steps are not quantized precisely at $2e^2/h$. The latter observation differs from a report by Wharam et al.[65] of quantized resistances for constrictions about 1 μm apart. The conductance found for lower index $N \approx 1$ can be well-quantized, however, even when the second constriction is very narrow. (See trace d in Fig. 10.)

The data of Figs. 10 and 11 demonstrate that a lead can be invasive and thereby influence the measurement of the resistance of a wire. The dashed lines in Fig. 11 depict the results of the three-terminal measurement $G_{12,13}$. We notice that the three-terminal conductance generally is larger than $G_{12,12}$ and deviates substantially from the quantized values except for low N in the forward constriction. For low N, $G_{12,13} \approx G_{12,12} \approx 2e^2N/h$. Under these circumstances, it follows from Eqs. (6a) and (7a) that $T_{31} \ll N$ for low N generally.

From Eq. (7b), we infer that the second constriction practically is a perfect conductor when N is small in the first constriction, i.e., $G_{12,32} \gg e^2/h$. The measured resistance of the second constriction was found to be less than the experimental error of 50 Ω. The resistance of the second series constriction vanishes because there is no contact resistance. There is no contact resistance because the first constriction focuses the electron wave on the second series constriction.[66] In addition, it is apparent from Fig. 11 that the resistance of two-series constrictions is not additive.[65] For example, when $V_{g1} = -1.6$ V and $V_{g2} = 0$ V, $G_{12,12} = 2e^2/h$; when $V_{g1} = 0$ V and $V_{g2} = -1.2$V, $G_{12,12} \approx 6e^2/h$; but when $V_{g1} = -1.6$ V and $V_{g2} = -1.2$ V, $G_{12,12} \approx 1.8e^2/h$, which is greater than the sum of the two resistances.

7. Discussion of the Three-Terminal Results

The collimation of the electron wave in the forward direction, which is a feature of the adiabatic approximation[43] for a tapered contact to the constriction, can produce a vanishing resistance in the second constriction and nonadditivity.[66] Because an electron wave in a low index subband in lead 1 does not propagate around the bend into the voltage contact 3, the constriction that defines lead 3 does not scatter or couple to all modes in the same way and a nonequilibrium distribution of the current among the available momenta develops. The widening of the constriction between leads 1 and 3 causes electrons injected into lead 1 from the 2DEG contact to travel through the constriction to the wide region of the junction while conserving the transverse quantum state. Consequently, there is a transfer of transverse momentum into the longitudinal momentum. Thus, the current injected from the 2D contact, which was equally distributed between the 1D modes of the constriction, reaches the wide junction between the three constrictions distributed only among the lowest-energy subbands of those allowed in the wider region of the junction. Ultimately, the junction widens abruptly so that the equipotential lines must turn through 90 degrees. From this point, the transverse momentum is conserved, not the number of occupied subbands, yielding a wave collimated in the forward direction.[67]

The reluctance of an electron wave to propagate around a bend, measured by T_{31}, is a manifestation of the collimation of the wave. The narrower the constriction is, the lower the index for the number of occupied subbands, and the more tightly focused the electron wave is. The wave radiates into the junction, within a cone of angle $\theta = \sin^{-1}[(W_i/W_f)]$. The cone is narrower

3. WHEN DOES A WIRE BECOME AN ELECTRON WAVEGUIDE? 143

and more tightly focused for the lower-index N modes, where W_i is small.

The nonadditivity of the resistance can be understood, on average, as a semiclassical ballistic effect,[66] but the quantization of the three-terminal resistance, the deviations from exact quantization, or the large reproducible fluctuations measured in $G_{12,13}$ and $G_{12,12}$ for small V_{g1}, which can be nearly as large as the conductance, cannot.[68] Conductance fluctuations as a function of the Fermi energy or magnetic field have been studied extensively in disordered metal and semiconductor wires, and attributed to quantum mechanical interference due to coherent scattering from impurities within the wire;[4] but fluctuations have been found in numerical evaluations of the conductance of junctions between electron waveguides, which have no impurities at all, as a function of the Fermi energy, too.[67,69] These fluctuations are attributed to scattering from the leads. Additional experimental support for this contention comes from periodic fluctuations found in the multi-terminal magnetoresistance of singly connected geometries.[12,70]

IV. Four-Terminal Resistance

8. THEORETICAL ESTIMATES OF THE FOUR-TERMINAL MAGNETORESISTANCE

The small probability for the low-index N modes to propagate around a bend, and the nonequilibrium distribution of the carriers among the subbands in the junction that develops as a consequence of it, have important implications for four-terminal resistance measurements of a constriction. Following Büttiker,[36] we can express the four-terminal resistance in terms of transmission probabilities by using Eq. (5). If we impose a current I between leads m and n, we then can solve Eq. (5) for the voltage difference, $V_k - V_l$, between leads k and l. The four-terminal resistance, $R_{mn,kl} = (V_k - V_l)/I$, can be expressed as

$$R_{mn,kl} = \frac{h}{2e^2} \frac{T_{km}T_{ln} - T_{kn}T_{lm}}{D}, \qquad (9)$$

where D always is positive and depends on the lead geometry, but is independent of permutations in the indices mn, kl. If a four-terminal junction like that shown in Fig. 8 is approximately symmetric, then we expect that $T_{31}, T_{24} \ll N$ and $T_{21}, T_{34} \approx N$ for low N; so $R_{14,32} = (T_{31}T_{24} - T_{34}T_{21})/D < 0$ and $R_{12,43} = (T_{41}T_{32} - T_{42}T_{31})/D \approx 0$ in the absence of a magnetic field.

Although Eq. (9) applies for an arbitrary magnetic field, the transmission coefficients vary with an applied magnetic field because of the Lorentz force.[71]

In a magnetic field, the Hamiltonian (Eq. (2)) of the waveguide becomes

$$-\frac{\hbar^2}{2m^*}\left(\left[\frac{\partial}{\partial x} - \frac{ieH}{\hbar c}y\right]^2 \Psi + \frac{\partial^2 \Psi}{\partial y^2}\right) + (V(y) \pm g\mu_B H)\Psi = E\Psi, \quad (10)$$

where H is the strength of the magnetic field along the z axis, the g factor is $g \leq 0.5$ in GaAs, and $\mu_B = e\hbar/2mc$ is the Bohr magneton. The $\pm$ sign in the Hamiltonian arises because the electronic spin component of the energy, $g\mu_B H$, depends on the orientation of the spin relative to the magnetic field. For economy, we will assume that $V(y)$ is harmonic. In this case, the form of the solution to Eq. (10) is similar to that for Eq. (2) except for a shift in the origin and a renormalization of the y coordinate.[72] The hybrid magneto-electric wave functions are $\Psi = [(\pi\alpha)^{1/2}r_o 2^n n!]^{-1/2} \exp[ik_x x]\exp[-\xi^2/2]\chi_N(\xi)$, where $\xi = (y - y_o)/\sqrt{\alpha}r_o$ with $\alpha = \omega_c/\omega$, and $\omega = \sqrt{\omega_c^2 + \omega_o^2}$. The cyclotron frequency is $\omega_c = eH/m^*c$, the magnetic length $r_o = \sqrt{\hbar c/eH}$ is the radius of the cyclotron orbit for the ground state, and the guiding center coordinate is $y_o = k_x \alpha^2 r_o^2$. The hybrid magneto-electric subbands are given by $E_N(y_o) = \hbar\omega(N + \frac{1}{2}) \pm g\mu_B H + (\hbar y_o \alpha r_o^2)^2/2m^* (1 - \alpha^2)$. The energy is an admixture of the oscillator frequency in the absence of a magnetic field, the Zeeman energy, and the cyclotron frequency. Moreover, it is a function of position along the transverse y direction because it depends upon the guiding center coordinate, y_o. Figure 12 illustrates the solutions to Eq. (10).

For $HeW^2 < 1$, the current is carried in the bulk of the wire and the resistance is due to scattering from either the device geometry, the leads, impurities, or other electrons, but for $HeW^2 > 1$, as illustrated by the solution to Eq. (10) and Fig. 12, the motion along the transverse and longitudinal

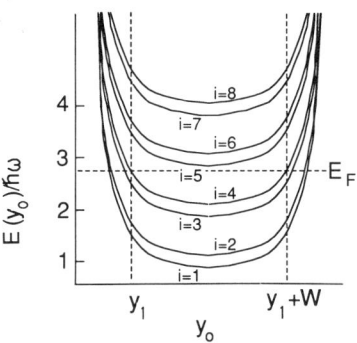

FIG. 12. Hybrid magneto-electric subbands of an electron waveguide extended along x and confined by a harmonic potential along the y direction. The width of the harmonic potential, W, is measured at the Fermi energy, E_F. The index i counts the spin-polarized subbands; y_o is the guiding center coordinate.

directions in the constriction is coupled, while the subband energy becomes a function of the position along the width of the wire and increases abruptly near the edges.[73,74] Since the Fermi energy crosses the subband energies near the edges of the wire, the net current is carried by states near the edges, which have a spatial extent $\sqrt{\alpha}\, r_o$. The current along the wire is proportional to the slope of the confining potential along the transverse direction. Thus, the driving force provided by the confining potential moves carriers along the edges of the wire with opposite edges carrying oppositely directed currents.

When the width of the wire becomes much larger than the extent of the wave function, a carrier along one edge of the wire cannot backscatter to the opposite edge. The absence of backscattering between the two edges results in dissipationless transport along the wire, i.e., $R_{14,32} = 0$.[75-79] While backscattering is suppressed, it is not necessarily so that $t_{jk} = \delta_{jk}$ as in Eq. (4). Scattering between edge states associated with the same edge still is allowed provided that

$$\sum_{j,k=1}^{N} t_{jk} = N. \qquad (11)$$

Figure 13 illustrates how the net current is carried through a cross when there is no backscattering between opposite edge states except at the 2D contact to the 1D constriction.

In an intense magnetic field, the Hall resistance, $R_{12,43}$, is quantized in steps of $h/e^2 i$, where i denotes the number of occupied spin-polarized edge states. For a magnetic field $\omega_c > \omega_o$, there are N hybrid-magneto-electric subbands below E_F, and the transmission probabilities between adjacent leads are exactly equal to the number of occupied edge states (times two because of the spin degeneracy), while the transmission between leads that are not adjacent vanishes. Thus, the Hall resistance measures the transmission probability between adjacent leads. Since $T_{13}, T_{24}, T_{41}, T_{32} = i$ while all other $T_{ij} = 0$, and $D = i^3$, we infer from Eq. (9) that $R_{14,32} \geq 0$ and $R_{12,43} = h/e^2 i$, where i denotes the number of (spin-polarized) edge states occupied in the 1D constriction.[78,79] In contrast with a 2DEG, a 1D constriction exhibits the integer-quantized Hall effect, $R_{12,43} = h/e^2 i$, only for $i \leq 2N$ because there are only a finite number of edge states to begin with. There is a one-to-one correspondence between the number of spin-degenerate edge states in a magnetic field and the number of 1D subbands occupied in the absence of a magnetic field. The correspondence is evident in the Hamiltonian of Eq. (10). In the absence of a magnetic field, $\alpha = 0$, $\omega = \omega_o$ and $2N$ subbands (counting the spin-degeneracy) are found below E_F, while for $\omega_c = \omega_o$, $\omega = \sqrt{2}\omega_o$, and $\alpha \approx 1/\sqrt{2}$ there still are $2N$ subbands occupied provided $N\hbar\omega$ remains below E_F.

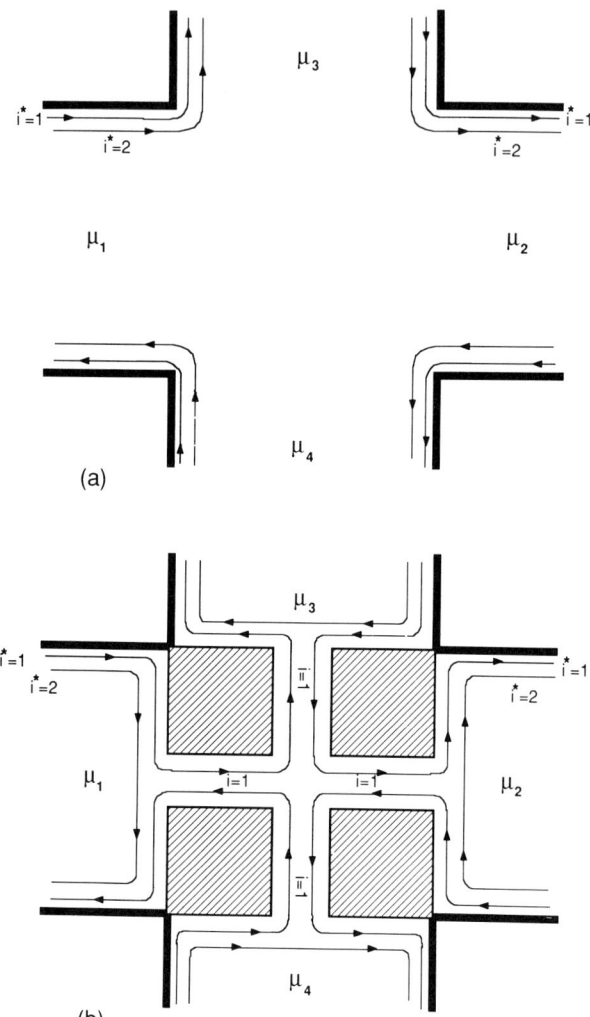

FIG. 13. A schematic representation of the net current distribution in a four-terminal junction in a magnetic field. (a) represents the net current in the junction without any constrictions in the 2DEG; (b) shows the net current with four constrictions in the 2DEG formed by applying a negative voltage to split-gate electrodes (indicated by the hashed lines). In each case, the net current follows the equipotential contours near the edges of the device, but in (b), the barrier between the 2D contact and the 1D constrictions reflects one edge state back into the 2D contact.

The integer quantization of the Hall resistance of a constriction is reminiscent of the quantization of the two-terminal resistance of a ballistic constriction as a function of width in the absence of a magnetic field. From Landauer's perspective, the two-resistance measurements are quantized for similar reasons.[80] In the absence of a magnetic field, within the adiabatic approximation for the contact, each transverse energy level within the constriction is associated with a current of $\Delta I = 2e/h \, \Delta\mu$ exactly, and $\Delta\mu$ is directly related to the measured voltage difference between the two contacts. In an intense magnetic field, the adiabatic approximation is even more appropriate than at $H = 0$. Each edge state in the constriction carries a current of exactly $\Delta I = (2e/h)\Delta\mu$, where $\Delta\mu = eV_H$ is the difference in chemical potential between edge states on opposite sides of the 2D electron gas, and V_H is the Hall voltage.

9. Four-Terminal Resistance Measurements for $HeW^2/hc < 1$

If the gate electrodes are biased similarly, then the potential distribution of the junction of Fig. 8 ideally resembles the cross geometry shown in Fig. 9. We can use the four constrictions as the current and voltage leads to measure the resistance. Figures 14a and b, respectively, show the calculated four-terminal resistances $R_{12,43}$ (dashed line) and $R_{14,32}$ (solid line) corresponding to an ideal junction with hard walls in the absence of a magnetic field, and the same resistances measured in a device like that shown in Fig. 8 made on heterostructure 4. In addition to the four-terminal resistances, we also show the two-terminal conductances $G_{12,12}$ associated with only one split-gate electrode biased (dot-dashed line) and with the two split-gate electrodes in series biased (dotted line). The calculated resistance is plotted as a function of the Fermi energy normalized by the ground state energy of the wire, while the measured resistance is plotted versus gate voltage. There is not necessarily a direct correspondence between the two abscissae[81] of Fig. 14.

While the calculated resistance $R_{12,43}$ is featureless because the junction was assumed to be perfectly symmetric, the calculation of $R_{14,32}$ reveals a negative resistance with sharp minima as a function of energy.[82,83] The resistance calculated using a harmonic confining potential instead of a hard wall exhibits similar features.[84] The negative resistance shown in Fig. 14 does not violate Onsager's relations.[85] Onsager showed that the *two-terminal* resistance is constrained to be greater than or equal to zero because the entropy increases with time; the four-terminal resistance is not so constrained.

The sharp minima found in $R_{14,32}$ correspond to the thresholds for occupation of 1D subbands in the wires comprising the junction.[82–84] As the energy increases from a value just above threshold for the occupation of a

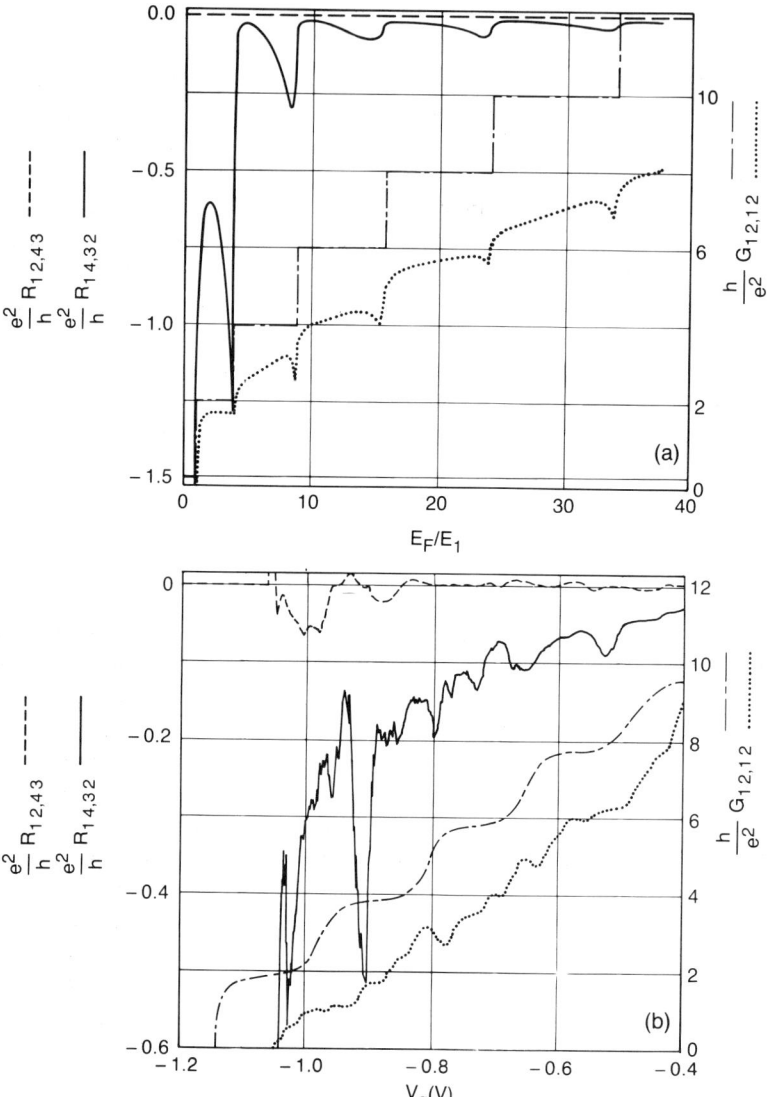

FIG. 14. (a) shows the calculated dependence on Fermi energy E_F of three characteristic resistance measurements made in a cross geometry with hard walls and sharp corners. The Fermi energy is normalized to the ground state energy of a square well, E_1. The solid line represents the energy dependence of the bend resistance $R_{14,32}$; the dashed line represents the dependence of the resistance $R_{12,43}$; and the dotted line represents the dependence of the two-terminal conductance $G_{12,12}$. The dashed-dotted line represents the energy dependence of the two-terminal conductance of only one constriction without any other constriction; it is used as a measure of the subband occupation. (b) shows the measured dependence on gate voltage V_g of the same three characteristic resistance measurements made in the cross geometry between two split-gate electrodes in series. The solid line represents the dependence on the voltage applied to two split-gate electrodes of the bend resistance $R_{14,32}$; the dashed line represents the dependence of the resistance $R_{12,43}$; and the dotted line represents the dependence of the two terminal conductance $G_{12,12}$. The dashed-dotted line represents the dependence of the two-terminal conductance on the gate voltage applied to only one split-gate electrode with the other grounded.

subband to a value below the threshold for occupation of another subband, all of the modes become lower in energy and the Fermi wavevector associated with each of the occupied modes becomes more forward-directed. Consequently, T_{21} increases relative to T_{31}, and the resistance $R_{14,32}$ becomes more negative. Globally, $R_{14,32}$ becomes less negative as the energy increases, however. The global change in $R_{14,32}$ occurs because the number of subbands increases. The sharp minima in $R_{14,32}$ are due to the quantum mechanical nature of the transport. Figure 15 compares a classical (dashed line), semiclassical (dotted line), and quantum mechanical (solid line) calculation of the resistance $R_{14,32}$ of a junction with hard walls and sharp corners.[69] The classical calculation of the resistance uses a Boltzmann transport equation assuming that the transport is ballistic without any quantization of the momenta or phase-coherent scattering. The injected particles have momenta that are distributed uniformly as a function of the angle. The semiclassical calculation also assumes ballistic transport without phase-coherent scattering, but *with* discrete transverse momenta. Consequently, only certain injection angles, of all those that are classically available, are allowed. The quantum mechanical calculation includes both the quantization of the transverse momenta and phase-coherent scattering. The importance of quantum mechanical reflections near the threshold for the occupation of a subband that develop from coherence is apparent from the dramatic differences between the classical and semiclassical calculations and the quantum mechanical calculation for low N. As the number of subbands increases, however, the differences between the results of the three calculations diminish.

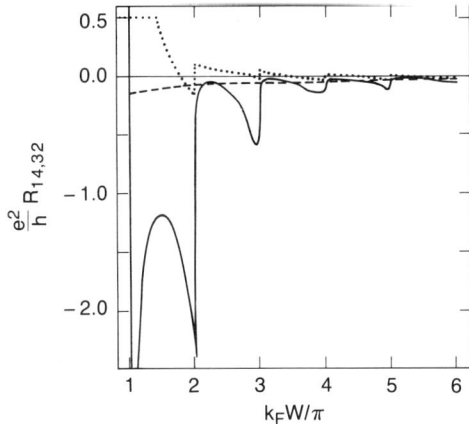

FIG. 15. Three calculations[69] of the dependence of the bend resistance $R_{14,32}$ of a junction with sharp corners and hard walls on the subband occupation measured by $k_F W/\pi$. The dashed, dotted, and solid lines depict classical, semiclassical, and quantum mechanical calculations of the resistance.

Sharp minima also are observed in a measurement of $R_{14,32}$ versus V_g as shown in Fig. 14b. As V_g becomes more negative, globally $R_{14,32}$ becomes more negative, but there are substantial fluctuations about the average at low temperature. As shown in Fig. 16a, the same features with the same magnitude are found in $R_{13,42}$ at the same gate voltages as well. We assume that the minima observed experimentally correspond to the thresholds for occupation of 1D subbands. The low-field Hall resistance is consistent with this interpretation.

There is a correspondence between the minima observed in $R_{14,32}$ in the absence of a magnetic field, and the value of the resistance associated with the first plateau found in the Hall effect with increasing magnetic field, when $N \leq 4$ in the constrictions comprising the junction. Since $G_{12,12}$, $G_{34,34}$, $G_{13,13}$ and $G_{14,14}$ all vanish beyond $V_g = -1.050$ V, we assume that the sharp minima shown in $R_{14,32}$ (and $R_{13,42}$) near $V_g = -1.025$ V in Fig. 16a correspond to the threshold for occupation of the $N = 1$ subband in all of the constrictions comprising the junction. This identification is supported by the Hall resistances shown in Fig. 16b, traces a, b, and c, where -1.035 V $\leq V_g \leq -0.98$ V. With an applied magnetic field, $R_{12,43}$ changes dramatically from a value near zero at zero magnetic field to a quantized value corresponding to the number of occupied subbands in the constriction. According to Fig. 16b, $R_{12,43}$ reaches a plateau at $(h/e^2)/(2.05 \pm 0.07)$ near $H = 2$ T, which we associated with the depopulation of a spin-polarized edge state, $i = 2$. There is no evidence of the depopulation of a higher-index edge state for $V_g < -1.0$ V. Traces d and e of Fig. 16b show the Hall resistance when the split-gate electrodes that define the junction each are biased near $V_g = -0.90$ V, where there is a sharp minimum in $R_{14,32}$. We presume that this range of gate voltage corresponds to the threshold for occupation of the second subband in all of the constrictions comprising the junction; i.e., $N = 2$ or $i = 4$ subbands are occupied. This identification is supported by the observation of a plateau in $R_{12,43}$ beyond $H = 0.5$ T at $(h/e^2)/(4.10 \pm 0.2)$ for -0.936 V $\leq V_g \leq -0.90$ V, which corresponds to the magnetic depopulation of the $i = 4$ edge state. The other minima found in $R_{14,32}$, e.g., near $V_g = -0.950$ V, we associate with the threshold for occupation of the $N = 2$ subband in at least one, but not all, of the constrictions comprising the junction.

While it is possible to identify the other minima found in $R_{14,32}$ and $R_{13,42}$ of Fig. 16a with the threshold for occupation of higher-energy subbands, the identification is dubious because of the possibility that the junction has at least one constriction with a different number of subbands occupied than the others for particular ranges of gate voltage, and because of the poor accuracy of the quantization of the low-field Hall effect for $N > 3$. For example, in the range $-0.820 < V_g < -0.70$ V, there are three minima in both

3. WHEN DOES A WIRE BECOME AN ELECTRON WAVEGUIDE?

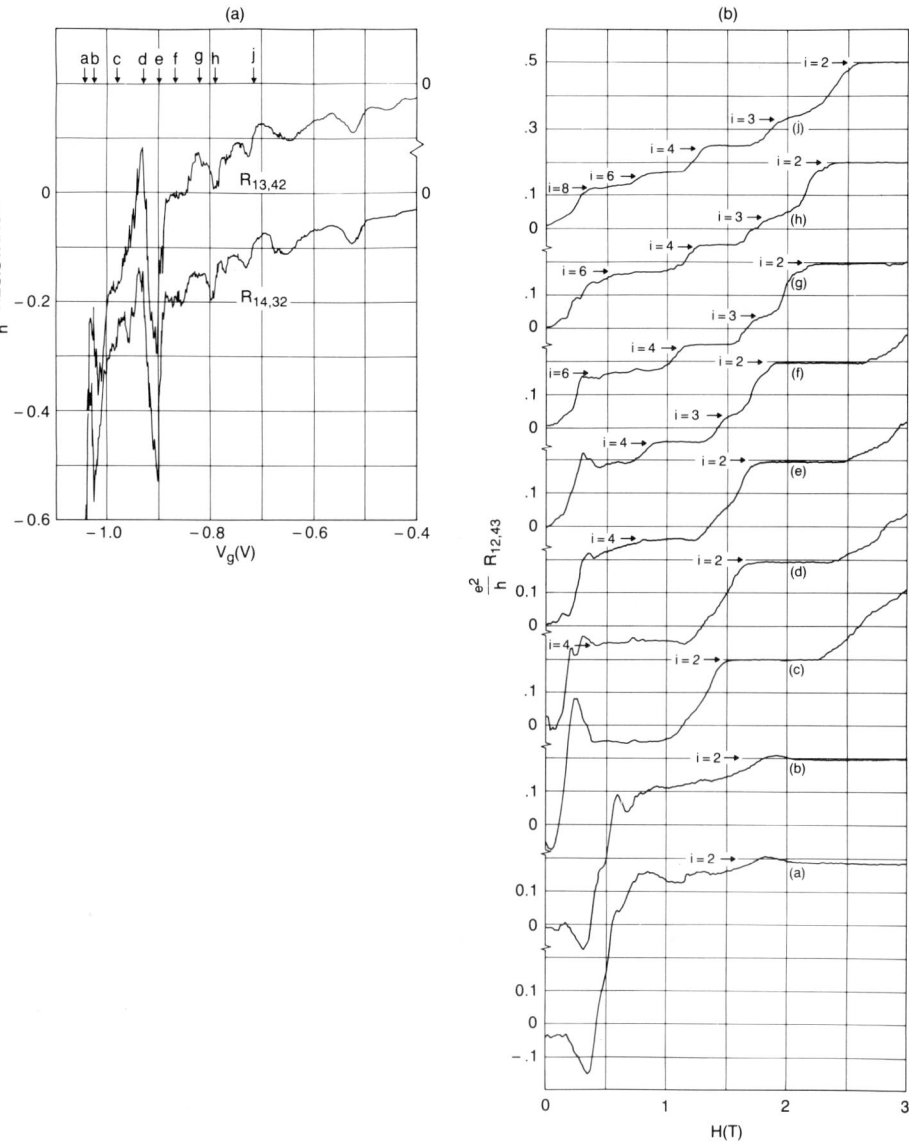

FIG. 16. (a) shows the bend resistances, $R_{14,32}$ and $R_{13,42}$, as a function of the gate voltage V_g applied to each of the series split-gate electrodes of a device like that shown in Fig. 8. $R_{13,42}$ is offset by $-0.2h/e^2$ from $R_{14,32}$ for clarity. (b) shows the Hall resistance $R_{12,43}$ versus magnetic field for nine different gate voltages (a–h,j) applied to the junction of (a). For traces, a–hj, $V_g = -1.035$ V, -1.025 V, -0.980 V, -0.936 V, -0.900 V, -0.871 V, -0.817 V, -0.795 V, and -0.712 V, respectively. These gate voltages are indicated by the arrows in (a) as well.

$R_{14,32}$ and $R_{13,42}$ as shown in Fig. 16a, but according to the Hall resistance, the subband index only changes from $N = 3$ to $N = 4$. Traces g and h of Figure 16b show that in the range -0.82 V $< V_g < -0.795$ V, the first plateau found in the Hall resistance, which occurs for fields beyond $H = 0.5$ T, is near $(h/e^2)/6$, corresponding to the magnetic depopulation of the $i = 6$ edge state; so, we assume that $N = 3$ in the constrictions comprising the junction for -0.820 V $< V_g < -0.780$ V. For -0.770 V $< V_g < -0.710$ V, $R_{12,43}$ shows two sharp minima. Yet, trace (j) of the Hall resistance, which corrrespond to the minima at $V_g = -0.712$ V, indicates that $N = 4$.

The temperature dependence is consistent with this interpretation of the sharp minima found in $R_{14,32}$, too.[13] As illustrated by Fig. 17, the minima are dramatically dependent upon temperature, and reappear after cycling the device to room temperature and cooling it again at nearly the same gate voltages. Features that are common to the traces taken before and after cycling to room temperature are numbered in Fig. 17. For example, the feature denoted by 2 in trace a corresponds to the feature numbered 2' in trace b. Since the resistance is determined by electronic states within $k_B T$ of the Fermi energy, where k_B is Boltzmann's constant, when the gate voltage is at the threshold for the occupation of a subband, only states within $k_B T$ of the $1/\sqrt{E_F - E_N}$ singularity in the 1D density of states in the constriction contribute to the resistance. As the temperature increases, however, other states, characterized by larger longitudinal wavevectors beyond the square root singularity, contribute and lower the magnitude of the resistance as indicated by trace c, which was obtained at 1.2 K. The sharp minima exhibit a different temperature dependence than the global negative resistance. The inset to Fig. 17 shows the temperature dependence of the magnitude of $R_{12,43}$ measured by Takagaki et al.[86] While the minima are attenuated and broadened beyond $T > 1$ K, the average negative resistance does not change appreciably until $T > 20$ K, a temperature indicative of the intersubband energy.[87]

The two-terminal conductance $G_{12,12}$, found when all of the split-gate electrodes are biased, is not consistent with this interpretation of the minima found in $R_{14,32}$. While both the measured and calculated conductances show a deterioration of the quantization for two constrictions in series, the minima in the calculated two-terminal conductance found near the threshold for the occupation of a subband cannot be unambiguously identified in the measured conductance. The lack of correspondence to the calculated two-terminal resistance may be related to disorder scattering in this junction or to asymmetry in the junction. The calculated resistance shown in Fig. 14a assumes that there is no disorder in the junction; however, we already have seen indications of disorder scattering in constrictions longer than 600 nm. (See Section 1.) Our measurements also reveal that, athough near zero, $R_{12,43}$ is not precisely zero, which we interpret as evidence of asymmetry in the

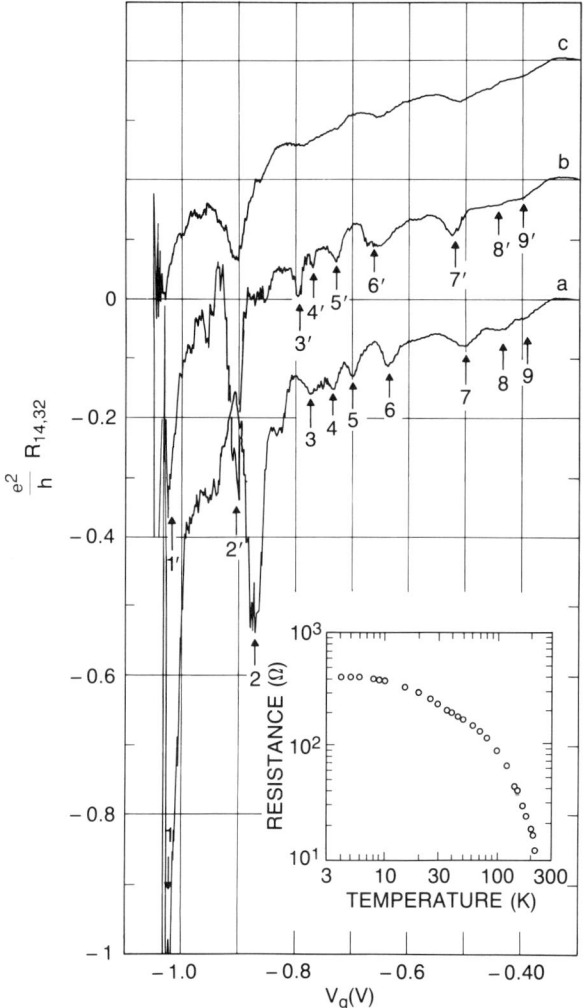

FIG. 17. The measured dependence of the bend resistance $R_{14,32}$ on the gate voltage applied to two split-gate electrodes in series. Traces a and b represent the same measurement on the same device at $T = 280$ mK, after the device had been cycled to room temperature and refrigerated again. The numbers 1–9 and 1'–9' are used to demonstrate the close correspondence between the two traces even after the device has been cycled to room temperature. Trace c shows the measured dependence on gate voltage found at 1.2 K. Traces b and c have been offset by $0.2h/e^2$ and $0.4h/e^2$ for clarity. The minima found in trace b at 280 mK are shallow and broad in trace c. The inset to the figure shows the temperature dependence of the magnitude of the bend resistance for $T > 3$ K measured by Takagaki et al.[86]

junction. The asymmetry evidently is small, judging from the size of the offset at $H = 0$ relative to h/e^2, however.[88]

Figures 18, 19, and 20 show the magnetoresistances found in the device of Fig. 14b for $V_g = -1.025$ V applied to each of the electrodes. Figure 18 shows that for $H < 200$ mT, the magnetoresistance $R_{14,32}$ is asymmetric and negative, and while generally suppressed below the conventional 2D Hall resistance, $R_{12,43}$ is nonzero at $H = 0$. As the applied magnetic field increases, the resistance changes dramatically. $R_{14,32}$ becomes positive and $R_{12,43}$ approaches a quantized value corresponding to twice the number of subbands

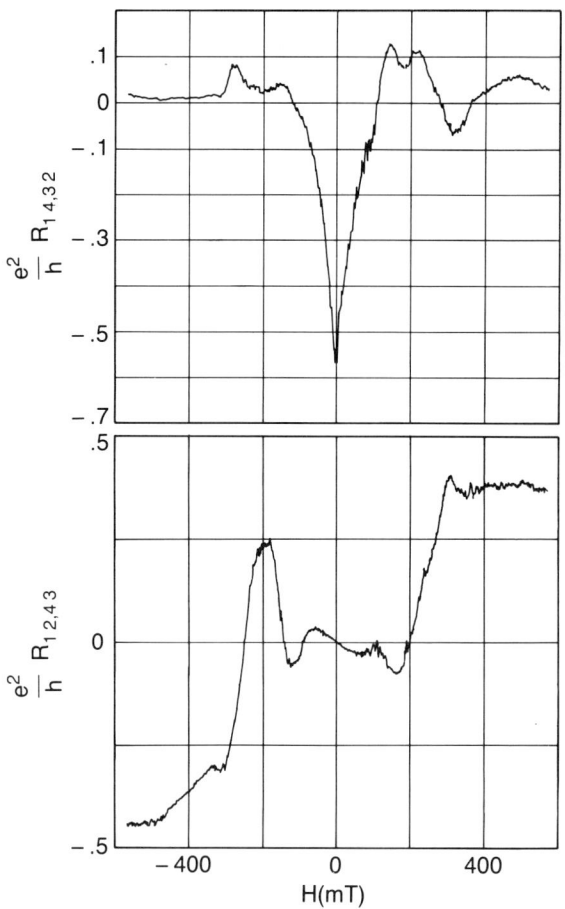

FIG. 18. The magnetoresistances $R_{14,32}$ and $R_{12,43}$ measured at 280 mK in a junction between two split-gate electrodes in series with each electrode biased at $V_g = -1.025$ V. With $V_g = -1.025$ V, the device has only one subband occupied, i.e., $N = 1 \pm 1$, in each of the constrictions comprising the junction.

in the constriction. For the data of Fig. 18, the zero field position was determined from a separate Hall probe in close proximity to the device, but it should be apparent that no translation of the field axis could result in $R_{14,32}(H) = R_{14,32}(-H)$. The asymmetry in the magnetoresistance, the negative resistance, and the finite Hall effect at $H = 0$ observed in these four-terminal measurements do not violate Onsager's symmetry relations because the four-terminal resistance is not trivially related to the coefficients to which Onsager relations apply. Because of the microscopic reciprocity of the transmission coefficients, however, it can be shown from Eq. (5) that the resistance $R_{mn,kl}(H)$ should be invariant under a field reversal when accompanied by an exchange of the current and voltage terminals, i.e.,[36]

$$R_{kl,mn}(H) = R_{mn,kl}(-H). \tag{12}$$

The decompositions $R^S = \frac{1}{2}[R_{kl,mn}(H) + R_{mn,kl}(H)]$ and $R^A = \frac{1}{2}[R_{kl,mn}(H) - R_{mn,kl}]$ are supposed to be symmetric and antisymmetric, respectively, as a direct consequence of this symmetry.

When the current and voltage leads are interchanged, we find generally that $R_{kl,mn}(H)$ is highly correlated with $R_{mn,kl}(-H)$ to within about 99%. The results of the decomposition of $R_{12,43}$ and $R_{14,32}$ of Fig. 18 are shown in Fig. 19a. The data of Fig. 19a represent a compilation of two separate resistance measurements taken with the current and voltage leads interchanged. Generally, R^S is approximately symmetric about $H = 0$, and R^A is approximately antisymmetric. While discrepancies of about $0.01 h/e^2$ from the perfect symmetry are observed, the reproducibility of the measurements is better than $0.002 h/e^2$ over two days time. Because of the reproducibility, it is unlikely that the asymmetry is due to magnetic impurities.

The origin of the symmetry in the symmetric part of the magnetoresistance is unknown, but it may be associated with the energy dependence of the transmission coefficients. The energy dependence of T_{jk} in Eq. (5) is evident from the nonlinearities in the resistance and second-harmonic generation.[89-94] Second- and higher-order harmonic generation has been observed by de Vegvar et al.[93] in submicron constrictions, and it was found that the second-harmonic signal did not exhibit the symmetry implicit in Eq. (12). Since the second-harmonic signal is not invariant under an exchange of current and voltage leads when the magnetic field is reversed, the second- and higher-harmonics could be a source of the asymmetry found in $R^S(H)$.

Figures 20a, b, c, and d show the symmetrized magnetoresistances found at $V_g = -1.035$ V, -0.900 V, -0.770 V, and -0.560 V, respectively, in the device of Fig. 16. Using our estimate for the number of subbands in the constriction and our estimate for the carrier density, and assuming a harmonic confinement potential, we can infer the width of the constrictions comprising the

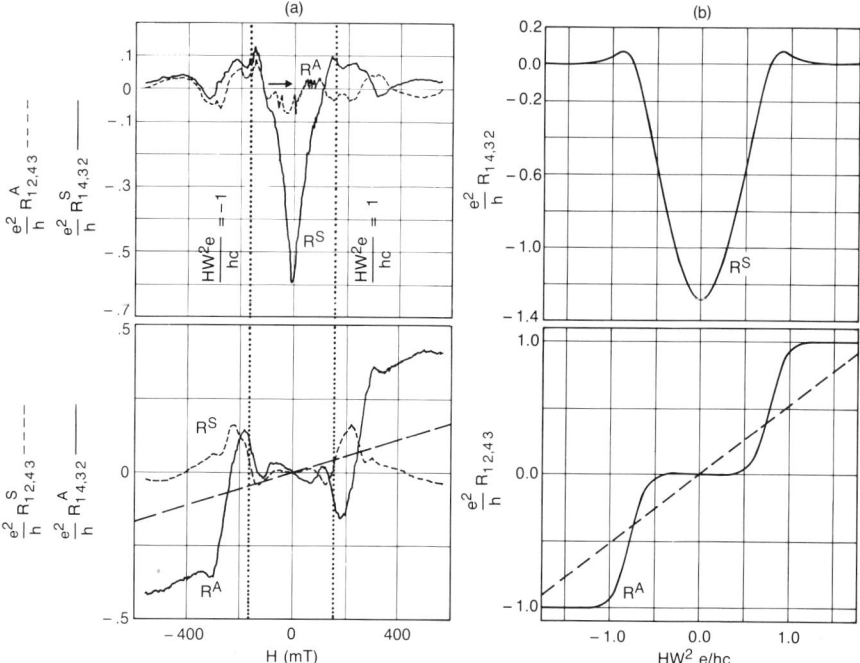

FIG. 19. (a) depicts the symmetric, R^S, and antisymmetric, R^A, components of the magnetoresistances $R_{14,32}$ and $R_{12,43}$ found at 280 mK for the device of Fig. 18. In a macroscopic junction with the same geometry, 2D mobility, and carrier density, $R_{14,32}$ is a positive number at zero field (as seen from the arrow in the top of (a)) and $R_{12,43}$ goes to zero linearly as a function of magnetic field (as indicated by the dashed line in the bottom of (a)). The dotted vertical line indicates where the condition, $HeW^2/hc = \pm 1$, is satisfied assuming a hard-wall confining potential. (b) shows the calculated magnetoresistance for a junction made from a hard-wall potential with sharp corners. Only one subband is occupied, and the spin degeneracy is ignored in the calculation. The essential features found in the measurement of the resistance are reproduced by the calculation. Beyond $HeW^2/hc = \pm 1$, both (a) and (b) show that $R_{14,32}$ is no longer negative, and the Hall resistance $R_{12,43}$ is not suppressed.

junction. The width of the constrictions is related to the carrier density and the number of 1D subbands through an integral over the 1D density of states; i.e.,

$$n = \frac{2}{\pi W} \sum_{j=0}^{N} \sqrt{2\frac{m^*}{\hbar^2}} \sqrt{E_F - E_j}. \tag{13}$$

We find that for the bias conditions, $V_g = -1.035$ V, -0.900 V, -0.770 V, and -0.560 V, which correspond, respectively, to the resistances shown in Figs. 20a–d, $W = 20 \pm 10$ nm, 25 ± 10 nm, 55 ± 10 nm and 65 ± 15 nm.

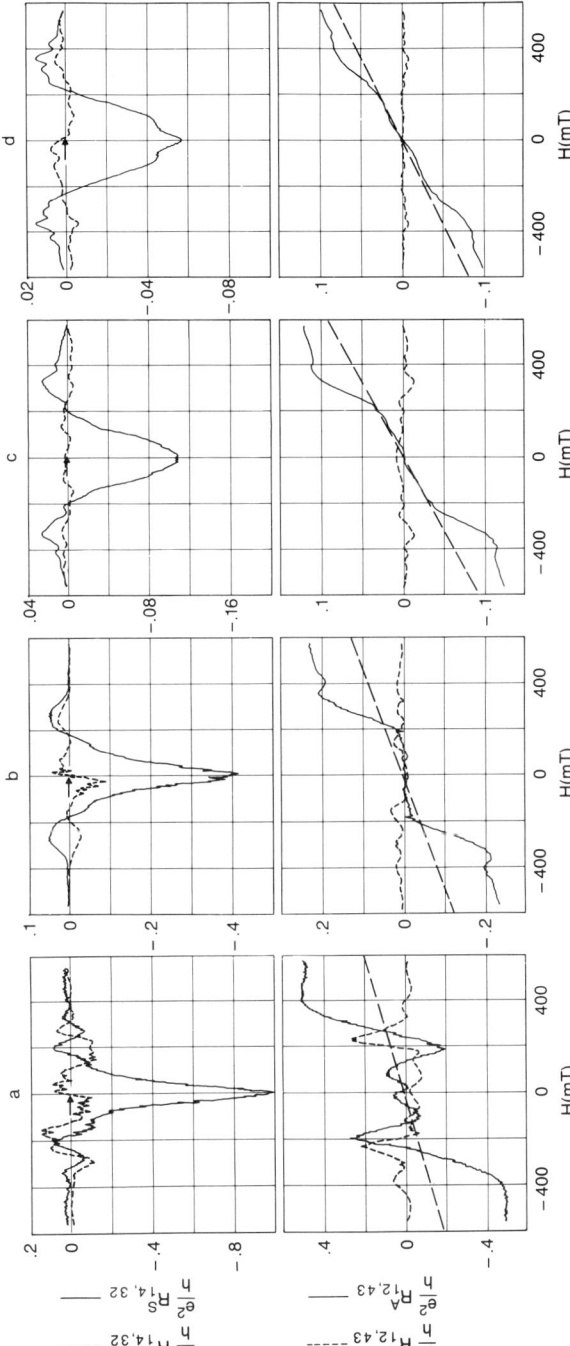

FIG. 20. The symmetric, R^S, and asymmetric, R^A, components to the magnetoresistances $R_{14,32}$ and $R_{12,43}$ when four different gate voltages are applied to a junction comprised of two split-gate electrodes in series. The data is obtained from the device of Fig. 16. In (a), $V_g = -1.035$ V and $N = 1 \pm 1$ in each of the constrictions comprising the junction; in (b), $V_g = -0.90$ V and $N = 2 \pm 1$; in (c), $V_g = -0.770$ V and $N = 4 \pm 1$; and in (d), $V_g = -0.560$ V with $N = 6 \pm 1$.

The arrows in the top row of Fig. 20 represent the conventional longitudinal resistance expected for a bend in a wide, diffusive junction with the same geometry, 2D mobility, and 2D carrier density. The long dashed lines in plots of $R_{12,43}$ represent the conventional 2D Hall resistance deduced from carrier density, which is estimated from the high-field Hall resistance. As Fig. 20 shows, the magnetoresistance $R_{14,32}$ is negative for -200 mT $< H < 200$ mT approximately. The range of field for which $R_{14,32}$ is negative is relatively independent of V_g. Figure 20 also shows that $R_{12,43}$ is suppressed below the conventional 2D Hall effect, but the range over which $R_{12,43}$ is suppressed changes from 270 mT for Fig. 20a to 0 in Fig. 20d. The discrepancy between the ranges of field for which the longitudinal resistance is negative and the Hall effect is suppressed is not understood. Figure 21a summarizes the initial slope of the Hall effect, $R^A_{12,43}$, and the range of magnetic field over which the Hall effect is suppressed from the conventional value found in the device of Fig. 16. Figure 21b shows the magnitude of $R^S_{14,32}(H = 0)$ and the range of magnetic field over which the resistance is negative for the same device.

It has been argued that the intrinsic Hall effect should not be suppressed,[80,95] and Imry has proposed that a measurement of the intrinsic Hall voltage could be accomplished using probes in which only one 1D subband is occupied, *provided the probes are weakly coupled*.[45] Using elementary perturbation theory to evaluate the effect of a small magnetic field on the wave function Ψ, it is possible to calculate the change in the electron density across the wire due to the field. This change is proportional to the electrochemical potential, i.e., the intrinsic Hall voltage across the wire. Thus, the polarization of the wave function in a 1D constriction yields a finite Hall effect.[80,96] Since our measurements show that the Hall effect still is suppressed when $N = 1$ in the constrictions comprising the junction in both the voltage and current probes as shown in Fig. 19, and when $N = 1$ only in the voltage probes,[13] we infer that the probes cannot be treated perturbatively.

Kirczenow,[97] Ravenhall,[98] Baranger and Stone,[67] and others[99] have explored the suppression of the Hall resistance numerically, treating the leads as an integral part of the resistance measurement, and have observed that the resistance is not generically suppressed for a junction with hard walls and sharp corners unless $N = 1$ in each of the constrictions comprising the cross. Figure 19 shows the magnetoresistance calculated according to such a model when $N = 1$ in each of the constrictions comprising the junction, beside the magnetoresistance measured when $N = 1 \pm 1$.[67,100] The field where $HeW^2/hc \approx 1$ is indicated in the experimental data to facilitate a comparison with the calculation. The general form and the magnitude of the measured resistance is reproduced by the calculation. (However, the agreement with the magnitude of the measured resistance $R_{14,32}$ achieved by this calculation may be coincidental.) For $N \geq 1$ in the constrictions comprising the junction, the models of Ravenhall and Kirczenow cannot reproduce all of

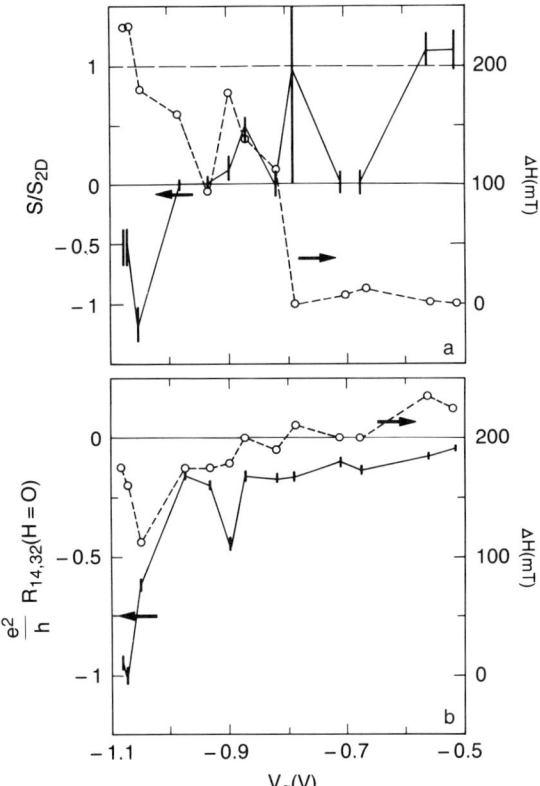

FIG. 21. (a) shows the measured slope S (solid line) of the resistance $R_{12,43}$ as a function of gate voltage V_g applied to the junction of Fig. 16. The slope is evaluated over the field range indicated by ΔH. This field range corresponds approximately to the condition $HeW^2/hc \approx 0.5$, where W is evaluated using a hard-wall confinement potential. This slope is normalized to S_{2D}, the slope associated with the Hall effect found in the constriction at 8 T. The Hall resistance generally is suppressed for $V_g < -0.8$ V with a few exceptions. For $V_g < -1.00$ V, the slope is negative, and for $V_g \approx -0.8$ V, the average normalized slope is near unity. The range of field over which the Hall resistance is suppressed changes with V_g as indicated by the open circles connected by dotted lines. The range is measured between zero field and the magnetic field where the Hall resistance touches the classical 2D value. (b) shows the bend resistance $R_{14,32}$ (solid line) versus V_g found at $H = 0$, and indicates the range in magnetic field over which the resistance is negative. The range is measured between $H = 0$, and the field where $R_{12,43}$ first becomes positive.

the features observed experimentally, however. The slope of $R_{12,43}$ found at low field as a function of Fermi energy is represented in Fig. 22a for such models.[67] A comparison with Fig. 21a reveals that the suppression of the Hall resistance is a much more generic phenomenon[101] applying at least to constrictions where $N \leq 4$.

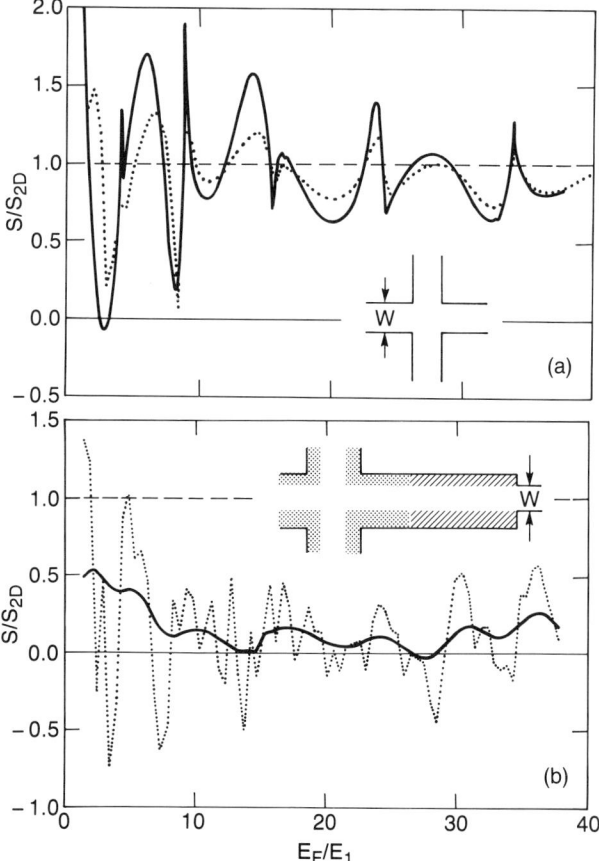

FIG. 22. The calculated slope of the Hall resistance, S, normalized to its classical 2D value, S_{2D}, as a function of Fermi energy E_F normalized by the bound state energy corresponding to ground state E_o. The junction is composed of four electron waveguides with hard walls and sharp corners. The suppression of the Hall resistance occurs only in narrow regions of energy. The slope at low field (solid line, $HeW^2/hc = 4 \times 10^{-7}$) agrees with that obtained by fitting $R_{12,43}(H)$ over a broad range of fields $[-H_1, H_1]$ (dotted line, $H_1 eW^2/hc = 0.6$ for $E_F < 14E_1$, $H_1 eW^2/hc = 0.8$ for $E_F > 14E_1$. (b) shows the slope obtained as in (a) in the same energy range for a junction in which the width of the waveguide is tapered (the cross-hatched region in the inset) into the junction. The suppression of the Hall resistance occurs over a broad range of energies ($T = 0$, dotted line; $T = E_1$, solid line). (From Ref. 22.)

Baranger and Stone[67] resolved this issue by considering more realistic potential contours for the junctions. In particular, they observed that for a junction in which the width of the wires is gradually tapered to meet the cross members, the Hall resistance is suppressed generally. Figure 22b demonstrates the generic suppression of the Hall resistance as a function of Fermi energy found for a tapered junction with hard walls. According to Baranger and Stone, the gradual widening of the wires comprising the junction causes the electron to travel adiabatically from the narrow to the wide region of the junction conserving the number of subbands. Thus, the current injected into the narrow wire, which is distributed equally between the subbands of the constriction, reaches the wide region distributed only among the low-lying modes. A collimation or focusing of the electron wave in the forward direction is a direct consequence of the adiabatic feed into the junction. Chang et al.[102] support this interpretation with experiments that claim to show that a uniform wire is not necessary to produce generic quenching; instead, a constriction on each end of a wide cross is sufficient for the suppression of the Hall resistance.

While the experimental conditions generally necessary for the suppression of the Hall effect still are controversial,[103] we can deduce two prerequisites from Figs. 20 and 21. When the $R_{14,32}$ is negative, the transmission probabilities T_{31}, T_{14}, T_{23}, and T_{42} all are small relative to T_{12} and T_{34}. We deduce from Figs. 20 and 21 and Eq. (9) that T_{31}, T_{14}, T_{23}, and T_{42} are comparable to T_{12} and T_{34} for $V_g > -0.9$ V or $N > 2$, since the magnitude of $R_{14,32}$ is small. Under these conditions, Figs. 20d and 21 show that $R_{12,43}^A$ no longer is suppressed below the conventional Hall resistance for magnetic fields beyond 20 mT. Therefore, the Hall resistance is suppressed below the 2D value when: (1) the coefficients T_{31}, T_{14}, T_{23}, and T_{42} all are small, and (2) $T_{31}T_{42} \approx T_{23}T_{14}$. Ford et al.[103] demonstrated this by changing the size of T_{31}, T_{14}, T_{23}, and T_{42} relative to T_{12} and T_{34} through a change in the geometry of the junction.

Recently, a semiclassical model for the resistance of a ballistic cross with tapered contacts, applicable for N large, has been shown to produce the suppression of the Hall resistance, the negative magnetoresistance found in $R_{14,32}$, and a feature in the Hall resistance resembling a plateau for magnetic fields such that the cyclotron diameter is greater than the width of the wire, i.e., $HeW^2/\hbar k_F < 1$.[2] Beenakker and van Houten assume in their semiclassical model that the angular distribution of the wavevectors of electrons injected into the tapered constrictions that comprise the cross is cosine-like and that an adiabatic taper focuses the classical electron trajectories. The semiclassical treatment is successful because the classical transmission probabilities are approximately the same as those calculated quantum mechanically, but there are exceptions.[100] For example, Beenakker and van Houten assert that a semiclassical description can produce the average negative resistance found as a function of width, but not the sharp minima shown in Fig. 14. The semiclassical model neglects quantum mechanical reflections that develop at the junction and so underestimates the resistance. In particular, a fit of the semiclassical theory to the data of Fig. 20b can account only for about half of the measured resistance, $R_{14,32}$, found at $V_g = -0.90$ V for $H = 0$, and cannot account for the well-quantized Hall effect[13] at $R_{12,43} = h/e^2 4$. (The general form of the magnetoresistance for $HeW^2/hc < 1$ is recovered, however.) The semiclassical Hall resistance calculated by Beenakker and van Houten is not quantized for $HeW^2/hc > 1$, as found experimentally for $N \leq 3$. (See, for example, Fig. 16.) However, the identification of a plateau in the measured resistance with the magnetic depopulation of the highest N subband becomes dubious for $N > 3$. The feature observed experimentally in the Hall resistance may be associated with that found in the semiclassical calculation when $N > 3$.

There is a third prerequisite for the suppression of the Hall effect. According to the interpretation by Baranger and Stone[67] and Beenakker and van Houten,[2] the suppression of the Hall resistance can occur only if the elastic mean free path is larger than the width of the junction. If an electron is scattered by an impurity from one subband into another, the collimation deteriorates because the distribution of the current among the available subbands is scrambled. The effects of *both* forward and backscattering are detrimental to the collimation effect, and the forward scattering rate in GaAs/AlGaAs is 10 times higher than the backscattered rate. To illustrate the short length scale over which the collimation of the electron wave can deteriorate, we examined the nonlocal bend resistance.

We have shown already that the average, symmetric magnetoresistance changes dramatically if the current path bends through a junction at the voltage terminals. The average resistance changes as well if the current path bends beyond the voltage terminals.[20,21] Figure 23 shows the changes in the four-terminal, symmetric magnetoresistances measured in a Hall bar geometry with voltage terminals 900 nm apart when the current path bends through a junction 700 nm from the voltage terminals ($R^S_{82,46}$), at one voltage terminal ($R^S_{32,46}$), and at both voltage terminals ($R^S_{37,46}$) relative to a geometry in which the current is injected through a junction along a straight line ($R^S_{12,46}$). The resistance in these devices always is observed to increase at zero field whenever the current path bends at a voltage probe (e.g., $R^S_{32,46}$), if the mobility of the starting heterostructure exceeds $\mu = 30$ m^2/Vs with $W \approx 50$ nm and $N \approx 5$. (No change is observed outside the amplitude of fluctuation when $\mu = 4.7$ m^2/Vs with $W \approx 75$ nm corresponding to $N \approx 8-10$.) The increase in the resistance is consistent with the observation that $R^S_{13,46}$ is negative, since $R_{32,46} - R_{12,46} = R_{31,46} > 0$, and since $R_{31,46} = -R_{13,46}$.

The average resistance consistently increases for $\mu \geq 87$ m^2/Vs and $N \approx 2-5$ even when the current bends through a junction separated from the voltage terminals by at least 700 nm; i.e., there is a nonlocal effect on the average resistance. The change in resistance due to a bend in the current path with respect to the resistance measured with no bend, ΔR^S, is plotted as a function of the distance between the nearest voltage terminal and the junction where the current path bends, ΔL, in the inset in the upper right corner of Fig. 23.

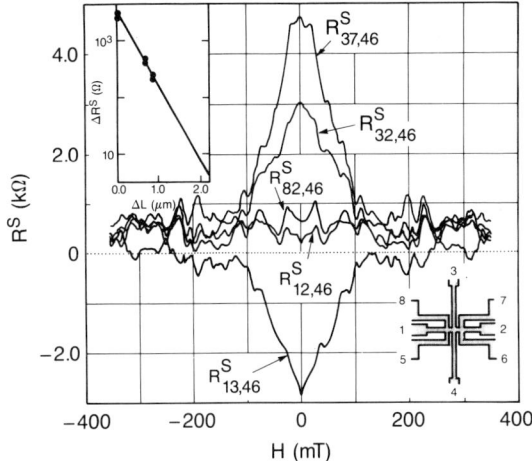

FIG. 23. The symmetric magnetoresistance R^s found at 280 mK using different current and voltage terminals in a device represented schematically by the inset in the lower right. The resistance is measured using different current terminals with the same voltage terminals, which are separated by 900 nm. The resistance generally increases when the current path bends through a junction. The inset in the upper left of the figure illustrates the decay of ΔR_s versus the distance between the bend in the current probe and the voltage terminal. (From Ref. 22.)

The changes observed are consistent with the law $\Delta R^S = R_0 \exp(-\Delta L/L_s)$ where R_0 is the magnitude of the resistance measured when the current bends through a junction at one voltage terminal (e.g., $|R_{13,46}(H = 0)|$), and L_s is a characteristic length less than 500 nm. Takagaki et al.[21] and Molenkamp et al.[104] recently have corroborated this short-length scale found in a device made in high-mobility GaAs/AlGaAs heterostructures.

We attribute the change in the nonlocal bend resistance, which occurs over a length of about 500 nm, to disorder scattering. In the absence of a magnetic field, elastic scattering of a carrier in one subband by disorder generally gives rise to about equal transmitted amplitudes into each of the occupied subbands. However, elastic scattering of a carrier in a subband, by a contact or lead within a junction with a definite symmetry, promotes scattering into particular subbands. The resulting nonequilibrium distribution of the current among the subbands, which gives rise to the change in resistance for a remote bend in the current path, is expected to decay as the bend gets further away from the voltage leads, since (elastic) disorder scattering scrambles the distribution. Theoretically, the decay of ΔR^S is consistent with an exponential dependence on the elastic mean free path.[69,100] The more rapid decay seen in the experiment ($L_e \approx 10 L_s$) may be due to preferential forward scattering in a constriction of the 2DEG in AlGaAs/GaAs heterostructures.[69] Although forward scattering by remote impurities is not measured by the mobility, it nevertheless can affect the distribution of the current among the subbands. If an impurity scatters a carrier, $\Sigma_{i,j=1}^{N} t_{ij}$ must be constant because the current is conserved within the wire, but the distribution of the current among the subbands is altered from what is characteristic of scattering only from the contacts.

10. FOUR-TERMINAL RESISTANCE MEASUREMENTS FOR $1 < HeW^2/hc < 10$—THE INTEGER QUANTIZED HALL EFFECT

Figures 24a, b, and c show the magnetoresistances, $R_{14,32}$ and $R_{12,43}$, observed at 100 mK when each of the constrictions comprising the junction are: (a) not biased ($V_g = 0$ V), (b) biased so that only the two lowest subbands ($N = 2 \pm 1$) are occupied in each of the constrictions, and (c) biased so that only the lowest subband ($N = 1 \pm 1$) is occupied. N is estimated using: (1) the quantization of the two-terminal resistances of each of the constrictions versus V_g at $H = 0$ T, (2) the peaks found in $R_{14,32}$ versus V_g at $H = 0$ T associated with the change in the number of occupied subbands, and (3) the quantization of $R_{12,43}$ versus H. Using a harmonic confinement potential, the carrier density deduced from $R_{12,43}$ at $H = 8$ T, and our estimate for N, we infer that $W \approx 30 \pm 10$ nm in Fig. 24b and $W \approx 20 \pm 10$ nm in Fig. 24c. Notice that $R_{14,32} \geq 0$ and $R_{12,43} = H/nec$ for $H < 100$ mT in the 2D junction of Fig. 24a. In contrast, $R_{14,32}$ is negative and $R_{12,43}$ is suppressed from the conventional 2D Hall resistance for $H < 100$ mT in Figs. 24b and c[11,20] in correspondence with our expectations for $HeW^2/hc < 1$.

A comparison of Fig. 24a with Figs. 24b and c shows that the four-terminal resistance of the junction is determined by the transmission through the constrictions and is not affected by the edge states reflected at the boundary with the 2DEG (shown in Fig. 13) and the corresponding discontinuity in the density there, or by the magnetoresistance in the 2D contact.[105] The 2D contact generally has a higher electron density than the constrictions that comprise the junction, and so for a particular magnetic field, the number of edge states occupied usually is larger in a 2D contact than in a constriction. If the edge states propagate adiabatically from the 2D contact to the constriction,[105] then no mixing of the states occurs, and edge states above the Fermi energy in the constriction are totally reflected. Thus, scattering from the constrictions completely determines the four-terminal resistance for fields $HW^2e/hc < 1$, and for fields where the Hall resistance is quantized. This was demonstrated explicitly in experiments by van Wees et al.,[106] which revealed that the Hall resistance of a wide 2D Hall bar may be quantized at anomalous values if both a voltage and current terminal are constricted to 1D.

3. WHEN DOES A WIRE BECOME AN ELECTRON WAVEGUIDE? 165

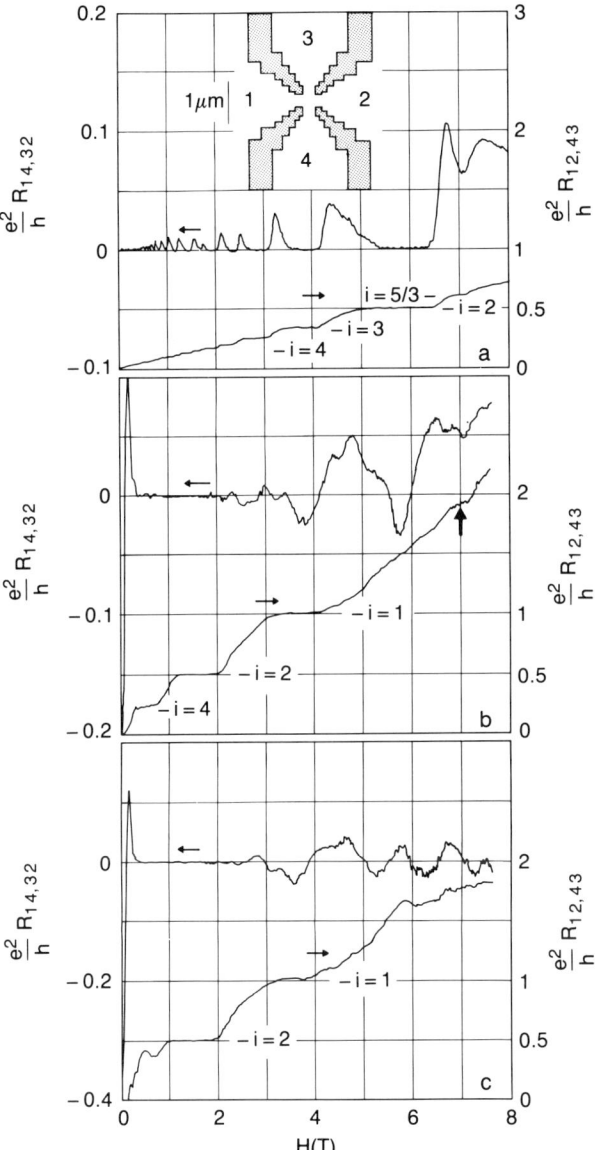

FIG. 24. The resistances, $R_{12,43}$, and $R_{14,32}$, versus magnetic field measured at 100 mK for a device made from heterostructure 4. In (a), the leads comprising the cross are 100 μm wide ($V_g = 0$ V), while in (b) and (c), the leads are about 30 nm with $N = 2 \pm 1$, and 20 nm wide with $N = 1 \pm 1$, respectively, assuming a harmonic confinement potential. The vertical arrow in (b) at $H = 7$ T indicates the position of an anomalous plateau in $R_{12,43}$ near $(h/e^2)/0.53$. The inset to (a) is a schematic representation of the junction; the convention for labeling the leads or the constrictions between the electrodes is given there. (From Ref. 14.)

As the magnetic field increases, the four-terminal Hall resistance, $R_{12,43}$, of the high-mobility 2DEG of Fig. 24a is quantized precisely in steps of $h/e^2 i$, where i is an integer (with 99.9% accuracy here) corresponding to the depopulation of spin-polarized edge states.[107] The four-terminal Hall resistance of a cross comprised of four 1D constrictions in a 2DEG also is quantized, but as shown in Figs. 24b and c, there is a finite number of steps corresponding to twice the number of spin-degenerate 1D subbands in the constriction in the absence of a magnetic field,[11] and moreover, the quantization is not precise.[108] For magnetic fields $HeW^2/hc > 1$, $R_{12,43}$ in Figs. 24b and c is quantized at approximately $(h/e^2)/i$ for integers $i \leq 2N$. In Fig. 24b, $R_{12,43}$ reaches a plateau at approximately 300 mT, i.e., $R_{12,43} = (h/e^2)/(4.10 \pm .15)$, corresponding to the depopulation of the highest spin-polarized 1D subband, $i = 4$. The plateau associated with $i = 3$ is not well-defined. However, with increasing magnetic field, other plateaus are observed in the resistance centered at $(h/e^2)/(1.97 \pm .03)$ and $(h/e^2)/(0.99 \pm .03)$ near $H = 1.6$ T and 4 T, respectively, corresponding to the $i = 2$ and $i = 1$ states. In Fig. 24c, $R_{12,43}$ reaches a plateau at approximately 1 T and 3.5 T, where $R_{12,43} = (h/e^2)/(2.02 \pm .03)$ and $(h/e^2)/(0.95 \pm 0.07)$, respectively, corresponding to $i = 2$ and $i = 1$.

The quantized Hall effect bound in a 2DEG is explained usually by invoking localization, but since the quantized Hall effect has been observed in junctions comprised of submicron ballistic constrictions, *shorter than the localization length* ($\approx L_e$), localization evidently is not a prerequisite.[12] Instead, it is the breakdown of translational invariance that is essential to observe the quantized Hall effect.[109] The breakdown of translational invariance can be provided by impurities or other imperfections (as in a macroscopic sample), but it also can be provided by contacts or a constriction, for example.

Furthermore, unlike a 2DEG where the filling factor $i = nhc/eH$, the magnetic field is not a direct measure of the number of occupied subbands in a 1D constriction. In a 1D constriction, the plateaus in the Hall resistance are not found at regular intervals in H^{-1} because the subbands are defined by both magnetic and electrostatic confinement, and because the carrier density in the constrictions is determined by the Fermi energy in the 2D contacts, which varies as a function of H. For example, for a harmonic confining potential, the magnetic field dependence of the filling factor is given by: $i = nhc/e(H^2 + H_0^2)^{1/2}$ with $H_0 = \omega_0 m^* c/e$. If $H_0 = 1$ T, then $i = 4$ should occur for $H = 0.26$ T, $i = 2$ for $H = 1.8$ T, and $i = 1$ at $H = 4.0$ T, similar to Fig. 24b.[32]

The value of the plateau resistance, $R_{12,43}$, is a measure of the number of occupied subbands. However, for 300 mT $< H <$ 5 T, $R_{12,43}$ is not exactly quantized, and moreover, $R_{14,32}$ can be negative. We attribute these observations to the overlap between edge states due to the narrow width of the constrictions. As shown next, for $H < 5$ T with $W \approx 20$ nm, the overlap between edge states is approximately $\exp[-(W/2r_0)^2] \approx 0.1$. Due to the overlap, there is about 10% backscattering between opposite edges, which gives rise to $<10%$ deviations in the transmission probability between adjacent leads, and ultimately to $<0.1h/e^2$ deviations from exact quantization in $R_{12,43}$ and $-0.1h/e^2 < R_{14,32} < 0$.

11. THE MAGNETORESISTANCE OF AN ANNULUS IN THE QUANTIZED HALL REGIME

To illustrate the dramatic change in current distribution and the backscattering that develops in an intense magnetic field, we examine the magnetoresistance of a prototypical scatterer in the quantized Hall regime—an annulus.[22-26] Following the analysis given by Jain[75] and Büttiker,[36,78,110] we assume that the magnetic field penetrates both the wire comprising the annulus and the annulus itself. For $HeW^2/hc < 1$, the magnetoresistance is periodic in the flux hc/e through the average area of the annulus due to the Aharonov–Bohm (AB) effect.[19] As represented in Fig. 25a, the AB effect occurs because two electron trajectories that encircle a magnetic flux acquire a relative phase shift proportional to the flux. As the flux changes, the

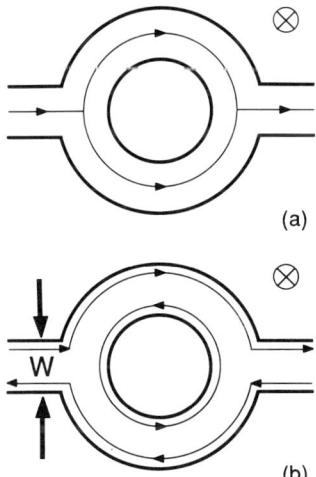

FIG. 25. A schematic representation of transport through an annulus near zero magnetic field, $W < 2r_o$ (a), and in the quantum Hall regime $W \gg 2r_o$ (b). The arrows indicate the direction of current. In the quantized Hall regime, a net current results from the difference between two oppositely directed currents associated with the edge states.

transmission probability of an electron through the annulus oscillates with a periodicity of hc/e. In contrast, in the quantized Hall regime, the current is carried around the annulus by edge states, as shown in Fig. 25b. The AB effect is suppressed in the quantized Hall regime because:

1. The outer edge states, which are connected to the leads and determine the resistance, do not enclose a flux.
2. The inner edge states that do enclose a flux are not coupled to the outer edge states.
3. Ideally, the edge states do not backscatter.

Since the magnetoresistance of the annulus oscillates periodically with magnetic field, the magnetoresistance provides an unambiguous periodic signature of the backscattering as a function of field. The observation of oscillations due to the AB effect has been reported before[22-26] in the magnetoresistance of annuli fabricated in high-mobility AlGaAs/GaAs heterojunctions. Under the conditions of the experiments, the magnetic field penetrated both the annulus and the 1D constrictions comprising the annulus. For magnetic fields where $HeW^2/hc > 1$, the oscillation amplitude was suppressed.[23,25] The oscillations, with a periodicity near zero field corresponding to a flux of hc/e through the average area of the annulus, decrease exponentially in intensity and shift to a lower frequency as the magnetic field increases.[22,26] The frequency of oscillation observed in a high magnetic field corresponds to a flux of hc/e through the area circumscribed by the inside diameter of the annulus, and the exponential decrease in the amplitude is indicative of the exponential decrease in the overlap between the outside and inside edges of the annulus. These observations show that the net current is carried by edge states in the quantized Hall regime, and that the suppression of the Aharonov–Bohm effect is due to the suppression of backscattering.

The fabrication procedures used for the annuli are described in detail elsewhere.[23,25] For economy, we focus on the results we obtained from devices with a starting mobility and carrier density of about $\mu = 50$ m^2/Vs and $n = 4.4 \times 10^{15}$ m^{-2}, respectively. The carrier density in the 1D constrictions deduced from the high-field Hall effect is approximately $n = 3.0 \times 10^{15}$ m^{-2}. We measured the four-terminal, magnetoresistance of nominally 2 μm-diameter annuli like that shown in the inset to Fig. 26. The current excitation was typically less than 1 nA. If the same lead configuration was used, the measured resistance was correlated from day to day to better than 95%.

Figure 26 shows the magnetoresistance $R_{25,16}$ of a nominally 2 μm- (actually 1.82 ± 0.05 μm) diameter annulus found for 70 mT intervals starting at magnetic fields of $H = 0.0$ T, 0.920 T and 2.470 T, respectively. The convention for numbering the leads is given in the inset. We observe that the magnetoresistance oscillates near 0.0 T with periodicity corresponding to a

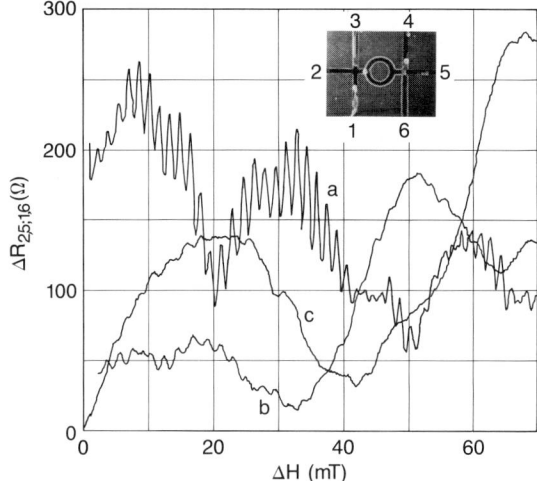

FIG. 26. The typical magnetoresistance $R_{25,16}$ of a nominally 2μm-diameter annulus obtained at $T = 280$ mK for H near (a) 0 T, (b) 0.92 T, and (c) 2.47 T. An electron micrograph of the device is shown in the inset and the convention for numbering the leads is indicated there. (From Ref. 22.)

flux of hc/e through the area of the annulus defined by the average diameter. As illustrated by traces b and c of Fig. 26, both the amplitude and frequency decrease dramatically as the magnetic field increases, however. Background fluctuations, which have lower-frequency aperiodic and periodic components,[111] also are found superimposed upon these oscillations. The background fluctuations beat with the periodic oscillations so that in short intervals of magnetic field, the amplitude can be reduced, e.g., near 45 mT in trace a of Fig. 26.

The changes observed in the Fourier power obtained from $R_{25,16}$ as a function of magnetic field are illustrated in Fig. 27. To avoid complications in the interpretation of the Fourier spectra due to the beating phenomenon mentioned earlier, the Fourier transforms were taken using a window approximately 220 mT wide, much larger than the typical periodicity associated with a beat (approximately 20 mT). Therefore, each transform represents an average over many beats. Near zero field, peaks in the Fourier spectra of the magnetoresistance of the annuli are observed at approximately 630 T^{-1} and 1,250 T^{-1}, corresponding to fluxes of hc/e and $hc/2e$, respectively, penetrating the area of the annulus defined by the average diameter. The width of the hc/e fundamental is assumed to be given approximately by $\Delta H^{-1} = 2\pi rW/(hc/e)$, where r is the radius, and W is the width of the constrictions that comprise the annulus. For the device of Fig. 26, we find that $W \approx 210$ nm.

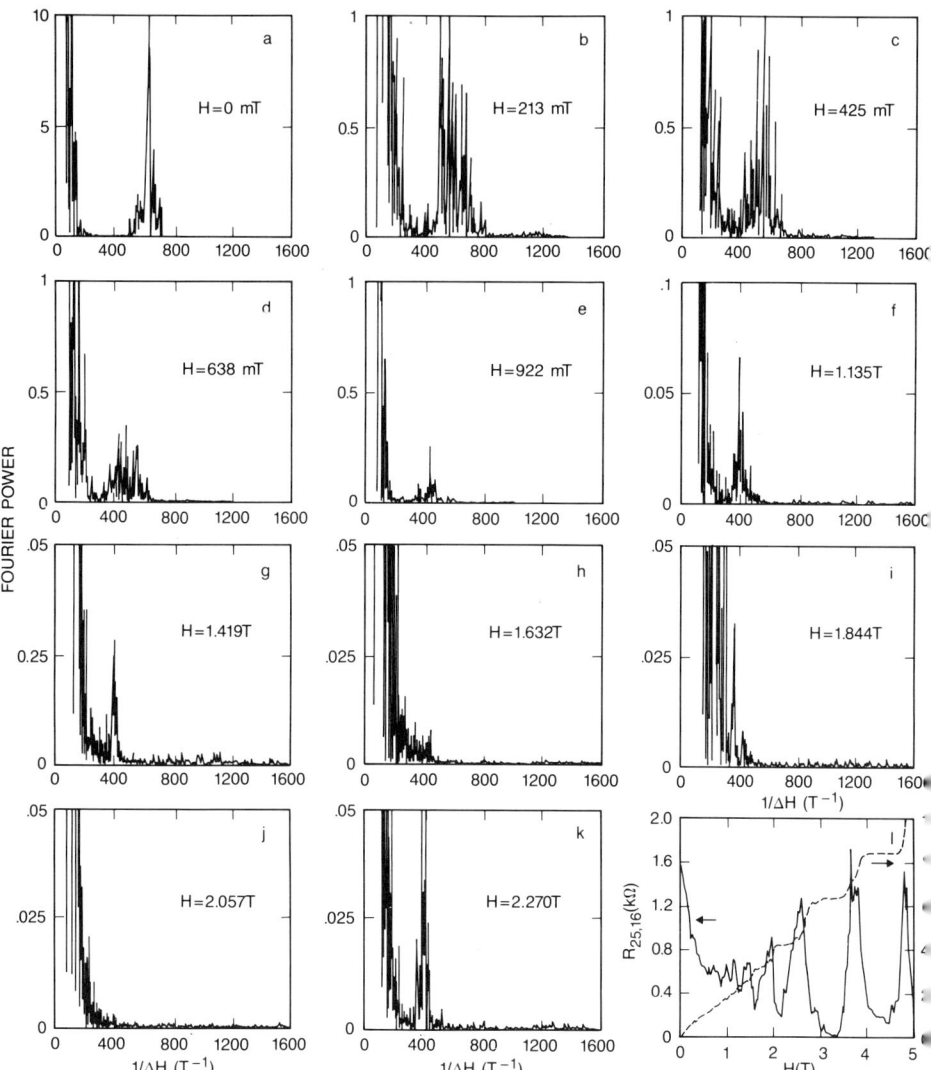

FIG. 27. (a)–(k) depict the Fourier power spectra found in $R_{25,16}$ as a function of magnetic field obtained over a 220 mT field range beginning at the field indicated. The peak in the Fourier spectra corresponds to a single flux quanta hc/e through the area of the annulus. The Fourier power is seen to decrease dramatically and shift to lower frequencies as the magnetic field increases into the quantum Hall regime. (1) shows the magnetoresistances $R_{25,16}$ and $R_{25,46}$ from $H = 0$ to 5 T. Notice that the hc/e peak in the spectra vanishes for $H = 1.632$ T and $H = 2.057$ T corresponding to minima in the resistance $R_{25,16}$ and plateaus in $R_{25,46}$, but reappears beyond the minima.

Figure 27 shows the shift in frequency and the decrease in Fourier power that is typical of Fourier spectra obtained from 12 annuli. As the field increases, the centroid of the power spectrum shifts monotonically from 630 T^{-1} to a lower frequency, eventually reaching approximately 420 T^{-1} near 1 T. The Fourier power in the band between 500–800 T^{-1}, the bandwidth associated with the width of the constrictions comprising the annulus at $H = 0$, is below the noise for $H \geq 1$ T, and no power above the noise ever is observed on the corresponding upper edge of the zero-field bandwidth beyond $H = 800$ mT. The shift in frequency can be appreciated readily by comparing Figs. 27a and f, or by examining Fig. 28. In Fig. 28, the spectrum obtained near $H = 1.20$ T (bottom) is juxtaposed with the spectrum obtained near zero field (top). It is apparent from Fig. 28 that the shift in the frequency of the hc/e peak corresponds to a shift to the lower-frequency edge of the hc/e band obtained near zero magnetic field. According to our interpretation of the spectrum obtained near zero field, the frequency observed at high field corresponds to a flux of hc/e enclosed by the area associated with the inside radius of the annulus.

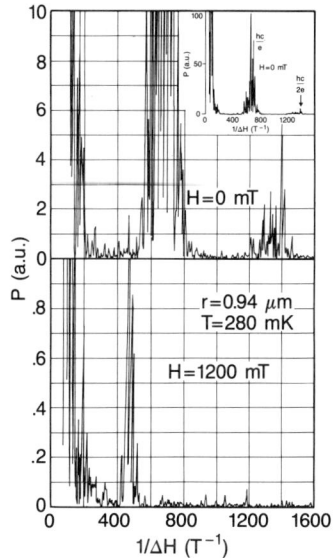

FIG. 28. The Fourier power spectra obtained near $H = 0$ T and near $H = 1.25$ T with a window 220 mT wide are shown for comparison. The zero field spectra is shown on a scale that exaggerates the power near the baseline. In the inset to the top figure, the spectra obtained near zero field are shown entirely. Notice that the centroid of the spectra for each device obtained near $H = 1.25$ T shifts to the inside edge of the spectra found at $H = 0$, indicative of conduction via the inside edge state. The power is reduced appreciably, by about a factor of 200.

On comparing Figs. 27a and b, we find that the Fourier power in the hc/e band of the spectrum obtained for $H \approx 220$ mT is below 10% of the peak value observed in the hc/e band near zero field. Generally, we find that $HeW^2/hc > 1$ for magnetic fields near the 10% attenuation of the AB effect. Beyond this field range, the Fourier power decreases exponentially with magnetic field. For 200 mT $< H <$ 1.6 T, the Fourier power decays exponentially with magnetic field according to law: $P = P_0 \exp(-L_0^2/4r_0^2)$, where P is the Fourier power and L_0 is a characteristic length of approximately 100 nm. The fit to the data of Fig. 27 (indicated by the dashed line in the bottom figure of Fig. 29) yields $L_0 = 95 \pm 7$ nm.

The spectral position of the peak does not depend on magnetic field for 1 T $< H <$ 2.5 T for the device of Fig. 26. This characteristic was observed in three of the four devices examined in detail. As the magnetic field increases, the peak in the spectrum near 420 T^{-1} vanishes near the resistance minima observed in $R_{25,16}$, corresponding to plateaus seen in $R_{25,46}$, but reappears beyond the minima. In Fig. 27l, minima in the longitudinal resistance, $R_{25,46}$, are observed near $H = 1.632$ T and $H = 2.057$ T, and no peak is found in the Fourier power spectra at the same fields as shown by Figs. 27h and j; but peaks are found in the spectra of Figs. 27g, i, and k, corresponding to the maxima found in $R_{25,46}$. The resistance $R_{25,46}$ is well-quantized to the values $h/e^2 i$ only for values of $i = 1, 2, 3, 4$, and 6 in the device of Fig. 26, and the resistance minima in $R_{25,46}$ below 3T are nonzero for the temperature range examined. The peak Fourier power is at most a factor of 200 lower (between the minima) at high field than that observed at zero field. The observation that the Fourier spectra vanish in the resistance minima was made on only two of 12 annuli examined.

The shift in the centroid of the hc/e band in the spectrum and the exponential decrease in the power with magnetic field are summarized in Fig. 29 for three devices. The devices represented in the top and center portions of the figure correspond to annuli with different radii, nominally $r = 0.50$ μm and 1 μm, fabricated in material with $\mu = 30$ m^2/Vs, while the device in the lower portion of the figure is the device of Fig. 26, which nominally has a 1 μm radius with $\mu = 50$ m^2/Vs. Generally, the centroid of the Fourier spectrum, represented by the solid triangles in Fig. 29, is shifted monotonically to a lower frequency as the magnetic field increases. The frequency observed at high field corresponds to the edge of the bandwidth (indicated by the error bar in Fig. 29) associated with the hc/e peak observed near $H = 0$ T, and a flux of hc/e through the area associated with the inside edge of the annulus.

We interpret these results to be indicative of the absence of backscattering in the quantized Hall regime.[75,78] For fields beyond $HeW^2/hc \approx 1$, the AB effect is suppressed because the current carried along the inner edge of the

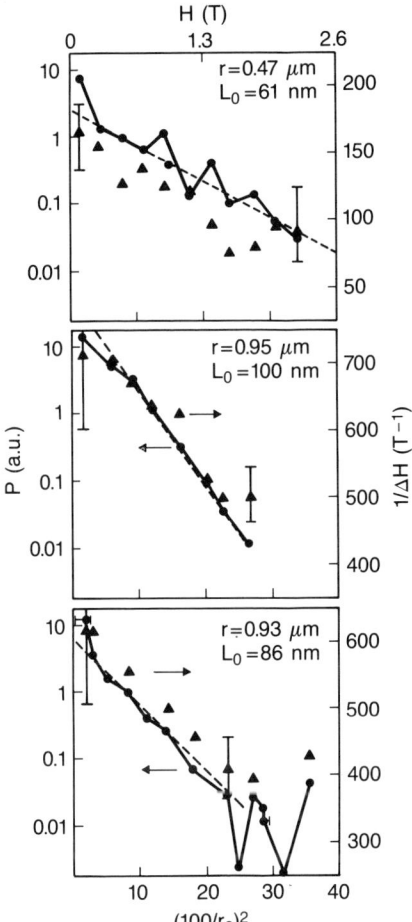

FIG. 29. A comparison of the results obtained for the peak Fourier power P associated with the hc/e peak (circle), and the centroid of the hc/e band (triangle) as a function of magnetic field and the reciprocal magnetic length squared for three different annuli. The exponential fit to the data $P = P_o \exp[-(L_o/r_o)^2]$ used to determine the parameter L_o is indicated by the dashed line. The changes in the slope of the Fourier power as a function of the r_o^{-2} are indicative of different width wires. The width of the Fourier spectra are indicated by the error bar on the centroid data.

annulus (as in Fig. 25b) backscatters only weakly to the outlying edge involved in the voltage measurement. The peak amplitude depends upon the backscattered intensity, and we presume that the backscattering depends upon the overlap between edge states. The exponential decay of the Fourier power with $r_o^{-2} \propto H$ between 200 mT and 1.6 T indicates the decrease in overlap with increasing magnetic field. The expected form for the overlap in a high magnetic field is proportional to $\exp(-\gamma W^2/4r_0^2)$,[112] where the parameter γ is a constant of order unity and depends upon the shape of the confining potential. The expected form with the exponential decay of the Aharanov–Bohm effect that we observe provided $\gamma \approx 0.25$, since $L_0 \approx W/2$.

We associate the peak in the Fourier spectrum, found at high field on the lower edge of the hc/e frequency band obtained near zero field, with the area circumscribed by the inner edge states. Realistically, the outer edge states are coupled to areas enclosed by localized states throughout the bulk of the constriction as well as states about the inner edge, but the latter states are more prominent in the Fourier spectra, presumably because that area, defined by the lithography, does not change appreciably with magnetic field. In the field range beyond the resistance minima and between the quantized Hall plateaus, we assume that the edge states along the inner edge of the annulus cannot tunnel directly to the outer edge, but rather scatter via potential fluctuations within the constriction. The radius corresponding to the 420 T^{-1} frequency is $r \approx 0.75$ μm. It is greater than the inside lithographic radius by about 70 nm and corresponds to $W \approx 360$ μm, assuming that half of the constriction width is given by the difference between the radii associated with the centroid of the hc/e peak in the Fourier spectrum for $H = 0$ mT (630 T^{-1}) and 1.2 T (420 T^{-1}), respectively. The estimate of W, deduced from the width of the hc/e fundamental at zero field, is not unequivocal and represents only a lower bound on the width of the constriction, especially in the limit where only a few 1D subbands carry the current. The width deduced from the zero-field Fourier spectrum systematically is smaller than that deduced from deviations of the SdH oscillations from $1/H$ periodicity.[25]

We suppose that the magnetic field dependence of the scattering from an annulus is indicative of the dependence of the scattering from any geometry. The implication is that scattering from a junction or a constriction also will decrease exponentially beyond $HeW^2 \approx 1$ where the overlap between the edge states decreases exponentially. The analysis of the suppression of the scattering due to the leads, like the suppression of the AB effect, in the field range below $HeW^2 \approx 1$ is expected to be complicated because the overlap is not a simple exponential function of magnetic field even in a ballistic constriction.

12. Four-Terminal Resistance Measurements for $HeW^2/hc > 10$—the Fractional Quantized Hall Effect

At low temperature, in an intense magnetic field when only *one* hybrid magneto-electric subband is occupied, the four-terminal Hall resistance of a 1D constriction in a two-dimensional electron gas (2DEG) also can be quantized at $(h/e^2)/i$, where i denotes certain simple fractions.[14,113,114] Beyond $H = 5$ T in Fig. 24, there are features in both $R_{14,32}$ and $R_{12,43}$ that we associate with the fractional quantized Hall effect (FQHE) in a 1D constriction. In Fig. 24b, there is a feature in $R_{12,43}$ centered near $(h/e^2)/(0.53)$ at $H = 7$ T and, corresponding to the plateaus, there is a minima in $R_{14,32}$. There also is a minima in $R_{14,32}$ near 5.7 T in Fig. 24b, but there is only an inflection at $R_{12,43} = (h/e^2)/(0.66)$. Figure 24c does not show features that we can associate unambiguously with the FQHE; however, we do observe minima in $R_{14,32}$ near $H = 5.25$ T, 6.2 T, and 7.2 T. Figure 30b shows the magneto-

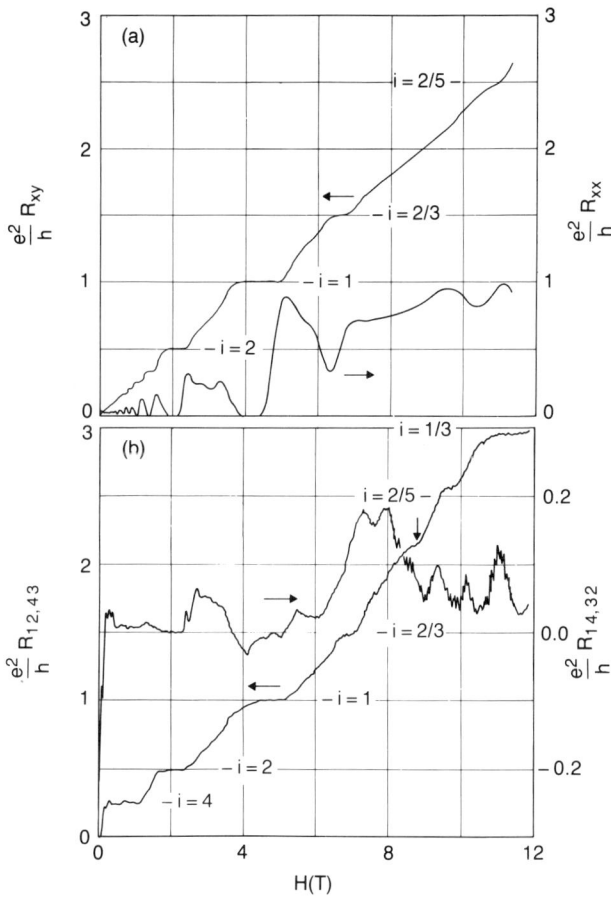

FIG. 30. (a) shows the magnetoresistances R_{xx} and R_{xy} of a high-mobility 2DEG, while (b) shows the corresponding magnetoresistances $R_{14,32}$ and $R_{12,43}$ found in a junction comprised of 1D constrictions with comparable carrier density. The arrow in (b) near $H = 8.7$ T indicates the position of an anomalous plateau in the Hall resistance near $(h/e^2)/(1/2)$. (From Ref. 14.)

resistance observed at 280 mK in the device of Fig. 14b. In Fig. 30b, the constrictions comprising the junction have $N = 2 \pm 1$ subbands occupied and are about $W \approx 25 \pm 10$ nm wide. In addition to the integer quantization observed below 5 T, we again find plateaus in $R_{12,43}$, at fractional occupation of the lowest subband. In Fig. 30b, there are plateaus centered near $(h/e^2)/0.340$, $(h/e^2)/0.395$, $(h/e^2)/0.47$, $(h/e^2)/0.685$, and $(h/e^2)/1.633$, at $H = 11.5$ T, 9.8 T, 8.7 T, 6.9 T, and 2.88 T, respectively, and, corresponding to the plateaus, there are minima in $R_{14,32}$. We associate these features with subband filling fractions, $i = 1/3$, $2/5$, $1/2$ or $6/13$, $2/3$, and $5/3$, respectively.

Figure 31 shows the dependence of the quantization of $R_{12,43}$ on the gate voltage applied to each of the split-gate electrodes comprising the device of Fig. 30b. At $V_g = 0$ V, the wires comprising the junction are 100 μm wide and $R_{12,43} = (h/e^2)/i$ precisely, where $i = 1$, $5/3$, and 2. When $V_g = -0.68$ V, $W \approx 60$ nm and $N = 4 \pm 1$, and a plateau is evident at $R_{12,43} = (h/e^2)/(0.667)$ near $H = 9.6$ T corresponding to $i = 2/3$. At $V_g = -0.87$ V, where $N = 3 \pm 1$ and $W \approx 50$ nm, a feature develops near $R_{12,43} = (h/e^2)/(0.52)$ at $H = 9.3$ T.

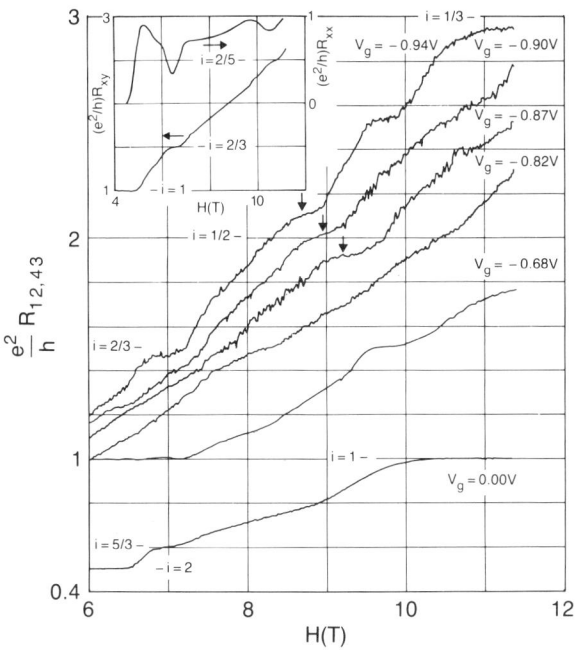

FIG. 31. The resistance $R_{12,43}$ versus magnetic field for $H > 6$ T found in the junction of Fig. 16 as a function of the gate voltage. The inset to the figure shows the magnetoresistance found in a 2D Hall bar with an areal density and mobility comparable to that found at $V_g \approx -0.9$ V. (From Ref. 14.)

The arrows in Fig. 31 follow this feature through $R_{12,43} = (h/e^2)/(0.50)$ as a function of the gate voltage. At $V_g = -0.90$ V, the feature appears centered near $R_{12,43} = (h/e^2)/(0.50)$, while for $V_g = -0.94$ V, with $N = 2 \pm 1$ and $W \approx 30$ nm, the feature occurs near $(h/e^2)/(0.47)$. The definition of the plateaus found about $(h/e^2)/(1/2)$ and the plateaus near $(h/e^2)/(0.66)$ and $(h/e^2)/(0.40)$ depends on V_g generally.

Figure 30a shows the Hall effect, R_{xy}, and the longitudinal resistance, R_{xx}, measured at 280 mK in a 2DEG configured in a Hall geometry comprised of 100 μm-wide wires in the same magnetic field range. The 2DEG is characterized by an areal density of $n = 1.0 \times 10^{15}$ m^{-2} and a mobility of $\mu = 87$ m^2/Vs at 280 mK, similar to the density found in Fig. 30b. In Fig. 30a, we observe that the Hall resistance is quantized at $R_{xy} = h/e^2 i$ where $i = 2/5$ and $2/3$, and corresponding to the plateaus in R_{xy}, there are minima in R_{xx} at the same magnetic fields. While R_{xx} in a high-mobility 2DEG routinely exhibits a broad minimum for magnetic fields near $i = 1/2$, no plateau has ever been observed in R_{xy}. The minimum found in R_{xx} in a 2DEG usually is attributed to a sequence of unresolved, odd-denominator fractions. Because of the poor accuracy of the quantization of the FQHE in a narrow constriction, we cannot unambiguously distinguish between $1/2$ and odd-denominator fractions such as $5/9$. However, the identification of the features found near $(h/e^2)/(1/2)$ with odd-denominator fractions is unlikely because the size of the gap for $i = 5/9, 4/9$, etc. is supposed to be smaller than 280 mK.

The fractional quantized Hall effect is an enigma. So far, exclusively *noninteracting* quantum mechanical models have been used to explain the experimental observations made on 1D constrictions in a 2DEG. These models are inadequate as an explanation for the fractional quantized Hall effect, however. The fractional quantized Hall effect (FQHE) in a 2DEG is attributed to the Coulomb interaction between electrons.[115-117] The many-electron 2D system in an intense magnetic field is described by a Hamiltonian of the form,

$$\sum_j \left[\frac{(-i\hbar \nabla_j - e\mathbf{A}_j/c)^2}{2m^*} + V_j \right] \Psi_m + \sum_{j<k} \frac{e^2}{z_j - z_k} \Psi_m, \qquad (14)$$

where $z_j = x_j - iy_j$ denotes the location of the jth electron in a complex x–y plane, and $\mathbf{A}$ is the vector potential associated with the applied field. Laughlin demonstrated that the best wave function for this multi-electron system, in a variational sense, is of the Jastrow-type,

$$\Psi_m = \prod_{j<k}^{N_e} (z_j - z_k)^m \prod_{j=1}^{N_e} \exp\left[\frac{-|z_j|^2}{4r_0^2} \right], \qquad (15)$$

where N_e denotes the number of electrons in the system, and a symmetric gauge is assumed, i.e., $\mathbf{A} = \frac{1}{2}H(\mathbf{yx} - \mathbf{xy})$. This wave function is comprised entirely of states equivalent to those deduced from Eq. (10) for the lowest Landau level. The ground-state wave function of the 2DEG in the fractional quantum Hall regime is not crystalline.[118-120] It is an incompressible quantum fluid because the positions z_i are not fixed.

If the ground-state wave function Ψ_m is required to obey Fermi statistics and be *antisymmetric* under particle exchange, then m must be an odd integer. For $m > 1$, the wave function vanishes as a high power of the two-electron separation; that is, there are m-fold degenerate nodes in the many-particle wave function. Whenever an electron is added to this system of N_e electrons, it resides at one of these nodes to minimize repulsion and satisfy the Pauli exclusion principle. Since each node can admit only one carrier, there is a maximum number of electrons that can be accommodated. If the wave function is expanded in powers of z_1, the highest power is $m(N_e - 1)$, which must be equal to the degeneracy of the Landau level, $L_x L_y/2\pi r_0^2$ minus one. For large N_e, we have

$$m \approx \frac{L_x L_y}{2\pi r_0^2 N_e} = \left(\frac{eH}{hc}\right)\frac{1}{n} = \frac{1}{i}, \qquad (16)$$

where i measures the number of flux quanta, hc/e, that thread the area associated with a node in the wave function (or an electron in lowest Landau level.)

The association between the nodes of the wave function and the flux quanta is the key element of our understanding of the FQHE in a 2DEG. When the number of carriers is less than $i = 1$, there is an excess number of flux quanta and nodes relative to the number of electrons added. When $i = 1/m$, these additional nodes and the associated flux quanta are juxtaposed onto the positions of the existing electrons to minimize the total energy of the multi-particle system. The m-fold vanishing of the wave function minimizes the repulsion between electrons and provides for exceptional stability when $i = 1/m$ and m is an odd integer. Laughlin demonstrated that a wave function of the type in Eq. (15) gives rise to downward cusps as a function of the fraction $i = 1/m$ in the total energy of the interacting system for odd integers m. The existence of a cusp in the energy indicates a discontinuity in the chemical potential there, which further implies that the filling fraction i is pinned for some range of chemical potential. (This situation is reminiscent of the integer quantized Hall effect, where the chemical potential is pinned in the gap between Landau levels.)

For finite temperatures or for a filling fraction i that is not a simple odd-denominator rational fraction, defects develop in the ground-state wave function that carry *fractional charge*. For example, if one electron is removed from a multi-electron system where $i = 1/m$, the ground state compensates for the deficit by releasing m flux quanta and the associated m-fold degenerate node in the many-electron wave function previously associated with that electron. A deficit of one electronic charge then amounts to exactly m flux quanta, each attached to a nondegenerate node with an effective charge $e^* = +e/m$.

The fractional quantized Hall effect is supposed to be a consequence of these mobile, fractionally charged defects, and of the existence of incompressible states of an electron gas. While there is not yet a theory of transport in the fractional quantized Hall regime, three different models have been proposed recently.[121-123] A naive extension of Büttiker's development of the Landauer formula, which exploits the phenomenological similarity to the integer quantized Hall effect, is possible if edge states associated with the defects or quasi-particles of fractional charge carry the current. In such a case, the charge e in Eq. (5) should be replaced by the quasi-particle charge e^*, i.e.,

$$I_j = -\left(\frac{2e^*}{h}\right) \sum_{k=1}^{N} T_{jk}\mu_k. \quad (17)$$

The chemical potential of the contact still is related to the measured voltage through $V_j = \mu_j/e$, however, because the electrons in the contact have charge e. Consequently, in the fractional quantized Hall regime, $R_{12,43} = h/ee^*i$, where $i = 1$ and the quantization of the resistance is a measure of the number and charge of the quasi-particles. Notice that, in contrast with the integer quantized Hall effect where a one-to-one correspondence exists between the number of edge states and the number of 1D subbands, in the FQHE there is, in principle, an infinite hierarchy of edge states, corresponding with the infinite hierarchy of incompressible states.[124]

Alternatively, Beenakker[121] has developed a model for transport in the fractional quantized Hall regime that exploits a more generalized edge state concept. According to Beenakker, the net current is carried at the edges similar to the integer quantized Hall effect, but by compressible states juxtaposed with the incompressible states as shown schematically in Fig. 32. According to Beenakker, the incompressible states do not contribute to the current because the current is proportional to the slope of the confining potential along the transverse direction, and this slope is proportional to $\partial i/\partial y$, which vanishes when $i = 1/m$. Instead, the contact injects current into the compressible bands between incompressible bands. The current injected into edge state j is $I_j = (e/h)\Delta\mu\,\Delta i_j$, so the total current I through a wire is $I = \Sigma_j I_j T_j$. $\Delta i_j = i_j - i_{j-1}$ measures the difference in the occupation number between two adjacent incompressible bands. Following Beenakker, Eq. (5) still applies, but

$$T_{jk} = \sum_\gamma T_{\gamma,jk}\Delta i_\gamma, \quad (18)$$

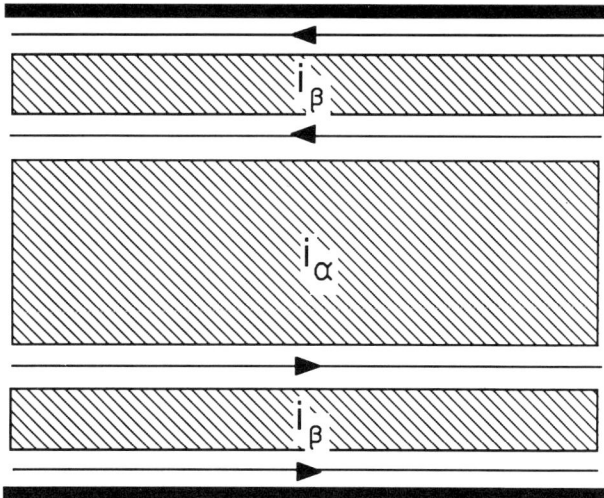

FIG. 32. A schematic representation of a two-terminal conductor in the fractional quantized Hall regime indicating the incomparable bands (hatched) of fractional filling factors i_α and i_β, alternating with the edge channels. (Arrows indicate the direction of current.)

where $T_{\gamma,jk}$ represents the fraction of the current injected from contact j into the generalized edge state γ to reach contact k. Equations (18) and (5) together then represent an alternative generalization of the Landauer formula for the conductance applicable to the fractional quantized Hall regime. It differs from Eq. (17) in the presence of the weights Δi_j and the absence of the quasi-particle charge e^*. According to Beenakker's interpretation, a resistance measurement in the fractional quantized Hall regime is not a measure of the quasi-particle charge e^* at all, but simply a measure of the occupation of the incompressible states. A multi-terminal generalization of this formula, applicable to the fractional quantized Hall regime, also was developed by Beenakker.[121] It is identical in form to Eq. (9), with the T_{jk} suitably modified as in Eq. (18). Thus, the resistance is quantized, $R_{12,43} = h/e^2 i$, where i is a rational fraction.

Recent experiments have been interpreted in support of these models;[15,125,126] but despite the phenomenological similarity to the integer quantized Hall effect, it is not experimentally obvious yet that edge states associated with the quasi-particles exist within the same Landau level, and that this extension of Büttiker's resistance formula is justified.

We suppose that the FQHE found in a 1D constriction in a 2DEG also is due to the Coulomb interaction between electrons. When the *attachment* of the electrons or the magnetic flux in the interacting system to the nodes of the many-electron wave function, Eq. (15), is disrupted, the quantization may deteriorate, however. The attachment can be disrupted if there is mixing between the energy levels with the same internal angular momentum, for example. This mixing can occur at finite magnetic fields in a 2DEG or in a narrow device. The observation of the FQHE in a constriction less than 40 nm wide, where 7.5 nm $< r_0 <$ 10.5 nm, is novel and it establishes an upper bound for the size of the current-carrying state at $i = 2/3$ for $H = 6.8$ T and at $i = 1/3$ for $H = 11$ T. In particular, if the FQHE is a measure of the quasi-particle charge, then the size of the quasi-particle must be less than 40 nm. We have not observed the FQHE for constrictions narrower than 37 ± 10 nm, but we cannot set a lower bound because we have observed minima in $R_{14,32}$ versus H in a junction in which $N = 1 \pm 1$ and $W \approx 20$ nm for fractional filling factors.

It is especially provocative that the four-terminal Hall resistance is quantized for $(h/e^2)/0.55 < R_{xy} < (h/e^2)/0.45$ in a narrow constriction. This is surprising because the Hall resistance of a macroscopic 2DEG usually is featureless near $(h/e^2)/(1/2)$.[27]

So far, only odd-denominator fractions i have been found in a 2DEG with only one edge state occupied. Following Laughlin, Halperin suggested that quantization at $(h/e^2)/i$ with $i = 1/2$ also might be observed, provided the Zeeman energy did not completely polarize the electronic spin.[127] Subsequently, Halperin and then Haldane and Rezayi[128] have proposed wave functions similar to Laughlin's, but without complete spin-polarization, that produce the FQHE when i is an *even-denominator* fraction. Experimental evidence,[129,130] associated with the first excited state of the 2DEG, has demonstrated the feasibility of an unpolarized ground state, and the FQHE at $(h/e^2)/(5/2)$ even has been observed.[131] However, fractional quantization of the Hall resistance at $(h/e^2)/(1/2)$ has not been observed yet in a 2DEG.[27]

We cannot associate unambiguously the observation of the feature in the Hall resistance of a junction composed of 40 nm constrictions in a 2DEG near $(h/e^2)/(1/2)$ with a spin-polarized state, but we cannot eliminate a spin-polarized state as a possibility either. The g factor in a 2DEG is small ($g \approx 0.5$), and the g factor in a constriction in a 2DEG is expected to be even smaller.[132] The small g factor for electrons within a constriction is evident in the data of Fig. 30b, since we cannot resolve a feature in the Hall resistance near $H = 3$ T at $R_{12,43} = h/e^2 3$, corresponding to the spin-polarized subband, $i = 3$. However, the data of Fig. 30a corresponding to a high-mobility 2DEG similarly does not show a plateau in the Hall resistance—as only an inflection at the filling factor corresponding to $i = 3$ is observed—and the 2DEG of Fig. 30a does not show a feature in the Hall resistance near $(h/e^2)/(1/2)$.

Based on numerical calculations, Chui[133] recently suggested that the Hall resistance of a constriction in a *spinless* 2DEG, for $HeW^2/hc < 100$, could be quantized at $(h/e^2)/(1/2)$. According to Chui, the spatial dependence of the multi-particle exchange interactions and the confining potential associated with a constriction produces a cusp in the energy near $i = 1/2$ and $i = 1/3$ for certain widths. The size of the gap at $i = 1/2$ can be larger even than that at $i = 1/3$, which suggests that plateaus near $(h/e^2)/(1/2)$ and $(h/e^2)/(1/3)$ would be observed at the same temperature. Moreover, the dependence of the plateau found near $(h/e^2)/(1/2)$ on V_g and the poor precision of the quantization may be indicative of the dependence of size of the calculated gap at $i = 1/2$ on the width of the constriction. To establish the correspondence between our experimental results and Chui's calculations, the results of the calculations must be reconciled with the four-terminal resistance measurements we have made. Since the size of the constriction typically is less than 40 nm, comparable to the size of the edge state $r_0^* = \sqrt{hc/e^*H}$, where e^* is the quasi-particle charge, strong scattering may occur between edge states. If the deviations observed from exact quantization in the FQHE are indicative of backscattering, then scattering from the leads for $HeW^2/hc \approx 10$ also may influence the resistance.

V. Summary and Outlook

Recent technological innovations in semiconductor materials engineering and device miniaturization have produced wires comparable in width and thickness to an electron wavelength, and shorter than the elastic and inelastic scattering lengths. The low-temperature transport through these small wires is determined by wave mechanics. The quantization of the resistance is just one manifestation of quantum mechanics in transport. The correspondence between calculations and the experimental observations of the resistance demonstrates that the resistance is due predominantly to scattering from the contacts or leads used for the measurement.

We have shown that the two-terminal conductance is quantized in steps of $2e^2/h$ in the absence of a magnetic field, in correspondence with quantization of the transverse momenta. The transverse momenta are quantized because of the lateral confinement of the 2DEG to a size of about 100 nm. The quantization of the conductance follows from the Landauer formula for the two-terminal resistance of a perfect conductor with contacts that adiabatically taper to the width of the constriction. Our measurements corroborate the discoveries of van Wees et al.[8] and Wharam et al.[9] for submicron lengths, but we find that the quantization depends upon the length of the constriction. Specifically, the resistance is not well quantized for a device about 1 μm long. The quantization is surprisingly resilient with respect to impurity scattering. We have observed quantization of the two-terminal resistance of a constriction in a 2DEG with a mobility of only about 15 m²/Vs.[13] The resilience develops from the restricted phase space for scattering in one dimension and the exponential insensitivity of the conductance to reflections at the contacts.

The precision of the quantization depends upon how it is measured. We have shown the three-terminal resistance measurements of two split-gate electrodes in series also can be quantized for low N, but a voltage terminal can cause the quantization to deteriorate for large N. The resistance of the two series constrictions is not additive,[65] but it is not necessarily quantized either. Depending upon how it is measured, the resistance of a constriction even can vanish. All of these observations are associated with the inability of low-index subbands to propagate around a bend. How an electron wave propagates around a bend has important implications for four-terminal resistance measurements as well. We have shown that the four-terminal resistance depends crucially on the disposition of the leads, and so, for example, the resistance can be negative. While the four-terminal resistance in the absence of a magnetic field is not quantized, there still is a quantum mechanical aspect to the transport. Specifically, the resistance of a bend shows sharp minima in the negative resistance that correspond to the threshold for occupation of another subband. The observation of a negative resistance does not imply that the local potential within the conductor increases in the direction of the current. The dependence of the four-terminal resistance on the leads used to measure it follows from the resistance formula developed by Büttiker.[36]

The four-terminal magnetoresistance of a constriction is quantized at $h/e^2 i$ for integers $i \leq 2N$ when $HeW^2/hc > 1$, but not precisely. Our study of the oscillations in the magnetoresistance of an annulus due to the Aharonov–Bohm effect revealed that the net current distribution shifts to the edges in this field regime. For $HeW^2/hc < 1$, the frequency of oscillation is associated with the area defined by the mean radius of the annulus because the current is carried in the bulk of the wire. However, for $HeW^2 > 1$, the frequency of oscillation corresponds to a flux of hc/e through the area defined by the inside radius of the annulus. The exponential decrease in the intensity of the oscillations found when $HeW^2/hc > 1$ is indicative of the exponential decrease in the overlap with increasing field between edge states outside and inside the annulus. Thus, our observations are interpreted to show that the net current for $HeW^2/hc > 1$ is carried by edge states, and that the suppression of the Aharonov–Bohm effect is due to the absence of backscattering between these states. We then associate the resilience of the quantization of the four-terminal Hall resistance of a ballistic constriction with the exponentially small overlap between the edge states and the absence of backscattering, which occurs even when the width of the constriction is comparable to the size of the edge state. The integer quantized Hall effect is amazingly resilient, since the quantization is observed even when only one 1D subband is occupied and the constriction is only about 20 nm wide.

Finally, we examined the fractional quantization of the four-terminal Hall resistance in the extreme magnetic quantum limit, which is due to the Coulomb interaction between electrons. We found evidence of the fractional quantum Hall effect in wires as narrow as 40 nm, and have observed quantization for $(h/e^2)/0.55 < R_{xy} < (h/e^2)/0.45$. The first result illustrates the importance of electron–electron interactions for transport in a narrow constriction and demonstrates that the scale of the correlated electronic state can be 40 nm or less when the magnetic length is 10 nm. Thus, our calculations of the resistance, which so far have focused naively on a noninteracting aspect, are incomplete and, while the correspondence with the available measurements of the resistance is striking, it may be coincidental. The observation of a plateau in the Hall resistance for $(h/e^2)/0.55 < R_{xy} < (h/e^2)/0.45$, also is novel and surprising. It is surprising because the Hall resistance of a 2D electron gas is featureless near $(h/e^2)/(1/2)$.

In this chapter, we have demonstrated unequivocally the quantum mechanical nature of the low-temperature transport in constrictions in a high-mobility 2DEG, which resemble electron waveguides. While the analogy to a waveguide is realistic, one important difference between an electron waveguide and its electromagnetic counterpart is that the electronic wave function can be manipulated easily with an external voltage[134–137] Harbingers of a new technology, which exploit this difference, have appeared recently. Specifically, Sols et al.[135] have suggested that transistor action is possible by manipulating the electronic wave function in a constriction using a gate electrode over one branch of a junction between three electron waveguides. The results of their calculation suggest that further development of semiconductor technology will make such a device practical. According to Sols, a "T" made from wires 10 nm wide and 100 nm long, might show transistor action at room temperature.

Transistor action from a single isolated junction between three electron waveguides may not be pertinent to ULSI (ultra-large scale integration), however. Currently, the system function associated with an integrated circuit is determined solely by metallized interconnections; but, as the features of an integrated circuit become smaller, device dynamics cannot be treated in isolation.[138] Coupling between devices in a high-density integrated circuit can occur via parasitic capacitances like it does in conventional devices or via wave function coherence and tunneling. These constraints are so severe in devices on the scale of the electron wavelength that they have to be included in the system theory of the architecture of a very dense integrated circuit.

Thus, a new convention for system design may be necessary. So far, scaling laws[139] have been used as a guideline to design small conventional integrated circuits, but the applicability of these laws is restricted. For example, as the scale of a transistor becomes smaller, it is desirable to reduce the supply voltage by the same factor so that all electric fields in the circuit remain constant. However, the voltage must be many times $k_B T/e$ to achieve nonlinearity in a conventional transistor. An electron waveguide can provide nonlinear operation whenever the voltage is smaller than the quotient of the intersubband energy and the electronic charge. (See Fig. 14, for example.) If the device is small enough, temperature may be irrelevant.

Acknowledgments

I gratefully acknowledge the scientific contributions of my collaborators in this work: H. Baranger, P. de Vegvar, R. Behringer, E. Westerwick, S. Sampere, R. Howard, P. Mankiewich, J. Cunningham, A. Chang, and J. Jain; I particularly would like to thank H. U. Baranger, R. E. Howard, and R. E. Behringer for critical readings of this manuscript. I also wish to acknowledge correspondence with M. Büttiker, J. Davies, L. Glazman, D. Khmel'nitskii, R. Landauer, Y. Takagaki, and S. Washburn.

References

1. C. W. Beenakker and H. van Houten, in *Electronic Properties of Multilayers and Low-Dimensional Semiconductor Structures* (J. M. Chamberlain, L. Eaves, and J. C. Portal eds.), *NATO Advance Study Institute Series*, Plenum, London (1990).
2. C. W. Beenakker and H. van Houten, *Phys. Rev. Lett.* **63**, 1857 (1989).
3. R. Landauer, in *Localization, Interaction, and Transport Phenomena* (G. Bergmann and Y. Bruynseraede, eds.), Springer-Verlag, New York, pp. 38–50 (1985).
4. Sean Washburn and Richard A. Webb, *Advan. Phys.* **35**, 375 (1986).
5. W. P. Kirk and M. Reed, eds., *Proc. Int. Symp. on Nanostructure Physics and Fabrication*, Academic Press, New York (1989).
6. W. Hansen, M. Horst, J. P. Kotthaus, U. Merkt, Ch. Sikorksi, and K. Ploog, *Phys. Rev. Lett.* **58**, 2586 (1987).
7. T. P. Smith, III, H. Arnot, J. M. Hong, C. M. Knoedler, S. E. Laux, and H. Schmid, *Phys. Rev. Lett.* 59, 2802 (1987).
8. B. J. van Wees, H. van Houten, C. W. J. Beenakker, J. G. Williamson, L. P. Kouwenhoven, D. van der Marel, and C. T. Foxon, *Phys. Rev. Lett.* **60**, 848 (1988).
9. D. A. Wharam, T. J. Thornton, R. Newbury, M. Pepper, H. Ahmed, J. E. F. Frost, D. G. Hasko, D. C. Peacock, D. A. Ritchie, and G. A. C. Jones, *J. Phys. C* **21**, L209 (1988).
10. G. Timp, R. E. Behringer, S. Sampere, J. E. Cunningham, and R. Howard, in *Proc. Int. Symp. on Nanostructure Physics and Fabrication* (W. P. Kirk and M. Reed, eds.), Academic Press, New York, p. 331 (1989).
11. M. L. Roukes, A. Scherer, S. J. Allen, Jr., H. G. Craighead, R. M. Ruthen, E. D. Beebe, and J. P. Harbison, *Phys. Rev. Lett.* **59**, 3011 (1987).
12. G. Timp, A. M. Chang, P. Mankiewich, R. Behringer, J. E. Cunningham, T. Y. Chang, and R. E. Howard, *Phys. Rev. Lett.* **59**, 732 (1987).

13. R. Behringer, G. Timp, H. U. Baranger, and J. E. Cunningham, *Phys. Rev. Lett.* **66**, 93 (1991).
14. G. Timp, R. E. Behringer, J. E. Cunningham, and R. E. Howard, *Phys. Rev. Lett.* **63**, 2268 (1989).
15. L. P. Kouwenhoven, B. J. van Wees, N. C. van der Vaart, C. J. P. M. Harmans, C. E. Timmering, and C. T. Foxon, *Phys. Rev. Lett.* **64**, 685 (1990).
16. A. M. Chang and J. E. Cunningham, *Solid State Commun.* **72**, 651 (1989).
17. R. G. Clark, J. R. Mallet, S. R. Haynes, J. J. Harris, and C. T. Foxon, *Phys. Rev. Lett.* **60**, 1747 (1988).
18. J. A. Simmons, H. P. Wei, L. W. Engel, D. C. Tsui, and M. Shayegan, *Phys. Rev. Lett.* **63**, 1731 (1989).
19. Y. Aharonov and D. Bohm, *Phys. Rev.* **115**, 485 (1959).
20. G. Timp, H. U. Baranger, P. de Vegvar, J. E. Cunningham, R. E. Howard, R. Behringer, and P. M. Mankiewich, *Phys. Rev. Lett.* **60**, 2081 (1988).
21. T. Kakuta, Y. Takagaki, K. Gamo, S. Namba, S. Takaoka, and K. Murase, *Phys. Rev B* **43**, 4321 (1991); Y. Takagaki, S. Takaoki, K. Gamo, K. Murase, S. Namba, K. Ishibashi, S. Ishida, and Y. Aoyagi, *Solid State Commun.* **69**, 811 (1989).
22. G. Timp, P. M. Mankiewich, P. de Vegvar, R. Behringer, J. E. Cunningham, R. E. Howard, H. U. Baranger, and J. K. Jain, *Phys. Rev. B* **33**, 6227 (1989).
23. G. Timp, A. M. Chang, J. E. Cunningham, T. Y. Chang, P. Mankiewich, R. Behringer, and R. E. Howard, *Phys. Rev. Lett.* **58**, 2814 (1987).
24. S. Datta, M. R. Mellock, S. Bandyopadhyay, R. Noren, M. Vazirir, M. Miller, and R. Reifenberger, *Phys. Rev. Lett.* **55**, 2344 (1985).
25. C. J. B. Ford, T. J. Thornton, R. Newbury, M. Pepper, H. Ahmed, C. T. Foxon, J. J. Harris, and C. Roberts, *J. Phys. C* **21**, L325 (1988).
26. C. J. B. Ford, T. J. Thornton, R. Newbury, M. Pepper, H. Ahmed, D. C. Peacock, D. A. Ritchie, J. E. F. Frost, and G. A. C. Jones, *Appl. Phys. Lett.* **54**, 21 (1989).
27. H. W. Jiang, H. L. Stormer, D. C. Tsui, L. N. Pfeiffer, and K. W. West, *Phys. Rev. B* **40**, 12013 (1989).
28. For a review, see *IBM J. Res. Develop.* **32**, 440–514 (1988).
29. Y. Hirayama, T. Saku, and Y. Horikoshi, *Phys. Rev. B* **39**, 5535 (1989); A. D. Wieck and K. Ploog, *Surf. Science* **229**, 252 (1990).
30. M. Pepper, *Microfabrication* **4**, 5 (1981).
31. C. C. Dean and Pepper, in *Localization, Interaction, and Transport Phenomena* (G. Bergmann and Y. Bruynseraede, eds.), Springer-Verlag, New York, p. 169 (1985).
32. K. F. Berggren, T. J. Thornton, D. J. Thornton, D. J. Newson, and M. Pepper, *Phys. Rev. Lett.* **57**, 1769 (1986).
33. This solution is due to J. H. Davies, private communication.
34. The self-consistent problem has been solved for a similar geometry to this by A. Kumar, S. E. Laux, and F. Stern, *Appl. Phys. Lett.* **54**, 1270 (1989); and for a geometry where $l \to \infty$, see S. E. Laux, D. J. Frank, and F. Stern, *Surf. Science* **196**, 101 (1988).
35. J. H. Davies, private communication.
36. M. Büttiker, *Phys. Rev. Lett.* **57**, 1761 (1986); *IBM J. Res. Develop.* **32**, 317 (1988).
37. A four-terminal resistance measurement $R_{12,34}$, where the leads and gate electrodes are disposed as in Fig. 8, and where only two diametrically opposite gate electrodes are used to form the constriction, has no series resistance at $V_g = 0$. Even though the series resistance was zero, the precision of the quantization did not improve.
38. Yu. V. Sharvin, *Zh. Eksp. Teor. Fiz.* **48**, 984 (1965) [*Sov. Phys. JETP* **21**, 655 (1965)].
39. Yu. V. Sharvin and N. I. Bogatina, *Zh. Eksp. Teor. Fiz.* **56**, 772 (1969) [*Sov. Phys. JETP* **29**, 419 (1969)].
40. V. S. Tsoi, *Pis'ma Zh. Eksp. Teor. Fiz.* **19**, 114 (1974) [*JETP Lett.* **19**, 701 (1974)].
41. R. Landauer, *J. Phys. Cond. Mat.* **1**, 8099 (1989).

42. Y. Imry, "Physics of Mesoscopic Systems" in *Directions in Condensed Matter Physics* (G. Grinstein and G. Mazenko, eds.), World Scientific Press, Singapore, p. 101 (1986).
43. L. I. Glazman, G. B. Lesovick, D. E. Khmel'nitskii, and R. I. Shekhter, *Pis'ma Zh. Eksp. Teor. Fiz* **48**, 218 (1988) [*JETP Lett.* **48** 238 (1988)].
44. A. Szafer and A. D. Stone, *Phys. Rev. Lett.* **62**, 300 (1989).
45. Y. Imry, in *Proc. Int. Symp. on Nanostructure Physics and Fabrication* (W. P. Kirk and M. Reed, eds.), Academic Press, New York, p. 379 (1989).
46. G. Kirczenow, *Solid State Commun.* **68**, 715 (1988).
47. E. G. Haanappel and D. van der Marel, *Phys. Rev. B* **39**, 5484 (1989); D. van der Marel and E. G. Haanappel, *Phys. Rev. B* **39**, 7811 (1989).
48. S. He and S. Das Sarma, *Solid State Electron.* **32**, 1695 (1989).
49. Y. Avishai and Y. B. Band, *Phys. Rev. B* **62**, 2527 (1989).
50. H. van Houten and C. W. J. Beenakker, in *Proc. Int. Symp. on Nanostructure Physics and Fabrication* (W. P. Kirk and M. Reed, eds.), Academic Press, New York, p. 347 (1989).
51. B. J. van Wees, L. P. Kouwenhoven, E. M. M. Willems, C. J. P. M. Harmans, J. E. Mooij, H. van Houten, C. W. J. Beenakker, J. G. Williamson, and C. T. Foxon, *Phys. Rev B* **43**, 2431 (1991).
52. P. L. McEuen, B. W. Alphenaar, R. G. Wheeler, and R. N. Sacks, *Surface Science* **229**, 312 (1990).
53. C. S. Chu and R. S. Sorbello, *Phys. Rev. B* **40**, 5941 (1989).
54. H. van Houten, C. W. J. Beenakker, J. G. Williamson, M. E. I. Broekaart, P. H. M. van Loosdrecht, B. J. van Wees, J. E. Mooij, C. T. Foxon, and J. J. Harris, *Phys. Rev. B* **39**, 8556 (1989); J. Spector, H. L. Stormer, K. W. Baldwin, L. N. Pfeiffer, and K. W. West, *Appl. Phys. Lett.* **56**, 1290 (1990).
55. J. H. Davies and J. A. Nixon, *Phys. Rev. B* **39**, 3423 (1988).
56. J. P. Harrang, R. J. Higgins, R. K. Goodall, P. R. Jay, M. Laviron, and P. Delescluse, *Phys. Rev. B* **32**, 8126 (1985).
57. F. F. Fang, T. P. Smith, and S. L. Wright, *Surf. Sci.* **196**, 310 (1988).
58. S. Das Sarma and F. Stern, *Phys. Rev. B* **32**, 8442 (1985).
59. H. Sakaki, *Jpn. J. Appl. Phys.* **19**, 1735 (1980).
60. D. E. Khmel'nitskii, private communication.
61. C. T. Rogers and R. A. Buhrman, *Phys. Rev. Lett.* **55**, 859 (1985).
62. W. J. Skocpol, L. D. Jackel, R. E. Howard, and C. A. Fetter, *Phys. Rev. Lett.* **49**, 951 (1982).
63. H. van Houten, B. J. van Wees, M. G. J. Heijman and J. P. Andre, *Appl. Phys. Lett.* **49**, 1781 (1986); P. M. Mankiewich, R. E. Behringer, R. E. Howard, A. M. Chang, T. Y. Chang, B. Chelluri, J. E. Cunningham, and G. Timp. *J. Vac. Sci. and Tech. B* **6**, 131 (1988).
64. A. Scherer, M. L. Roukes, H. G. Craighead, R. M. Ruthen, E. D. Beebe, and J. P. Harbison, *Appl. Phys. Lett.* **51**, 2133 (1987).
65. D. A. Wharam, M. Pepper, H. Ahmed, J. E. F. Frost, D. G. Hasko, D. C. Peacock, D. A. Ritchie, and G. A. C. Jones, *J. Phys. C* **21**, L887 (1988).
66. C. W. Beenakker and H. van Houten, *Phys. Rev. B* **39**, 10445 (1989).
67. H. U. Baranger and A. D. Stone, *Phys. Rev. Lett.* **63**, 414 (1989).
68. P. H. Beton, B. R. Snell, P. C. Main, A. Neves, J. R. Owers-Bradley, L. Eaves, M. Henini, O. H. Hughes, S. P. Beaumont, and C. D. W. Wilkinson, *J. Phys. Condens. Mat.* **1**, 7505 (1989).
69. H. U. Baranger, *Phys. Rev. B* **42**, 1479 (1990).
70. Y. Takagaki, K. Gamo, S. Namba, S. Takaoka, and K. Murase, *Solid State Commun.* **75**, 873 (1990).
71. H. U. Baranger and A. D. Stone, *Phys. Rev. B* **40**, 8169 (1989).
72. L. D. Landau and E. M. Lifshitz, eds., *Quantum Mechanics*, vol. 3 of *Course of Theoretical Physics*, Pergamon, New York, p. 457 (1981).

73. B. I. Halperin, *Phys. Rev. B* **25**, 2185 (1980).
74. R. E. Prange and R. Joynt, *Phys. Rev. B* **25**, 2943 (1982).
75. J. K. Jain, *Phys. Rev.* **60**, 2074 (1988).
76. J. K. Jain and S. A. Kivelson, *Phys. Rev. Lett.* **60**, 1542 (1988); *Phys. Rev. B* **37**, 4276 (1988).
77. P. Streda, J. Kucera, and A. M. MacDonald, *Phys. Rev. Lett.* **59**, 1973 (1987).
78. M. Büttiker, in *Proc. Int. Symp. on Nanostructure Physics and Fabrication* (W. P. Kirk and M. Reed, eds.), Academic Press, New York, p. 319 (1989); *Phys. Rev. B* **38**, 9375 (1988).
79. H. van Houten, C. W. J. Beenakker, P. H. M. van Loosdrecht, T. J. Thornton, H. Ahmed, M. Pepper, C. T. Foxon, and J. J. Harris, *Phys. Rev. B* **25**, 2185 (1988).
80. G. Kirczenow, *Phys. Rev. B* **38**, 10958 (1988).
81. Notice that the threshold for the measured conductance changes when both constrictions are biased. The threshold shifts because the electrostatic potential associated with the second split-gate electrode, within 300 nm of the first split-gate electrodes, affects the potential of the first.
82. Y. Avishai and Y. Band, *Phys. Rev. Lett.* **62**, 2527 (1989).
83. H. U. Baranger and A. D. Stone, in *Science and Engineering of One and Zero Dimensional Semiconductors*, (S. P. Beaumont and C. M. Sotomayor-Torres eds.) Plenum, New York, p. 121 (1990).
84. G. Kirczenow, *Solid State Commun.* **71**, 469 (1989).
85. H. G. B. Casimir, *Rev. Mod. Phys.* **17**, 3 (1945).
86. Y. Takagaki, K. Gamo, S. Namba, S. Takaoka, K. Murase, and S. Ishida, *Solid State Commun.* **71**, 809 (1989).
87. G. Timp, in *Mesoscopic Phenomena in Solids* (B. L. Altshuler, P. A. Lee, and R. A. Webb, eds.), Elsevier, Amsterdam (1989).
88. The threshold for each of the four constrictions differed by as much as 50 mV for the device of Fig. 14b when biased independently. We could not evaluate how this difference affects the symmetry of the junction when all four constrictions are biased similarly simultaneously.
89. R. Landauer, in *Nonlinearity in Condensed Matter* (A. R. Bishop, D. K. Campbell, R. Kummaravid, and S. E. Trullinger, eds.), Springer, Heidelberg, p. 2 (1987).
90. B. L. Al'tschuler and D. E. Khmel'nitskii, *Pis'ma Eksp. Teor. Fiz.* **42**, 291 (1985) [*JETP Lett.* **42**, 359 (1985)].
91. A. I. Larkin and D. E. Khmel'nitskii, *Zh. Eksp. Teor. Fiz.* **91**, 1815 (1986) [*Sov. Phys. JETP* **64**, 1075 (1986)].
92. R. A. Webb, S. Washburn, and C. P. Umbach, *Phys. Rev. B* **37**, 8455 (1988).
93. P. G. N. de Vegvar, G. Timp, P. M. Mankiewich, J. E. Cunningham, R. Behringer, and R. E. Howard, *Phys. Rev. B* **38**, 4326 (1988).
94. L. P. Kouwenhoven, B. J. van Wees, C. J. P. M. Harmans J. G. Williamson, H. van Houten, C. W. J. Beenakker, C. T. Foxon, and J. J. Harris, *Phys. Rev. B* **39**, 8040 (1989).
95. F. Peeters, *Phys. Rev. Lett.* **61**, 580 (1988).
96. A. Devenyi and Y. Imry, unpublished.
97. G. Kirczenow, *Phys. Rev. Lett.* **62**, 2993 (1989).
98. D. G. Ravenhall, H. W. Wyld, and R. L. Schult, *Phys. Rev. Lett.* **62**, 1780 (1989).
99. H. Akera and T. Ando, *Phys. Rev. B* **39**, 5508 (1989).
100. H. U. Baranger, private communication.
101. C. J. B. Ford, T. J. Thornton, R. Newbury, M. Pepper, H. Ahmed, D. C. Peacock, D. A. Ritchie, J. E. F. Frost, and G. A. Jones, *Phys. Rev. B* **38**, 8518 (1988).
102. A. M. Chang, T. Y. Chang, and H. U. Baranger, *Phys. Rev. Lett.* **63**, 996 (1989).
103. C. J. B. Ford, S. Washburn, M. Büttiker, C. M. Knoedler, and J. M. Hong, *Phys. Rev. Lett.* **62**, 2724 (1989).

104. L. W. Molenkamp, A. A. M. Staring, C. W. J. Beenakker, R. Eppenga, C. E. Timmering, J. G. Williamson, C. J. P. M. Harmans, and C. T. Foxon, *Phys. Rev. B* **41**, 1274 (1990).
105. Y. Zhu, J. Shi, and S. Feng, *Phys. Rev. B* **41**, 8509 (1990).
106. B. J. van Wees, E. M. M. Willems, C. J. P. M. Harmans, C. W. J. Beenakker, H. van Houten, J. G. Williamson, C. T. Foxon, and J. J. Harris, *Phys. Rev. Lett.* **62**, 1181 (1989).
107. Marvin E. Cage, in *The Quantum Hall Effect* (R. E, Prange and S. M. Girvin, eds.), Springer-Verlag, New York, p. 37 (1987).
108. A. M. Chang, G. Timp, J. E. Cunningham, P. M. Mankiewich, R. E. Behringer, and R. E. Howard, *Solid State Commun.* **76**, 769 (1988).
109. R. E. Prange, in *The Quantum Hall Effect*, (R. E. Prange and S. M. Girvin, eds.), Springer-Verlag, New York, p. 1 (1987).
110. Another prototype scatterer is a simple barrier. See, for example, S. Washburn, A. B. Fowler, H. Schmid, and D. Kern, *Phys. Rev. Lett.* **61**, 2801 (1988); R. J. Haug, A. H. MacDonald, P. Streda, and K. von Klitzing, *Phys. Rev. Lett.* **61**, 2794 (1988).
111. G. Timp, A. M. Chang, P. de Vegvar, R. E. Howard, R. Behringer, J. E. Cunningham, and P. Mankiewich, *Surf. Science* **196**, 68 (1988).
112. B. I. Shklovskii and A. L. Efros, *Electronic Properties of Doped Semiconductors*, Springer, Berlin (1984).
113. T. P. Smith, K. Y. Lee, C. M. Knoedler, J. M. Hong, and D. P. Kern, *Phys. Rev. B* **38**, 1558 (1988).
114. J. A. Simmons, H. P. Wei, L. W. Engel, D. C. Tsui, and M. Shayegan, *Phys. Rev. Lett.* **63**, 1731 (1989).
115. T. Chakraborty and P. Pietilainen, eds., *The Fractional Quantum Hall Effect*, Springer-Verlag, New York (1988).
116. D. C. Tsui, H. L. Stormer, and A. C. Gossard, *Phys. Rev. Lett.* **48**, 1559 (1982).
117. R. B. Laughlin, *Phys. Rev. Lett.* **50**, 1395 (1983); R. B. Laughlin, in *The Quantum Hall Effect*, (R. E. Prange and S. M. Girvin, eds.), Springer-Verlag, New York, p. 276 (1987).
118. D. J. Yoshioka, B. I. Halperin, and P. A. Lee, *Phys. Rev. Lett.* **50**, 1219 (1983).
119. R. B. Laughlin, *Phys. Rev. B* **27**, 3383 (1983).
120. F. D. M. Haldane and E. H. Rezayi, *Phys. Rev. Lett.* **54**, 237 (1985).
121. C. W. J. Beenakker, *Phys. Rev. Lett.* **64**, 216 (1990).
122. A. H. MacDonald, *Phys. Rev. Lett.* **64**, 220 (1990).
123. A. M. Chang, *Solid State Commun.* **74**, 871 (1990).
124. F. D. M. Haldane, in *The Quantum Hall Effect* (R. E. Prange and S. M. Girvin, eds.), Springer-Verlag, New York, p. 303 (1987).
125. A. M. Chang and J. E. Cunningham, *Solid State Commun.* **72**, 651 (1989).
126. R. G. Clark, J. R. Mallet, S. R. Haynes, J. J. Harris, and C. T. Foxon, *Phys. Rev. Lett.* **60**, 1747 (1988).
127. B. I. Halperin, *Helv. Phys. Acta* **56**, 75 (1983).
128. F. D. M. Haldane and E. H. Rezayi, *Phys. Rev. Lett.* **60**, 956 (1988).
129. J. P. Eisenstein, R. L. Willett, H. L. Stormer, D. C. Tsui, A. C. Gossard, and J. H. English, *Phys. Rev. Lett.* **61**, 997 (1988).
130. R. G. Clark, S. R. Haynes, A. M. Suckling, J. R. Mallet, P. A. Wright, J. J. Harris, and C. T. Foxon, *Phys. Rev. Lett.* **62**, 1536 (1988).
131. R. Willett, J. P. Eisenstein, H. L. Stormer, D. C. Tsui, A. C. Gossard, and J. H. English, *Phys. Rev. Lett.* **59**, 1776 (1987).
132. P. Streda, private communication.
133. S. T. Chui, *Phys Rev. Lett.* **56**, 2395 (1986); *Phys. Rev. B* **36**, 2806 (1987).
134. S. Datta and S. Bandyopadhyay, *Phys. Rev. Lett.* **58**, 717 (1987).

135. F. Sols, M. Macucci, U. Ravaioli, and K. Hess, *Appl. Phys. Lett.* **54**, 350 (1989).
136. P.G. N. de Vegvar, G. Timp, P. M. Mankiewich, R. Behringer, and J. Cunningham, *Phys. Rev. B* **40**, 3491 (1989).
137. C. J. B. Ford, A. B. Fowler, J. M. Hong, C. M. Knoedler, S. E. Laux, J. J. Wainer, and S. Washburn, *Surf. Science* **229**, 307 (1990).
138. D. K. Ferry, in *The Physics of Submicron Semiconductor Devices* (H. L. Grubin, D. K. Ferry, and C. Jacoboni, eds.), *NATO ASI Series*, Plenum, New York, p. 503 (1988).
139. Carver Mead and Lynn Conway, *Introduction of VLSI Systems*, Addison-Wesley, Menlo Park, CA.

CHAPTER 4

The Quantum Hall Effect in Open Conductors

M. Büttiker

IBM RESEARCH DIVISION
THOMAS J. WATSON RESEARCH CENTER
YORKTOWN HEIGHTS, NEW YORK

I. INTRODUCTION	192
II. BASIC ELEMENTS OF ELECTRICAL CONDUCTION	195
1. Introductory Remarks	195
2. The Scattering Problem	197
3. The Current Response due to Differing Chemical Potentials	199
4. Resistance and Transmission	202
5. Experimental Tests of Reciprocity	204
6. Conductors with an Arbitrary Number of Probes	205
7. Alternative Derivations and Applications	206
III. QUANTIZATION AND INTERFERENCE IN THE ABSENCE OF A FIELD	208
8. Quantized Transmission at a Saddle-Shaped Potential	208
9. Four-Terminal Measurement of Constriction Resistances	212
10. On the Nature of Phase-Sensitive Four-Terminal Resistances	215
IV. MOTION IN HIGH MAGNETIC FIELDS: THE TWO-TERMINAL RESISTANCE	221
11. Motion in High Magnetic Fields	221
12. The Two-Terminal Conductance	227
13. The Constriction in a Magnetic Field	233
14. The Aharonov–Bohm Effect and Backscattering	237
V. THE QUANTUM HALL EFFECT IN OPEN CONDUCTORS	238
15. Conductor with Ideal Contacts and without Backscattering	238
16. Simultaneously Quantized Longitudinal Resistances and Hall Resistances	240
17. Contacts with Internal Reflection Alternating with Ideal Contacts	244
18. The Anomalous Quantum Hall Effect due to Adjacent Contacts with Internal Reflection	248
19. Backscattering and Contacts with Internal Reflection	251
20. The Long-Range Nature of Non-Equilibrium Populations	253
VI. RESONANT DEPARTURES FROM THE HALL RESISTANCE	257
21. Two-Terminal Conductance	257
22. Effect of Resonant States on the Hall Effect	261
23. Intrinsic Resonances in Strongly Confined Systems	269
VII. DISCUSSION	271
ACKNOWLEDGMENT	272
REFERENCES	273

I. Introduction

In this chapter, we are chiefly concerned with transport in a high magnetic field in conductors that confine a two-dimensional electron gas. Such a conductor is outlined schematically in Fig. 1. It is almost a decade now, since it was discovered by von Klitzing et al.[1] that the Hall effect is quantized in such systems. The Hall resistance at these quantized values is given by the fundamental unit of resistance $h/e^2 = 25812.8$ ohm divided by an integer. The development preceding this discovery is reviewed by Ando et al.[2] The accuracy of the quantization is so high that the quantum Hall effect recently was adopted as a resistance standard. The general nature of this effect has given rise to the belief that it must have a fundamental explanation that is independent of the geometrical shape of the conductor and microscopic details. Laughlin considered a ribbon of two-dimensional electron gas wrapped into a cylinder and has studied the response of the electron gas to an Aharonov–Bohm flux through the opening of the cylinder.[3] Such purely topological considerations have attracted considerable attention.[4–7]

Surprisingly, over the past two years, a different picture of the quantum Hall effect has emerged. Instead of the closed conductors, implicit in these topological explanations of the quantum Hall effect, we consider a conductor as shown in Fig. 1. A number of contacts are made to the conductor that serve as carrier source and carrier sink and serve to attach voltmeters. The source serves as an injector of carriers and the sink as a receptor of carriers. The voltage contacts are carrier detectors. Do we observe a quantum Hall effect quite independent of the properties of the injector and the detectors? The answer to this question is: No! The distribution of incident carriers into states of the two-dimensional electron gas depends on the properties of the source contact.[8] The memory of this initial distribution can be lost only through the action of inelastic events. Elastic scattering alone can lead to a

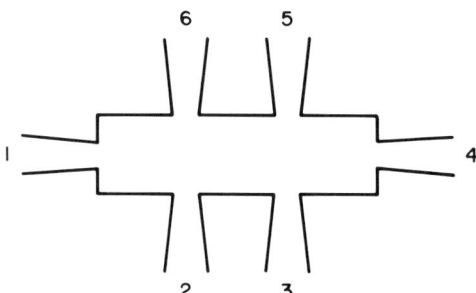

FIG. 1. Conductor with a number of probes permitting carriers to enter and leave. (From Ref. 8.)

modification of the incident distribution but not to an equilibration. Therefore, if the injector and detector are *close enough*, the Hall resistance is not quantized unless the injection and the detection proceeds according to very stringent conditions. This effect has been demonstrated in a clear manner in experiments by van Wees et al.[9] and Komiyama et al.[10]

Each of the contacts is connected to an electron bath. A voltmeter connected to two of the probes measures the difference in electrochemical potentials between two of these reservoirs. The voltage difference, which together with the current defines the measured resistance, is an electrochemical potential difference (not an electrostatic potential difference between two arbitrary chosen points somewhere in the interior of the conductor or somewhere at the boundary of the conductor). The consideration of conductors with contacts therefore, leads, first of all to an unambiguous characterization of what we mean by a *resistance measurement*. This is not a trivial point. Halperin,[11] commenting on Laughlin's argument,[13] writes: ".. the argument handles immediately the case where the voltage difference between the two edges is due in part to a chemical potential difference in addition to an electrostatic potential difference across the sample." Next, we stress that voltages measured at contacts permit carrier exchange with the conductor are *chemical potential differences*. To find answers that are in agreement with experiments, we must treat a multi- terminal conductor. Hall resistances and longitudinal resistances are measured in a four-terminal set-up: a carrier flux I is driven from the current source contact k to the current sink contact l and the measured voltage is determined by connecting a pair of probes m and n to a voltmeter. The resistance measured in the configuration $mnkl$ is

$$R_{kl,mn} = \frac{V_m - V_n}{I}. \tag{1}$$

If an imaginary line is drawn from the carrier source contact to the carrier sink contact, the measured resistance is called a *Hall resistance* if the two contacts are on opposite sides of this line and is called a *longitudinal resistance* if the contacts are on the same side of this line. In Eq. (1), the potentials eV_m characterize chemical potentials (Fermi energies) of electron baths connected to contacts.

The resistances in Eq. (1) can be calculated if it is known how the carrier fluxes incident on the conductor are related to the applied voltages. This relationship can be expressed with the help of conductances G_{ij}, which characterize transport from probe j to probe i. The incident current at probe i is

$$I_i = \sum_{j \neq i} G_{ij}(V_i - V_j). \tag{2}$$

The conductances in Eq. (2) are defined for carrier transport from one probe to another. They not only are a consequence of the bulk properties of the sample, but in general also depend on the properties of the contact. The resistances defined in Eq. (1) are obtained from Eq. (2) under the conditions $I = I_k = -I_l$ and $I_m = I_n = 0$. (All probes that are not used directly in the four-probe measurement also are assumed to carry zero net current.) The quantum Hall effect is a consequence of the quantization of the conductances G_{ij}, which are (ideally) either equal to $(e^2/h)N$ or zero. A derivation of Eq. (2) and evaluation of the resistances as defined by Eq. (1) is the subject of Section II of this chapter.

That some conductances need to be zero is as important for the occurrence of the quantum Hall effect as are the conductances that are proportional to an integer. Expressed in experimental terms,[12] this means that in addition to the Hall resistances, the longitudinal resistances are (ideally) also quantized (are zero). This equivalence is demonstrated best in situations in which, through application of a gate across the conductor, the sample is made strongly nonuniform. In a range of gate voltages, Eqs. (1) and (2) then predict that both Hall resistances and longitudinal resistances can be quantized simultaneously at non-trivial integer values.[8] By non-trivial integer values, we mean integers that do not characterize the property of the sample far away from the gate. This simultaneous quantization was demonstrated in experiments by Washburn et al.[13] in a submicron conductor with a submicron gate structure and in experiments by Haug et al.[14] The comparison of the simple and direct discussion of the experimental results by Washburn et al. and Haug et al. with the discussion of similar experiments by Syphers and Stiles[15] best illustrates the progress made.

Carriers lose memory of the particular way in which they are injected due to inelastic events. One might suspect, therefore, that over distances of a few microns, this memory is lost and that a voltage probe a distance larger than a few microns away from the carrier-injecting probe could not be sensitive to the particular injection mechanism. However, Komiyama et al.[10,16] and van Wees et al.[17] reported experiments in which a memory loss does not occur over distances larger than 250 μm. Alphenaar et al.[18] have investigated the degree of equilibration between edge channels over similar distances. These experiments suggest that inelastic events, at least with regard to some of the current carrying states, are strongly suppressed in high magnetic fields. These experiments clearly indicate that the properties of the contacts are relevant not only in small systems but even in very large samples.

The experimental demonstration of the long distances taken for equilibration invalidates theoretical descriptions of high-field transport in terms of *local conductivities* σ_{xx} and σ_{xy}. As a consequence of this long-range nature, the longitudinal resistances away from plateaus do not scale as $\rho_{xx}L/W$, where L and W are the length and width of the sample even in conductors that are very large compared to a magnetic length. For recent experimental results on this scaling behavior, we refer the reader to Haug and von Klitzing,[19] who for an explanation point explicitly to the role of unequally populated states.

The need to put forth a discussion of the quantum Hall effect based on Eqs. (1) and (2) became apparent with the work of Streda et al.[20] This work used a method of calculating voltages that was proposed originally in Ref. 21, but which we subsequently have criticized.[22] A discussion similar to that of Streda et al.[20] but restricted to one Landau level only was presented by Jain and Kivelson.[23] Refs. 8 and 24 criticize the discussions of Streda et al.[20,25] for not using an approach that treats all contacts on an equivalent footing. Sivan et al.[26] defend the discussion of Streda et al.,[20] but Streda et al.,[20] in view of experiments,[27] eventually adopted our approach.[8,22] It is clear that the theoretical task is to investigate the transport coefficients in Eq. (2) that are *global* conductances. These coefficients describe transport from one probe to another. Therefore, a theory of electrical conductance that gives answers about these global conductances is needed. We discuss such an approach in Section II. Section III illustrates this approach by discussing a few phenomena in the absence of a magnetic field. Section IV starts with a simple discussion of transport in high magnetic fields and discusses the high-field two-terminal conductance for structures of various shapes. The phenomena mentioned in this introduction are discussed in Section V. Finally, in Section VI, we deal with situations where the phase of the quantum mechanical wave functions is of importance.

II. Basic Elements of Electrical Conduction

1. Introductory Remarks

In this section, we discuss the basic approach to electrical conduction that we employ later on in this chapter. Figure 2 shows a conductor with several probes that permit the entrance or exit of carriers. We assume that the motion of carriers from the entrance to the exit can be treated as a purely elastic scattering problem. Dissipation is removed to the electron baths, which are connected to the probes. (See Fig. 2.) An electron wave incident in probe j typically is associated with waves transmitted into all the other probes $i \neq j$ and a reflected wave in probe i. The conduction process thus is viewed as a transmission and reflection problem. The description of wave motion in terms of reflection and transmission coefficients is familiar from the propagation of electromagnetic waves in waveguides and has found wide application in this field. It also is familiar, of course, in tunneling theory, which calculates the resistance of tunneling devices in terms of tunneling matrix

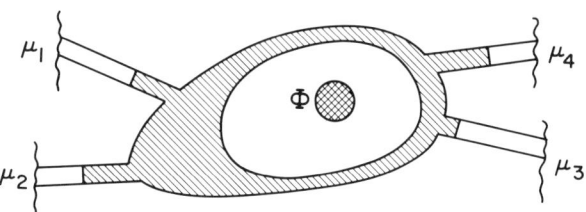

FIG. 2. Conductor with four probes connected to electron reservoirs at chemical potentials μ_i, $i = 1, 2, 3, 4$ and an Aharonov—Bohm flux Φ. (From Ref. 22.)

elements. Yet both of these stimuli did not lead to a transport theory of electric conduction in terms of transmission and reflection coefficients in the 1930s or 1940s. Papers that treat the conduction not only of a single scatterer but of an entire sample in terms of transmission and reflection probabilities seem to have appeared only with the work of Landauer in 1957 and in a more explicit way in 1970. Landauer's 1970 paper treated a one-dimensional sample. Subsequent to the work of Anderson et al.[28] on the scaling behavior of localization and conductance, a number of works appeared that aimed to find more general expressions for the resistance in terms of transmission and reflection probabilities.[29-32] Landauer's discussion of the resistance[33,34] emphasizes the local electrostatic potential that arises due to charge accumulation and depletion at the scattering centers of the sample. More precisely, the voltage across a scatterer is determined in portions of the sample a few screening lengths away from the scatterer. This leads to an electrostatic potential difference that is determined by a *charge neutrality condition*. The author, together with Imry and Landauer, reestablished in 1985 a generalized resistance formula[21] that had been derived earlier by Azbel.[32] None of these papers addressed the role of contacts. The distinction between open and closed systems was emphasized in the early discussions of sample-specific interference effects in normal loops.[35-38] However, only the probes connected to the current source and the sink were treated explicitly. Eventually, it was recognized that a voltage probe, which permits no *net* carrier flux,[39] nevertheless allows carrier exchange into and out of the conductor. The consequence is that a voltage probe permits phase randomization and, in the presence of transport, dissipation of energy.[38] Thus, voltage probes provide a (conceptual) model of a dissipative scatterer for electrical conduction.[21,40-46,161] These notions, amplified by a set of experiments by Benoit et al.,[47] led to a formulation of resistances that treats all probes connected to a conductor on an equivalent footing.[22,48] It is this formulation that is the subject of this section and that will be used in this chapter. In contrast to the electrostatic potential at points inside a conductor, which essentially is determined by a charge neutrality condition, the consideration of a conductor including the probes leads to a formulation of resistances in terms of the chemical potentials of the electron baths. The measurement of a chemical potential is conveniently expressed in terms of a *zero-current condition*.

2. THE SCATTERING PROBLEM

To provide a mathematical underpinning for the formulation of a scattering problem, *asymptotic regions* are needed that permit the definition of incident and outgoing waves. This is achieved by assuming that each probe of the conductor in Fig. 2 eventually widens into a perfect lead. For mathematical purposes, we assume such a perfect lead to extend uniformly for an infinite distance away from the conductor. This mathematical device clearly has no counterpart in the real physical world. Perfect wires do not exist. Therefore, this mathematical device makes sense only if our final result is largely independent of these assumptions. As we will show, that indeed is the case. In the perfect conductors, the Hamiltonian is separable into a part that describes motion along the perfect wire and a part that describes motion transverse to the perfect wire. Therefore, in an infinitely extended perfect wire, the wave functions are of the form,

$$\psi_{n,\pm k} = e^{\pm ikx} f_{n,\pm k}(y). \tag{3}$$

The factor $e^{\pm ikx}$ is due to translational invariance along the conductor, $f_{n,\pm k}$ is the transverse wave function. For fixed k, the spectrum consists of a ladder of discrete states with energy $E_n(k)$. The specific k dependence of the dispersion $E_n(k)$ depends on the type of perfect conductor considered and depends on whether or not a magnetic field is present. The specific form of the dispersion, as we shall see, is irrelevant, however, and our discussion in this section is completely general. It applies in particular also if a magnetic field is present. To help the reader's intuition, we will, for simplicity, appeal to the zero-magnetic-field case and repeat part of the following discussion for the case when a magnetic field is present in Section IV. In the absence of a magnetic field, the dispersion is the sum of two energies, $E_n(k) = \hbar^2 k^2/2m + E_n(0)$, the energy associated with longitudinal motion along the perfect wire and an energy $E_n(0)$ due to quantization of the transverse motion. Each of the dispersions $E_n(k)$ commonly is referred to as a quantum channel. (See also Fig. 3.)

Suppose now that we have defined the properties of the asymptotic regions in all the probes connected to the conductor. Consider a wave of unit amplitude of the form given by Eq. (3), which is incident in probe j and channel n. A solution of the scattering problem then consists of this incident wave and of outgoing waves in (typically) all the other probes. The outgoing waves, in

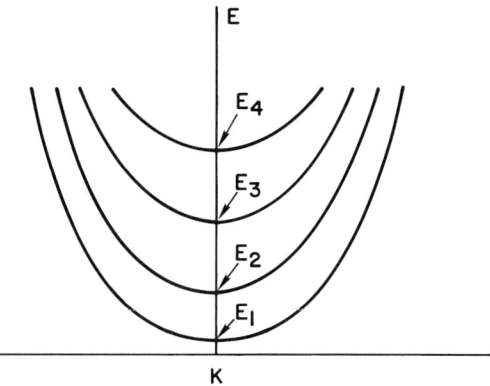

FIG. 3. Energy dispersion $E_n(k)$ of the quantum channels of a perfect wire. k is the wavevector along the wire and $E_n(0)$ is the threshold of the nth quantum channel. (From Ref. 165.)

the asymptotic region again are of the form given by Eq. (3) multiplied by a factor $s_{ij,mn}$, which accounts for the fact that current is conserved. The outgoing wave in channel m in probe i due to a wave of unit amplitude incident in probe j in channel n is

$$\left(\frac{v_{jn}}{v_{im}}\right)^{1/2} s_{ij,mn} e^{-ik_{im}x} f_{im}(y). \qquad (4)$$

Here,

$$v_{jn} = \left(\frac{1}{\hbar}\right)\left(\frac{dE_{jn}(k)}{dk}\right)\bigg|_{E_F}, \qquad (5)$$

is the velocity of carriers in probe j and quantum channel n at the Fermi energy E_F. The velocity factors in Eq. (4) are chosen such that the absolute square of the amplitude $s_{ij,mn}$ is the transmission probability $T_{ij,mn}$ (or the reflection probability $R_{ii,mn}$) of a carrier incident in lead j and channel n to exit in lead i and channel m, (or incident in lead i in channel n and reflected into lead i into channel m),

$$T_{ij,mn} = |s_{ij,mn}|^2, \; R_{ii,mn} = |s_{ii,mn}|^2. \qquad (6)$$

The incident wave function has an amplitude that is normalized to one and thus gives rise to an incident current $I_{jn} = v_{jn}$. The outgoing wave in probe i and channel m is associated with a current $I_{im} = v_{im}(v_{jn}/v_{im})|s_{ij,mn}|^2 = v_{jn} T_{ij,mn}$. Alternatively, if the incident current is $I_{jn} = |\alpha_{jn}|^2$, then the outgoing currents are determined by the square of the amplitudes $s_{ij,mn}\alpha_{jn}$. The scattering matrix **S** with the elements $s_{ij,mn}$ thus provides a linear relationship between the current amplitudes of the incident waves and the outgoing waves. Because current is conserved, **S** must be a unitary matrix. Denoting the Hermite conjugation by †, we must have $\mathbf{S}^\dagger = \mathbf{S}^{-1}$. Since the Hamiltonian describing the scattering at the sample also is invariant, if we reverse the momenta and the magnetic field simultaneously, the scattering matrix also must have the property $\mathbf{S}^*(-B) = \mathbf{S}^{-1}(B)$. Here, the asterisk denotes complex conjugation. Taken together, these two conditions imply that the scattering matrix has the symmetry $\mathbf{S}^T(B) = \mathbf{S}(-B)$. The amplitudes of the **S** matrix obey a microreversibility condition, $s_{ij,mn}(B) = s_{ji,nm}(-B)$. The transmission (or reflection) amplitude of a carrier incident in lead j in channel n in a magnetic field to exit in lead i in channel m in the presence of a magnetic field is equal to the amplitude of the reciprocal process of a carrier incident in lead i in channel n to exit in probe j in channel m if the magnetic field is reversed. Thus, as a consequence of microreversibility, the transmission and reflection probabilities also obey the relations,

$$T_{ij,mn}(B) = T_{ji,nm}(-B), \quad R_{ij,mn}(B) = T_{ji,mn}(-B). \tag{7}$$

The microreversibility properties, Eq. (7), of the transmission and reflection probabilities are necessary to provide the reciprocity of resistances, which we will derive shortly; but the microreversibility of the transmission probabilities is not a sufficient condition to arrive at the reciprocity of resistances. The reciprocity of resistances is a consequence of both microreversibility and equilibration provided by the electron baths connected to the conductor. We now will discuss this in more detail.

3. THE CURRENT RESPONSE DUE TO DIFFERING CHEMICAL POTENTIALS

The mathematical formulation of the scattering problem given earlier needs to be augmented by physical considerations that specify the role of the reservoirs attached to the probes of the conductor. First, it is assumed that a carrier that approaches a reservoir from inside the conductor once it has reached the *asymptotic regions* discussed before will enter the reservoir with probability one. Scattering that the carrier undergoes in the reservoir completely randomizes the energy and phase of the carrier such that a carrier entering the conductor bears no phase and energy relationship to an exiting carrier. Entering and leaving the reservoir is treated as an irreversible event.[34] Since the reservoirs connected to the probes are a source of irreversibility, it follows that waves incident from differing reservoirs cannot exhibit quantum interference effects. This allows us to treat waves incident from differing probes as incoherent. Second, we must specify how a reservoir attached to a probe populates the quantum channels. In part of the theory of electric conduction, it sometimes is assumed that the incident carrier stream is populated according to a shifted Fermi distribution. The result is that the quantum channels with low threshold energy, which have a total momentum parallel to the probe, are populated up to higher energies than quantum channels with high threshold energies, which have a small longitudinal momentum component. Such a non-equilibrium distribution of incident carriers does not give rise to the reciprocity of resistances. Rather, we assumed that the reservoirs are at equilibrium and fill all the quantum channels of probe i according to the Fermi function of reservoir i. The reservoir populates all quantum channels equivalently.[21] Next, we shall assume, for simplicity, that the temperature is so small that the Fermi function of reservoir i can be described in terms of the chemical potential (the Fermi energy) μ_i of reservoir i. The equilibrium population of incident quantum channels simplifies the calculation considerably, and together with the microreversibility of the scattering matrix, ensures the reciprocity of the resistances.

Since the reservoirs feed all channels equally and up to their chemical potential, it is only the total transmission probabilities that are relevant. If each incident channel in probe j supports a unit current $|\alpha_{jn}|^2 = 1$, the total current in probe i is $\Sigma_{nm}|s_{ij,mn}|^2|\alpha_{jn}|^2 = \Sigma_{nm}|s_{ij,mn}|^2$. Here, the sum is over all M_j incident channels in probe j and over all outgoing channels M_i in probe i. Hence, transmission from probe j into probe i can be characterized by a total transmission probability,[22]

$$T_{ij} = \sum_{m=1}^{m=M_i} \sum_{n=1}^{n=M_j} T_{ij,mn}. \tag{8}$$

Similarly, the total reflected current in lead j due to carriers incident in lead j can be characterized by a total probability for reflection,

$$R_{jj} = \sum_{m=1}^{m=M_j} \sum_{n=1}^{n=M_j} R_{jj,mn}. \tag{9}$$

Equations (8) and (9) are the relevant transport coefficients as we now will demonstrate.

Next, we have to evaluate the currents driven through a sample in the presence of small differences between the chemical potentials. To find the relation between the currents and the chemical potentials, we proceed as follows: Denote the lowest of the chemical potentials μ_i by μ_0. At energies smaller than μ_0, all states are fully occupied. These states cannot contribute to a net current flow. Thus, the considerations can be limited to energies larger than μ_0. Reservoir j feeds all quantum channels in the energy range $\mu_j - \mu_0$. The current incident from this reservoir in channel n in this energy interval is $I = ev_{jn}(dn_{jn}/dE)(\mu_j - \mu_0)$. Since the density of states in a one-dimensional quantum channel is $dn/dk = 1/2\pi$, we obtain for the density of states in channel n in probe j, $dn_{jn}/dE = (dn/dk)(dk/dE_{jn}) = 1/hv_{jn}$. Hence, the reservoir j feeds a current,

$$I = \frac{e}{h}(\mu_j - \mu_0), \tag{10}$$

into each of the M_i incident quantum channels. Note that this current is independent of the properties of the channel (density of states, velocity, effective mass...). The universality expressed by Eq. (10) is basic for the occurrence of quantized resistances.

It now is a simple matter to calculate the net currents flowing into the probes of a conductor. The total incident carrier flux in probe i is $M_i(\mu_i - \mu_0)/h$. Of this flux, a portion $R_{ii}(\mu_i - \mu_0)/h$ is reflected. The incident flux is diminished further by carriers that are incident in the other probes and are transmitted into probe i. These fluxes are proportional to $T_{ij}(\mu_j - \mu_0)/h$. Thus, the current in probe i is[22]

$$I_i = \frac{e}{h}\left[(M_i - R_{ii})\mu_i - \sum_{j \neq i} T_{ij}\mu_j\right]. \tag{11}$$

The reference chemical potential does not appear. Due to current conservation, we have

$$M_i = R_{ii} + \sum_{j \neq i} T_{ij}. \tag{12}$$

Using current conservation, we also can express Eq. (11) in the form,

$$I_i = \frac{e}{h}\sum_{j \neq i} T_{ij}(\mu_i - \mu_j). \tag{13}$$

Equation (12) and (13) are the important equations of this section. They represent a quantum mechanical version of the Kirkhoff laws. These equations provide a linear relationship between the chemical potentials of the electron reservoirs connected to the conductor and the currents at the connections of the conductor and the reservoirs. It is a linear response relation.

The transmission and reflection coefficients in Eqs. (12) and (13) are evaluated *at equilibrium at the Fermi energy*. The linear response coefficients T_{ij} and R_{ii} have the symmetry that, according to Onsager and Casimir,[49] is required. The diagonal coefficients R_{ii} are symmetric under flux reversal and the off-diagonal coefficients obey a reciprocity relation,

$$R_{ii}(B) = R_{ii}(-B), \qquad T_{ij}(B) = T_{ji}(-B). \tag{14}$$

The symmetry of the total transmission and reflection coefficients is a consequence of the symmetry of the individual transmission and reflection probabilities as stated by Eq. (8).

4. Resistance and Transmission

Using the relations between currents and chemical potentials, Eq. (11), we now can determine the resistance for a given configuration of current sources and sinks and voltage probes. We discuss here two cases explicitly: the case of a two-probe conductor and the case of a resistance measurement in a four-probe setup.

If the conductor is connected only to two reservoirs with M_1 and M_2 quantum channels, current conservation as stated by Eq. (12) requires $M_1 = R_{11}(B) + T_{12}(B)$ and $M_2 = R_{22}(B) + T_{21}(B)$. Now since the total reflection coefficients are symmetric under field reversal, it follows that $M_1 = R_{11}(B) + T_{12}(-B)$ and $M_2 = R_{22}(B) + T_{21}(-B)$. Consequently, the two-terminal transmission probability $T_{21} = T_{12} \equiv T$ is also symmetric under field reversal. Thus, Eqs. (11) and (13) yield a two-probe resistance,

$$R_{12,12} = \frac{\mu_1 - \mu_2}{eI} = \left(\frac{h}{e^2}\right)\left(\frac{1}{T}\right). \tag{15}$$

Equation (15) has been obtained by Fisher and Lee[30] in a linear response calculation that did not appeal to reservoirs and is contained in a number of papers as a limiting result.[21,28] The physical content of Eq. (15), which gives the resistance for current flow between two equilibrium reservoirs, was comprehended by Imry.[50] The comparison of the two-probe resistance expression, Eq. (15), with the resistances of multi-probe conductors,[22] further helped to clarify Eq. (15).

Consider a four-probe conductor next. Let probe k be the carrier source and probe l be the carrier sink. Probes m and n are voltage probes. The voltmeter connected to a probe is taken to have infinite impedance. Hence, the voltage measured at a probe is obtained by adjusting the chemical potential of this probe such that the net current through the measurement probe is zero. To find the measured resistance in such a configuration, we have to solve Eq. (11) with the conditions $I \equiv I_k \equiv -I_l$ and $I_m = I_n = 0$. We cannot directly invert Eq. (11) to find the chemical potentials as a function of the current. If all chemical potentials are equal, all the resulting currents are zero. That implies that Eq. (11) has an eigenvector with eigenvalue zero. Hence, the determinant of the matrix of transport coefficients in Eq. (11) is zero. A simple way to proceed is to subtract from each chemical potential in Eq. (11) the potential μ_n. Writing the equations for I_k, I_l, and I_m gives a system of three equations with three unknown variables, $\mu_i - \mu_n$, $i = k, l, m$. Solving for $\mu_m - \mu_n$ yields

$$\mu_m - \mu_n = \frac{h}{e} I \frac{(T_{mk} T_{nl} - T_{ml} T_{nk})}{D_{kn}}. \qquad (16)$$

Here, D_{kn} is a subdeterminant of the matrix of transport coefficients in Eq. (12) of rank 3 (with row k and column n deleted from the full matrix). All subdeterminants of rank 3 of the matrix of transport coefficients are equal, $D_{kn} \equiv D$. This is a consequence of current conservation. To show this, let us denote the determinant of the full matrix by "det." It is zero. Now, expanding it in terms of subdeterminants yields

$$0 = \det = (M_i - R_{ii})D_{ii} - \sum_{j \neq i} T_{ij} D_{ij}. \qquad (17)$$

This is just Eq. (11) with zero current in all the probes and with the subdeterminants in place of the chemical potentials; but the only solution of Eq. (11) that yields zero current in all the probes is the equilibrium solution consisting of identical chemical potentials. Hence, all subdeterminants are equal, $D_{ij} \equiv D$. With the voltage drop, $V_{mn} = (\mu_m - \mu_n)/e$ as determined by Eq. (16), we find for the four-terminal resistance, $R_{kl,mn} = V_{mn}/I$,[22]

$$R_{kl,mn} = \left(\frac{h}{e^2}\right) \frac{(T_{mk} T_{nl} - T_{ml} T_{nk})}{D}. \qquad (18)$$

Equation (18) has the property that the two resistances measured in two configurations in which the current source and current sink are exchanged (exchange of k an l) are equal in magnitude but differ in their sign. Similarly, two configurations that differ only in the voltage probes (exchange of m and n) are equal up to a sign. The important symmetry of the four-terminal resistance, Eq. (17), is the reciprocity relation, which states the following. The two resistances measured in the four-terminal configurations, in which exchange of the role of the current and voltage probes is accompanied by a reversal of the magnetic field, are identical. The reciprocity of resistances,

$$R_{kl,mn}(B) = R_{mn,kl}(-B), \qquad (19)$$

follows from Eq. (18) on account of the microreversibility property of the transmission probabilities, Eq. (8), and the invariance of the subdeterminant D under field reversal. $D(B) = D(-B)$ follows from the fact that the subdeterminants also obey a reciprocity relation, $D_{ij}(B) = D_{ji}(-B)$, and that all subdeterminants are equal, as we have shown in the preceding.

5. Experimental Tests of Reciprocity

The reciprocity of the resistances is the experimentally observed symmetry of electric conduction. A very interesting test of reciprocity symmetry demonstrating the phase-sensitive nature of voltage measurements is provided in an experiment by Benoit et al.[47] They investigated the symmetry of the Aharonov–Bohm oscillations of a metallic four-probe conductor at milli Kelvin temperatures. In metallic loops,[36–38,52–54] the Aharonov–Bohm oscillations manifest themselves in a small oscillatory contribution to the total resistance. For reviews on Aharonov–Bohm effects in normal conductors, we refer the reader to Imry,[50] Washburn and Webb,[53] and Aronov and Sharvin.[54] According to Eq. (19), the h/e oscillations need not be symmetric if the flux Φ through the loop is reversed, but can appear with an arbitrary phase ϕ. (See Fig. 4.) However, reciprocity requires that if the oscillatory contribution to $R_{14,23}$ is $\Delta R \cos(2\pi\Phi/\Phi_0 - \phi)$, where $\Phi_0 = h/e$ is the elementary flux quantum, then the oscillatory component in $R_{23,14}$ is precisely $\Delta R \cos(2\pi\Phi/\Phi_0 + \phi)$. Just such a behavior was observed in the experiments of Benoit et al.[47] An added significance of these experiments is the demonstration of the phase sensitivity of the voltage measurement in small conductors. The observation of a phase different from zero or π needs quantum mechanical phase coherence.[22] Only if the probes are close to the loop or, as in Fig. 4, directly connected to the loop can a non-trivial phase in the oscillations be observed. DiVincenzo and Kane[55] have calculated the probability distribution of the phase ϕ as function of the phase-breaking length for a conductor in the metallic diffusive limit. Many additional tests of reciprocity have been performed and for references to this work, we refer the reader to Ref. 48. Of special interest in the context of this chapter are the reciprocity tests in the electron-focusing geometry reported by van Houten et al.[56] and reciprocity tests in high magnetic fields reported by Komiyama et al.[57]

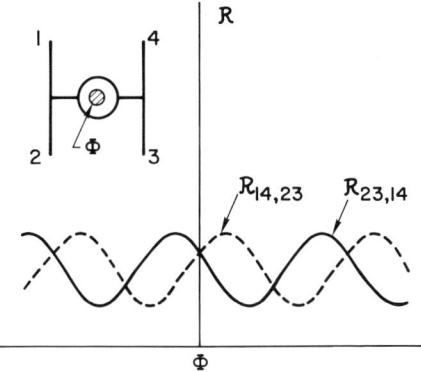

FIG. 4. h/e-Aharonov–Bohm resistance oscillations for the conductor shown in the inset for two measurements that are related by reciprocity. (From Ref. 165.)

6. Conductors with an Arbitrary Number of Probes

In Section 4, we discussed the two-probe resistance and four-terminal resistance measurements. For completeness, we briefly discuss here the three-probe resistance measurement on a three-probe conductor. Subsequently, we discuss two probe, three-probe, and four-probe measurements on conductors with many probes. In a three-terminal measurement, there is one voltage measurement that is performed at the voltage probe n, and the second voltage measurement is performed either at the carrier source probe k or the carrier sink probe l. Here, k, l, n are any permutations of the probe indices of Eq. (11) for $i = 1$, 2, and 3. The calculation gives[48,58]

$$R_{kl,kn} = \frac{h}{e^2} \frac{T_{nl}}{D} \tag{20}$$

if one of the voltage measurements is made at the carrier source, and gives

$$R_{kl,nl} = \frac{h}{e^2} \frac{T_{nk}}{D} \tag{21}$$

if one of the voltage measurements is made at the carrier sink. Here, D is a subdeterminant of rank 2 of the 3 × 3 matrix of transport coefficients. It can be expressed, for instance, as $D = T_{lk}T_{nl} + T_{nk}(T_{ln} + T_{lk})$. Note that these resistances are not independent of one another.

Consider a conductor with K probes. Only N probes are used for the actual measurement, $N < K$. Label the current source and current sink probes as probes 1 and 2 and let probe 3 (and probe 4) be the voltage terminal(s) in the case of a three-probe (or four-probe) measurement. Let the chemical potential of the remaining $K-N$ probes be adjusted such that no current flows at these probes, $I_i = 0$ for $i = K-N + 1,..,K$. We can use the condition $I_K = 0$ to obtain the chemical potential μ_K as a function of all the remaining chemical potentials. We then eliminate μ_K in all the $i = 1, ... K-1$ equations. (See Eq. (11).) The new transport coefficients $\hat{T}_{ij}$ and $\hat{R}_{ii}$, which appear in the $K-1$ equations, are functions of the original transmission and reflection probabilities T_{ij} and R_{ii} and satisfy the microreversibility conditions from Eq. (14) and the current conservation property, Eq. (12). Thus, elimination of a chemical potential of a voltage probe maps a conductor of K terminals onto a conductor of $K-1$ terminals.[48] Therefore, if we have a conductor of K probes and want to calculate a measured four-terminal resistance, we first eliminate all $K-N$ probes that are not used to make the measurement. For the transport coefficients found in this way, we finally use Eqs. (15), (18), (20) or (21) to obtain the desired resistance. There of course are many alternative ways to arrive at these results.

7. ALTERNATIVE DERIVATIONS AND APPLICATIONS

A natural question is the relation of Eqs. (11) and (18) to the linear response formalism. We already have mentioned the connection between Eq. (8) and the linear response approach established by Fisher and Lee.[30] Proceeding in a similar way, Stone and Szafer[59] have given a linear response derivation of Eqs. (25)–(28) for conductors without a magnetic field. The inclusion of a magnetic field is more difficult and only recently has such derivation been found by Baranger and Stone.[60] A major by-product of such a derivation is the expression of the transmission coefficients in terms of Green's functions. This is important for the computation of these coefficients starting from a given Hamiltonian. However, it should be stressed that linear response formalism cannot be used to determine which conductance formula is *correct*. There is no reason why a linear response approach together with a charge neutrality condition for a local potential cannot lead to the formulae of Azbel[32] and Büttiker et al.[35] if the assumptions in these papers are followed. As an example of such a discussion, we mention the derivation of Landauer's 1970 result by Thouless.[61] These formulae are a consequence of a number of physical assumptions, and to redrive them with a linear response approach, the same assumptions need to be made.

The discussion given in the preceding, which determines resistances in terms of chemical potentials of reservoirs, must be self-consistent. By this, we mean the following: Suppose that $V(r)$ is the electrostatic potential in the presence of transport. $V(r)$ can be obtained by calculating the accumulated charges and by solving a Poisson equation. Let E_i be the Fermi energy of reservoir i. Our problem is self-consistent if deep inside reservoir i, $eV(r) + E_i = \mu_i$. If a reservoir that acts as a current source has to obey this relation with a certain accuracy, the current density must nearly vanish. This point has been stressed repeatedly by Landauer[62,63] and Büttiker,[48] and has been a focal point of a paper by Levinson.[64] A narrow lead with only a few quantum channels is not a reservoir and does not permit self-consistency. This is of special importance in a two- or three-terminal measurement, since in this case, a current source or sink also serves as a voltage terminal. In such cases, we always must require that the number of channels M_i, which characterize a reservoir, largely exceeds all transmission probabilities into this terminal. On the other hand, if we use a terminal only as a voltage probe, the density of states is of no importance. (The net current is zero, anyway.) Sufficiently far into the voltage probe, we have $eV(r) + E_i = \mu_i$ regardless of the density of states of the probe.[48] The four-terminal discussion is more general in the following sense: The μ_i can be assumed to characterize the incident carriers. In a four-terminal measurement, self-consistency imposes less restrictive conditions. The discussion presented in this section also is valid if the *incident* states in the current source and current sink probe can be characterized by a *common* chemical potential. In such a case, the relation $eV(r) + E_i = \mu_i$ does not hold for the current probes, but only for the voltage probes. The four-terminal approach is more general for exactly the reason it is the preferred experimental technique of measuring resistances: *Series resistances* in the current probes and voltage probes leave the four-terminal resistance unchanged. In this sense, Fig. 2 can be viewed as a model of a portion of a much larger conductor, i.e., a conductor with long leads.

A number of applications of the resistance formulae given in the preceding are presented in the remaining part of this chapter. Here, we mention only one field, which recently has been a testing ground for this approach. The discovery by Roukes et al.[65] of the vanishing of the low-field Hall effect in narrow submicron ballistic junctions and the closely related negative longitudinal resistances seen by Takagaki et al.[66,67] have generated considerable interest. At helium temperatures, electron scattering in these structures is due predominantly to the confinement and the junction geometry. In these structures, the Hall resistance can be manipulated geometrically to be zero, enhanced over the classical value of a boundary-free two-dimensional electron gas, or can be made to be negative as shown in experiments by Ford et al.[68,69] The role of the junction geometry also is elucidated in experiments by Chang et al.[70] The extent to which a recent set of theoretical papers[71–74] using Eq. (18) has been able to reproduce the experimental observations is discussed by Roukes et al.[75] and by Thornton et al.[162] In this chapter, we will address ballistic junctions only briefly in Section VI, focusing on high-field behavior.[68,76,77]

III. Quantization and Interference in the Absence of a Field

In this section, we illustrate the application of the resistance formulae given in Section II by treating several problems of electron motion in the absence of a field. Our first example relates to the recent experimental discovery of quantized conductance steps through narrow constrictions formed by split gates over a two-dimensional electron gas. The technique, which achieves narrow conduction channels, was pioneered by Thornton et al.,[78] and quantized conductance steps have been reported Wharam et al.[79] and van Wees et al.[80] The same technique can be used to build multi-probe samples with controllable contacts, which provides geometries for many interesting experiments[56,81] The discussion given next only treats a simple situation and is advanced here only to demonstrate some of the key elements needed to observe a quantized conductance. This is followed by a discussion of a multi-terminal measurement of a such a constriction resistance. Finally, to illustrate the importance of phase-coherent quantum effects in a four-terminal measurement, we discuss a simple single-channel conduction problem.

8. Quantized Transmission at a Saddle-Shaped Potential

The experimental discovery of the conductance steps in the absence of a magnetic field in narrow constrictions of a two-dimensional electron gas was preceded by a number of more or less clearly formulated theoretical hints. Ref. 21 discussed the opening of new quantum channels in terms of a conductance formula that expressed resistance as the ratio of an electrostatic potential drop calculated in narrow perfect wires connected to a barrier in a narrow conductor. In this case, the opening of a new channel does not lead to a conductance that is a monotonic function of the gate voltage, but shows an oscillatory behavior. Imry[50] made the connection between Eq. (15) and contact resistances. In his discussion, Imry assumes a (uniform) perfect wire coupled without scattering to reservoirs. In an investigation of the addition of series resistances in Ref. 40, it was noted that the resistance of a perfect wire increases by $(h/e^2)(1/N)$ if an inelastic scatterer is inserted into the perfect wire, which randomizes both the phase and the velocity of the carriers. The clearest anticipation known to us came in informal discussions with Garcia,[82] who pointed to oscillations in a monotonically increasing conductance. None of these works, however, specified conditions for the observation of the effect. The need to consider in detail how carriers emanating from a reservoir reach the constriction is spelled out by Landauer.[62] Since the experimental discovery, a rather large number of theoretical papers have appeared. Various geometries have been treated mostly by solving the Schrödinger equation numerically, typically assuming that the confining potential is a hard-wall potential with sharp or, at best, tapered corners.[25,83–91] In contrast, a set of papers by Glazman et al.,[92] Kawabata[93] and Yacoby and Imry[94] consider the same geometry as Landauer[62] and assume that the potential is smooth along the constriction, but still assume hard walls. Only a few papers are concerned with the calculation of the real shape of the potential.[95–98] Clearly, due to screening effects, the potential must be a smooth function in every direction. Charge is depleted not only underneath the gates but also some distance from the gates. Locally at the bottleneck, therefore, we can expand the potential and obtain, in appropriate coordinates, an electrostatic saddle point potential,

$$V(x, y) = V_0 - \left(\frac{1}{2}\right) m\omega_x^2 x^2 + \left(\frac{1}{2}\right) m\omega_y^2 y^2. \tag{22}$$

Here, we have expressed the curvature of the saddle-point potential in terms of the frequencies ω_x and ω_y. We assume now that we can neglect scattering far away from the saddle and first treat only the potential specified by Eq. (22). To the extent that higher-order corrections in $V(x, y)$ can be neglected, this is a simple, illustrative problem that can be solved exactly.[99,100] (It is, as we discuss later on, exactly solvable even in the presence of a magnetic field.[101] The total energy is given by Eq. (22), supplemented by a kinetic energy term $p^2/2m$. This Hamiltonian is separable for transverse motion and longitudinal motion. The transverse motion gives rise to quantized eigenvalues $\hbar\omega_y(n + 1/2)$. This generates an effective potential,

$$V_0 + \hbar\omega_y\left(n + \frac{1}{2}\right) - \left(\frac{1}{2}\right) m\omega_x^2 x^2, \tag{23}$$

for longitudinal motion. This energy can be viewed as the band bottom of the nth quantum channel. (See also Ref. 92.) Classically, i.e., if quantum tunneling is neglected, the nth channel has a threshold energy at the saddle given by

$$E_n = V_0 + \hbar\omega_y\left(n + \frac{1}{2}\right). \tag{24}$$

Channels with energy $E_n < E_F$ are open and channels with energy $E_n > E_F$ are closed. Due to quantum tunneling, this threshold is not sharp. The result for the transmission probability for carriers incident in channel n with energy E can be expressed with the help of the variables,

$$\varepsilon_n = \frac{2(E - \hbar\omega_y(n + 1/2) - V_0}{\hbar\omega_x}, \qquad (25)$$

in the simple form,

$$T_{nm} = \delta_{nm}\frac{1}{1 + e^{-\pi\varepsilon_n}}. \qquad (26)$$

Because of the harmonic form of the potential, there is no channel mixing. Figure 5 shows the transmission probabilities T_{nn} for $n = 0, 1, 2, 3, 4, \ldots$ for the case $\omega_y/\omega_x = 3$, and shows the total transmission probability $T = \Sigma_n T_{nn}$ as a function of $(E - V_0)/\hbar\omega_x$. Figure 5 makes it evident that the appearance of plateaus in the total transmission probability T (and hence in the conductance, Eq. (15)) depends on the rapid rise of the transmission probabilities T_{nn} from zero to one over an energy range that is small compared to the distance between the plateaus. The energy derivative of the total transmission

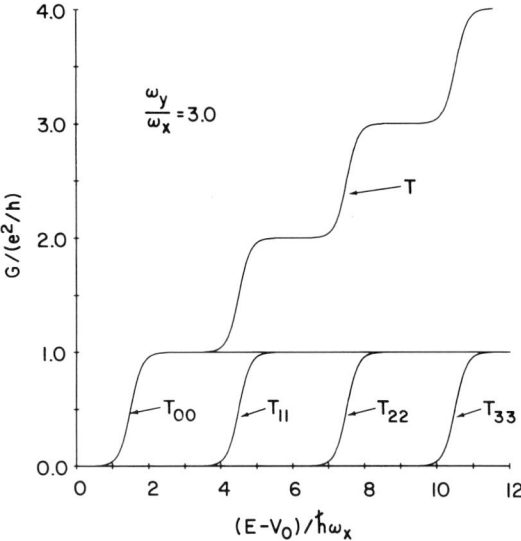

FIG. 5. Transmission probabilities of a saddle-point potential given by Eq. (22) as a function of $(E - V_0)/\hbar\omega_x$ for a ratio $\omega_y/\omega_x = 3$. T_{nn}, $n = 0, 1, 2, 3, \ldots$ are the transmission probabilities of the individual quantum channels. $T = \Sigma_n T_{nn}$ is the total transmission probability that, when multiplied by (e^2/h), yields the conductance of the saddle-point constriction. (From Ref. 102).

probability,

$$\frac{dT}{dE} = \sum_{n=0}^{n=\infty} \frac{2\pi}{4\hbar\omega_x} \frac{1}{\cosh^2(\pi\varepsilon_n/2)}, \qquad (27)$$

is maximal at the energies at the classical channel threshold and is minimal at the energies,

$$E_n = V_0 + \hbar\omega_y(n + 1). \qquad (28)$$

The nearly flat portions of the total transmission probability are centered around the energies determined by Eq. (28). The distance between the points of minimal slope is $\hbar\omega_y$. Hence, quantization occurs if the transmission probabilities T_{nn} rise from zero to one in an energy interval that is small compared to $\hbar\omega_y$. Since the energy sensitivity of the transmission probability is governed by $\hbar\omega_x$, we find that the conductance shows a series of marked steps if[102]

$$\omega_y \geq \omega_x. \qquad (29)$$

Thus, the potential has to vary slowly along x compared to the variation along y. Evaluated at the center of the plateau, Eq. (27), yields a slope,

$$\frac{dT}{dE} = \left(\frac{4\pi}{\hbar\omega_x}\right) e^{-2\pi\omega_y/4\omega_x}, \qquad (30)$$

up to corrections of order $(e^{-\pi\omega_y/4\omega_x})^2$. Thus, for large ratios ω_y/ω_x, the slope at the center of the plateaus is exponentially small. Figure 6 shows a series of conductance traces for a sequence of saddle-point potentials determined by the ratio ω_y/ω_x that is varied in the interval from 0 to 5 in increments of 0.25. It is seen already that for ratios that are moderately larger than 1, the conductance shows a considerable structure, and if the ratio approaches the maximum value shown, the plateaus are entirely flat.

The local scattering problem investigated here is a complete solution of reservoir-to-reservoir transmission only under a set of very restrictive conditions: It is a complete solution if every carrier that has been transmitted past the local region of the saddle has a probability of one to leave the constriction in the forward direction, and if every carrier that is reflected leaves the constriction in the backward direction with a probability of one. That does not necessarily mean that carriers are not scattered as they leave the saddle, but it requires that carriers do not reverse momentum. This can be achieved by gradually widening the constriction away from the bottleneck as discussed by

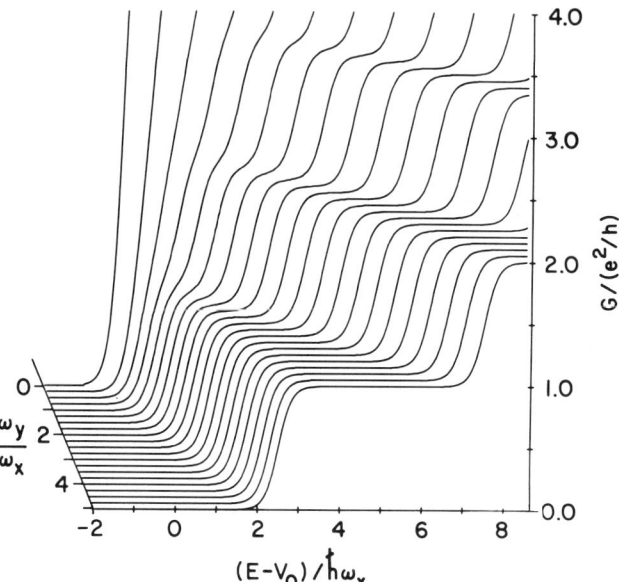

FIG. 6. Conductance of a saddle-point constriction as a function of $(E - V_0)/\hbar\omega_x$ for differing saddle potentials characterized by ω_y/ω_x. (From Ref. 102.)

Landauer[62] and Glazman et al.[92] Multiple crossings of the bottleneck due to scattering can lead to resonances.[83,86,90] Furthermore, we have in the discussion just given assumed that the channels arriving at the saddle are completely filled up to the Fermi energy. Scattering away from the saddle, which reverses momenta, will give rise to channels that are not completely filled. For these reasons, experiments do not give rise to plateaus that are flat with an accuracy as predicted by Eq. (30), but can show considerable deviations. Furthermore, all two-terminal measurements have the feature that additional sources of resistance add in series and thus, the plateaus seen in experiments are typically displaced away from the quantum values in a systematic way. In view of this, it is clear that the quantization of ballistic contact resistances is unlikely to be a competitor of the quantum Hall effect as far as accuracy is concerned.

9. FOUR-TERMINAL MEASUREMENT OF CONSTRICTION RESISTANCES

Some experiments on constriction resistances invoke a four-probe geometry as shown in Fig. 7. Van Houten et al.[56] have studied the four-probe resistance for weak magnetic fields and have presented both experimental

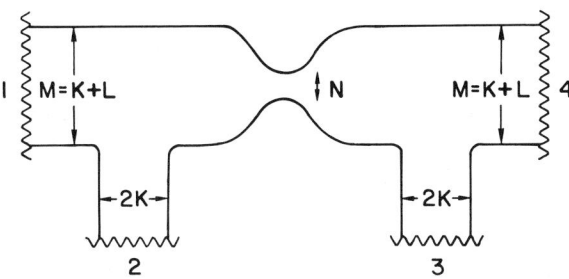

FIG. 7. Four-probe geometry with constriction. The number of perfectly transmitted quantum channels of the constriction is N. Probes 1 and 4 admit $M = K + L$ quantum channels and probes 2 and 3 admit $2K$ quantum channels. Carriers incident in probes 1 or 4 in L quantum channels do not enter probes 2 and 3.

data and theoretical considerations. Landauer[63,103] investigates the problem by calculating a local electric potential using a charge neutrality condition. We augment these discussions by taking all four probes into account to obtain an *estimate* for the measured four-terminal resistance. In a strict sense, we go beyond the formalism of Section II here, since we want to incorporate at least some inelastic relaxation inside the conductor.

We make the following assumptions: The constriction permits transmission of N channels with the lowest subband energies. If carrier flow is from left to right, this has the consequence that immediately past the constriction, the subbands are populated unequally, an effect that is discussed under the label of *collimation*.[60,98,104,161,163] On the left-hand side of the constriction, the modes with the lowest N subband energies are filled. On the right-hand side of the constriction, the N modes with the lowest thresholds are not fully occupied by reflected carriers. We shall now assume that weak inelastic forward scattering equilibrates channels that carry carriers in the same direction. That means the current that has been transmitted through the junction and is proportional to N eventually will be equilibrated such that all M channels in the wide lead are equally populated with a population proportional to N/M per channel. Similarly, the reflected current that is proportional to $M - N$ eventually populates all M channels, permitting carriers to move away from the junction equally and proportional to $(M - N)/M$. Let us assume that probes 2 and 3 are sufficiently far away from the constriction for this equilibration process to take place. Therefore, the fact that the probes inject carriers up to differing chemical potentials will have no effect on the quantization of the constriction resistance.

Let us specify the junction properties (the interconnection of probes 1 and 2 and of probes 3 and 4) by assuming the following: If the M incident channels in lead 1 are full, the total probability of carriers to go past the junction without entering probe 2 is L, and the total probability of carriers entering probe 2 is K. We further assume, for simplicity, that the junction exhibits no reflection regardless from which branch carriers are incident. Finally, we want to assume that the carriers incident from probe 2 are scattered equally to the left and to the right with a total probability K. For the junction of probes 3 and 4, we make the same assumptions. Thus, probes 1 and 4 have $M = K + L$ conducting channels and probes 2 and 3 each have $2K$ conducting channels. Note that here, K, L, and M merely estimate the width of the probes and are not indicative of quantized transmission. With these specifications, it now is easy to calculate all the global transmission and reflection probabilities. Consider R_{11}. Of the total incident flux M, a flux L is transmitted past the junction, and at the constriction, carriers are reflected with probability $(M - N)/M$. Thus, the flux reflected at the constriction is $L(M - N)/M$. Of this flux, a portion L/M makes it past probe 2 back into probe 1. Therefore, we find $R_{11} = L^2(M - N)/M^2$ for the total reflection probability at probe 1. As a second example, consider the probability T_{21} for transmission of carriers incident in probe 1 and outgoing in probe 2. There is a total probability K for carriers from probe 1 to reach probe 2 directly. In addition, there is indirect transmission into probe 2 from carriers that at first pass probe 2 and subsequently are reflected at the constriction. A portion of these reflected carriers enters probe 2 on their return. The contribution to the total probability of this process is $L((M - N)/M)(K/M)$. Thus, the total probability for transmission from probe 1 to probe 2 is $T_{21} = K + L((M - N)/M)(K/M)$. In this fashion, all the transmission and reflection coefficients can be calculated. The result for the four-terminal resistances is

$$R_{14,23} = R_{23,14} = \frac{h}{e^2}\left(\frac{1}{N} - \alpha\frac{1}{M}\right), \quad (31)$$

where the factor α is

$$\alpha = \frac{2L}{2L + K}. \quad (32)$$

Van Houten et al.[105] found $\alpha = 1$ and understood Eq. (31) as the difference of two resistances associated with the N open quantum channels of the constriction and the M channels of the wide lead. (The wide lead also forms a constriction with regard to the still larger reservoirs 1 and 4). Landauer found a factor $\alpha = 2/\pi$. Our result stresses the dependence of the four-terminal resistance on the width of all the probes. Suppose that every carrier that approaches probe 2 from either the left-hand side or the right-hand side enters probe 2. This corresponds to $L = 0$. Hence, in this case, α is zero. Probe 2 provides complete equilibration and the four-terminal resistance is the same as for a constriction between two reservoirs in a two-terminal configuration.

If probes 1 and 4 nominally have the same width as probes 2 and 3, $L = K$, and we find $\alpha = 2/3$. Finally, when $K \ll L$, the factor α is proportional to $1 - (K/2L)$ and approaches 1. The reason that the four-terminal resistance depends on the relative width of the probes is a purely *classical* effect. Carriers enter the conductor from all four terminals and with maximum energies equal to the chemical potential of the contact. Clearly, the result we found depends on the details of our assumptions. The results discussed in the preceding are a consequence of weak inelastic scattering in the forward direction, and are a consequence of our assumption that there is no reflection at the junction. Even for this purely *classical* scattering problem, there is no shortcut to the answer. If a magnetic field is switched on, our assumption that carriers injected in probe 2 are transmitted equally to the left and to the right breaks down. We will reconsider the geometry of Fig. 7 in our discussion of high-field transport. In the quantum Hall regime, there are situations where Eq. (31), with $\alpha = 1$, is not merely an approximation but *exact*.[8] Next, we consider a four-terminal single-channel problem where we focus specifically on quantum-coherence effects.

10. ON THE NATURE OF PHASE-SENSITIVE FOUR-TERMINAL RESISTANCES

To elucidate the role of phase coherence in a voltage measurement, we consider the simple situation shown in Fig. 8. A conductor in which electrons populate only one subband contains a barrier with reflection probability R and transmission probability T. A third probe is very weakly coupled to this conductor via a tunnel barrier. This third probe could be the tip of a tunneling microscope (or could be a very narrow constriction). A similar geometry was considered by Engquist and Anderson.[39] In their discussion, they disregard the possibility of a phase-sensitive voltage measurement based on the notion that the reservoir to which the voltage probe couples is wide and has many modes. The use of the tunneling miscroscope as a voltage probe has been demonstrated,[106] and also has been the subject of theoretical investigations by Chu and Sorbello.[107,108] The calculation[109] presented shortly makes no pretense of being realistic but serves to demonstrate phase-sensitive effects associated with voltage measurements. Levinson[64] has made a similar calculation and reaches conclusions similar to those of Ref. 109. An experimental demonstration of voltage measurements that are sensitive to the phase of the electron wave function, but use voltage probes that are strongly coupled to the conductor, was given by Benoit et al.[47] In this experiment, it was shown that the phase of the Aharonov–Bohm oscillations depends on the multi-probe character of the conductor.[22] Subsequent experiments, which investigated the fluctuations at voltage probes spaced closely

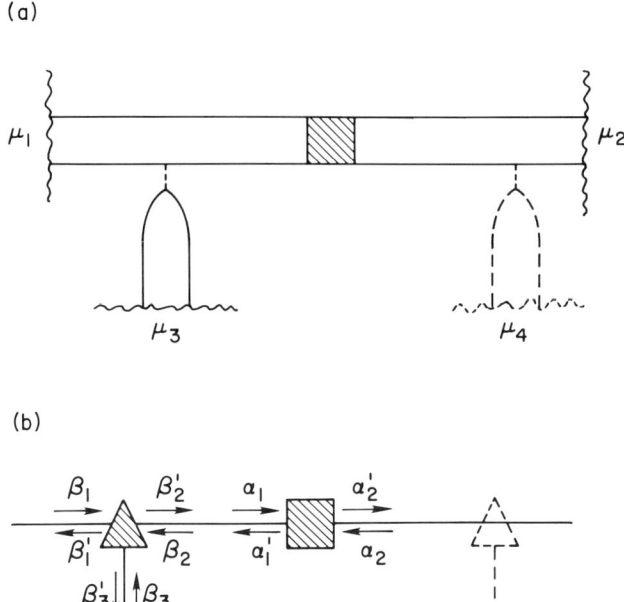

FIG. 8. (a) Multiprobe conductor with one or more weakly coupled probes. Aside from the probes, the only scattering process considered is that due to a single barrier (shaded area). (b) Schematic representation of the conductor depicted in (a). α and β are the current amplitudes that need to be calculated to determine the overall transmission probabilities.

compared to a phase-coherence length (reviewed in Ref. 110), attracted more attention to these phenomena and led to theoretical efforts to describe these phenomena in detail.[55,58,111,112] In contrast to this work, the situation investigated here uses voltage probes that only are weakly coupled to the conductor. An additional very direct demonstration of phase-sensitive effects in four-probe measurements came with the electron focusing experiments of van Houten et al.[56]

The coupling of the probe to the conductor is described by a 3×3 scattering matrix with three input and output channels corresponding to incoming and outgoing electron waves in the three branches leading up to the junction. (See Fig. 8b.) The matrix elements s_{13} and s_{23}, which describe transmission from the probe to the conductor, are $|s_{13}|^2 = |s_{23}|^2 = \varepsilon$, where the parameter ε is very small compared to 1. Consider now the transmission probabilities for a carrier incident in the weakly coupled probe to reach probe 1. There are two processes: Carriers incident from the junction that go to the left give rise to an amplitude s_{13}; carriers that enter the conductor to the right have a probability amplitude r to be reflected at the barrier. These carriers travel to the left past the probe and give rise to an additional amplitude $s_{23}r \exp(-i2k_F d)$. Here, d is the distance between the barrier and the probe and r is the reflection amplitude of the barrier. The total transmission probability amplitude t_{13} is the sum of the amplitudes of both of these processes. The two processes are coherent, since they are a consequence of a single incident wave. Therefore, the transmission probability T_{13} contains a phase-sensitive term and is given by[109]

$$T_{13} = \varepsilon(1 + R + 2R^{1/2}\cos(\chi_3)). \tag{33}$$

$R = |r|^2$ is the reflection probability of the barrier, and $\chi_3 = 2k_F d + \Delta\phi + \phi_{13} - \phi_{23}$. Here, $\Delta\phi$ is the phase associated with reflection and ϕ_{13} and ϕ_{23} are the phases associated with the matrix elements s_{13} and s_{23}. Consider next the transmission amplitude for carriers incident in probe 3 and reaching probe 2. To order $\sqrt{\varepsilon}$, there is no interference effect and this amplitude simply is $t_{23} = s_{23}t$, where t is the transmission amplitude of the barrier. The transmission probability is just $T_{23} = \varepsilon T \equiv \varepsilon|t|^2$. In the absence of a magnetic field, the case of interest here, $T_{13} = T_{31}$ amd $T_{23} = T_{32}$ Furthermore, current conservation also determines $R_{33} = 1 - T_{31} - T_{32}$.

Let us now consider the flux from probe 1 to probe 2 and study the chemical potential measured at probe 3. Considering this as a three-terminal problem, Eq. (11), with $I_3 = 0$, yields[48]

$$\mu_3 = \frac{T_{31}\mu_1 + T_{32}\mu_2}{T_{31} + T_{32}}. \tag{34}$$

Using the transmission probabilities as determined in the preceding, we find

$$\mu_3 = \frac{\mu_1 + \mu_2}{2} + \frac{(R + R^{1/2}\cos\chi_3)}{1 + R^{1/2}\cos\chi_3}\frac{(\mu_1 - \mu_2)}{2}. \tag{35}$$

A repetition of the calculation for the case that the probe is to the right of the barrier yields a chemical potential,

$$\mu_4 = \frac{\mu_1 + \mu_2}{2} - \frac{(R + R^{1/2}\cos\chi_4)}{1 + R^{1/2}\cos\chi_4}\frac{(\mu_1 - \mu_2)}{2}, \tag{36}$$

where $\chi_4 = 2k_F d_R + \Delta\phi' + \phi_{23} - \phi_{13}$. Here, d_R measures the distance of the probe from the scatterer and $\Delta\phi'$ is the phase associated with reflection at the barrier for carriers incident from the right. Figure 9 shows the chemical potentials μ_3 and μ_4 as a function of the distance of the probe from the scatterer. The chemical potentials are on oscillatory function of the probe distance. The oscillation period for this simple one-channel problem is

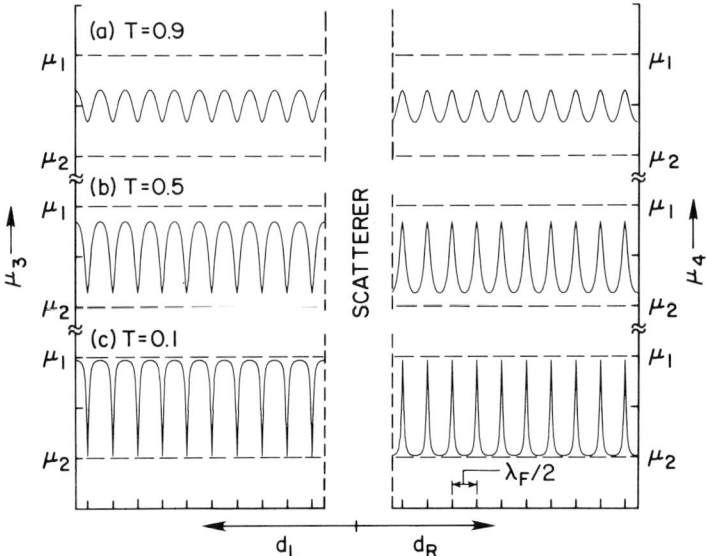

FIG. 9. Chemical potential oscillations as a function of the distance d_L and d_R of the probes away from the scatterer for the transmission probabilities, $T = 0.9$, $T = 0.5$, $T = 0.1$. Note that the amplitude of the oscillations is the same on either side of the barrier. (From Ref. 109.)

$\lambda_F/2 = \pi/k_F$, reminiscent of Friedel oscillations. (Note, however, that in a multi-channel conductor, the fluctuations are multiply-periodic with long periods of the order of the width of the conductor.) The remarkable feature of these results is the fact that, as shown in Fig. 9, the peak amplitude of the oscillations of the chemical potentials is the same on both sides of the barrier. Suppose that we make a four-terminal measurement of the resistance. Two weakly coupled probes as described before measure a chemical potential difference of $\mu_3 - \mu_4$. Since the amplitudes of the oscillations of the chemical potential are the same on either side of the barrier, it follows that the chemical potential difference can be either positive or negative, depending on the location of the probes. The resistance is calculated easily. The current from probe 1 to probe 2 is in the lowest-order independent of the coupling parameter ε and given by $I = (e/h)T(\mu_1 - \mu_2)$. Thus, the four-probe resistances are[109]

$$R_{12,34} = R_{34,12} = \frac{\mu_3 - \mu_4}{eI}$$

$$= \left(\frac{h}{e^2}\right)\left(\frac{1}{2}\right)\frac{2R + (1 + R)R^{1/2}(\cos\chi_3 + \cos\chi_4) + 2R\cos\chi_3\cos\chi_4}{(1 + R^{1/2}\cos\chi_3)(1 + R^{1/2}\cos\chi_4)}.$$

(37)

We could have used Eq. (18) equally well and, by taking into account that two probes are weakly coupled, also would arrive at Eq. (37). This resistance is shown in Fig. 10 for three different phases, $\chi_3 = 1, 0, -1$ as a function of χ_4 for a barrier with a transmission probability, $T = 0.5$. The upper bound for these resistances is $R^{1/2}/T$ and the lower bound is $-R^{1/2}/T$. Let us compare this with the resistance we would measure if both probes are, say, to the left of the barrier. One measurement is given by μ_3 with a phase χ_3; the other measurement also is determined by Eq. (35) but with a different phase which we denote by χ_5. Denote the potential associated with this phase by μ_5. We then find, instead of Eq. (37),

$$R_{12,35} = R_{35,12} = \frac{\mu_3 - \mu_5}{eI} = \left(\frac{h}{e^2}\right)\left(\frac{1}{2}\right)\frac{R^{1/2}(\cos\chi_3 - \cos\chi_5)}{(1 + R^{1/2}\cos\chi_3)(1 + R^{1/2}\cos\chi_5)}, \quad (38)$$

which again has an upper bound of $R^{1/2}/T$ and a lower bound of $-R^{1/2}/T$. Hence, from a single four-terminal resistance measurement, we could not determine if the barrier in fact is between the probes or to either side of the two probes. To make this distinction, a lot of measurements are needed to examine the likelihood of positive and negative resistances.

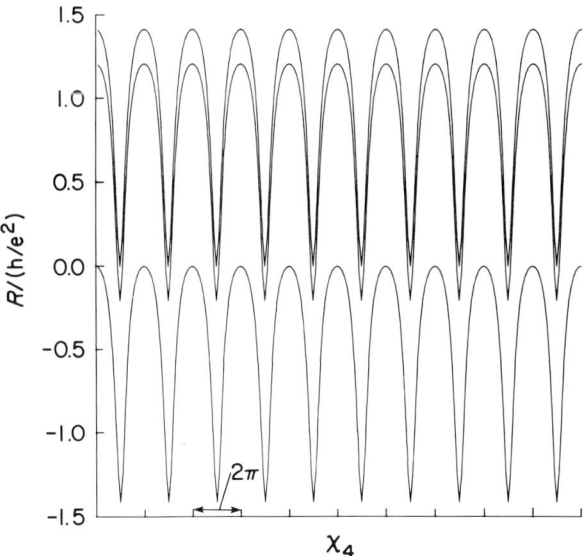

FIG. 10. Phase-sensitive four-terminal resistance, Eq. (37), as a function of the phase angle χ_4 for fixed values of $\chi_3 = 1$ (top), 0 (middle), and -1 (bottom). The transmission probability of the barrier is $T = 0.5$. (From Ref. 109.)

Suppose we move the probes over distances that are large compared to a Fermi wavelength and determine the phase-averaged chemical potentials. A simple calculation yields

$$\langle \mu_{3,4} \rangle = \frac{\mu_1 + \mu_2}{2} \pm \frac{1}{2}(1 - \sqrt{T})(\mu_1 - \mu_2), \qquad (39)$$

where the brackets denote a phase average and the plus sign applies for μ_3 and the minus sign for μ_4. Consequently, the phase-averaged resistance is

$$\langle R \rangle = \frac{h}{e^2} \frac{1 - \sqrt{T}}{T} \qquad (40)$$

if the barrier is between the probes. Since μ_5 differs from μ_3 only with respect to the phase over which we average, we have $\langle \mu_5 \rangle = \langle \mu_3 \rangle$ and consequently, the phase-averaged resistance, Eq. (38), is zero. The result for the phase-averaged resistance with the barrier between the probes is interesting. It depends not only on the transmission probability but also on the square root of the transmission probability. This points to its origin, which is an interference effect.

So far, we have assumed that there is no inelastic scattering except in the reservoirs to which the probes are connected. It is for this reason that the oscillations in the chemical potential persist over arbitrary distances away from the barrier. A more realistic description should take into account that phase-randomizing events can take place throughout the conductor. Let us denote the length over which phase coherence is maintained by l_ϕ. Sufficiently far away from the barrier, the oscillations of the chemical potential are reduced exponentially, i.e., the interference terms proportional to $R^{1/2}$ acquire a factor $\exp(-2d_L/l_\phi)$ for the probe to the left of the barrier and a factor $\exp(-2d_R/l_\phi)$ for the probe to the right of the barrier. Instead of the resistance, Eq. (40), we obtain after phase-averaging,

$$\langle R \rangle = \frac{h}{e^2} \frac{1}{T} \left\{ 1 - \frac{T}{2} \left(\frac{1}{\sqrt{1 - Re^{-4d_L/l_\phi}}} + \frac{1}{\sqrt{1 - Re^{-4d_R/l_\phi}}} \right) \right\}. \qquad (41)$$

For $l_\phi \gg d_L$ and $l_\phi \gg d_R$, Eq. (41) reduces to the phase-averaged resistance just calculated. For $l_\phi \ll d_L$ and $l_\phi \ll d_R$, Eq. (41) gives Landauer's result,[34]

$$\langle R \rangle = \frac{h}{e^2} \frac{1 - T}{T}. \qquad (42)$$

Not all junctions necessarily give rise to the interference effect discussed in the preceding. The simple junctions as specified in Refs. 40 and 41 have the property that a wave approaching a junction from the left enters one probe channel and a wave approaching from the right enters another probe channel. Certain matrix elements are taken to be zero. That simplifies calculations but hardly can be expected to be a generic property of small junctions. Levinson[64] reaches similar conclusions concerning the range of validity of Eq. (42). He points out that temperature averaging can be another reason for loss of phase sensitivity. The phase $2k_F d$ is a strong function of energy if d is large. The phase sensitivity is lost if $kT > \hbar v_F/d$. For a Fermi velocity of $v_F = 3 \times 10^7$ cm/sec and $kT = 1$K, the distance over which probes can be phase-sensitive is still of the order of 2 μm. It is interesting to note that the Friedel-like oscillations in the chemical potential do not occur only in single-mode wires but as has been shown now by Chu and Sorbello,[108] also near grain boundaries in metallic films. With regard to the main interest of this chapter, electron transport in high fields, we mention here only that interference effects at voltage probes are detrimental to the observation of quantized resistances. It is interesting, therefore, to study the circumstances in high magnetic fields under which such effects are absent.

IV. Motion in High-Magnetic Fields: The Two-Terminal Resistance

In this section, we extend the concept of quantum channels to high magnetic fields. It is shown that these channels have the properties that were used in Section II to derive the resistance formulae. This then guarantees that the formulae presented in Section II also are valid in high magnetic fields. This is followed by a discussion of the quantization of the two-terminal resistance in high magnetic fields for conductors of differing aspect ratios and for constrictions.

11. Motion in High Magnetic Fields

We begin our discussion with an investigation of the geometry of Fig. 11. A strip of two-dimensional electron gas is connected to metallic contacts to the right and left. Let us assume at first that the two-dimensional strip is a perfect, impurity-free conductor extended along x and with a transverse coordinate y. (Later on, we will consider a disordered potential.) The Hamiltonian describing motion in the perfect strip is

$$H = \frac{1}{2m}\left(p - \frac{eA}{c}\right)^2 + V(y), \tag{43}$$

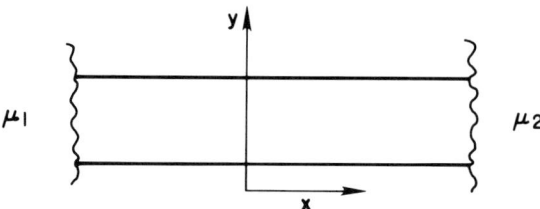

FIG. 11. Two-dimensional electron gas strip connected to the right and left to metallic contacts.

where $V(y)$ is the confining potential and where the vector potential in the Landau gauge is taken to be $A = (By, 0)$. This Hamiltonian is invariant under translations along x and the wave functions thus are of the type,

$$\psi_{j,k} = e^{ikx} f_{j,k}(y), \tag{44}$$

where k is the momentum along x and f_k is a function of y only. The Schrödinger equation for f becomes

$$Ef = -\frac{\hbar^2}{2m}\frac{d^2f}{dy^2} + V(y)f + \frac{1}{2}m\omega_c^2(y - y_0)^2 f. \tag{45}$$

Hence, $\omega_c = |eB/mc|$ is the cyclotron frequency, and

$$y_0 \equiv -kl_B^2, \tag{46}$$

with $l_B = (\hbar c/|eB|^{1/2})$ the magnetic length. This is just the Schrödinger equation for a particle moving in the one-dimensional potential $V(y)$ supplemented by a magnetic energy,

$$E_M = \frac{1}{2}m\omega_c^2(y - y_0)^2, \tag{47}$$

which creates a harmonic potential centered at y_0.

First, let us consider a confining potential $V(y)$ that is zero in the interior region of the electron gas, and at its boundary, $y_{1,2} = \pm W/2$ rises to infinity. (Potentials for real samples are rather smooth functions and will be discussed shortly.) In the interior of the sample, where $V(y)$ is zero, Eq. (45), is the eigenvalue problem for a harmonic oscillator and, consequently, the eigenvalues are

$$E_{jk} = \hbar\omega_0\left(j + \frac{1}{2}\right), \tag{48}$$

with $j = 0, 1, 2, 3, \ldots$ These eigenvalues are independent of k and the coordinate y_0. The eigenvalues of Eq. (48) give the bulk Landau levels of a perfect conductor. Near the boundary of the sample, we no longer can neglect the confining potential, and the energy levels depend on y_0 through the distance to the wall, $y_1 - y_0$ or $y_2 - y_0$. Because the confining potential adds to the magnetic energy potential, the energy levels rise as the wall is approached.[4,113,114] Thus, the energy of a state is determined by

$$E_{jk} = E(j, \omega_c, y_0(k)), \quad (49)$$

which is represented graphically in Fig. 12. The velocity (drift) of the carriers is

$$v_{jk} = \frac{1}{\hbar} \frac{dE_{jk}}{dk} = \frac{1}{\hbar} \frac{dE_{jk}}{dy_0} \frac{dy_0}{dk}. \quad (50)$$

In the bulk of the sample, where for the perfect conductor of Fig. 11 the Landau levels are independent of position y_0, the velocity is zero. The states in the bulk correspond to carriers describing a cyclotron motion with a guiding center that is stationary. Near the wall of the sample, each Landau level is associated with a state that is a strong function of position. These states correspond to carriers skipping along the boundary of the sample and are

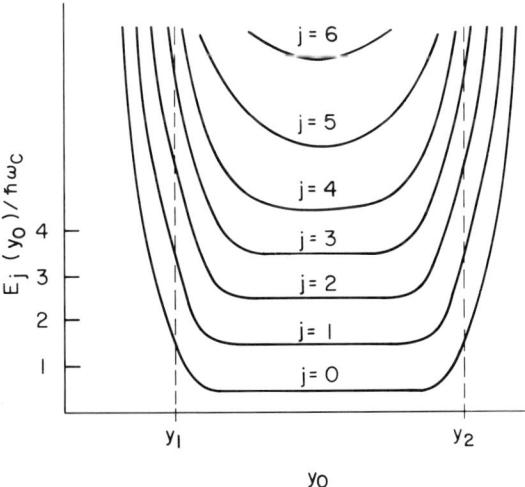

FIG. 12. Quantum channels in high magnetic field for a perfect conductor. $y_0 = -kl_B^2$ is the center of the harmonic oscillator wave functions. Hard walls are located at y_1 and y_2. Each bulk Landau level (flat portion of the dispersion in the center of the strip) gives rise to a branch of edge states near the sample boundary. (From Ref. 144.)

called *edge states*. The dispersions of Fig. 12 define the quantum channels for high magnetic transport much as the simpler $E(k)$ relations shown in Fig. 3 define the quantum channels in the absence of a magnetic field. For each of these quantum channels, the appropriate density of states is $dn/dk = 1/2\pi$, and thus the density of states with respect to energy is

$$\frac{dn_j}{dE} = \frac{dn}{dk}\frac{dk}{dE_j} = \frac{1}{hv_{jk}}. \tag{51}$$

Hence, the density of states is inversely proportional to the velocity of the carriers, as we found already in the absence of a field. Therefore, the current that can be injected into an edge state is

$$I = v_j \frac{dn_j}{dE}(\mu_1 - \mu_2) = \frac{e}{h}(\mu_1 - \mu_2). \tag{52}$$

Thus, the current that can be fed into an edge state (quantum channel) by a reservoir is quite independent of any particular properties of the system (magnetic field, effective mass...). Equation (52) is the essential ingredient in the derivation of the resistance formulae given in Section II. It follows, therefore, that the resistance formulae of Section II hold for the entire range of magnetic fields.[8] In view of the history of linear response formalism in high magnetic fields,[60] it is very remarkable that Eq. (11) provides a simple expression that is valid over the entire range of magnetic fields.

Let us next consider a confining potential $V(y)$, which varies slowly compared to the magnetic length l_B. Suppose y_A is a point near the edge of the sample. We then can expand $V(y)$ in a Taylor series away from y_A,

$$V(y) = V(y_A) + eF(y - y_A) + \frac{1}{2}\frac{d^2V}{dy^2}(y - y_A)^2 + O((y - y_A)^3). \tag{53}$$

Here, we have written the first derivative dV/dy as an electric field F characteristic of the slope of the potential at y_A. Let us, for the single purpose of a more concise notation, neglect the second order derivative of $V(y)$. The eigenvalue problem, Eq. (45), then can be rewritten with a magneto-electric energy that is centered not at $y_0 \equiv -kl_B^2$ but at $Y = y_0 - eF/m\omega_c^2$, giving rise to eigenenergies,

$$E + \frac{m}{2}v_B^2 = \hbar\omega_c\left(j + \frac{1}{2}\right) + eF(y_0(k_j) - y_A) + V(y_A). \tag{54}$$

The second term on the left-hand side is the kinetic energy of the guiding center in crossed magnetic and electric fields. These fields give rise to a drift velocity $v_B = cF/B$. (Since the kinetic energy is proportional to $(F/B)^2$ and the field eF is assumed to be small, we neglect this term next.) Thus, up to small corrections, the edge states have energies that follow the slope of the confining potential. A very important consequence of this behavior is that neighboring edge states at the Fermi energy can be very far apart in real space. Let $y_0(k_j)$ be the guiding center coordinate for carriers in the jth edge state at the Fermi energy. Using Eq. (53), we find that the distance between two states, $\Delta y = y_0(k_j) - y_0(k_{j+1})$, is

$$\Delta y = \frac{\hbar \omega_c}{eF}. \tag{55}$$

The transverse functions f_{jk} are harmonic oscillator wave functions and are proportional to $H_j((y - y_0)/l_B)\exp(-(y - y_0)^2/2l_B^2)$, where H_j is a Hermite polynomial of degree j. The width of these states is $\sigma = (\langle j|(y - y_0)^2|j\rangle)^{1/2} = l_B(j + 1/2)^{1/2}$. The ratio of the separation of edge states to the width of the edge states is

$$\frac{\Delta y}{\sigma} = \frac{\hbar \omega_c}{eFl_B} \frac{1}{\left(j + \frac{1}{2}\right)^{1/2}}. \tag{56}$$

This ratio is much larger than unity if the energy that a carrier gains in an electric field over the distance of a magnetic length eFl_B is much smaller than the cyclotron energy. This offers the important possibility that differing edge states can be rather well isolated from one another with little elastic or inelastic scattering between them.

Let us now assume that the confining potential $V(y)$ is supplemented with an impurity potential $V_i(x, y)$ such that the total potential is $V(x, y) = V(y) + V_i(x, y)$. If the impurity potential also is a smooth function over large distances compared to the magnetic length, then we can repeat the analysis given earlier at each point (x, y) to find that energy levels in the disordered region are determined by

$$E = \hbar \omega_c \left(j + \frac{1}{2}\right) + V(x, y). \tag{57}$$

The states of the system at a given energy are given by a set of equipotential lines, $E_G = V(x, y)$, where $E_G = E - \hbar \omega_c(j + \frac{1}{2})$ is the guiding center energy. If it is assumed that the fluctuations in the impurity potential are small compared to the cyclotron energy $\hbar \omega_c$, therefore, near the boundaries of the two-dimensional region, the confining potential is the dominant energy. It follows that there always are equipotential lines along the wall of the sample connecting one contact with another. A typical pattern of states at the Fermi energy for a conductor with three contacts is shown in Fig. 13. We reserve the name *edge states* or *open states* for solutions of Eq. (57) that connect contacts. Away from the edges of the sample, there are, due to the impurity potential, solutions of Eq. (57) that close on themselves, and we call such states *localized states*. Both the edge states and the localized states provide uni-directional paths for the carriers, as is evident when we think of them as analogs of classical skipping orbits. Each solution thus can be assigned an arrow that indicates the direction of carrier flow along the state. Figure 13 shows a typical configuration of edge states and localized states at the Fermi energy for the case when the Fermi energy lies between the second and third bulk Landau level.

Equation (57) is not exact. There might be regions of strong potential fluctuations in which the assumptions of a smoothly varying potential do not hold. One of these regions in Fig. 13, labeled A, is near the sample edge. A strongly fluctuating potential can cause scattering from one edge state to another. Such scattering, however, still permits perfect transmission along the edge of the sample, as the following simple consideration shows. Suppose scattering from one edge of the sample to the other edge of the sample is impossible. We then can encase a region along the edge that exhibits deviations from the adiabatic behavior described by Eq. (57) with a box that has N incoming states and N outgoing states corresponding to the number N of edge states on that side of the sample. Now if each of the incoming edge states

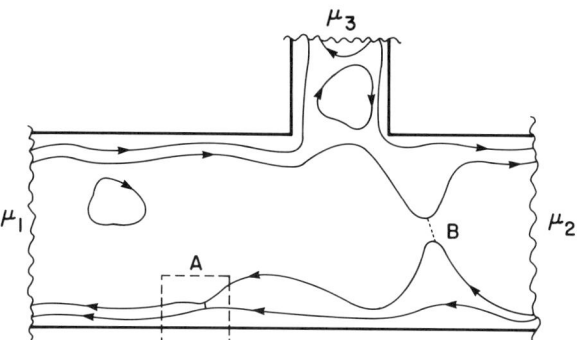

FIG. 13. Typical configuration of edge states at the Fermi energy. The local inter-edge state scattering process A allows for quantized transmission: If the incident states are fully occupied up to the same energy, all the outgoing states are fully occupied up to the same energy. The scattering process B allows carriers to be scattered back into the reservoir from which they are incident. Backscattering brings the edge state populations out of equilibrium. Hence, scattering near A is irrelevant only if processes of type B are absent. Contact 1 is ideal, contact 2 in the presence of backscattering, and contact 3 exhibit internal reflection.

is completely filled (carriers a unit current), then all the outgoing states are completely filled (carry a unit current). Hence, the total probability for transmission through the box is $T = N$ equal to the number of edge states. As long as all the edge states are completely filled, scattering along an edge causing mixing among edge states is irrelevant. Therefore, as long as there are no scattering paths from one side of the sample to the other, the edge states provide perfect conduction paths regardless of impurity scattering or roughness of the edge.

The processes of the type labeled B in Fig. 13 permit scattering of carriers from one sample side to the other. As we will discuss shortly, it is the absence of such backscattering processes which leads to the quantum Hall effect. Such processes can be expected to be absent if the Fermi energy lies between two Landau levels in a region where there are a few localized states present at the Fermi energy. We can achieve such a situation, since tunneling from an edge state to the localized state is proportional to $\exp(-d^2/l_B^2)$ due to the harmonic oscillator nature of the wave functions. Therefore, a separation d of a few magnetic lengths between edge states on opposite sides of the sample or between edge states and intervening localized states is sufficient to suppress backscattering to physically undetectable limits.

12. THE TWO-TERMINAL CONDUCTANCE

Suppose that the the two reservoirs contacting the strip of two-dimensional electron gas in Fig. 11 fill the edge states completely. Suppose also that there are no backscattering processes. Then the edge states at the Fermi energy permit perfect transmission from one reservoir to another. Thus, the total transmission probability in the presence of N edge states is $T = N$, and the conductance is

$$G = \frac{e^2}{h} T = \frac{e^2}{h} N. \tag{58}$$

We emphasize here that the quantization is a consequence of the equilibrium filling of the edge states. (A more detailed discussion of this point is given in Section 19.) The quantization of the two-terminal conductance in high magnetic fields is demonstrated nicely in a series of experiments by Fang and Stiles.[115,116] To describe the transition from one quantized value of the two-terminal conductance to another quantized value, we show in Fig. 14 the states at the Fermi energy for a sequence of four differing magnetic fields. The situation that corresponds to a quantized conductance $G = 3e^2/h$ is shown in Fig. 14a. We assume that the impurity potential is a smooth function of position[117-119] such that the states are described almost everywhere by Eq. (57).

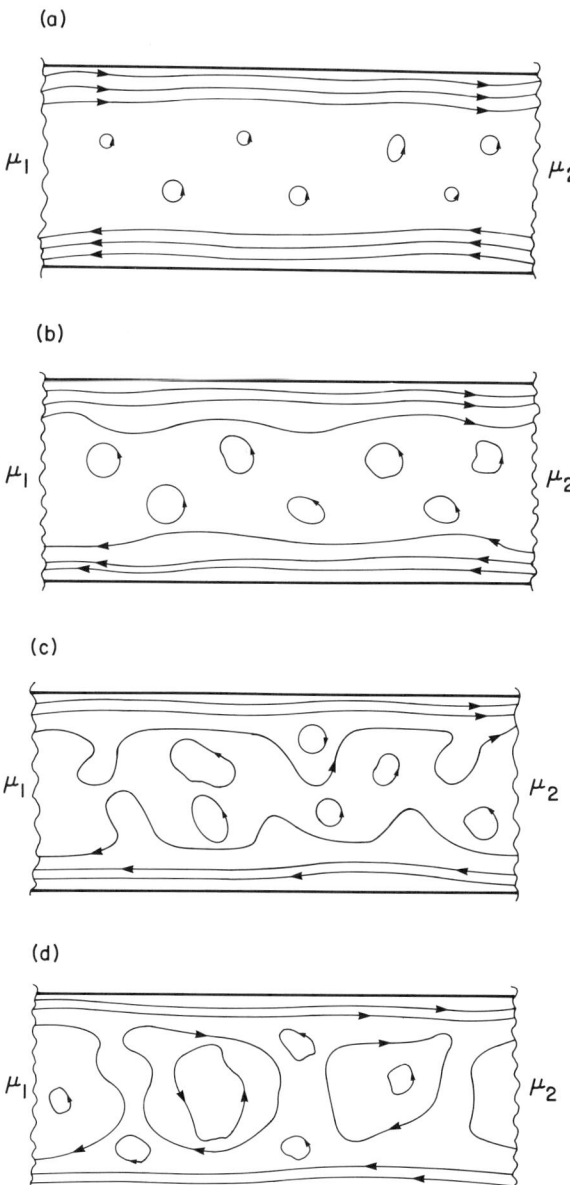

FIG. 14. States at the Fermi energy for a sequence of magnetic fields depicting the transition from a quantized two-terminal conductance plateau due to three edge states to the plateau determined by two edge states. With increasing magnetic field, the innermost edge state moves away from the sample edge and eventually is broken up into localized states. (a) Center of $N = 3$ plateau. (b) High magnetic field side of the plateau. (c) Onset of backscattering. (d) For fields such that the Fermi energy is below the center of the $N = 3$ Landau level, the innermost edge state is completely broken up into localized states.

4. THE QUANTUM HALL EFFECT IN OPEN CONDUCTORS

Let us denote the maximum slope of the impurity potential by $|\nabla V_i|_{\max}$. The potential is a smooth function if it fluctuates weakly over a length scale larger than the width σ of the wave functions, i.e., if $\sigma |\nabla V_i|_{\max} \ll \hbar \omega_c$. Furthermore, we assume that the confining potential $V(y)$ also is a smooth function. If $eF = -\nabla V(y)$ near the Fermi energy is small compared to $\hbar \omega_c/\sigma$ (as seen in Eq. (56)), the edge states are much further apart than a magnetic length. If the confining potential at the Fermi energy has a slope such that the electric field is in the window,

$$\frac{\hbar \omega_c}{\sigma} \gg eF \gg |\nabla V_i|_{\max}, \tag{59}$$

a particularly simple situation arises. The edge states are far apart from one another and experience no impurity potential near the sample boundary. Obviously, Eq. (59), cannot be correct under all circumstances. Especially, it cannot hold for the innermost edge state when the center of the Landau level is approached. For the innermost edge state, the relevant electric field must approach zero if the Fermi energy approaches the center of a Landau level. The innermost edge state then is very far apart from the outer edge states but experiences the impurity potential. Therefore, as the Fermi energy approaches a Landau level, the innermost edge state moves away from the sample edge into the bulk of the sample. (See Fig. 14b.) The density of localized states increases as shown in Fig. 14b. The localized states correspond to peaks in the impurity potential and hence the direction of carrier flow of the localized states is clockwise. If the field is increased further, the innermost edge state undergoes very large excursions into the bulk of the sample. (See Fig. 14c.) Processes of type B described in Fig. 13 now become possible and lead to backscattering. The specific process that gives rise to backscattering is quantum tunneling at very low temperatures if two states in the bulk approach very closely and are Mott-hopping processes at elevated temperatures. Eventually, if the field increases further and the Fermi energy falls below the center of the third Landau level, the edge state breaks up into disconnected loops (localized states). We then have reached a situation where there are two edge states along the sample boundaries disconnected from localized states in the bulk of the sample, and the conductance again is quantized but now with a value $N = 2$.

This picture differs from that given by the percolation theory[118,119] in that we deal with a confined sample with boundaries, and therefore, have, edge states. In a sample with a width W and length L such that $L \gg W$, the relevant threshold must be identified with the energy E_N, which marks the onset of an edge state connecting the two reservoirs. (Due to finite size effects, this threshold in general will be different from the percolation threshold calculated for a system without boundaries). For $E_F > E_N$, there are N edge states, and for $E_F < E_N$, there are $N - 1$ edge states.

For a sample with a width W and length L such that $L \ll W$, the situation is different. Again, such a sample will exhibit quantized conductance steps; but if the field is increased and the Fermi energy approaches a bulk Landau level, several paths might exist now that percolate from one contact to the other through the entire sample. The metallic contacts now feed current not only in the states near the edges of the sample, but also into the states associated with the bulk Landau level. Therefore, the conductance, instead of decreasing as in the previous example, now can increase and become much larger than the conductance at the plateau. It will reach a maximum when the Fermi energy lies at the center of the Landau level and decreases as the Fermi energy starts to fall below the bulk Landau level. Therefore, the transition from one quantized conductance plateau to another need not be monotonic but can occur via an intervening conductance maximum. Figure 15 shows the two-terminal resistance of a Si MOSFET as a function of gate voltage. The length of the conductor is 10 μm and the width 200 μm. The integers indicate the quantized values of the conductance. The data is from unpublished work by Fang.[120] A strong non-monotonic behavior of the conductance between the quantized values is seen.

This behavior can be described in more detail. We are dealing with an impurity potential that is correlated over a distance ζ, which is large compared to the magnetic length. Let us first consider the simple case where the sample length L is even shorter than the correlation length of the impurity potential but is still large compared to a magnetic length $l_B \ll L < \zeta$.

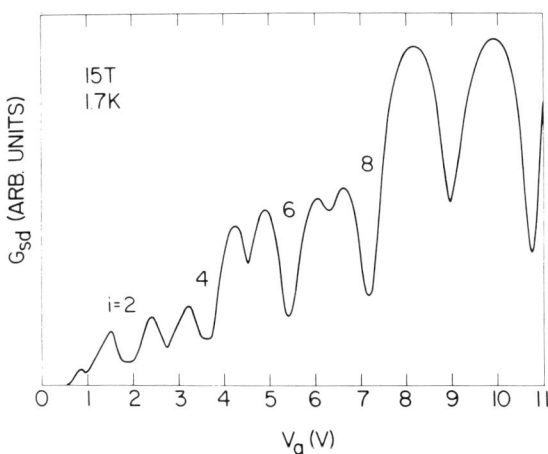

FIG. 15. Two-terminal conductance of a Si MOSFET with an aspect ratio of $W/L = 20$, and $L \approx 10$ μm as a function of gate voltage. (From Ref. 120.)

The behavior of the fluctuating potential along the width of the conductor can be described by a one-dimensional random function with zero average, $\langle V_i(y) \rangle = 0$, and a Gaussian distribution with a mean-square value equal to $V_0^2 < (\hbar\omega_c)^2$. We further assume that $V(y)$ is a smooth function with correlation $\langle V_i(y)V_i(y+d) \rangle = V_0^2 \exp(-d^2/\zeta^2)$, where ζ is the correlation length. The quantity of interest is the density n per unit length of states that percolate from one side to the other. We obtain a percolating state at the Fermi level each time the fluctuating potential exceeds the Fermi energy. (See Fig. 16.) The average number of percolating paths per unit length is equal to the average density of excursions of the potential $V(y)$ above the energy value E_F. The average density for a smooth potential is[121]

$$n(E) = \frac{1}{\zeta} e^{-E^2/2V_0^2}. \tag{60}$$

Here, the energy E is measured away from the average bulk Landau level energy. Thus, as the Fermi energy approaches a bulk Landau level, we obtain an additional contribution to the conductance given by

$$\Delta G = \frac{e^2}{h} n \left(E_F - \left(j + \frac{1}{2}\right)\hbar\omega_c \right) W. \tag{61}$$

This contribution is proportional to the width W of the sample and is maximal when the Fermi energy coincides with the center of the bulk Landau

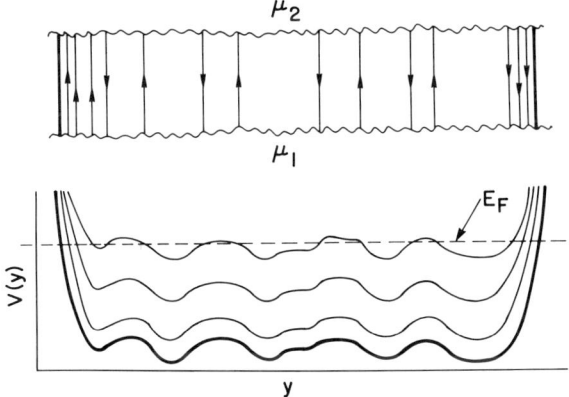

FIG. 16. Two-dimensional electron gas in high magnetic fields that is much wider than long. The states at the Fermi energy are shown in the top part. The potential fluctuation along the width of the sample (bottom) gives rise to many edge states in the interior of the sample at the Fermi energy (top). The two-terminal conductance of such a conductor is a non-monotonic function of magnetic field (or Fermi energy).

level. If W/ζ is much larger than the number of edge states N, then the contribution to the conductance given by Eq. (61) dominates the contribution due to the edge states. In this case, the conductance changes from one plateau to the next one in a non-monotonic fashion, and is maximal at an intervening energy (magnetic field).

In some two-terminal measurements,[115] it is seen that the conductance between the $(N + 1)$th plateau and the Nth plateau falls below the Nth plateau value. Instead of an *overshoot*, a depression is observed. It is possible that such depressions are not intrinsic, but are a consequence of additional contacts on the peripheries of the sample that electrically short the edge state and bulk states. A two-terminal measurement on a two-probe conductor and a two-terminal measurement on a conductor with additional contacts are not equivalent.

Suppose that we now make longer the sample depicted in Fig. 15. As we lengthen the sample, more and more of the initial percolating states will merge and form an orbit on which carriers are reflected back to the contact from which they are incident. Therefore, the number of percolating states $N_0 = n(E)W$ decreases as we increase the length of the sample. In a random potential, the orbits can be viewed as a realization of a diffusion process with a diffusion length characterized by the correlation length ζ. Hence, the number of trajectories that percolate in a sample of length L is $N_0\zeta/L$. Hence, at the center of a Landau level, $E = 0$, the number of states that percolate over a distance W is $(W/\zeta)(\zeta/W) \approx 1$. Only one of the many incident states survives if the sample length is equal to the width W, in agreement with percolation theory.[118,119]

Thus, for $L < W$, we can expect an overshoot of the conductance between steps, and for $L > W$, a monotonic behavior between steps. For $L \gg W$, the transmission problem acquires a one-dimensional character. Assuming scattering is only elastic, we can describe all the scattering processes in terms of 2×2 scattering matrices, with the incoming and outgoing states given by the innermost edge state. The sequence of these scattering matrices is random, corresponding to the detailed nature of the impurity potential. Such a problem can be mapped onto a one-dimensional transmission problem through a random sequence of barriers of differing heights and distances. The solution to this problem is well-known[34,122] and gives rise to a transmission probability that falls off exponentially with a localization length λ. The uppermost Landau level is characterized by a transmission probability,

$$\langle T_{33} \rangle \cong \exp\left(\frac{-L}{\lambda}\right). \tag{62}$$

The localization length λ is a function of the Fermi energy. It is infinite as long as the onductance is quantized (i.e., for the situation depicted in Fig. 14a), and decreases and approaches a finite value of the order of W as the Fermi energy approaches the center of the Landau level.

The localization length λ, which we discuss here to characterize the transition from one quantized value to another, is a very different quantity then the localization length usually discussed in connection with the quantum Hall effect. Many papers discuss a localization length that characterizes the localized states of a bulk system without boundaries. Mil'nikov and Sokolov[123] write about this localization length that although it "does not directly determine the transport properties of the system, it must be calculated if only because it is the only characteristic of the system which can be determined very accurately in numerical simulations." In contrast, the λ in Eq. (58) characterizes transmission along a disappearing edge state and is directly related to the transport properties of the system. It allows not only comparison among differing theoretical work but also with experiments! Chalker and Coddington,[124] at the very end of their paper, do discuss a localization length for a system with confining boundaries. They also make the distinction between a localization length that characterizes bulk states and a localization length that characterizes a system with boundaries. Some work[125,126] considers systems of finite width in one direction but takes periodic boundary conditions in the other direction. It is the study of a scattering problem that can be related most directly to experiments.

The key point of the admittedly very preliminary discussion given in the preceding is the following: For values of the magnetic field (or the gate voltage) at which the conductance is quantized, the conductance is independent of the length and width of the conductor. This is a consequence of the equilibrium filling of the edge states. Between plateaus, the conductance depends crucially on the geometry of the conductor. Next, we consider the two-terminal resistance of a conductor that is not very much larger than a magnetic length.

13. THE CONSTRICTION IN A MAGNETIC FIELD

Let us now subject a split-gate constriction discussed in Section 8 to a magnetic field. The two-terminal resistance of a split-gate constriction in a magnetic field was measured by van Wees et al.[127] Again, we assume that the constriction can be described by an electrostatic potential of the form, $V(x, y) = V_0 - (1/2)m\omega_x^2 x^2 + (1/2)m\omega_y^2 y^2$. With the help of the abbreviation, $\Omega^2 = \omega_c^2 + \omega_y^2 - \omega_x^2$, the energies that govern the transmission and the reflection at the saddle in a magnetic field are[101]

$$E_1 = \frac{1}{2}\frac{\hbar}{\sqrt{2}}([\Omega^4 + 4\omega_x^2\omega_y^2]^{1/2} - \Omega^2)^{1/2}, \qquad (63)$$

$$E_2 = \frac{\hbar}{\sqrt{2}}([\Omega^4 + 4\omega_x^2\omega_y^2]^{1/2} + \Omega^2)^{1/2}. \qquad (64)$$

The transmission probability is a function of

$$\varepsilon_n = \frac{E - E_2(n + 1/2) - V_0}{E_1}. \qquad (65)$$

The transmission probability then is (as in the absence of field, seen in Eq. (26)) given by

$$T_{mn} = \delta_{mn}\frac{1}{1 + \exp(-\pi\varepsilon_n)}. \qquad (66)$$

Note that there is no channel mixing. It is only the diagonal transmission probabilities that are nonzero.

In the limit of zero applied fields, these formulae reduce to the results discussed in Section 8. At high fields, when ω_c exceeds ω_x and ω_y, the expressions given before simplify considerably. Fertig and Halperin[101] find

$$E_1 \approx \frac{\hbar\omega_x\omega_y}{2\omega_c} \approx \left(\frac{m}{2}\right)\omega_x\omega_y l_B^2, \qquad (67)$$

where we have used the magnetic length, $l_B^2 = |\hbar c/eB|$, and

$$E_2 \approx \hbar\omega_c \qquad (68)$$

At high fields, we can study the trajectories that follow the equipotential lines of the saddle, using Eq. (57). Figure 17 gives a (qualitative) picture of the equipotential lines and the trajectories of the guiding center. Edge states with a guiding center energy $E_G = E_F - \hbar\omega_c(n + 1/2)$ are transmitted with probability 1 if $E_G \gg V_0$. Edge states with a guiding center energy $E_G = E_F - \hbar\omega_c(n + 1/2)$ are reflected with probability 1 if $E_G \ll V_0$. At most, one edge state with a guiding center energy $E_G \approx E_F - V_0$ is partially reflected and partially transmitted.[56] This can be seen in the following way: Denote the closest distance between the $(n + 1)$th edge state at the saddle for $\varepsilon_n < 0$

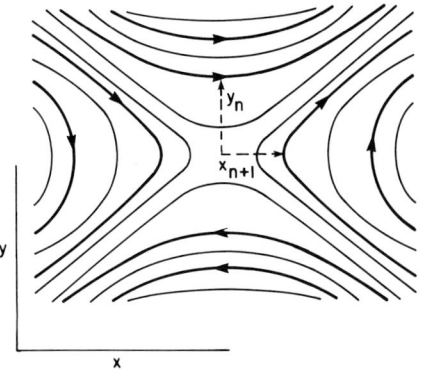

FIG. 17. Saddle-point potential of the constriction (faint lines). The edge states (full lines) follow equipotential lines. Scattering between edge states on one side of the saddle to the edge states along the other side of the saddle is determined by the length x_{n+1} if the guiding center energy is below the energy of the saddle point V_0 and by y_n for the edge state that crosses the saddle at energies above V_0.

by x_{n+1}. (See Fig. 16.) This coordinate is obtained by solving Eq. (57) for $y = 0$. The tunneling probability then is[101]

$$T_{mn} = \delta_{mn} \frac{1}{1 + e^{\pi(\omega_x/\omega_y)(x_n/l_B)^2}}. \tag{69}$$

For states with $\varepsilon_n > 0$, we similarly can introduce the distance y_n, which is a solution of Eq. (57) for $x = 0$, and obtain

$$T_{mn} = \delta_{mn} \frac{1}{1 + e^{-\pi(\omega_y/\omega_x)(y_n/l_B)^2}}. \tag{70}$$

For high magnetic fields, the difference of the distances x_n and x_{n+1} to the saddle point is large compared to the magnetic field, since $(x_{n+1}^2 - x_n^2)/l_B^2 = 2\omega_c^2/\omega_x^2 \gg 1$. A similar condition applies for the distances in the y direction. It is at most, therefore, one edge that has a transmission probability that differs appreciably from zero or one. Hence, at high magnetic fields, the conductance of a saddle point always is quantized.

To study the transition from small magnetic fields to high magnetic fields, we must return to the general equations, (61)–(64). A rapid opening of quantum channels as a function of energy E occurs if $E_1 \leq E_2$. This requires $\Omega^2 \geq 4\omega_x^2\omega_y^2$ or[102]

$$\omega_c + \omega_y \geq \omega_x. \tag{71}$$

Therefore, we see that the addition of a magnetic field helps to bring about a quantization of the conductance. Even saddle potentials that for zero field are not quantized (since ω_y might be less than ω_x) are quantized eventually in high enough magnetic fields. Such an example is shown in Fig. 18, which shows the two-terminal resistance as a function of (Fermi) energy for cyclotron energies in an interval ranging from zero to $\hbar\omega_c = 5\hbar\omega_x$ for the case that $\omega_x = \omega_y$. In this example, the quantization is poor at zero field and improves dramatically once the cyclotron energy exceeds the energies associated with the curvature of the saddle. The flatness of the plateau again is best studied by considering the derivative of the total transmission probability with respect to the (Fermi) energy. (Experimentally, it is the derivative of the conductance with respect to gate voltage that needs to be investigated.) We find that the slope at the center of the plateaus in high magnetic fields is determined by[102]

$$\frac{dT}{dE} \approx \left(\frac{2\pi}{E_1}\right) e^{-\pi E_2/4E_1} \approx \left(\frac{4\pi\omega_c}{(\omega_x\omega_y)^{1/2}}\right) e^{-\pi\omega_c^2/\omega_x\omega_y}. \tag{72}$$

Wee see that a cyclotron energy dominating the geometrical frequencies of

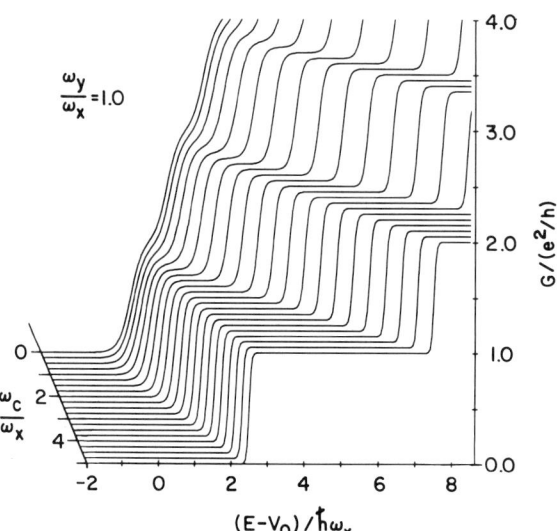

FIG. 18. Two-terminal conductance of a point-contact constriction in a magnetic field as a function of the Fermi energy. The ratio of the cyclotron frequency to ω_x, which parameterizes the magnetic field, is increased in steps of .25 from 0 to 5. The shape of the saddle is determined by the ratio ω_y/ω_x, which is taken to be equal to 1. (From Ref. 102.)

the saddle can give rise to very flat plateaus. The experimental results[127] are more complex because of the spin degree of freedom of the electrons that we have not included.

14. THE AHARONOV–BOHM EFFECT AND BACKSCATTERING

The search for sample-specific Aharonov–Bohm effects in small normal conductors[35–38,50,53,110] was a major stimulus at the beginning of the 1980s for the investigation of transport in small structures. Figure 19 shows a conductor with a hole subject to a high perpendicular magnetic field. The edge states are shown for the case that only the lowest Landau level is populated. If the picture we have developed is correct, it has the following interesting consequence. The hole of the conductor represents, for the two-dimensional gas, a huge potential fluctuation in the interior of the sample. A localized state around the hole of the sample develops. Electrons in this state are sensitive to the flux through the hole of the loop if phase coherence is maintained along the circumference. The edge states that form at the outer sample edge, however, are not sensitive to the flux through the hole of the loop. In the conductor of Fig. 19, these outer edge states are open states that originate at the transition between the metallic contact and the two-dimensional gas and end at a metallic contact. If the reservoirs are phase-randomizing, as we assume

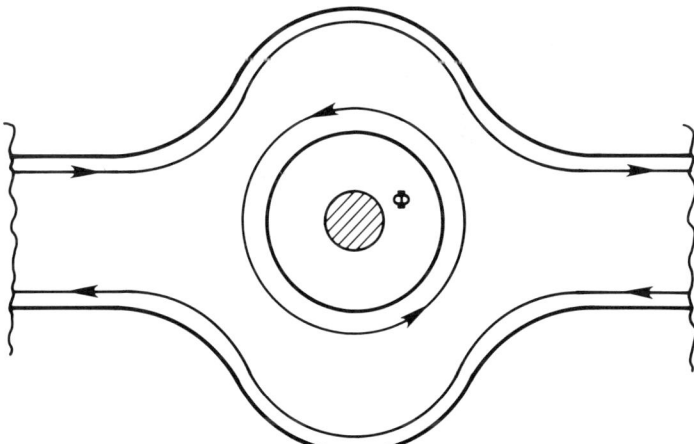

FIG. 19. Suppression of the Aharonov–Bohm effect in high magnetic fields. Only the edge state encircling the hole of the conductor is sensitive to the flux through the hole. The states along the outer sample boundaries start and terminate at reservoirs and are insensitive to the flux through the hole of the conductor.

throughout this chapter, electrons in these states are not sensitive to a flux through the hole of a conductor; but it is the carriers in these states that determine the net carrier flow from the source to the sink. Hence, in a situation where backscattering is suppressed, the carriers on the outer edge states do not interact with the flux-sensitive localized state. Therefore, the conductance in such a situation is quantized but not sensitive to an Aharonov–Bohm flux. Conversely, if we want to have an Aharonov–Bohm effect, we must have backscattering,[8,128] but in the presence of backscattering, the conductance cannot be quantized. We see, therefore, that the Aharonov–Bohm effect and the quantization of resistance are mutually exclusive! In this sense, the gauge argument provided by Laughlin[3] is—strictly speaking—not correct. Many discussions of the quantum Hall effect invoke close geometries and also invoke Aharonov–Bohm fluxes. (See, for instance, Hajdu et al.,[129] Pook and Hajdu,[130] and Ohtsuki and Ono.[125] In many cases, a quantized conductance is obtained after averaging the conductance over several flux periods. Clearly, such procedures are not to the point: Here, we are not interested whether or not the conductance is quantized in some averaged sense, but more strictly, whether a given sample for a given flux exhibits a quantized conductance or not. Halperin[4] also starts his discussion by investigating the response of the electron gas in a Corbino geometry to an Aharonov–Bohm flux. Ultimately, in his explanation of the quantum Hall effect, he neglects all phase-sensitive effects and, for this reason, finds quantization.

Away from the quantized conductance plateau, the innermost edge state moves away from the sample boundary and backscattering becomes possible. Such a behavior consisting of Aharonov–Bohm oscillations between plateaus and a complete suppression of the Aharonov–Bohm effect in the plateaus has been observed in many experiments.[9,146,131–136] We return to the subject of Aharonov–Bohm oscillations in high magnetic fields in connection with our discussion of very simple models[26,128,137] that permit the study of this effect.

V. The Quantum Hall Effect in Open Conductors

15. Conductor with Ideal Contacts and without Backscattering

Let us now discuss how the quantum Hall effect is established in an open conductor with current probes and Hall bars.[8,24,137] Figure 20 shows a conductor where the Hall bars are of the same width as the probes connected to the current source and sink. In the previous section, we have shown that

4. THE QUANTUM HALL EFFECT IN OPEN CONDUCTORS

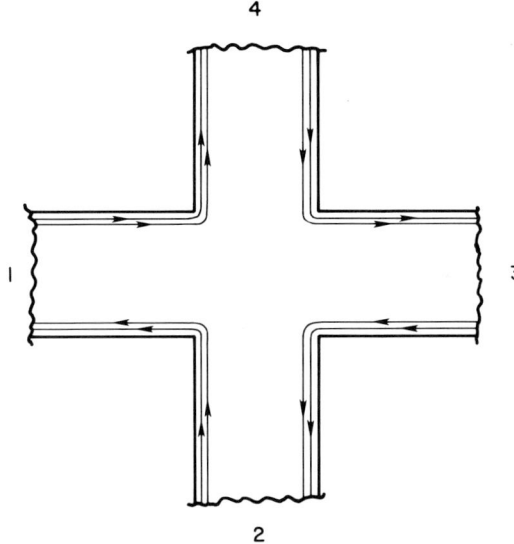

FIG. 20. Conductor with Hall probes. (From Ref. 143.)

edge states are established along the sample boundary. In the conductor of Fig. 20, N edge states connect the four contacts in a cyclical fashion. We assume that all the contacts are ideal; i.e., the contacts populate all edge states equally up to the chemical potential of the source. Further, since all contacts are taken to be ideal, carriers that reach a contact leave the sample with probability of 1. We further assume that processes of type B shown in Fig. 13 are absent; i.e., the conductor is assumed to be so wide that scattering across the sample from one edge to the other does not occur. Under these circumstances, the edge states provide perfect transmission channels for carriers and lead to transmission probabilities, $T_{41} = N$, $T_{34} = N$, $T_{23} = N$, and $T_{12} = N$. All the other transmission probabilities are zero! The reflection probabilities are determined by current conservation. If contact i is described by a (large) number of states M_i characterizing the metallic contact, current conservation requires $R_{ii} = M_i - N$. With the transmission and reflection probabilities as specified in the preceding, it now is easy to calculate the four-terminal resistances, from Eq. (18). The Hall resistance $R_{13,42}$ is determined by $T_{41}T_{23} - T_{43}T_{21}$, which is equal to N^2. Evaluation of the subdeterminant D in Eq. (18) yields $D = N^3$. All Hall resistances of the conductor of Fig. 20 are quantized and given by $\pm(h/e^2)(1/N)$. The *longitudinal* resistances, for example $R_{12,43}$ are zero, since in the products forming the expression, $T_{41}T_{32} - T_{42}T_{31}$, at least one transmission probability always is zero.

Since there is no backscattering in the conductor of Fig. 20, it not only is the four-terminal resistances that either are quantized or zero, but all the three-terminal resistances also are zero or quantized. Furthermore, all the two-terminal resistances also are quantized and given by $R_{kk,mm} = (h/e^2)(1/N)$. This is because with carrier flow as shown in Fig. 20, the chemical potential at contact 4 is equal to that of the carrier source $\mu_1 = \mu_4$, and the chemical potential at contact 2 is equal to that of the carrier sink, $\mu_2 = \mu_4$.

It is easy to extend the preceding discussion to a conductor with an arbitrary number of leads, with the result that all generalized Hall resistances are quantized and all generalized longitudinal resistances are zero. Similarly, all three- and two-terminal resistances are quantized or zero. An interesting experiment on a conductor with these properties has been performed by Fang and Stiles.[116] They use metallic wires (with resistances much smaller than the quantized Hall resistance) to interconnect various Hall probes. The interconnections are made in such a way that only a two-probe conductor remains. Depending on the way the interconnections are arranged, Fang and Stiles obtain two-terminal resistances, $R_{12,12} = (h/e^2)(p/q)(1/N)$, with p and q integers. If, for instance, terminals 2 and 4 of the conductor in Fig. 20 are connected, the two-terminal resistance is proportional to $2/N$. It is a simple matter to check that with the assumptions made before, and completing the equation (11) according to the Kirkhoff laws, we do indeed reproduce the quantized values observed by Fang and Stiles.[116]

In the preceding, we assumed contacts without internal reflection and assumed that backscattering is completely absent. These assumptions obviously are sufficient to obtain the quantum Hall effect. Are they necessary? To find out, we next consider a few simple examples, where we introduce backscattering at singular locations and where we also assume contacts with internal reflection.

16. Simultaneously Quantized Longitudinal Resistances and Hall Resistances

In the conductor of Fig. 20, quantization of the Hall voltage comes about due to transport along N edge states that connect the contacts in a cyclical fashion. We can prevent such a cyclical connection of contacts by assuming that a barrier (or a narrow constriction) diverts a number K of edge states as shown in Fig. 21. Such a situation will be established over a certain range of gate voltage, which creates a saddle-like potential such that $N-K$ edge states are transmitted and K edge states are reflected. The transmission probabilities now are as follows: $T_{41} = N$, $T_{43} = N - K$, $T_{41} = K$, $T_{23} = N$, $T_{32} = K$, $T_{12} = N - K$, and all other transmission probabilities are zero. Inserting this

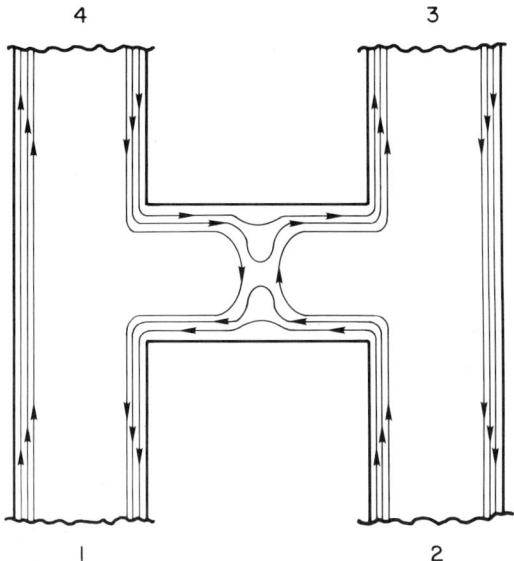

FIG. 21. Conductor with barrier reflecting K edge states. (From Ref. 143.)

into Eqs. (11) and (18) yields Hall resistances,[8,24]

$$R_{13,42} = \left(\frac{h}{e^2}\right)\frac{1}{(N-K)}, \tag{73}$$

$$R_{42,13}(B) = R_{13,42}(-B) = -\left(\frac{h}{e^2}\right)\frac{N-2K}{N(N-K)}, \tag{74}$$

and *quantized* longitudinal resistances that are symmetric in the field,[8,24]

$$R_{12,43}(B) = R_{12,43}(-B) = R_{43,12}(B) = \left(\frac{h}{e^2}\right)\frac{K}{N(N-K)}. \tag{75}$$

All other four-terminal resistance measurements on the conductor of Fig. 21 are zero. The plateaus predicted by Eqs. (75)–(77) have been observed in strikingly clear experiments by Washburn et al.[13] and Haug et al.[14,27] Recently, Snell et al.[138] carried out an experiment using a narrow constriction and measured all the resistances predicted by Eqs. (73)–(75). Equation (75) was found independently by van Houten et al.,[105] who applied this result to magnetic depopulation studies of point contacts and found good agreement with the experiment down to fairly small magnetic fields. This experiment suggests that whatever the value of the factor α in Eq. (31) is at zero field, already moderate magnetic fields (cyclotron radius smaller than the wire width) are sufficient to drive α towards 1 (at least as long as inelastic equilibration occurs between the constriction and the voltage probes).

Experiments with gated structures similar to those of Washburn et al.[13] and Haug et al.[14] were performed earlier by Syphers and Stiles.[15] The progress since this earlier work is chiefly one of *understanding*. The explanations by Syphers and Stiles of their observations are complex; the explanation of the observations by Washburn et al. and Haug et al. are simple and direct. Another distinction (perhaps not an important one) is the size of the systems. The gates employed in the Syphers and Stiles work are of the order of $(10 \, \mu m)^2$. The gates used in the Haug et al. work are even larger. Such large gate structures lead to a non-monotonic behavior in the transition from one plateau to another. The sample of Washburn et al.[13] is very small, with a lithographical channel width of 2 μm and a gate width of the order of 0.1 μm. The topology of the conductor is the same as that shown in Fig. 21, with an overall distance between contacts of the order of 10 μm. Figure 22 shows the experimental data from the experiment by Washburn et al.[13] The lowest trace shows the longitudinal resistance $R_{12,43}$ in the absence of a magnetic field as a function of gate voltage. For a field of $B = 5.16$ tesla, the sample exhibits for zero gate voltage a quantum Hall effect with a Hall resistance

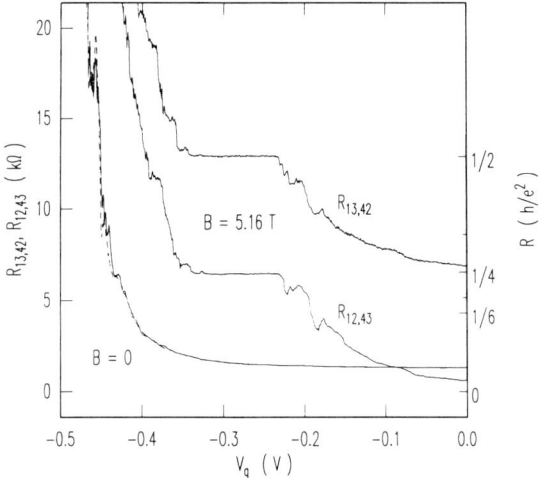

FIG. 22. Generalized longitudinal resistance and generalized Hall resistance measured at a four-probe conductor with the same topology as that shown in Fig. 21 as a function of gate voltage. The magnetic field is fixed at a value such that for zero applied gate voltage (no barrier), the Hall resistance is quantized at $(h/e^2)(1/4)$. As the gate voltage increases, two (spin-degenerate) edge states are reflected at the barrier. The resulting plateaus in the longitudinal resistance and the Hall resistance correspond, therefore, to $N = 4$ and $K = 2$ in Eqs. (92) and (94). (From Ref. 13.)

$R_{12,42}$ equal to $(e^2/h)(1/4)$. With increasing voltage, there eventually is a range where two spin-degenerate edge states are reflected. Because the spin degeneracy is not resolved, $K = 2$ in Eq. (73). Thus, with $N = 4$ and $K = 2$, Eq. (73) predicts a new plateau in the Hall resistance to occur at $(h/e^2)(1/2)$, which is just what is observed in the experiment. Similarly, for the longitudinal resistance $R_{12,43}$, we find from Eq. (74) a new plateau predicted at $(h/e^2)(1/4)$, which also is observed. The experiment by Washburn et al. demonstrates in a clear way, therefore, the *simultaneous* quantization of the longitudinal resistance and the Hall resistance.

It is possible to use the gated structures to increase the density of carriers beneath the gate. Instead of a reflection of edge states as discussed before, we have in this case additional edge states beneath the gate that form closed loops. Interestingly, as shown by Haug et al.,[139] there are situations where the presence of these additional edge states leaves the longitudinal resistance at zero. From this, we can conclude that the additional edge states do not couple to the edge states penetrating the gated region, even in gated regions of macroscopic size. Such decoupling effects will be discussed later on in more detail.

Equations (73)–(75) can be extended for the case that transmission through the gated area is not quantized. If instead of an integer number K of reflected edge states, we consider a total reflection probability R and instead of $N-K$ transmitted edge states, we consider more generally a transmission probability T, we obtain Hall resistances,

$$R_{13,42} = \left(\frac{h}{e^2}\right)\frac{1}{T}, \tag{76}$$

$$R_{42,13}(B) = R_{13,42}(-B) = -\left(\frac{h}{e^2}\right)\frac{T-R}{NT}. \tag{77}$$

The longitudinal resistances are symmetric in the field and given by

$$R_{12,43}(B) = R_{12,43}(-B) = R_{43,12}(B) = \left(\frac{h}{e^2}\right)\frac{R}{NT}. \tag{78}$$

We emphasize that Eqs. (76)–(78) are strictly valid only if backscattering occurs at the gate. They are not applicable if the magnetic field is varied away from a range in which the bulk of the sample is quantized.

The generalized Eqs. (76) and (77) are given in Ref. 24, using a different notation. The longitudinal resistance has a close resemblance to the single-channel Landauer formula R/T, i.e., Eq. (42), as has been pointed out by Haug et al.,[14,27] Jain and Kivelson,[140] and Stone and Baranger[60] This is because in Landauer's derivation of his result, it is implicitly assumed that the voltage measurement is phase-insensitive as we have emphasized in Section 10. In the conductor of Fig. 21, the voltage measurement is not phase-sensitive because there is no interference at the contacts. Equation (78) is not a consequence of piled-up charges near the barrier, but is a consequence of ideal contacts coupled to equilibrium electron reservoirs. The contacts as discussed here provide complete equilibration and provide complete phase-randomization. Next, we consider more general situations and show how this result changes if we do not assume contacts that are ideal.

17. Contacts with Internal Reflection Alternating with Ideal Contacts

In the preceding example, we have assumed that backscattering occurs in the interior of the sample such that it completely or partially diverts edge states from one contact to another. Suppose now that backscattering occurs in a branch of the conductor in Fig. 22. This can be brought about again by making a gate over a branch or by making a constriction. Contacts that are made by alloying provide another possibility of contacts with internal reflection. It is possible that the metallic diffusion occurs not deep enough to reach the edge of the two-dimensional electron gas. Entrance and exit of the carriers occurs effectively then through a tunneling barrier. Another possibility is that the two-dimensional electron gas is reached but not near the edge of the sample. In such a situation, the contact again is reached only by tunneling or Mott hopping. A transition region between the two-dimensional electron gas and a metal is depicted to the left in Fig. 23. In the conductor of Fig 23, two of the branches are subject to backscattering and two intervening contacts are assumed to be ideal. We show next that regardless of the scattering properties of the non-ideal contacts, the Hall voltages of this conductor are quantized.

The properties of a contact with scattering can be described by a scattering matrix with transmission and reflection amplitudes, as for a two-terminal resistor. Let μ_i be the chemical potential of the contact under consideration. Let μ_0 be the lowest chemical potential of all the terminals connected to the conductor. The flux from the probe into the nth edge of the sample is

$$I_n = \left(\frac{e}{h}\right) \sum_{m=1}^{m=M} T'_{nm}(\mu_i - \mu_0) \tag{79}$$

where m refers to the quantum channels in the (metallic) contact and n designates the edge states. The prime is used to denote waves incident from the

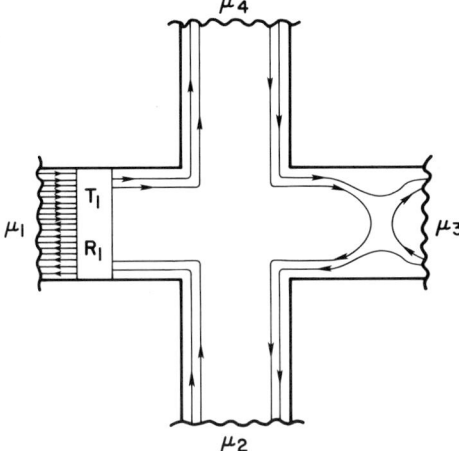

FIG. 23. Conductor with two contacts with internal reflection and two ideal contacts. Along the perimeter of the conductor, contacts with internal reflection and ideal contacts are alternating. Contact 1 is a schematic depiction of a metallic contact. Contact 3 exhibits internal reflection due to a cross-gate or a split-gate constriction.

reservoir. Since all the reservoir states are fully occupied, the total probability for tunneling from the reservoir into the nth edge state is

$$T'_n \equiv \sum_{m=1}^{m=M} T'_{nm} \leq 1. \tag{80}$$

If $T'_n < 1$, the nth edge state is only partially filled by the reservoir. If $T'_n \neq T'_k$, the reservoir fills different edge states to a differing degree.[8] Thus, a contact with internal reflection leads to a non-equilibrium injection of carriers.[8] In the simple case that the scattering properties of a branch are determined by a quantized constriction resistance or an appropriately biased gate, the T'_n either are zero or one. This is a simple limiting case of the more general situation in which all the T'_n are different from one another. The total transmission probability of a contact is $T \equiv \Sigma_n T_n$. The total transmission probability of the contact is independent of the direction of incidence of the carriers (as seen from the discussion of Eq. (15)), and for contact 1 is denoted by T_1. The probability of reflection for carriers approaching the contact from the metallic contact (the reservoir) for contact 1 is $R'_1 = M_1 - T_1$, where M_1 is the number of states characterizing the reservoir. The reflection probability for carriers approaching the contact from the region of two-dimensional gas is $R_1 = N - T_1$, where N is the number of edge states. Since the number of reservoir states is very large compared to only a few edge states typically in the two-dimensional gas, the reflection probability R'_1 always is large, independent of the properties of the contact. This *external* reflection is not of further interest. The reflection probability R_1 is zero for an ideal contact, i.e., if carriers approaching the contact can leave the sample with probability 1. Then $T_1 = N$ and $R_1 = 0$. Thus, an ideal contact is characterized by the absence of *internal* reflection. A contact with scattering $T_1 < N$ exhibits internal reflection $R_1 \neq 0$. A fraction of the carriers approaching the contact from the interior of the sample along the lower side of the sample skips along the contact and reaches the upper edge of the sample.

In the conductor of Fig. 20 with ideal contacts, the chemical potentials of the voltage probes are equal to the chemical potential of the reservoir that emits the carriers. The chemical potentials of the voltage probes are locked to the chemical potentials of the current source and current drain contact. In contrast, all chemical potentials in the conductor of Fig. 22 are different from one another if carrier flow is from contact 1 to contact 3. Application of Eq. (11) to the conductor of Fig. 22 shows that despite the two contacts with internal reflection, all Hall voltages still are quantized and all longitudinal resistances still are zero. That the Hall resistance is quantized in the presence of two weakly coupled contacts ($T_1 \ll 1$ and $T_3 \ll 1$) also is shown in the calculation of Peeters.[141] The Hall resistance is quantized because the conductor contains two ideal contacts, which provide complete equilibration. If carrier flow is from contact 1 to contact 3, the carrier source feeds the different edge channels in a non-equilibrium fashion as described previously. At the voltage contact, inelastic relaxation occurs. Carriers incident into probe 4 have energies up to μ_1; but contact 4, since it is ideal, re-injects carriers into all edge states emanating from it up to the chemical potential $\mu_4 < \mu_1$. Contact 4 dissipates energy. An explicit discussion of the energy dissipated at contacts is given by Komiyama and Hirai.[142]

The additional scattering in branches 1 and 3 is reflected in the two-terminal resistance $R_{13,13}$, which no longer is quantized but given by[8]

$$R_{13,13} = \frac{h}{e^2} \frac{N^2 - R_1 R_2}{N T_1 T_2}. \tag{81}$$

It is instructive to note that with a little algebra, the resistance given by Eq. (81) can be expressed in the form,

$$R_{13,13} = \frac{h}{e^2} \left[\frac{1}{N} + \frac{N - T_1}{N T_1} + \frac{N - T_2}{N T_2} \right].$$

The total two-terminal resistance is equal to the quantized resistance augmented by a contribution from each backscattering process proportional to $(N - R_i)/N T_i$. We emphasize that this result rests on the particular assumptions made about the equilibration of edge channel populations between successive backscattering processes. It is not a general law for the addition of resistances in the quantum Hall regime. (A similar result for $N = 1$ was found in Ref. 40 at zero field and also is a consequence of particular assumptions about equilibration.) The three-terminal resistances of the conductor depicted in Fig. 20 also are not quantized. We find

$$R_{13,12} = \frac{h}{e^2} \frac{1}{T_1} \tag{82}$$

and

$$R_{13,23} = \frac{h}{e^2} \frac{R_3}{NT_3}. \tag{83}$$

For the three-terminal resistances measured with respect to probe 4, we find

$$R_{13,14} = \frac{h}{e^2} \frac{R_1}{NT_1} \tag{84}$$

and

$$R_{13,43} = \frac{h}{e^2} \frac{1}{T_3}. \tag{85}$$

On the other hand, if carrier flow is from contact 4 to contact 2, all edge states are equally populated by the carrier source contact and the chemical potentials of voltage probe 3 is equal to that of the source, $\mu_3 = \mu_4$, and the chemical potential of contact 1 is equal to that of the sink, $\mu_1 = \mu_2$. The two-terminal conductance is quantized, $R_{42,42} = (h/e^2)(1/N)$, the three-terminal resistances $R_{42,43}$ and $R_{42,12}$ are zero, and the three-terminal resistances $R_{42,32}$ and $R_{42,41}$ are quantized and equal to $(h/e^2)(1/N)$.

Equation (81) also can be regarded in the following way: It describes a two-terminal conductor with disordered contacts that are so far apart that inelastic relaxation occurs and equilibrates the differing populations of the edge channels. Thus, Eq. (81) provides a simple demonstration that scattering at contacts is sufficient to bring about deviations from quantization of the two-terminal conductance. As stated in Section 12, the quantization of the two-terminal conductance hinges on an equilibrium population of edge states. Modeling inelastic scattering with the help of contacts, it is possible to build up samples that are large compared to an inelastic length. In Fig. 23, a sample is shown in which all physical contacts are disordered (exhibit internal reflection). The equilibration of the edge states is indicated by the wavy line.

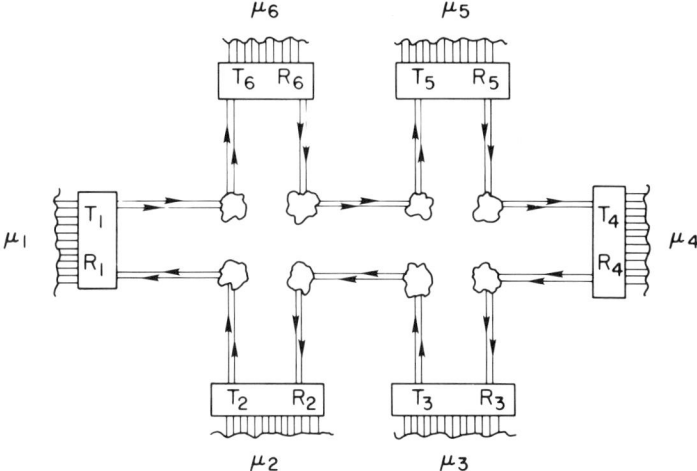

FIG. 24. Conductor with contacts with internal reflection. Such a conductor exhibits the quantum Hall effect only if inelastic scattering is sufficient to equilibrate edge states between contacts. (From Ref. 8.)

Mathematically, this conductor is described by a sequence of contacts along the perimeter of the sample, which has the property that contacts with internal reflection (the physical contacts) alternate with contacts that are ideal (the equilibrating devices). Such a conductor exhibits the quantum Hall effect: All generalized Hall resistances are quantized, and all generalized longitudinal resistances are zero; but all two-terminal measurements at physical contacts and all three-terminal measurements are not quantized, but are determined by expressions similar to Eqs. (82)–(85). The important point made with the help of Fig. 24 is the following: Four-terminal resistances can be quantized even in situations where two- and three-terminal resistances are not quantized; but the quantization for this conductor is a consequence of inelastic relaxation.

18. THE ANOMALOUS QUANTUM HALL EFFECT DUE TO
    ADJACENT CONTACTS WITH INTERNAL REFLECTION

In the previous section, we have shown that the quantum Hall effect prevails if contacts with internal reflection and contacts without internal reflection alternate along the perimeter of a sample. However, if two contacts with internal reflection are adjacent to one another, that is not the case. At least one Hall measurement exhibits a resistance that is not quantized at the bulk value. As pointed out in Ref. 8, this is so because carriers injected through a disordered contact (a contact with internal reflection), streaming towards a voltage contact that also is disordered, give rise to a chemical potential at this contact that depends on the detailed fashion with which the voltage contact couples to the system. Despite the intense research on the quantum Hall effect for almost a decade, such an effect had not been predicted previously, nor were there to our knowledge any experimental results published indicating such behavior. Within less than a year, there were two very interesting experiments by Komiyama et al.,[10] and van Wees et al.[9] demonstrating this effect. The experiment by van Wees et al.[9] employs point contacts and provides a conceptually very simple example. We discuss this experiment first and later discuss the experiments by Komiyama et al.[10]

Figure 25 shows a particular situation, where $N-K$ edge states transmit at contact 1 and $N-L$ edge states transmit at contact 2. The Hall resistance $R_{13,42}$ is determined by the number N of bulk edge states, since carrier flow is from contact 4, and contact 4 provides equilibration; but if carrier flow is from contact 2 to contact 4, the Hall resistance is

$$R_{24,13} = \left(\frac{h}{e^2}\right)\frac{1}{(N-M)}, \qquad (86)$$

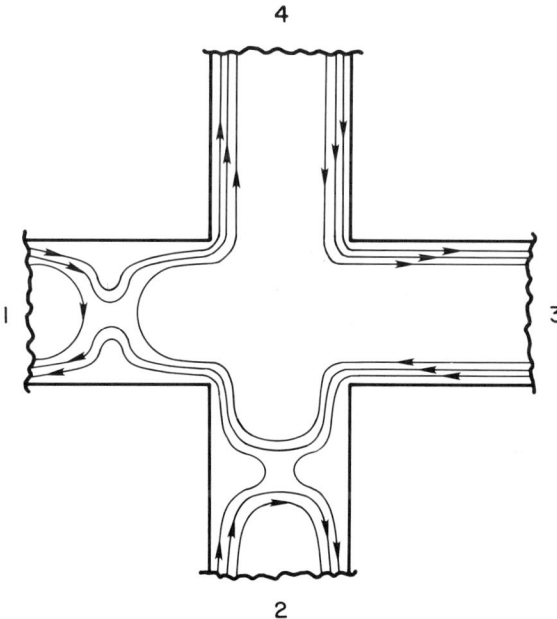

FIG. 25. Conductor with contact 1 reflecting K edge states and contact 2 reflecting L edge states. (From Ref. 143.)

where $M = \min(K,L)$. It is the contact that provides less reflection that determines the outcome of the resistance measurement.

Equation (86) is valid if the transmission probabilities of contacts 1 and 2 are quantized. It is instructive to study the result for arbitrary contacts characterized by arbitrary scatterers. Denote the transmission matrix (the matrix that yields the current amplitudes of the edge states in terms of the current amplitudes of the reservoir states) by t'_2 and denote the transmission matrix of contact 1, which yields the current amplitudes of the reservoir states in terms of the amplitudes of the edge states by t_1. The transmission matrix from contact 2 to contact 1 is given by $t_{12} = t_1 t'_2$, and the total transmission probability from contact 2 to contact 1 thus is

$$T_{12} = Tr(t'^{\dagger}_2 t^{\dagger}_1 t_1 t'_2). \tag{87}$$

With the help of T_{12} and the total transmission probabilities $T_1 = Tr(t^{\dagger}_1 t_1)$ and $T_2 = Tr(t^{\dagger}_2 t_2)$, which characterize the individual contacts, the Hall resistance is found to be[9,56]

$$R_{24,13} = \frac{h}{e^2} \frac{T_{12}}{T_1 T_2}. \tag{88}$$

Let us now discuss several limiting cases of this simple result. If the contact conductances are quantized, i.e., $T_1 = N - K = N_1$ and $T_2 = N - L = N_2$ and there is no interaction between the edge states, then as is evident from Fig. 25, the transmission probability from contact 2 to contact 1 is determined by the smaller of the two transmission probabilities, $T_{12} = \min(N - K, N - L)$. Thus, the Hall resistance is quantized and given by Eq. (86).

Next, let us consider the case that one of the edge states is transmitted only partially at the contacts, as shown in Fig. 17 for a single contact. The analysis for this case has been presented by Refs. 9 and 56. The transmission probabilities of the individual contacts are $T_1 = N_1 + \Delta T_1$ and $T_2 = N_2 + \Delta T_2$, where N_1 and N_2 are the numbers of completely transmitted edge states and $\Delta T_i \leq 1$. Three different situations occur: (a) For $N_1 < N_2$, the transmission probability is $T_{21} = N_1 + \Delta T_1$; (b) for $N_1 > N_2$, we find $T_{21} = N_2 + \Delta T_2$; and (c) for $N_1 = N_2 \equiv N_c$, we find $T_{21} = N_c + \Delta T_1 \Delta T_2$. Thus, for case (a), the Hall resistance is

$$R_{24,13} = \frac{h}{e^2} \frac{1}{N_2 + \Delta T_2}. \tag{89}$$

Therefore, for case (a), the Hall resistance is quantized if $\Delta T_2 = 0$. Similarly, for $N_1 > N_2$, the Hall resistance is quantized and determined by N_1 if ΔT_1 is zero independent of the value of ΔT_2. In case (c), the Hall voltage is quantized only if both ΔT_1 and ΔT_2 are zero, i.e., if both contacts are quantized. The experiments of van Wees et al.[9] and van Houten et al.[56] are in complete agreement with these conclusions.

Next let us consider the case where contact 1 and contact 2 exhibit internal reflection, but without making the simplifying assumptions that apply to saddle-like point contacts. In general, all the transmission amplitudes of the t matrix can be nonzero. In the case that the two contacts are close together such that the total transmission probability T_{12} has to be evaluated phase-coherently, Eq. (87) applies without simplification. Phase-coherent effects can come about if the contacts mix channels, i.e., if the channel number is not conserved at the contacts. Phase-coherent effects also can occur as a consequence of elastic scattering between edge states between the two contacts if the edge states are populated initially in a non-equilibrium manner. If the two contacts are mixing but are so far apart that phase memory is not preserved, the evaluation of T_{12} becomes simpler. In this case, it is only the population of the edge channels that needs to be considered, i.e., the currents but not the amplitudes.[142] The transmission probability T_{21} then can be expressed in the following way: Contact 2 populates the ith edge states with probability $T'_{2,i} = \Sigma_j T'_{2,ij}$. Carriers in the ith edge state penetrate contact 1 with probability $\Sigma_i T'_{1,ki} T_{2,i}$. The total transmission probability is

$$T_{12} = \sum_i T_{1,i} T'_{2,i}, \qquad (90)$$

where $T_{1,i} = \Sigma_k T_{1,ki}$. The situation discussed before with complete transmission and reflection of edge states and possibly a partial transmission of one edge state is a special case of Eq. (90), with all the $T_{n,k}$, $T'_{n,k}$ either zero or one except for possibly one edge state. The more general case is analyzed in detail by Komiyama and Hirai[142] for a number of conductors with differing configurations of contacts with and without internal reflection.

19. BACKSCATTERING AND CONTACTS WITH INTERNAL REFLECTION

In Section 18, we have discussed how backscattering affects the Hall effect and the longitudinal resistance if backscattering occurs in the bulk. All the contacts were taken to be ideal. In the previous section, we have discussed conductors where backscattering processes occur only at the contacts. A number of interesting phenomena can be expected if both backscattering and internal reflection at contacts occur simultaneously. This is the case in the experiments of Komiyama et al.,[10,16] and also is the case in an experiment by van Wees et al.,[135] which shows that the Shubnikov–de Haas oscillations can be suppressed if measured with a voltage contact with internal reflection. Next, we discuss a very simple example that demonstrates the interaction of bulk backscattering processes and contacts with internal reflection. Our particular example relates to the quantized backscattering process discussed in Section 18. How are these results changed if the contacts are not ideal? Let us return briefly to the conductor of Fig. 21. Clearly the anomalous quantization found in the conductor of Fig. 21 also hinges on the properties of the contacts (at least as long as the contacts are within an equilibration length of the barrier). To show this, consider the conductor in Fig. 26,[143,144] where two of the contacts are separated by barriers from the main conducting channel. If this barrier forms a smooth saddle, the contacts interact only with the outermost edge state. Let the probability for transmission from this contact to the outermost edge state be $T_1 < 1$ at contact 1 and $T_3 < 1$ at contact 3. Let us, as in Section 18, associate a transmission probability T for transmission through the gated region (or a point contact). We assume that there are $N-K$ edge states that are completely transmitted and possibly one edge state that is partially transmitted with probability ΔT. Thus, the gated region has a total transmission probability, $T = N - K + \Delta T$. How are the results presented in Section 18 affected by the differing properties of the contacts

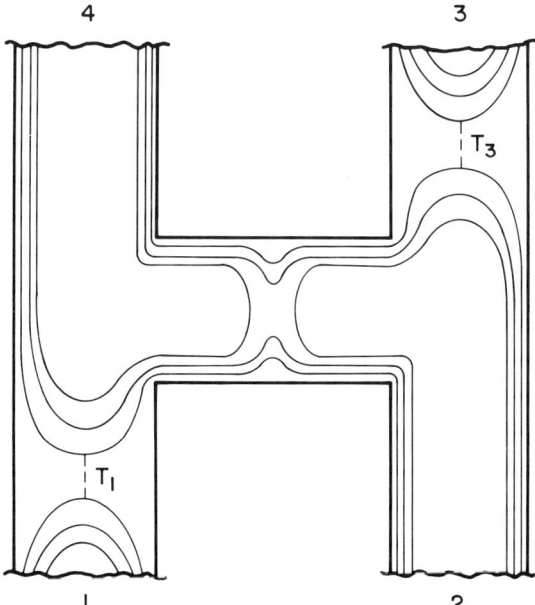

FIG. 26. Conductor with barrier and two weakly coupled contacts. (From Ref. 143.)

in Fig. 26? Two cases need to be distinguished. If there is only partial transmission via *one* edge state, i.e., if $N = K$, the results given in Section 18 still apply with $T = \Delta T$ and $R = 1 - \Delta T$. If, however, $N - K \geq 1$, application of the multi-terminal formula, Eq. (18), to the conductor of Fig. 26 yields an *anti-symmetric* Hall resistance,

$$R_{13,42}(B) = -R_{13,42}(-B) = \left(\frac{h}{e^2}\right)\frac{1}{T}, \qquad (91)$$

independent of T_1 and T_3. An even more drastic change occurs in the longitudinal resistance. Instead of Eqs. (75) or (78), we obtain $R_{12,43} = 0$. This is due to the fact that two of the contacts couple only to the outermost edge state. The backscattering process does not affect the outermost edge state and, with the contacts as shown in Fig. 26, is simply not detected! The key point is that Eq. (78) is not in any way a general expression. Equation (78) is valid for a singular backscattering process in an otherwise perfect sample if the measurement is made with ideal contacts.

20. The Long-Range Nature of Non-Equilibrium Populations

A dramatic set of experiments that can be understood as a combination of backscattering processes and contacts with internal reflection have been performed by Komiyama et al.[10,16] (Preprints of these works reached the author in late 1988 and early 1989. The fact that neither one of their two experimental papers had reached publication as of the initial date of this writing is not to the glory of our editorial and refereeing procedures.) The first of these papers drew attention to deviations from Eqs. (73)–(75) that can be understood as being due to contacts with internal reflection, as discussed in connection with Fig. 26. The second experiment perhaps is more dramatic and is described next in more detail.

The experimental setup of the Komiyama et al. experiment[16] is shown in the inset of Fig. 27. The dimensions of the sample are large, with a distance between contacts of the order of 50 μm. Again, we deal with a gated structure and consider a situation where, in the absence of a barrier (zero gate voltage), the Hall resistances are quantized. The key effect that we address is shown in the longitudinal resistance $R_{57,68}$. In the absence of a voltage at the gate, this resistance is nearly zero. As the gate voltage is increased, an edge state eventually is reflected at the barrier created by the gate, as in the Washburn et al.[13] and Haug et al.[14] experiments. The reflection of this edge state is accompanied by a considerable rise in the resistance $R_{57,68}$. We stress here, without appealing to any interpretation of this effect, that this a very striking

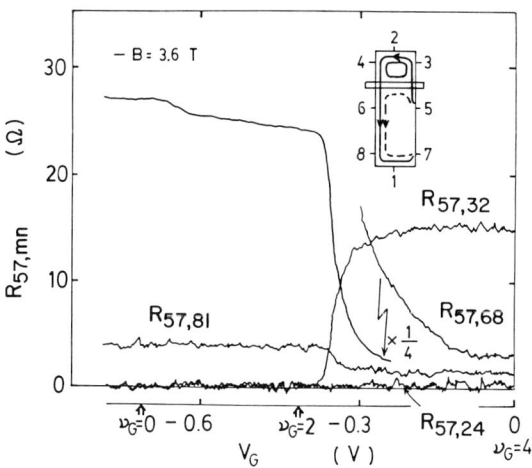

FIG. 27. Gate voltage dependence of various longitudinal resistances in a GaAs–AlGaAs heterostructure with a cross gate. The sample is 30 μm wide and 400 μm long. All contacts exihibit internal reflection. The dramatic effect shown in this experiment is the rise of the four-terminal longitudinal resistance $R_{57,68}$ as the innermost edge state is reflected by the gate, which is a macroscopic distance away from voltage contacts 6 and 8. If the edge states would equilibrate before they reach contact 6, this longitudinal resistance would be zero. (Reprinted with permission from *Solid State Commun.* **73**, 91, S. Komiyama, H. Hirai, S. Sasa, and F. Fujii, Copyright 1990, Pergamon Press plc.)

phenomenon: The change in the longitudinal resistance occurs due to a change in a potential that is located more than 50 μm away from the contacts that are used to measure a generalized longitudinal resistance. In a drastic way, this experiment, therefore, points to the long-range and non-local nature of electron transport in high magnetic fields. How can we understand this effect? In the absence of a barrier, both edge states follow the sample boundary. In the presence of a barrier, one edge state is reflected and reaches the two contacts that are used to measure $R_{57,68}$ along a much shorter path. From the discussion given previously, it is clear that the longitudinal resistance $R_{57,68}$ would be zero if the contacts were ideal, and would be zero even if the contacts exhibit internal reflection and the edge states are equally populated due to inelastic relaxation. Now Komiyama *et al.* also measured the contact resistances of their samples (the three-terminal resistances) and found indeed that they are not zero (or quantized). Therefore, since the contacts provide for internal reflection, we know that there is an initial non-equilibrium population of edge states. Thus, the Komiyama experiment demonstrates the following: In the absence of a barrier, the two edge states follow the outer sample perimeter and over a long distance, therefore, are *parallel*. Inelastic processes, even if weak, can bring about a partial equilibration between these edge states and, therefore, the measured resistance $R_{57,68}$ is small. (It is possible that in this case, the equilibration is due mainly to the contacts at the periphery of the sample between the current source contact and the contacts used to measure the voltage difference.) As the gate voltage is raised and the innermost edge state is reflected toward contacts 6 and 8 on a shorter path, equilibration between edge states is far less effective and, therefore, the measured resistance $R_{57,68}$ rises dramatically. (Note that in this configuration, there are no additional contacts that can equilibrate the edge states.) This experiment demonstrates, therefore, that even over very long distances, edge states that initially have been populated unequally do not equilibrate. These results point to a crucial role of the contacts for high-precision measurements of the quantum Hall effect. Since edge states do not equilibrate even in large samples, it is necessary even for large samples to inject (or detect) carriers in an equilibrium fashion if one wants to observe quantized resistances.

Van Wees et al.,[17] in a fascinating experiment, have demonstrated the absence of inter-edge scattering by showing that the Shubnikov–de Haas resistance oscillations are suppressed if carriers are injected and detected selectively. The experiment in principle is similar to Fig. 26, where we have shown that a localized backscattering process is not detected if two of the contacts couple only to the outermost edge state. Van Wees et al. have adjusted the Fermi energy such that it lies in the center of the highest populated Landau level. If the voltage contact is ideal, the normal Shubnikov–de Haas oscillations are detected. If the voltage contact is made narrower such that it successively couples to fewer edge states, the Shubnikov—de Haas oscillations gradually disappear. The distance between the current and voltage contact is 200 μm.

Alphenaar et al.[18] undertook an experiment to determine the amount of equilibration between various edge states. The experiment is performed in a geometry as shown in Fig. 22. There are three (spin-degenerate) edge states. The current injecting contact 2 is adjusted such that one edge state is completely transmitted. Thus, $T_2 = 1$ in Eq. (89). Contact 1, the voltage probe, is adjusted such that it permits successively transmission of 1, 2, and 3 edge states. The transmission probability T_1 in Eq. (89) thus is $T_1 = N_1$, where N_1 is the number of edge states transmitted at contact 1. The distance between contacts 2 and 1 is very large. Therefore, due to scattering, a portion of the carriers injected into the outermost edge state eventually will be distributed to some degree among the other channels[8,142] Let the population in the edge states at the entrance to the voltage contact be denoted by α_n, where $n = 1, 2, 3, \ldots$ indicates the edge state. The transmission probability T_{21} thus is equal to the population in the outermost edge state, $T_{21} = \alpha_1$ if $N_1 = 1$, and is $T_{21} = \alpha_1 + \alpha_2$ if $N_1 = 2$. According to Eq. (88), the measured Hall resistance thus is

$$R_{24,13} = \frac{h}{e^2} \frac{\sum_{k=1}^{N_1} \alpha_k}{N_1}. \tag{92}$$

Thus, by measuring the Hall resistance in this configuration, one obtains direct information on the degree to which carriers injected in the outermost edge state have been scattered into other edge states (or into the bulk). At a temperature $T = 0.45$ K and with two contacts separated by 80 μm, Alphenaar et al.[18] measure $\alpha_1 = 0.48$ and $\alpha_2 = 0.44$. Therefore, the population injected in the outermost edge state has almost equilibrated with the second edge state, but only 8% of the injected carriers are lost to the innermost edge state! Alphenaar et al.[18] conclude quite generally that the $N - 1$ outermost edge states equilibrate over long distances but that the $N - 1$ outermost edge states do not equilibrate with the innermost edge state. We stress that these long equilibration lengths apply to inter-edge state scattering. Scattering in the bulk of the sample can be expected to be characterized by different length scales. Also, it is worth emphasizing that the intra-edge state scattering rates probably are not suppressed. The distance over which carriers maintain phase-coherence is limited by intra-edge state scattering and thus is not any longer than a phase-coherence length in the absence of a field.

In Section IV, we discussed the possibility that edge states in a soft confining potential are far away from one another. The separation between adjacent edge states can be large compared to the width of the wave function. Moreover, if the field is not too weak, the outer edge states simply will not experience a fluctuating impurity potential. Martin and Feng[145] have calculated the inter-edge state scattering rate for elastic scattering and for acoustic phonons and have found that these scattering rates indeed are exponentially reduced in high fields. If r_0 is the elastic scattering rate for zero magnetic field, then Martin and Feng have found that in high fields, the inter-edge state scattering rate is $r_{n,n+1} \approx r_0 \exp(-(\Delta y_{n,n+1})^2/l_B^2)$. Here, $\Delta y_{n,n+1}$ is the distance between the nth and $(n+1)$th edge state and l_B is the magnetic length. Since the edge states (at the Fermi surface) can be far apart (as seen in Eq. (55)), the inter-edge state scattering rates are suppressed exponentially. This result was obtained for delta function impurity potentials, which is not very realistic. If it is taken into account that the impurity potential is smooth, an even stronger reduction of the elastic scattering rate can be expected. For acoustic phonon scattering, Martin and Feng found a similar result, but the exponential reduction factor now, in addition, is multiplied by $\exp(-\Delta E/kT)$, where ΔE is the energy exchanged between electrons and phonons.

The experiments discussed previously point to an extremely weak interaction between different edge states. Clearly, this weak interaction also will have an effect on the magnitude of the Shubnikov–de Haas oscillations, i.e., on resistances measured away from the plateaus. In Fig. 14, it is shown that only the innermost edge state is affected in the transition from one plateau to another. The longitudinal resistances away from a plateau also will depend on the way current is carried from one contact to another. Haug and von Klitzing[19] point out that the longitudinal resistances do not scale as expected for bulk conduction. The fact that bulk conduction is dominant in a short and wide sample, but edge state conduction is dominant in a long and narrow sample, also must be manifested in the measured longitudinal resistances. It should be emphasized that away from plateaus, all contacts invariably will exhibit internal reflection and, in the terminology used here, are not ideal. It seems that there hardly is an aspect of the quantum Hall effect that can be understood without taking the detailed role of edge states and contacts into account.

VI. Resonant Departures from the Hall Resistance

In this section, we address resonant tunneling processes in high magnetic fields. Resonant tunneling processes requires phase coherence. Recent experiments in GaAs structures[69,131,134,135,146] suggest that the phase-coherence length is only of the order of a few μm even at temperatures as small as half a Kelvin. The phase-coherence length at high fields thus is not very different from its low-field value. As suggested already, this might be due to the fact that the phase-coherence length is limited by intra-edge state scattering.[134] The discussion in this section thus is limited to length scales of a few μm. Our key interest is the effect of localized states on the Hall resistance.[26,71,76,77,137] We start, however, by analyzing very simple situations of resonant tunneling in two-terminal conductors.

21. Two-Terminal Conductance

Figure 28 shows a single resonant state in a two-terminal conductor. Figure 28a shows a localized state formed between two barriers analyzed by Sivan et al.[26] In Fig. 28b, the localized state arises due to a fluctuation in the impurity potential. The localized state permits interaction of one edge state of the sample with another edge state on the other side of the sample.[23,128] Since motion in high magnetic fields is uni-directional, these situations are particularly simple to analyze. At low magnetic field, the analysis is more

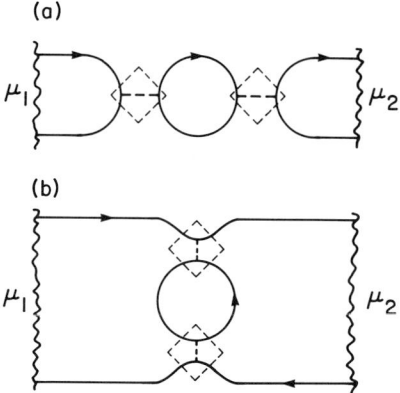

FIG. 28. Two-terminal conductor with (a) a localized state between two barriers (from Ref. 26), (b) a localized state permitting scattering from one edge state to another across the width of the sample. (from Ref. 23).

complex because each current path allows propagation of carriers in both directions. As stressed by Sivan et al., the high-field models presented next can be viewed simply as limiting cases of models investigated a number of years ago at low fields.[35–38] From a scattering point of view, the configurations in Figs. 28a and b are equivalent: They correspond to a simple permutation of the scattering matrix. This permutation[137] exchanges the roles of transmission and reflection coefficients.

The configurations in Fig. 28 are easy to analyze because transmission from the edge state to the localized state is determined by a 2×2 scattering matrix, which is specified by r_i, r'_i, t_i and $t'_i \equiv t_i$. The saddle points described by these scattering matrices are enclosed by dashed squares in Fig. 28. For the saddle point to the left connecting the edge state and the localized state, these scattering amplitudes have the following meaning: t_1 gives the probability of a carrier incident from the localized state to reach the edge state and r_1 is the probability that a carrier remains on the localized state. r'_1 is the probability that a carrier incident in the edge state in the upper left-hand side is reflected into the edge state in the lower left-hand side. t'_1 is the probability that a carrier is transmitted from the upper edge state into the localized state. Microreversibility implies that $t_1 \equiv t'_1$. Similarly, there is a set of probabilities, r_2, r'_2, $t_2 \equiv t'_2$, that characterizes transmission from the edge states to the right into and out of the localized state. Let the phase accumulated during traversal along the upper branch of the localized state be ϕ_a and the phase accumulated during traversal of the lower branch of the localized state be ϕ_b. For a smooth potential, the phases ϕ_a and ϕ_b also can be determined at least in a WKB approximation. Denote the equipotential trajectory of

Eq. (57) by $Y_\pm(E, x)$, then [147,148]

$$\phi_a = \int_{x_1}^{x_2} dx' \frac{Y_+(E_G, x)}{l_B^2} \qquad (93)$$

$$\phi_b = \int_{x_1}^{x_2} dx' \frac{Y_-(E_G, x)}{l_B^2}, \qquad (94)$$

where $E_G = E - \hbar\omega(j + 1/2)$. The sum of the two phases ϕ_a and ϕ_b is equal to the total area A enclosed by the orbit divided by the magnetic length squared, $\phi_a + \phi_b = A/l_B^2$. The area $A(E_G)$ enclosed by an orbit at energy E_G is determined by Eq. (57). Quantum mechanically, a localized state exists if the total flux through the area enclosed by the localized state is equal to a multiple of an elementary flux quantum,

$$A(E_G)B = \Phi_0\left(n + \frac{1}{2}\right), \qquad (95)$$

or if $\phi_a + \phi_b = 2\pi(n + 1/2)$. For a fixed magnetic field, the quantization condition, Eq. (95), determines a sequence of energies $E_{j,n}$ at which the local potential well (or potential peak) supports a quantum state with energy $E_{j,n}$. This happens whenever the total area enclosed by the orbit admits a flux that is an $(n + 1/2)$ multiple of the elementary flux quantum. Equation (95) is approximate, it measures the flux enclosed by an orbit by following the guiding center, and does not take the actual width of the wave function into account. Corrections to this semiclassical quantization rule are discussed by Azbel[147] and Glazman and Jonson.[149] In contrast to Glazman and Jonson,[149] we emphasize here, as in our discussion of the point contact resistances, [102] the case of a smooth potential; i.e., there are no hard walls. Even with the semiclassical quantization rule, the flux periodicity of the quantization rule, Eq. (45), does not give rise to a periodicity in the magnetic field: The increment in magnetic field ΔB needed to increase the flux through the area A by a flux quantum is[135,136] $\Delta B = \Phi_0/(A + (dA/dE)(dE_G/dB))$. Thus, a periodicity in magnetic field results only if the area is weakly dependent on energy. In addition, we also should keep in mind that the Fermi energy in reality is not fixed with respect to the potential, but adjusts in such a way as to keep the electron density constant.

A simple calculation gives for the overall transmission amplitude, denoted by s, the result,

$$s = \frac{t_1 t_2 e^{i\phi_a}}{1 - r_1 r_2 e^{i(\phi_a + \phi_b)}}. \qquad (96)$$

Next, we parameterize the reflection coefficients in terms of their amplitudes and phases, $r_n = R_n^{1/2} e^{i\Delta\phi_n}$, $i = 1, 2$. This gives a transmission probability,

$$|s|^2 = \frac{T_2 T_1}{1 + R_1 R_2 - 2 R_1^{1/2} R_2^{1/2} \cos \chi}, \tag{97}$$

where the overall phase is determined by $\chi = \phi_a + \phi_b + \Delta\phi_1 + \Delta\phi_2$. The two-terminal conductance of the conductor of Fig. 28a is given simply by $G = (e^2/h)T$ with $T \equiv |s|^2$. Now it is easy to see that transmission in the conductor of Fig. 28b is equivalent to reflection in the conductor of Fig. 28a. Hence, for the conductor of Fig. 28b, we have $R \equiv |s|^2$. Thus, the conductance of Fig. 28b simply is $G = (e^2/h)T$, with $T = 1 - R = 1 - |s|^2$.

Suppose now that the coupling of such a localized state to the edge states is very weak; i.e., T_1 and T_2 are small compared to 1. The localized state then is very long-lived and we can expect resonant transmission through the localized state. If the saddle only is weakly transparent, the phase $\Delta\phi_n$, $n = 1, 2$ is given by $\Delta\phi_n = -\pi/2$. Therefore, the total phase χ is equal to $\chi = A/l_B^2 - \pi$. The transmission probability $|s|^2$ given by Eq. (97) is maximal if $\cos(\chi) = 1$. Thus, if the localized state is weakly coupled to the external edge states, the condition for maximal transmission is just the quantization condition for the closed orbits. Namely, $\cos(\chi) = 1$ implies, $A/l_B^2 = 2\pi(n + 1/2)$, which is Eq. (95). Now consider the total phase χ as a function of energy E. Let the resonant energy E_r, determined by $\cos \chi(E) = 1$, be one of the solutions of Eq. (95). Typically, in the vicinity of the resonant energy, the phase χ is a linear function of the energy. The constant of proportionality is determined by the attempt frequency ν. Hence, for energies close to the resonant energy E_r, we find that the phase is $\chi = 2\pi n + h\nu(E - E_r)$. We now can express Eq. (97) in the Breit–Wigner form,

$$|s|^2 = \frac{\Gamma_1 \Gamma_2}{(E - E_r)^2 + \Gamma^2/4}, \tag{98}$$

where $\Gamma_i = h\nu T_i$, $i = 1, 2$ are the partial decay widths, and $\Gamma = \Gamma_1 + \Gamma_2$ is the total decay width.

Equations (97) and (98) are trivially extended to the case where there are N edge states that are totally reflected (Fig. 28b) and completely transmitted (Fig. 28b), i.e., when these additional edge states are well separated from the edge state that interacts with the localized edge state. This is the case in recent experiments by Brown et al.[134] and van Wees et al.,[135] who have considered transmission through a cavity.

Now we expand on the resonant Breit–Wigner behavior of the transmission probability. The power of the Breit–Wigner approach lies in the fact that it can treat even the case where additional edge states interact with one another and with the localized state. Consider the conductors as shown in Fig. 28a, but with N edge states to the left and N edge states to the right, all of which might interact with the localized state if the potential is not smooth. The Breit–Wigner approach yields a transmission probability from edge state j to the left to edge state i to the right that is given by

$$S_{ij} = \frac{\Gamma_i \Gamma_j}{(E - E_r)^2 + \Gamma^2/4}, \qquad (99)$$

where Γ_i is the partial decay width associated with transmission from the localized state into the edge state i and $\Gamma = \Sigma_i \Gamma_i$. The total transmission probability is

$$S = \frac{\Gamma_l \Gamma_r}{(E - E_r)^2 + \Gamma^2/4}, \qquad (100)$$

where Γ_l and Γ_r are the total decay width due to transitions from the localized state into edge states to the left and to the right, respectively. Therefore, the two-terminal conductance of the conductor in Fig. 28a is determined by $T = S$,

$$G = \left(\frac{e^2}{h}\right) \frac{\Gamma_l \Gamma_r}{(E - E_r)^2 + \Gamma^2/4}, \qquad (101)$$

and by permutation of the S matrix, the two-terminal conductance of the conductor in Fig. 28b is determined by $R = S$, and $T = N - R = N - S$,

$$G = \left(\frac{e^2}{h}\right)\left(N - \frac{\Gamma_l \Gamma_r}{(E - E_r)^2 + \Gamma^2/4}\right). \qquad (102)$$

Again, these expressions can be generalized if there is an additional set of edge states that does not interact with the other edge states and with the localized state. Equations (99)–(102) are very general, since they also apply when the localized state interacts with many edge states, as would be the case for a potential that varies strongly on a magnetic length scale.

22. EFFECT OF RESONANT STATES ON THE HALL EFFECT

Consider now the four-terminal conductors[137] in Figs. 29a and 29b. In the conductor of Fig. 29a, a central localized state can be entered if a barrier in the leads is surmounted. Figure 29a depicts a quantum dot with

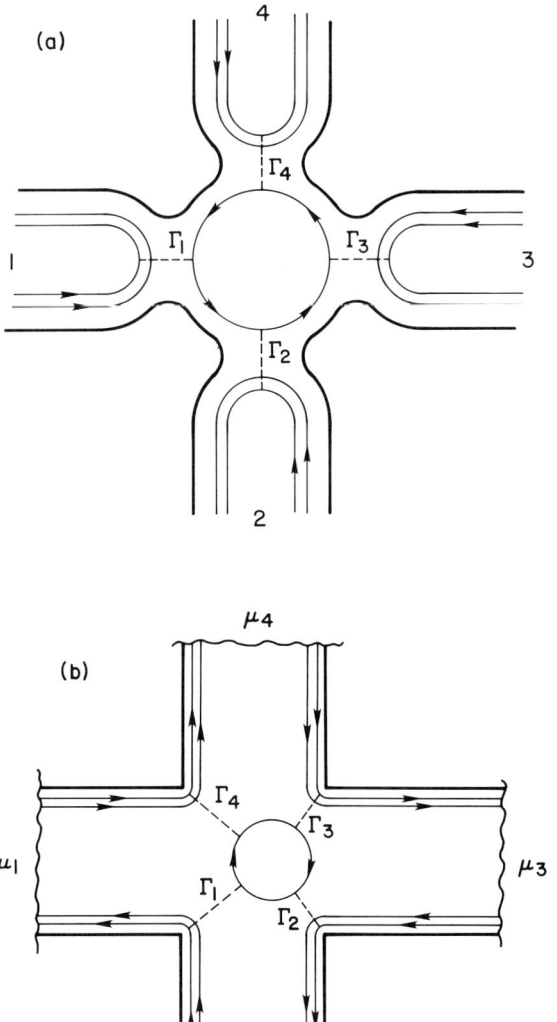

FIG. 29. Four-probe conductor with a resonant state in the center. (a) The localized state is separated from states in the probes by tunneling barriers. (b) The localized state interacts with edge states that without this interaction permit perfect transmission through the sample. Near resonances, the interaction of the localized state with edge states can be described by decay widths Γ_i. (From Ref. 137.)

narrow leads weakly coupled to it. In the conductor of Fig. 29b, a central localized state interacts with edge states. We start again by considering the simple case when only one edge state is present. Again, we can characterize the interaction of the localized state and the edge state in the leads with a 2×2 scattering matrix with amplitudes r, $t = t'$, and r'. Here, r_i' is the reflection probability of a carrier incident in lead i of the conductor and r_i is the probability of a carrier in the localized state to continue on the localized state past the most probable tunneling path connecting it to the lead i. Let us denote the phase that is accumulated by a carrier on the localized state during traversal from one possible escape path to the next escape path by ϕ. Let us calculate the transmission probabilities from one lead to another. Consider the amplitude s_{41}. This amplitude consists of the following processes: There is direct transmission with probability,

$$t_4 e^{i\phi_4} t_1. \tag{103}$$

Next, there is the possibility that a carrier completes n full turns on the localized state before escaping, and the amplitude for that is

$$t_4 e^{i\phi_4} (r_1 e^{i\phi_1} r_2 e^{i\phi_2} r_3 e^{i\phi_3} r_4 e^{i\phi_4})^n t_1. \tag{104}$$

Summing over all these amplitudes, we obtain a total transmission amplitude,

$$s_{41} = \frac{t_4 e^{i\phi_4} t_1}{1 - r_1 r_2 r_3 r_4 e^{i(\phi_1 + \phi_2 + \phi_3 + \phi_4)}}. \tag{105}$$

Hence, the resulting transmission probability is

$$S_{41} = |s_{41}|^2 = \frac{T_4 T_1}{1 + R_1 R_2 R_3 R_4 - 2(R_1 R_2 R_3 R_4)^{1/2} \cos(\chi)}, \tag{106}$$

with a total phase,

$$\chi = \sum_{i=1}^{i=4} (\phi_i + \Delta\phi_i). \tag{107}$$

For the transmission probability S_{31}, we obtain

$$S_{31} = \frac{T_3 R_4 T_1}{1 + R_1 R_2 R_3 R_4 - 2(R_1 R_2 R_3 R_4)^{1/2} \cos(\chi)}, \tag{108}$$

and for S_{21},

$$S_{21} = \frac{T_1 R_4 R_3 T_2}{1 + R_1 R_2 R_3 R_4 - 2(R_1 R_2 R_3 R_4)^{1/2} \cos(\chi)}. \tag{109}$$

The reflection probability then is found by current conservation, $S_{11} = 1 - S_{21} - S_{31} - S_{41}$. In a similar fashion, we can determine all the other transmission and reflection probabilities.

For the conductor in Fig. 29a, the transmission and reflection probabilities are given by $T_{ij} = S_{ij}$ and $R_{ii} = S_{ii}$. Equations (106)–(109) are completely general and allow the discussion of many different situations. Since we deal with one edge state only, the Hall resistance remains quantized as long as *one* of the transmission probabilities is equal to 1.[8] Sivan et al.[26] have studied the case where two of the leads only are weakly coupled (say, $R_2 \to 1$ and $R_4 \to 1$). Their discussion of the voltage measurement (which is in terms of piled-up charges and neglects phase-coherence) is correct in this simple case, since we deal with one Landau level only and since the transmission probabilities of Eqs. (106)–(109) are phase-sensitive only in the denominator (in contrast to the zero-field problem discussed in Section 10). Next, we focus chiefly on the case where all the leads are identical, i.e., on the case of a conductor that is fourfold-symmetric. The Hall resistance is calculated with the help of Eq. (18) and the transmission probabilities as determined previously. For the case of a fourfold-symmetric conductor, $T_i \equiv T$, $R_i \equiv R$, $\phi_i = \phi$ and $\Delta\phi_i \equiv \Delta\phi$, $i = 1, 2, 3, 4$, we find

$$R_H = \frac{h}{e^2} \frac{1}{T^2} \frac{(1 - R^2)(1 + R^4 - 2R^2 \cos\chi)}{(1 + R^2)(1 + R)^2}. \tag{110}$$

The Hall resistance, Eq. (110), for the fourfold symmetric junction with scattering in the leads is shown in Fig. 30 as a function of χ. It is, of course, a periodic function of χ. The periodicity in χ does not necessarily translate, however, into a Hall resistance that is periodic in B for reasons already discussed after Eq. (95), and since the phases $\Delta\phi_n$ and the reflection probabilities R_n also depend on B. Let us now discuss Fig. 30 in more detail. In the absence of backscattering in the leads, $R = 0$, the conductor in Fig. 29a exhibits no localized state. We have direct transmission along the edges of the conductor connecting the contacts in a cyclical fashion. The Hall resistance is quantized. For small reflection probabilities, corresponding to a *localized* state that is strongly coupled to the leads, weak oscillations develop around the quantized value (Hall plateau). With increasing reflection probability R, the localized state becomes long-lived. The Hall resistance depends strongly on the phase χ and shows large excursions above the quantized value. For

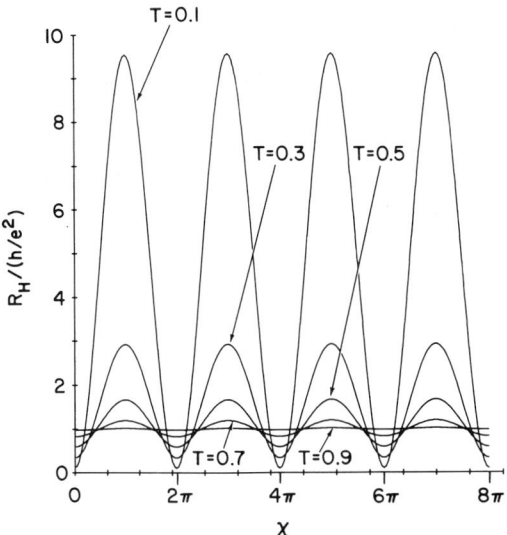

FIG. 30. Hall resistance of a conductor with leads separated by a barrier with transmission probability T and reflection probability R from a central region (current pattern as in Fig. 28a with one edge state only).

$\chi = 0$, i.e., at resonance, the Hall resistance becomes very small, and for reflection probabilities R close to 1, it is zero. As pointed out,[137] it is possible, therefore, to quench the Hall effect even at very high fields. We will return to this interesting effect next.

A curious *sum rule* holds for these models. If the Hall conductance, $G_H = 1/R_H$, is averaged over a full period of χ, the average is given by the quantized value of e^2/h. This was noted by Sivan et al.[26] for the case of two weakly coupled contacts. Such a sum rule also is at the heart of discussions of the quantum Hall effect based on an analysis of the flux sensitivity of a closed two-dimensional electron gas.[125,129] From an experimental point of view, this sum rule is not relevant, since a conductance is measured as a function of the magnetic field and not the phase χ.

Next, consider the conductor in Fig. 29b with perfect contacts. The pattern of current carrying states at the Fermi energy of this conductor again can be mapped onto that of the conductor in Fig. 29a. The scattering matrix for the conductor of Fig. 29b can be obtained by a permutation of the scattering matrix of the conductor in Fig. 29a. In particular, for the transmission and reflection probabilities, we find

$$T_{ij} = \mathbf{S}_{i,j+3}, \qquad R_{ii} = \mathbf{S}_{i,i+3}, \qquad (111)$$

where the matrix **S** is as defined previously and where the second index of **S** is taken to be modulo 4.

A number of different current patterns of the type shown in Fig. 29b have been discussed in Ref. 137. A version of the current pattern of Fig. 29b can be used to show that it is possible even in high magnetic fields to obtain four-terminal longitudinal resistances that are absolutely negative as observed in the experiments by Chang et al.[150] Here, we focus on the Hall resistance. For the case of a fourfold symmetric conductor, with a current pattern at the Fermi energy as in Fig. 29b, the Hall resistance is

$$R_H = \frac{h}{e^2} \frac{1 - F(1 + R)^2}{1 - 2FR(1 + R) + F^2(1 + R)^2(1 + R^2)}, \qquad (112)$$

where F is the function,

$$F = \frac{T^2}{1 + R^4 - 2R^2 \cos \chi}. \qquad (113)$$

Figure 31 shows this resistance as a function of the total phase χ for different

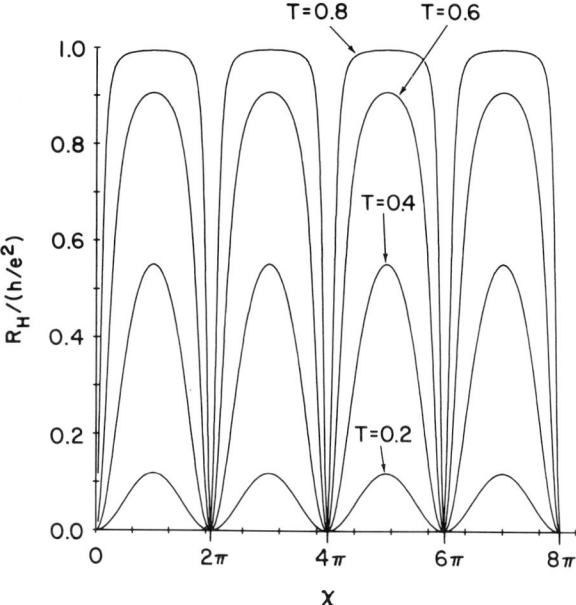

FIG. 31. Hall resistance of a fourfold-symmetric conductor with a resonant state in the center and perfect probes (current pattern as in Fig. 28b with one edge state only.)

transmission probabilities T. There is no interaction between the edge states and the localized state if $T = 0$. Even very small interactions (small transmission probabilities) lead to sharp deviations from the quantized value for $\chi = 0$. Note that at these values of the phase, the Hall resistance is zero (completely quenched). With increasing transmission probabilities, the interaction between the edge states and the localized states increases and the Hall resistance for all χ lies below the quantized value. Finally, if the interaction of the edge states and the localized states becomes strong ($T \approx 1$), the Hall resistance becomes small over the entire range of χ.

Let us look once again at the special case where the interactions between the localized states and the edge states in the leads are weak. We assume that all the transmission probabilities are small, $T_i \ll 1$. The transmission amplitudes then again become of the Breit–Wigner form. For the conductor of Fig. 29a, the transmission probabilities are given by[137]

$$S_{ij} = \frac{\Gamma_i \Gamma_j}{(E - E_r)^2 + \Gamma^2/4}, \quad (114)$$

where $\Gamma_i = \hbar v T_i$ is the decay rate into lead i, and

$$\Gamma = \sum_{i=1}^{i=4} \Gamma_i, \quad (115)$$

is the total decay width of the localized state. The formal expressions, Eqs. (114) and (115), are valid independent of the number of edge states in the leads if it is understood that Γ_i is the sum of all the partial decay widths into edge states of lead i. The only amplitude that is explicitly dependent on the number of edge states is the reflection probability for which we obtain

$$S_{ii} = N - \left(\frac{\Gamma_i(\Gamma - \Gamma_i)}{(E - E_r)^2 + \Gamma^2/4}\right). \quad (116)$$

For the conductor in Fig. 29a, we have $T_{ij} = S_{ij}$ and $R_{ii} = S_{ii}$. It is easy to show that the Hall resistance near resonance vanishes. (The zero Hall resistance described by the Breit–Wigner formulae corresponds to the minima of the Hall resistance shown in Fig. 30.) This consideration shows that the Hall resistance is quenched near resonance quite independently of the symmetry of the problem if only all the transmission probabilities are small enough such that the localized state becomes long-lived. (A simple calculation shows that the longitudinal resistances also vanish near resonance.) This is a consequence of the factorization of the transmission probabilities, $T_{ij} \propto \Gamma_i \Gamma_j$. Because of the factorization, each of the two products, $T_{mk}T_{nl} - T_{ml}T_{nk}$, whose difference determines according to Eq. (18) the four-probe resistance, is proportional to $\Gamma_1\Gamma_2\Gamma_3\Gamma_4$ regardless of the sequence of indices $klmn$. Physically, this result can be understood in the following way: If the localized state becomes long-lived, the current along the localized state is conserved along the closed orbit.

Let us next consider the conductor of Fig. 29b. The transmission and reflection probabilities again are obtained from Eqs. (114) and (115) using the permutation Eq. (111). Now the transmission probabilities differ strongly from one another even at resonance. For simplicity, let us return to the fourfold-symmetric conductor. In the fourfold-symmetric case, all decay widths are equal, $\Gamma_i \equiv \Gamma_0$, $i = 1, 2, 3, 4$, and the total decay width is $\Gamma = 4\Gamma_0$. Only three transmission probabilities remain to be determined. Introducing for the resonant denominator the abbreviation $\Delta = (E - E_r)^2 + \Gamma^2/4$, which is equal to $\Delta = (E - E_r)^2 + 4\Gamma_0^2$, we find from the transmission probabilities, Eq. (114), when permuted according to Eq. (111),

$$T_+ = N - \frac{3\Gamma_0^2}{\Delta}, \tag{117}$$

$$T_- = T_d = \frac{\Gamma_0^2}{\Delta}. \tag{118}$$

Here, $T_+ = T_{41} = T_{34} = T_{23} = T_{12}$ is the transmission probability for scattering from the incident probe into a side probe in the direction favored by the Lorentz force, and $T_- = T_{14} = T_{43} = T_{32} = T_{21}$ is the transmission probability for scattering against the Lorentz force. $T_d = T_{13} = T_{31} = T_{42} = T_{24}$ is the transmission probability for direct transmission without entering a side probe. Away from resonance, the probability for transmission into the direction favored by the Lorentz force is quantized. All other transmission probabilities vanish. Consequently, the Hall resistance is quantized. At resonance, the transmission probability T_+ is reduced and the other transmission probabilities are equal and nonzero. Interestingly, as in the case of a two-terminal resonant transmission problem, where the transmission can become perfect if the decay rates to the left and right are equal, the probabilities for the fourfold-symmetric conductor Eqs. (117) and (118) given previously also become just numbers at resonance. This is a multi-terminal analog of the well-known fact that transmission through a single resonant state in a two-port conductor is equal to 1 if the decay rate to the right and to the left are equal. At resonance, we have $\Delta = 4\Gamma_0^2$ and hence,

$$T_+ = N - \frac{3}{4} \tag{119}$$

and

$$T_- = T_d = R = \frac{1}{4}, \quad (120)$$

independent of the confining potential and other physical parameters! Thus, the Hall resistance is predicted to decrease from a well-quantized plateau at $R_H = (h/e^2)(1/N)$ to a minimum value,

$$R_H = R_{13,42} = \left(\frac{h}{e^2}\right)\frac{N-1}{N^2 - N + 1/2}, \quad (121)$$

at resonance. For $N = 1$, Eq. (121) predicts a complete suppression of the Hall resistance, $R_H = 0$, as shown in Fig. 31.

23. Intrinsic Resonances in Strongly Confined Systems

Transition of carriers from one side of the sample to another side of the sample can occur not only because of impurities and deliberately introduced barriers, but also for purely geometrical reasons. Consider the conductor shown in Fig. 29b. The energy spectrum in the branches of the conductor consists of magneto-elastic subbands as shown in Fig. 12. The channels with low threshold energy might be determined by purely magnetic energies if the width of the wire is large compared to the magnetic length. The threshold energy of the higher-lying modes typically is determined both by the confining potential and the magnetic energy. However, in the center of the cross, the elastic confinement is relaxed, and therefore, it is possible to have states in the cross that have lower energies than would be predicted from consideration of the branches of the conductor alone. This was noticed by Peeters,[151] who calculates the bound-state energies using a variational approach, and Schult et al.,[152] who present numerical computations for zero field. The effect on transport has been discussed by Ravenhall et al.[76] and Kirczenow.[71,77] Kirczenow[71] has mapped these states as a function of magnetic field for a Hall conductor with quadratic confinement. The resonant states are somewhat below the threshold of the nth magneto-elastic subband in the branches of the cross. In other words, as the magnetic field is lowered, the $(n + 1)$th magneto-elastic subband has a precursor that shows up as a resonant state in the center of the cross. It is clear that this is a universal phenomenon that is to be expected whenever the magneto-elastic confinement plays a role. (If the magnetic field is so strong that the spectrum in the center of the wire is independent of confinement, then, of course, this phenomenon does not occur.) Such resonant features are seen clearly in the computations of Ravenhall et al.[76] and Kirczenow.[71,77] A more detailed comparison between these analytic predictions and the computational results is the subject of Kirczenow.[153]

Interestingly, experiments on very narrow junctions do reveal a sequence, indeed, of resonant-like suppressions of the Hall resistance.[69] Figure 32 from Ford et al.[69] shows the high-field Hall resistance of two junctions of differing geometry. The junctions are $GaAs/Al_xGa_{1-x}As$ heterostructures with an Au gate that allows the density and channel width to be varied as a function of gate voltage. At high fields, these junctions exhibit a quantized Hall effect. As the junction is narrowed, large dips and spikes appear between Hall plateaus. As the junction is narrowed further, these strong fluctuations move into the $i = 4$ plateau as shown in Figs. 32a and b. The junction in Fig. 32a is a nominally straight Hall cross and the junction in Fig. 32b has a central area that is widened as shown in the inset. The measurements are made at $T = .022$ K. The measurements that are shown are for a gate voltage at which the strongest suppression of the Hall resistance is seen. Comparing these data with the model calculations of Figs. 30 and 31, it is evident that neither of these two models can account fully for the observed data. The model of Fig. 29a, which gives oscillations in the Hall resistance that exceed the quantized resistance, is inadequate because it predicts amplitudes above

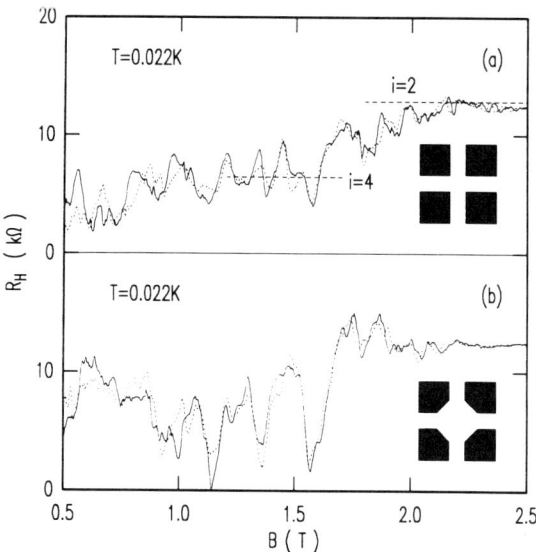

FIG. 32. High-field, low-temperature Hall resistance of two narrow ballistic junctions of differing geometries. (From Ref. 69.)

the plateau that are much larger than the excursions below the plateau. The model of Fig. 29b possibly can account for the large dips, but does not give excursions above the plateau. It thus is likely that a more complicated scattering model is needed that is a combination of Figs. 29a and b. Indeed, experimentally, we always must expect some scattering in the leads, especially when the Fermi energy approaches a threshold of a subband. Variation in the width and variations in the electron density along the leads are likely to produce—even in these small conductors—localized states at the threshold of the subbands.[69,154] It thus is possible that by augmenting Fig. 29b with backscattering processes in the leads (instead of sharp channel thresholds), theoretical results can be obtained that closely match the experimental data.

VII. Discussion

In this chapter, we have emphasized that to understand the quantum Hall effect in open conductors, it is necessary to consider in detail the properties of the contacts. It is the conductances calculated with respect to the potentials at the contacts that lead to results that are in agreement with experiments. The emphasis on contacts has led to the appreciation that longitudinal resistances and Hall resistances can be quantized simultaneously. The emphasis on contacts and the understanding of the role of the contacts also has led to the striking discovery of the extremely long distances taken to equilibrate edge channel populations. These experiments make the discussion of contacts relevant not only for small samples but also for macroscopic Hall conductors. It is likely, therefore, that in the future, contacts will play an essential role in assessing the accuracy of the quantum Hall effect. By stressing the role of contacts, we have taken a considerable step away from the discussion of this effect based on topological arguments or on discussions that treat motion in high-magnetic fields as a bulk effect without regarding the shape and boundaries of the conductor.

Motion in high magnetic fields is sample-specifc and the quantum Hall effect appears only if a set of conditions is obeyed. Even if these conditions are obeyed, the chemical potential at interior contacts[143] can be expected to reveal a complicated pattern. The sample-specific nature of high magnetic field transport must be especially important for the discussion of the Shubnikov–de Haas oscillations, since in this case, we deal with electron transport both in the bulk of the sample and along edge states.

We focused in this chapter on the integer quantum Hall effect. A similar picture, however, also can be expected to apply to the fractional quantum Hall effect. Interesting experiments on gated structures, similar to the ones used by Washburn et al.[13] and Haug et al.,[14] have been performed by Chang and Cunningham[155] for the fractional quantum Hall effect. Kouwenhoven et al.[156] have repeated the van Wees et al.[9] experiment with two adjacent contacts that exhibit internal reflection in the fractional quantum Hall regime and also have found quantization at anomalous fractional values. Timp et al.[132] have observed a plateau for a filling factor of one-half. Theoretical discussions of transport in the fractional quantum Hall regime near sample edges have been presented by Beenakker[157] and MacDonald.[158] These two discussions give rather different answers for the nature of the edge states. Beenakker finds that transport, even in the fractional quantum Hall regime, is only through compressible regions that have charge e. MacDonald, on the other hand, finds that there are edge states with fractional charges. The resolution of these questions is important. It bears on the question of whether in a dc measurement we can ultimately measure fractional charges.[159] It is related furthermore to the question of the periodicity of the Aharonov–Bohm effect in the fractional regime. An experiment by Simmons et al.[136] finds a marked difference for the field needed to produce a sample-specific resistance fluctuation in the integer regime and the fractional regime.

In this chapter, we only have discussed phenomena that occur if the chemical potential differences at the terminals of the conductor are small. The behavior under large applied potential differences and the calculation of critical currents requires a detailed knowledge of the potential distribution in the sample. It is clear that a number of questions related to high-field transport will occupy physicists for some time to come. This chapter thus likely contains only some initial and preliminary results that hopefully, in a few years, can be complemented with a more rounded view of this fascinating field.

Acknowledgment

I have had the benefit of meeting and talking with many of the authors of the works referenced in this article. Special thanks are due to F. Fang for allowing me to publish Fig. 15 and to S. Washburn, who prepared Figs. 22 and 32. I also thank R. Haug and C. W. J. Beenakker for critical comments on early versions of this chapter.

Note Added in Proof

Since the submission of this chapter in December of 1989, the topics discussed here have remained areas of active research. It is impossible to allude to all the important work. Instead we refer only to a few papers that can guide the interested reader to the more recent literature. Progress has been made toward a quantitative description of the transition region between Hall plateaus and the associated Shubnikov–de Haas peaks.[166–168] Structures with many gates have been fabricated[169] and structures with interior and exterior contacts[168] have been used to investigate non-equilibrium effects. Substantial theoretical and experimental progress has been made in the investigation of non-linear effects.[169–170] Much of the discussion given in the last section of this chapter has independently been carried out in Ref. 171 and has, in comparison with experiments,[172] been further developed in Ref. 173. We mention an experiment[174] that directly measures the electric field distribution in a Hall bar (also not on a mesoscopic scale) and an experiment that directly observes the Joule heat generated in the contacts.[175] Finally, we allude to recent work on thermal noise and shot noise in the quantized Hall regime.[176]

References

1. K. von Klitzing, G. Dorda, and M. Pepper, *Phys. Rev. Lett.* **45**, 494 (1988).
2. T. Ando, A. B. Fowler, and F. Stern, *Rev. Mod. Physics* **54**, 437 (1982).
3. R. B. Laughlin, *Phys. Rev.* **B23**, 5632 (1981).
4. B. J. Halperin, *Phys. Rev.* **B25**, 2185 (1982).
5. Q. Niu and D. J. Thouless, *J. Phys.* **A17**, 2453 (1984).
6. Q. Niu and D. J. Thouless, *Phys. Rev.* **B35**, 2188 (1987).
7. J. E. Avron, A. Raveh, and B. Zur, *Rev. Mod. Phys.* **60**, 873 (1988).
8. M. Büttiker, *Phys. Rev.* **B38**, 9375 (1988).
9. B. J. van Wees, E. M. M. Willems, C. J. P. M. Harmans, C. W. J. Beenakker, H. van Houten, J. G. Williamson, C. T. Foxon, and J. J. Harris, *Phys. Rev. Lett.* **62**, 1181 (1989).
10. S. Komiyama, H. Hirai, S. Sasa, and S. Hiyamizu, *Phys. Rev.* **B40**, 12566 (1989).
11. B. J. Halperin, *Helveta Physica Acta*, **56**, 75 (1983).
12. K. von Klitzing, *Rev. of Mod. Phys.* **58**, 519 (1986).
13. S. Washburn, A. B. Fowler, H. Schmid, and D. Kern, *Phys. Rev. Lett.* **61**, 2801 (1988).
14. R. J. Haug, A. H. MacDonald, P. Streda, and K. von Klitzing, *Phys. Rev. Lett.* **61**, 2797 (1988).
15. D. A. Syphers and P. J. Stiles, *Phys. Rev.* **B32**, 6620 (1985).
16. S. Komiyama, H. Hirai, S. Sasa, and F. Fujii, *Solid State Commun.* **73**, 91 (1990).
17. B. J. van Wees, E. M. M. Willems, L. P. Kouwenhoven, C. J. P. M. Harmans, H. G. Williamson, C. T. Foxon, and J. J. Harris, *Phys. Rev.* **B39**, 8066 (1989).
18. B. W. Alphenaar, P. L. McEuen, R. G. Wheeler, and R. N. Sacks *Phys. Rev. Lett.* **64**, 677 (1990).
19. R. J. Haug and K. von Klitzing, *Europhysics Lett.* **10**, 489 (1989).
20. P. Streda, J. Kucera, and A. H. MacDonald, *Phys. Rev. Lett.* **59**, 1973 (1987).
21. M. Büttiker, Y. Imry, R. Landauer, and S. Pinhas, *Phys. Rev.* **B31**, 6207 (1985).
22. M. Buttiker, *Phys. Rev. Lett.* **57**, 1761 (1986).
23. J. K. Jain and S. A. Kivelson, *Phys. Rev. Lett.* **60**, 1542 (1988).
24. M. Büttiker, *Phys. Rev. Lett.* **62**, 229 (1989).
25. P. Streda, *J. Phys.* **C1**, L12025 (1989).
26. U. Sivan, Y. Imry, and C. Hartzstein, *Phys. Rev.* **B39**, 1242 (1989).
27. R. J. Haug, J. Kucera, P. Streda, and K. von Klitzing, *Phys. Rev.* **B39**, 10892 (1989).
28. P. W. Anderson, D. J. Thouless, E. Abrahams, and D. S. Fisher, *Phys. Rev.* **B22**, 3519 (1980).

29. E. N. Economou and C. M. Soukoulis, *Phys. Rev. Lett.* **46**, 618 (1981).
30. D. S. Fisher and P. A. Lee, *Phys. Rev.* **B23**, 6851 (1981).
31. D. C. Langreth and E. Abrahams, *Phys. Rev.* **B24**, 2978 (1981).
32. M. Ya. Azbel, *J. Phys.* **C14**, L225 (1981).
33. R. Landauer, *IBM J. Res. Dev.* **1**, 223 (1957).
34. R. Landauer, *Phil. Mag.* **21**, 863 (1970).
35. M. Büttiker, Y. Imry, and R. Landauer, *Phys. Lett.* **A96**, 365 (1983).
36. Y. Gefen, Y. Imry and M. Ya. Azbel, *Phys. Rev. Lett.* **52**, 129 (1984).
37. M. Büttiker, Y. Imry, and M. Ya Azbel, *Phys. Rev.* **A30**, 1982 (1984).
38. M. Büttiker, *Phys. Rev.* **B32**, 1846 (1985). For an extended account, see M. Büttiker, in *New Techniques and Ideas in Quantum Measurement Theory*, Annals of the New York Academy of Sciences, Vol. 480, p. 194 (1986).
39. H. L. Engquist and P. W. Anderson, *Phys. Rev.* **B24**, 1151 (1981).
40. M. Büttiker, *Phys. Rev.* **B33**, 3020 (1986).
41. M. Büttiker, *IBM J. Res. Develop.* **32**, 63 (1988).
42. H. F. Cheung, Y. Gefen, and E. K. Riedel, *IBM J. Res. Develop.* **32**, 359 (1988).
43. D. Sokolovski, *Phys. Lett.* **A123**, 381 (1988).
44. C. W. J. Beenakker and H. van Houten, *Phys. Rev.* **B39**, 10445 (1989).
45. S. Datta, *Phys. Rev.* **B40**, 5830 (1989).
46. J. L. D'Amato and H. M. Pastawski, *Phys. Rev.* **41**, 7411 (1990).
47. A. D. Benoit, S. Washburn, C. P. Umbach, R. P. Laibowitz, and R. A. Webb, *Phys. Rev. Lett.* **57**, 1765 (1986).
48. M. Büttiker, *IBM J. Res. Develop.* **32**, 317 (1988).
49. H. B. G. Casimir, *Rev. Mod. Phys.* **17**, 343 (1945).
50. Y. Imry, in *Directions of Condensed Matter Physics*, Vol. 1 (G. Grinstein and G. Mazenko, eds.), World Scientific Publishing Co., Singapore, p. 101 (1986).
51. R. A. Webb, S. Washburn, C. Umbach and R. A. Laibowitz, *Phys. Rev. Lett.* **54**, 2996 (1985).
52. A. D. Stone, *Phys. Rev. Lett.* **54**, 2692 (1985).
53. S. Washburn and R. A. Webb, *Advances in Physics*, **35**, 375 (1986).
54. A. G. Aronov and Yu. V. Sharvin, *Rev. Mod. Phys.* **59**, 755 (1987).
55. D. P. DiVincenzo and C. L. Kane, *Phys. Rev.* **B38**, 3006 (1988).
56. H. van Houten, C. W. J. Beenakker, J. G. Williamson, M. E. I. Broekaart, P. H. M. van Loosdrecht, B. J. van Wees, J. E. Moij, C. T. Foxon, and J. J. Harris, *Phys. Rev.* **B39**, 8556 (1989).
57. S. Komiyama, H. Hirai, S. Sasa, and F. Fujii, *J. Phys. Soc. Japan* **58**, 4086 (1990); *Surface Science* **229**, 224 (1990).
58. M. Büttiker, *Phys. Rev.* **B35**, 4123 (1987).
59. A. D. Stone and A. Szafer, *IBM J. Res. Develop.* **32**, 384 (1988).
60. H. U. Baranger and A. D. Stone, *Phys. Rev.* **B40**, 8169 (1989).
61. D. J. Thouless, *Phys. Rev. Lett.* **47**, 972 (1981).
62. R. Landauer, *Z. Phys.* **68**, 217 (1987).
63. R. Landauer, in *Analogies in Optics and Microelectronics* (W. van Haeringan and D. Lenstra, eds.), Kluwer Academic Publishers, Dordrecht, Federal Republic of Germany p. 243 (1990).
64. I. B. Levinson, *Sov. Phys. JETP* **68**, 68 (1989).
65. M. L. Roukes, A. Scherer, S. J. Allen, Jr., H. G. Craighead, R. M. Ruthen, E. D. Beebe, and J. P. Harbison, *Phys. Rev. Lett.* **59**, 3011 (1987).
66. Y. Takagaki, K. Gamo, S. Namba, S. Ishida, S. Takaoka, K. Murase, K. Ishibashi, and Y. Aoyagi, *Solid State Communic.* **68**, 1051 (1988).
67. Y. Takagaki, K. Gamo, S. Namba, S. Takaoka, K. Murase, and S. Ishida, *Solid State Communic.* **71**, 809 (1989).

68. C. J. B. Ford, S. Washburn, M. Büttiker, C. M. Knoedler, and J. M. Hong, *Phys. Rev. Lett.* **62**, 2724 (1989).
69. C. J. B. Ford, S. Washburn, M. Büttiker, C. M. Knoedler, and J. M. Hong, *Surface Science* **229**, 298 (1990).
70. A. M. Chang, T. Y. Chang, and H. U. Baranger, *Phys. Rev. Lett.* **63**, 2268 (1989).
71. G. Kirczenow, *Phys. Rev. Lett.* **62**, 2993 (1989).
72. Y. Avishai and Y. B. Band, *Phys. Rev. Lett.* **62**, 2527 (1989).
73. H. U. Baranger and A. D. Stone, *Phys. Rev. Lett.* **63**, 414 (1989).
74. C. W. J. Beenakker and H. van Houten, *Phys. Rev. Lett.* **63**, 1857 (1989).
75. M. L. Roukes, T. J. Thorton, A. Scherer, and B. P. Van der Gaag, in *Electronic Properties of Multilayers and Low Dimensional Structures*, (J. M. Chamberlain, L. Eaves, and J. C. Portal, eds.), Plenum New York, p. 95 (1990).
76. D. G. Ravenhall, H. W. Wyld, and R. L. Schult, *Phys. Rev. Lett.* **62**, 1780 (1989).
77. G. Kirczenow, *Solid State Commun.* **71**, 469 (1989).
78. T. J. Thornton, M. Pepper, H. Ahmed, D. Andrews, and G. J. Davies, *Phys. Rev. Lett.* **56**, 1198 (1986).
79. D. A. Wharam, T. J. Thornton, R. Newbury, M. Pepper, H. Ahmed, J. E. F. Frost, D. G. Hasko, D. C. Peacock, D. A. Ritchie, and G. A. C. Jones, *J. Phys.* **C21**, L209 (1988).
80. B. J. van Wees, H. van Houten, C. W. J. Beenakker, J. G. Williamson, L. P. Kouwenhouven, D. van der Marel, and C. T. Foxon, *Phys. Rev. Lett.* **60**, 848 (1988).
81. C. G. Smith, M. Pepper, H. Ahmed, J. E. F. Frost, D. G. Hasko, R. Newbury, D. C. Peacock, D. A. Ritchie, and G. A. C. Jones, *J. Phys. C* **1**, 9035 (1989).
82. N. Garcia, private communication (1987).
83. D. van der Marel and E. G. Haanappel, *Phys. Rev.* **B39**, 7811 (1989).
84. R. Johnston and L. Schweitzer, *J. Phys.* **C21**, L861 (1988).
85. G. Kirczenow, *Solid State Commun.* **68**, 715 (1988).
86. G. Kirczenow, *Phys. Rev.* **B39**, 10452 (1989).
87. A. Szafer and A. D. Stone, *Phys. Rev. Lett.* **62**, 300 (1989).
88. L. Escapa and N. Garcia, *J. Phys.* **C1**, 2289 (1989).
89. Y. Avishai and Y. B. Band, *Phys. Rev.* **B62**, 2527 (1989).
90. E. Tekman and S. Ciraci, *Phys. Rev.* **B39**, 8772 (1989); *Phys. Rev.* **B43**, 7145 (1991).
91. Song He and S. Das Sarma, *Phys. Rev.* **B40**, 3379 (1989).
92. L. I. Glazman, G. B. Lesovik, D. E. Khmel'nitskii, and R. I. Shekter, *JETP Lett.* **48**, 239 (1988).
93. A. Kawabata, *J. Phys. Soc. Jap.* **58**, 372 (1989).
94. A. Yacoby and Y. Imry, *Phys. Rev.* **B41**, 5341 (1990).
95. A. Kumar, S. E. Laux and F. Stern, *Appl. Phys. Lett.* **54**, 1271 (1987).
96. J. A. Davies and J. A. Nixon, *Phys. Rev.* **B39**, 3423 (1989).
97. N. D. Lang, *Phys. Rev.* **B37**, 10395 (1988).
98. N. D. Lang, A. Yacoby, and Y. Imry, *Phys. Rev. Lett.* **63**, 1499 (1989).
99. J. N. L. Connor, *Mol. Phys.* **15**, 37 (1968).
100. W. M. Miller, *J. Chem. Phys.* **48**, 1651 (1968).
101. H. A. Fertig and B. I. Halperin, *Phys. Rev.* **B36**, 7969 (1987).
102. M. Büttiker, *Phys. Rev.* **B41**, 7906 (1990).
103. R. Landauer, *J. Phys. C* **1**, 8099 (1989).
104. H. van Houten and C. W. J. Beenakker, in *Nanostructure Physics and Fabrication* (M. A. Read and W. P. Kirk, eds.), Academic Press, Boston, p. 347 (1989).
105. H. van Houten, C. W. J. Beenakker, P. H. M. van Loosdrecht, T. J. Thornton, H. Ahmed, M. Pepper, C. T. Foxon, and J. J. Harris, *Phys. Rev.* **B37**, 8534 (1988).
106. J. R. Kirtely, S. Washburn, and M. J. Brady, *Phys. Rev. Lett.* **60**, 1546 (1988).

107. C. S. Chu and R. S. Sorbello, *Phys. Rev.* **B40**, 3409 (1989).
108. C. S. Chu and R. S. Sorbello, *Bull. Am. Phys. Soc.* **35**, 642 (1990).
109. M. Büttiker, *Phys. Rev.* **B40**, 3409 (1989). An extended account of this work is given in *Analogies in Optics and Microelectronics* (W. van Haeringen and D. Lenstra, eds.), Kluwer Academic Publishers, Dordrecht, Federal Republic of Germany p. 105 (1990).
110. R. A. Webb and S. Washburn, *Physics Today* **41**, 46 (1988).
111. H. U. Baranger, A. D. Stone, and D. P. DiVincenzo, *Phys. Rev. B* **37**, 6521 (1988).
112. C. L. Kane, P. A. Lee, and D. P. DiVincenzo, *Phys. Rev. B* **38**, 2995 (1988).
113. R. E. Prange and T. Nee, *Phys. Rev.* **168**, 779 (1968).
114. A. H. MacDonald and P. Streda, *Phys. Rev.* **B29**, 1616 (1984).
115. F. F. Fang and P. J. Stiles, *Phys. Rev.* **B27**, 6847 (1983).
116. F. F. Fang and P. J. Stiles, *Phys. Rev.* **B29**, 3749 (1984).
117. R. E. Prange and R. Joynt, *Phys. Rev.* **B25**, 2943 (1982).
118. S. L. Luryi and R. F. Kazarinov, *Phys. Rev.* **B27**, 1386 (1983).
119. S. A. Trugman, *Phys. Rev.* **B27**, 7539 (1983).
120. F. F. Fang (unpublished) (1983).
121. R. L. Stratonovich, *Topics in the Theory of Random Noise*, Vol. 1, Gordon and Breach, New York (1963).
122. P. Erdös and R. C. Herndon, *Adv. Phys.* **31**, 65 (1982).
123. G. V. Mil'nikov and I. M. Sokolov, *JETP Lett.* **48**, 537 (1989).
124. J. T. Chalker and P. D. Coddington, *J. Phys.* **C21**, 2665 (1988).
125. T. Ohtsuki and Y. Ono, *Solid State Communications* **65**, 403 (1988).
126. Y. Ono, T. Ohtsuki, and B. Kramer, *J. Phys. Soc. Japan* **58**, 1705 (1989).
127. B. J. van Wees, L. P. Kouwenhoven, H. van Houten, C. W. J. Beenakker, J. E. Mooij, C. T. Foxon, and J. J. Harris, *Phys. Rev.* **B38**, 3625 (1988).
128. J. K. Jain, *Phys. Rev. Lett.* **60**, 2074 (1988).
129. J. Hajdu, M. Janssen, and O. Viehweger, *Z. Phys.* **B66**, 433 (1987).
130. W. Pook and J. Hajdu, *Z. Phys.* **B66**, 427 (1987).
131. C. J. B. Ford, T. J. Thornton, R. Newbury, M. Pepper, H. Ahmed, D. C. Peacock, D. A. Ritchie, J. E. F. Frost, and G. A. C. Jones, *Appl. Phys. Lett.* **54**, 21 (1989).
132. G. Timp, R. Behringer, J. E. Cunningham, and R. E. Howard, *Phys. Rev. Lett.* **63**, 2268 (1989).
133. P. H. M. van Loosdrecht, C. W. J. Beenakker, H. van Houten, J. G. Williamson, B. J. van Wees, J. E. Mooij, C. T. Foxon, and J. J. Harris, *Phys. Rev.* **B38**, 10162 (1988).
134. R. J. Brown, C. G. Smith, M. Pepper, R. Newbury, H. Ahmed, D. G. Hasko, J. E. F. Frost, D. C. Peacock, D. A. Ritchie, and G. A. C. Jones, *J. Phys. Cond. Matt.* **1**, 6291 (1989).
135. B. J. van Wees, L. P. Kouwenhoven, C. J. P. M. Harmans, J. G. Williamson, C. E. Timmering, M. E. I. Broekaart, C. T. Foxon, and J. J. Harris, *Phys. Rev. Lett.* **62**, 2523 (1989).
136. J. A. Simmons, H. P. Wei, L. W. Engel, D. C. Tsui, and M. Shayegan, *Phys. Rev. Lett.* **63**, 1731 (1989).
137. M. Büttiker, *Phys. Rev.* **B38**, 12724 (1988).
138. B. R. Snell, P. H. Beton, P. C. Main, A. Neves, J. R. Owers-Bradely, L. Eaves, M. Henini, O. H. Hughes, S. P. Beaumont, and C. D. W. Wilkinson, *J. Phys.* **C1**, 7499 (1989).
139. R. J. Haug, K. von Klitzing, K. Ploog, P. Streda, *Surface Science* **229**, 229 (1990).
140. J. K. Jain and S. A. Kivelson, *Phys. Rev. Lett.* **60**, 1542 (1988).
141. F. M. Peeters, *Phys. Rev. Lett.* **61**, 589 (1988).
142. S. Komiyama and H. Hirai, *Phys. Rev.* **B40**, 7767 (1989).
143. M. Büttiker, in *Nanostructure Physics and Fabrication* (M. A. Read and W. P. Kirk, eds.), Academic Press, Boston, p. 319 (1989).
144. M. Büttiker, *Surface Science* **229**, 201 (1990).
145. T. Martin and S. Feng, *Phys. Rev. Lett.* **64**, 1971 (1990).

146. G. Timp, P. M. Mankiewich, P. de Vegar, R. Behringer, J. E. Cunningham, R. E. Howard, H. U. Baranger, and J. K. Jain, *Phys. Rev.* **B33**, 6227 (1989).
147. M. Ya. Azbel, *Soviet Physics JETP* **12**, 891 (1961).
148. H. A. Fertig, *Phys. Rev.* **B38**, 996 (1988).
149. L. I. Glazman and M. Jonson, *J. Phys.* **C1**, 5547 (1989).
150. A. M. Chang, G. Timp, J. E. Cunningham, P. M. Mankiewich, R. E. Behringer, and R. E. Howard, *Solid State Commun.* **76**, 769 (1988).
151. F. M. Peeters, *Supperlattices and Microstructures* **6**, 217 (1989).
152. R. L. Schult, D. G. Ravenhall, and H. W. Wyld, *Phys. Rev.* **B39**, 5476 (1988).
153. G. Kirczenow, *Phys. Rev.* **B42**, 5357 (1990).
154. M. Ya. Azbel and O. Entin-Wohlman, *J. Phys.* **A22**, L957 (1989).
155. A. M. Chang and J. E. Cunningham, *Solid State Comm.* **72**, 651 (1989); A. M. Chang, *Solid State Comm.* **74**, 871 (1990).
156. L. P. Kouwenhoven, B. J. van Wees, N. C. van der Vaart, C. J. P. M. Harmans, C. E. Timmering, and C. T. Foxon, unpublished (1989).
157. C. W. J. Beenakker, *Phys. Rev. Lett.* **64**, 216 (1990).
158. A. H. MacDonald, *Phys. Rev. Lett* **64**, 220 (1990).
159. S. A. Kivelson and V. L. Prokovski, *Physical Rev.* **B40**, 1373 (1989).
160. C. W. J. Beenakker and H. van Houten, in *Electronic Properties of Multilayers and Low Dimensional Semiconductor Structures* (J. M. Chamberlain, L. Eaves, and J. C. Portal, eds.), Plenum, New York (in press) (1989).
161. P. H. Beton, B. R. Snell, P. C. Main, J. R. Owers-Bradley, L. Eaves, M. Henini, O. H. Hughes, S. P. Beaumont, and C. D. W. Wilkinson, *J. Phys. C* **1**, 7505 (1989).
162. T. J. Thornton, M. L. Roukes, A. Scherer and B. P. Van de Gaag, *Phys. Rev. Lett.* **6**, 2128 (1989).
163. L. W. Molenkamp, A. A. Staring, C. W. J. Beenakker, R. Eppenga, C. E. Timmering, J. G. Williamson, Harmans, and C. T. Foxon, *Phys. Rev.* **B41**, 1274 (1990).
164. R. Johnston and L. Schweitzer, *Z. Phys.* **B72**, 217 (1988).
165. M. Büttiker, in *Electronic Properties of Multilayers and Low Dimensional Structures*, (J. M. Chamberlain, L. Eaves, and J. C. Portal, eds.), Plenum New York, p. 51 (1990).
166. P. L. McEuen, A. Szafer, C. A. Richter, B. W. Alphenaar, J. K. Jain, A. D. Stone, R. G. Wheeler, and R. N. Sacks, *Phys. Rev. Lett.* **64**, 2062 (1990).
167. P. C. van Son, F. W. de Vries, and T. M. Klapwijk, *Phys. Rev.* **B43**, 6764 (1991).
168. J. Faist, P. Gueret, and H. P. Meier, *Phys. Rev.* **B43**, 9332 (1991); J. Faist, *Europhys. Lett.* **15**, 331 (1991).
169. G. Müller, D. Weiss, S. Koch, K. von Klitzing, H. Nickel, W. Schlapp, and R. Lösch, *Phys. Rev.* **B42**, 7633-7636 (1990).
170. P. C. van Son and T. M. Klapwijk, *Europhys. Lett.* **12**, 429 (1990); P. C. van Son, G. H. Kruithof, and T. M. Klapwijk, *Phys. Rev.* **42**, 11267 (1990).
171. L. W. Molenkamp, M. J. P. Brugman, H. van Houten, C. W. J. Beenakker, C. T. Foxon, *Phys. Rev.* **B43**, 12118 (1991).
172. C. J. B. Ford, S. Washburn, R. Newbury and C. M. Knoedler, and J. M. Hong, *Phys. Rev.* **B43**, 7339 (1991).
173. G. Kirczenow and E. Castano, *Phys. Rev.* **B43**, 7343 (1991)
174. P. F. Fontein, J. A. Kleinen, P. Hendriks, F. A. P. Blom, J. H. Wolter, H. G. M. Lochs, F. A. J. M. Driessen, L. J. Giling, C. W. J. Beenakker, *Phys. Rev.* **B43**, 12090 (1991).
175. U. Klass, W. Dietsche, K. von Klitzing, and K. Ploog, *Z. Phys.* **B82**, 351 (1991).
176. M. Büttiker, *Phys. Rev. Lett.* **65**, 2901, (1990); in *The Physics of Semiconductors* (E. M. Anastassakis and J. D. Joannopoulos, eds.), World Scientific, Singapore, p. 49, (1991).

CHAPTER 5

Electrons in Laterally Periodic Nanostructures

W. Hansen and J. P. Kotthaus

SEKTION PHYSIK
UNIVERSITÄT MÜNCHEN
FEDERAL REPUBLIC OF GERMANY

U. Merkt

INSTITUT FÜR ANGEWANDTE PHYSIK
UNIVERSITÄT HAMBURG
FEDERAL REPUBLIC OF GERMANY

I. INTRODUCTION	279
II. REALIZATION OF LATERALLY PERIODIC NANOSTRUCTURES	286
III. TRANSPORT IN PERIODIC NANOSTRUCTURES	295
1. Density-Modulated Systems.	296
2. Arrays of Quasi-One-Dimensional Systems	307
3. Quasi-Zero-Dimensional Systems	326
IV. FAR-INFRARED SPECTROSCOPY OF ELECTRONIC EXCITATIONS	334
4. Plasmons in Density-Modulated Systems	336
5. Excitations in Confined Systems	345
V. CONCLUSIONS AND PERSPECTIVES	372
ACKNOWLEDGMENTS	373
REFERENCES	374

I. Introduction

The outstanding role of artificially created superlattices for the development of semiconductor physics and devices was envisioned initially by Esaki and Tsu in their famous paper on superlattices and negative differential conductivity in semiconductors.[1] Following their definition, we mean by *superlattices* periodic potentials in solids with periods well above the interatomic spacings but still below the elastic mean free path of conduction electrons or other quasi-particles of interest. In semiconductors, this puts us in the nanometer range, i.e., in the range of periods typically between 10 nm and several 100 nm. In their original work, as well as in subsequent explorations, Esaki and Tsu considered possible effects of a one-dimensional superlattice created by layer-by-layer molecular beam epitaxy (MBE) on the electrical and optical properties of a three-dimensional (3D) semiconductor.[1-3] This field now is well-developed and an essential topic of all semiconductor conferences.[4]

Here, we want to focus on the effects of artificially generated periodic potentials on quasi-two-dimensional (2D) electron systems in semiconductors resulting in superlattice phenomena or, in the extreme case of tight-binding superlattice potentials, in arrays of quasi-one-dimensional (1D) quantum wires or quasi-zero-dimensional (0D) quantum dots. Various ways to realize such laterally periodic nanostructures were proposed more than a decade ago by Sakaki et al.,[5] Bate,[6] and Stiles.[7] All these concepts make use of the advantageous properties of 2D electron systems that have been studied extensively now for over 20 years and are well-documented in the proceedings of the biannual conference on electronic properties of two-dimensional electron systems (recent proceedings in Refs. 8 and 9) and a well-known review by Ando, Fowler, and Stern.[10] One essential advantage of 2D electron systems is that the donor charge can be spatially removed from the plane of mobile charge, thus making possible high mobilities at relatively high carrier densities as well as easy tunability via field effect. This is one of the main reasons for the success of the metal-oxide semiconductor field-effect transistor (MOSFET), where the active channel in essence is a 2D electron system. The modulation-doping technique used in epitaxially grown heterojunctions[11] makes possible electron mobilities in semiconductor devices that today exceed values of 10^7 cm^2/Vs. Another advantage is the relative ease with which a 2D electron system can be subjected to lateral structuring by employing high-resolution lithography and planar technology. This is best evidenced by the steady advance of MOS technology. A third advantage of 2D systems in comparison to 3D systems that becomes important here is that a further reduction in the dimensionality of an electron system by a tight-binding superlattice potential is expected to affect the electronic properties more if the original dimensionality of the system is lower. This becomes obvious immediately if one looks at the dependence of the density of states $D(E)$ of an electron system on its dimensionality as sketched in Fig. 1. If one subjects a 3D electron system to a sufficiently strong superlattice potential periodic in one direction, the resulting electron system will be 2D-like and strong modifications of the electronic density of states occur only in the vicinity of the quantization energies that are formed in the wells of the superlattice potential, i.e., at the respective bottoms of the 2D subbands. The effect of a tight-binding 1D superlattice potential on a 2D electron system is stronger, since the transition from a 2D to a 1D electron system modifies the

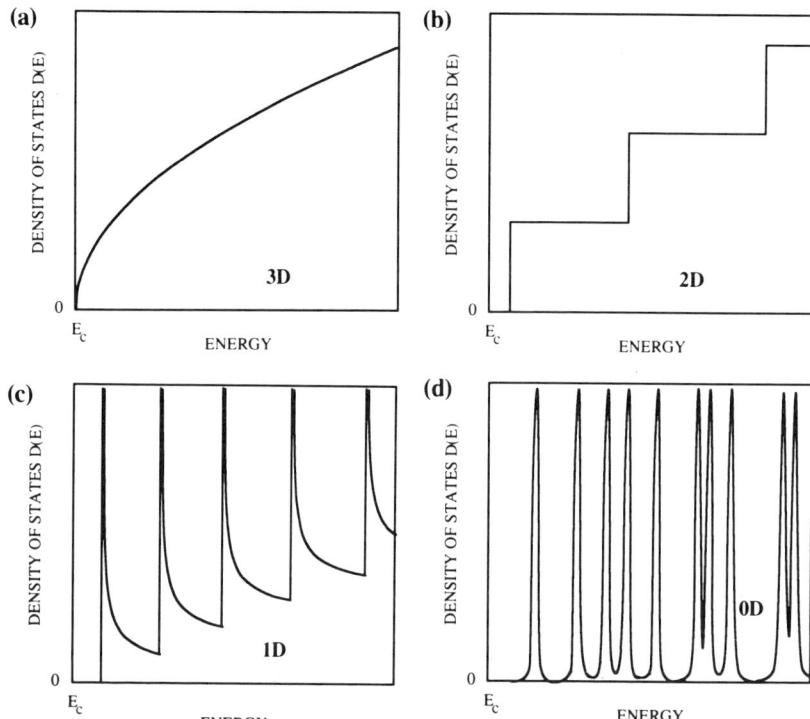

FIG. 1. Sketch of the densities of states in ideal three-dimensional (a), quasi-two-dimensional (b), and quasi-one-dimensional (c) electron systems. In the case of a quasi-zero-dimensional electron system (d), the discrete energy levels are assumed to be broadened.

electronic density of states at all relevant energies. Even more pronounced effects are expected if we expose a 2D electron system to a tight-binding superlattice potential in both lateral dimensions and effectively reduce the density of states from the 2D to the discrete 0D case, i.e., to a series of quasi-atomic levels.

As for the corresponding 3D case,[1] there exist basically two routes to create lateral superlattices. One is to modulate spatially the atomic structure of the underlying solid. Then the superlattice potential may become visible as a modulation of the band gap as sketched in Fig. 2a and/or a modulation of the electron effective mass.[12] The other is to generate electrostatically the superlattice, either by periodic doping as suggested by Esaki and Tsu[1] or by employing periodically nanostructured field electrodes as proposed by Bate.[6] In the latter cases, a potential modulation is generated in which the band gap remains spatially constant but the band edges are periodically modulated

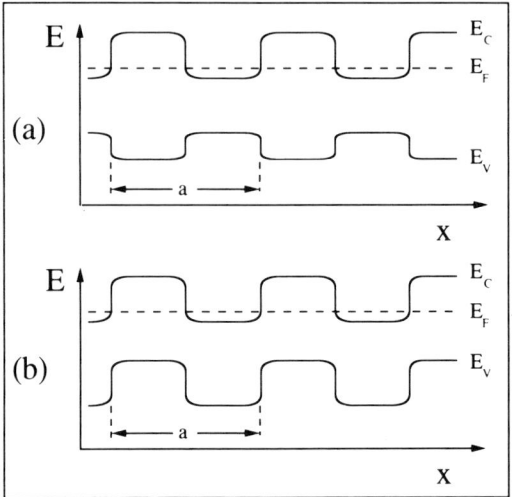

FIG. 2. Superlattice formed by (a) band-gap modulation and (b) electrostatic potential modulation. E_c and E_v denote the conduction and valence band edge, respectively, and E_F is a typical Fermi level.

as sketched in Fig. 2b. The use of surface-corrugated thin films suggested by Sakaki et al.[5] may be understood as another way of electrostatic potential modulation.

For the 3D case, layer-by-layer growth with molecular beam epitaxy (MBE) has proved to be extremely successful to achieve various types of band-gap-modulated superlattices.[13] Doping superlattices as an example of potential modulation have the disadvantage that the statistical distribution of the donors in the active layers usually causes a relatively low electron mobility. Nevertheless, they are very useful for optical applications.[14] For the 2D case, nearly all experiments reported so far have used some type of potential modulation to generate lateral superlattices. Lateral confinement to arrays of dots or wires by periodically removing semiconductor material via etching[15–17] does not really constitute a method of band-gap modulation, since in all reported experiments, surface charges cause sidewall depletion so that the effective confinement essentially is electrostatic. Real band-gap modulation of a 2D system seems to require rather difficult regrowth techniques that so far have not been able to produce true lateral band-gap superlattices. Alternative promising methods to create band-gap-modulated lateral superlattices are provided by the strain-induced confinement reported recently by Kash et al.[18] and the growth on a terraced high-index plane.[19–22] Because of their important role in experiments reported so far, electrostatically generated lateral superlattices will be the main subject of our review.

A simple way, at least conceptually, to electrostatically generate a 2D electron system with a lateral superlattice is illustrated in Fig. 3. Here, the gate oxide of a MOSFET is periodically thickness-modulated and then covered with a metal gate. With a gate bias above inversion threshold, one generates a 2D-like inversion layer in which the 2D electron density is laterally modulated with period a as sketched schematically in Fig. 4a. Such density-modulated nanostructures first were realized on Si by Mackens et al.[23] for the study of superlattice effects on the 2D plasmon dispersion. They are not

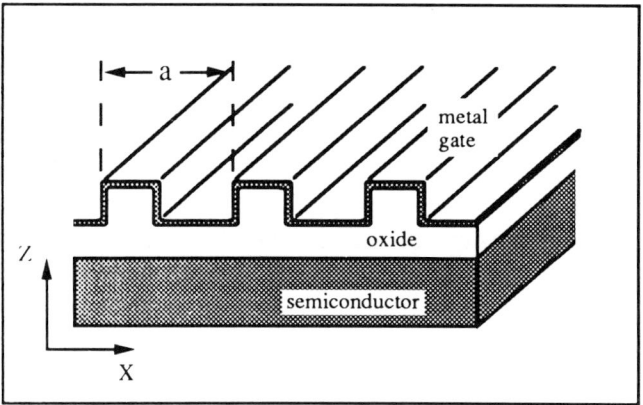

FIG. 3. Metal-oxide semiconductor (MOS) structure with periodically thickness-modulated oxide insulator between metal gate and semiconductor substrate.

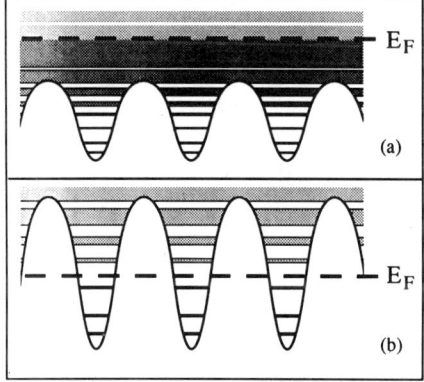

FIG. 4. Sketch of the effect of a superlattice onto the subband structure of a 2D electron system. (a) Weak potential modulation causing a periodic density modulation, (b) strong potential modulation generating a periodic array of isolated quantum wires or quantum dots.

well-suited, however, to achieve a tight-binding superlattice with which a transition from 2D layers to 1D quantum wires can be induced as sketched in Fig. 4b. The latter now can be achieved by dual-gate MOS devices similar to the ones originally proposed by Bate.[6] The operation of dual-gate MOS devices will be discussed in detail in the following section.

It may appear surprising that it took nearly a decade to realize lateral superlattices that were proposed conceptually in the mid-1970s and for which quantitative predictions existed for many years.[7,24] One reason is that it was necessary to refine the lithograhic tools required to generate periodic nanostructures.[25] Another more important reason, however, is that it proved quite difficult to find nanostructure processing technologies that could be used to artificially create laterally periodic nanostructures without destroying the mobility of the electron system that was to be subjected to the lateral superlattice. Even today there have been relatively few laterally periodic structures realized as in Fig. 4a that exhibit single-electron superlattice effects directly reflecting the superlattice period. Because of the fabrication problems still associated with such systems, most experiments have been concentrated on periodic arrays of essentially isolated quantum wires and dots similar to the systems sketched in Fig. 4b.

The effects of lateral superlattices on the electronic properties of 2D systems nevertheless were investigated extensively about 10 years ago, both experimentally and theoretically. This work makes use of natural superlattice effects occurring in a 2D electron system confined near a high-index surface, i.e., a surface vicinal to a high-symmetry surface of the underlying semiconductor crystal. Such natural superlattice effects first were discovered in pioneering work by Cole et al.[26] in which the existence of minigaps in the subband structure of 2D inversion channels of MOSFETs fabricated on high-index surfaces of Si was shown by magneto-transport experiments. Though such natural superlattice effects are not the purpose of our review, we briefly want to address them, since they reveal some of the basic physical phenomena that also are expected in artificially created laterally periodic nanostructures. For the case of MOS devices on high-index surfaces, the mechanisms that cause the superlattice potential are provided by the particular surface boundary conditions at high-index surfaces and are described either by the so-called valley projection model put forward by Sham et al.[27] or by a staircase-like model proposed by Volkov et al.[28,29] In either model, the period of the superlattice is determined by the surface orientation alone, whereas the strength of the superlattice potential depends on the particular surface boundary condition and usually is found to be a relatively weak perturbation.

If we consider a high-index surface with its normal tilted by an angle θ with respect to the surface normal of the adjacent high-symmetry surface, one obtains the superlattice period G in reciprocal space simply as $G = Q \sin \theta$, where Q denotes the extension of the first Brillouin zone for the direction normal to the high-symmetry surface. Correspondingly, the real space period of the superlattice is $L = 2\pi/G = 2\pi/(Q \sin \theta)$. For electrons on Si, the surface superlattice effects are complicated somewhat by the fact that one has to take into account the many-valley conduction band structure, and we refer the interested reader to the literature.[26,27,30–32] For a single-valley conduction band such as that of InSb,[33] the surface superlattice band structure becomes as simple as illustrated in Fig. 5a. At the boundaries and the center of the 1D

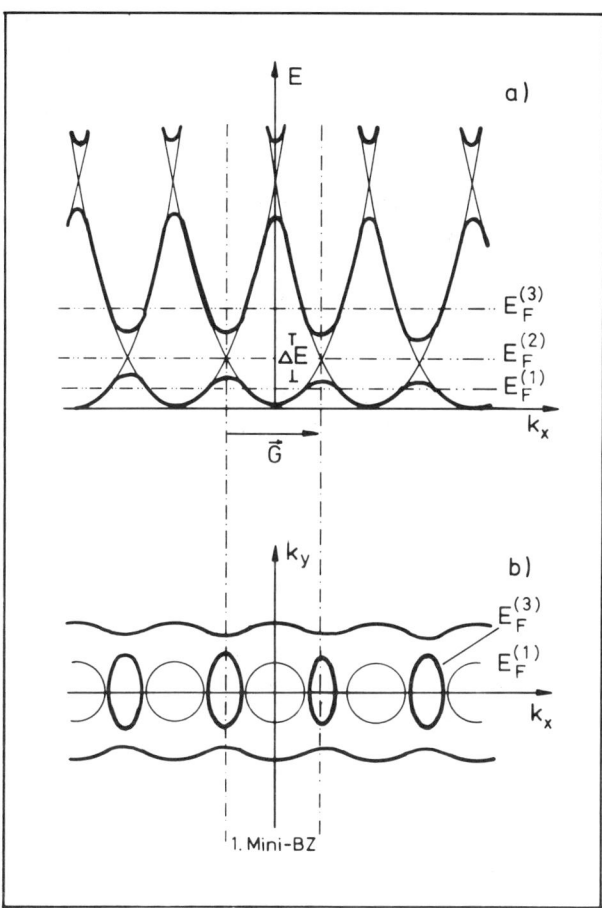

FIG. 5. Schematic representation of minigaps in the subband structure of a single-valley 2D electron system in a superlattice potential periodic in one dimension with reciprocal lattice vector **G** in x direction. (a) Energy dispersion in an extended minizone scheme for zero (thin line) and finite (thick line) strength of the superlattice potential. The Fermi levels for the cases of a partly occupied lowest miniband $E_F^{(1)}$, a fully occupied miniband $E_F^{(2)}$, and a partly occupied second miniband $E_F^{(3)}$ are indicated. (b) Fermi contours in the k_x–k_y plane for the cases $E_F = E_F^{(1)}$ (thin line) and $E_F = E_F^{(3)}$ (thick line). (From Ref. 33.)

minizones, the superlattice potential causes the appearance of 1D minigaps in the 2D subband structure. Tuning the 2D electron density via field effect, the Fermi energy can be moved from below to above a minigap. Fermi contours corresponding to different positions of the Fermi energy are illustrated in Fig. 5b. In transport experiments, the observation of such 1D minigaps becomes possible either by their effect on the Fermi contours, which are revealed via Shubnikov–de Haas oscillations of the magneto-resistance,[26,32,33] or by their effect on the density of states, which becomes visible as structure in the conductance when the Fermi energy is swept across a minigap.[26,30,33] As has been demonstrated in experiments on high-index surfaces of Si, the appearance of minigaps also can be observed by far-infrared spectroscopy that detects the excitations across the minigaps.[27,31,34] All these techniques that have proved to be quite valuable in the experimental investigations of 1D superlattice effects in MOS devices on high-index surfaces also can be applied to artifically created lateral superlattices.

The essential disadvantage of the aforementioned natural superlattices is that neither their superlattice potential nor their period can be tuned over a wide range. Typical superlattice potentials achieved are of the order of 1 meV and are destroyed easily by disorder at the semiconductor interface. Superlattice periods realized on high-index surfaces are relatively short and of order 10 nm corresponding to tilt angles θ of a few degrees. Most likely, longer-period natural superlattices have not been observed because of interface disorder destroying long-range correlation. However, the recently reported MBE growth on vicinal planes[21,22] eventually may combine some of the features of such natural superlattices with artificial band-gap engineering provided by epitaxial technologies.

In the following, we will restrict our discussion to artificially generated lateral superlattices imposed on originally 2D electron systems. After a short section on the realization of such laterally periodic nanostructures, we first will discuss their static transport properties and then the spectroscopy of intraband excitations. In each case, we will treat weakly modulated superlattices as sketched in Fig. 4a separately from tight-binding superlattices as described in Fig. 4b.

II. Realization of Laterally Periodic Nanostructures

As stated in the introduction, proper preparation techniques are crucial for precise control of the superlattice potential that patterns the originally 2D electron system in the semiconductor. Since L. V. Keldysh's suggestion in 1962 to generate a potential modulation by means of an intense ultrasonic wave,[35] an enormous variety of methods to implement a superlattice potential in a crystal have been proposed in publications and patents. Many of them suggest laterally patterning of 2D electron systems, because of their high electron mobility and the unequivocal advantage that the electron density can be controlled. Furthermore, the 2D electron system in Si MOSFETs is the basis of very large-scale integration of industrial electronic devices. The common need for high-resolution patterning techniques in device technology as well as fundamental research lead to a very rapid development in this area and can be noticed in the proceedings of conferences devoted to this area.[36–39]

Methods of lateral patterning may be classified into lithographic techniques and maskless techniques. *Litographic techniques*, in which the periodic structures are defined as a replica of a mask, take advantage of resist materials sensitive to exposure with particle or radiation beams. Here, we would like to explain briefly a specific optical exposing method, namely *holographic lithography*, that proved very successful in producing simple periodic structures as model systems for fundamental research. Two overlapping coherent laser beams produce a grating-type interference pattern with which the photoresist is exposed. Figure 6a is a schematic of the optical setup for holographic lithography. A laser beam of wavelength λ is expanded and split into two beams of equal intensity. With two plane mirrors, the beams are brought to

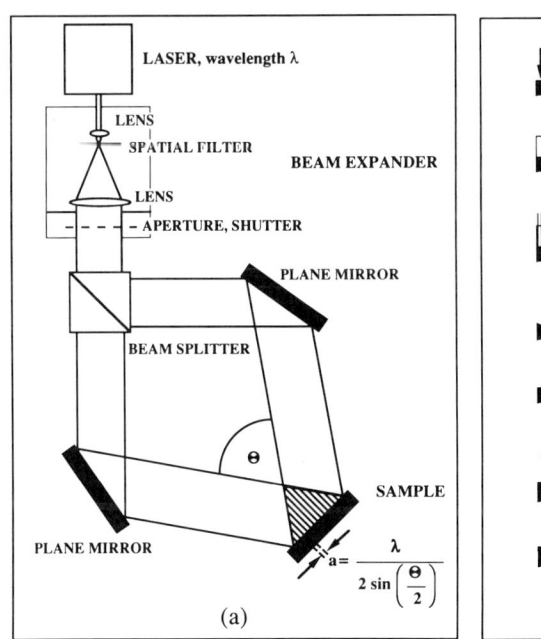
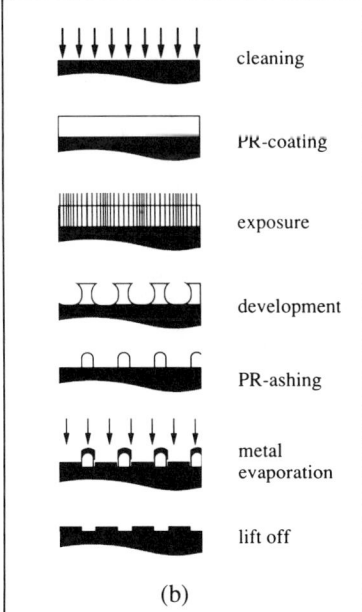

FIG. 6. (a) Optical setup for exposure with holographic lithography. (b) Processing steps for microstructuring with lithographic techniques.

interfere at the sample surface, which is coated with a thin (thickness $d \leq \lambda/4$) layer of photoresist. The dose of the exposure controls the solubility of the resist in a developing solution. Using a positive photoresist, the exposed stripes of the resist are dissolved. The remaining grating of undissolved resist serves as a mask in subsequent preparation processes, such as the one depicted in Fig. 6b. Examples of photoresist patterns defined by holographic lithography are depicted in Fig. 7. Holographic lithography is advantageous because proximity effects and exposure beam-induced damage—important issues in electron beam and ion beam exposure—are negligible.[25] Also, the exposure time, which typically is below one minute, is relatively small and large areas of the order of cm^2 can be patterned with excellent homogeneity

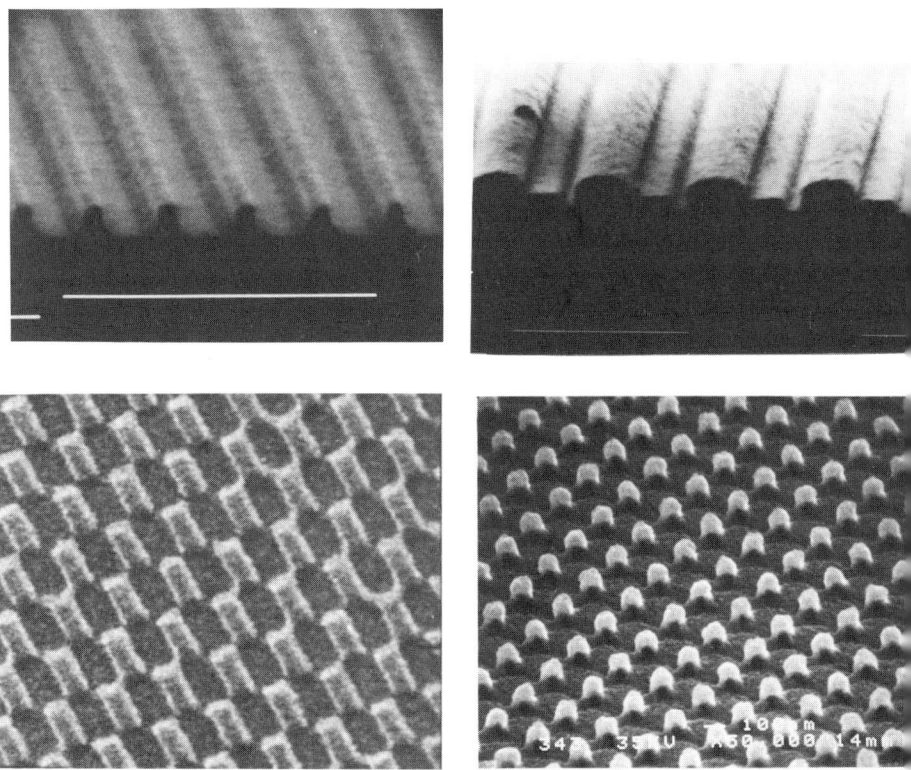

FIG. 7. Scanning electron micrographs of photoresist masks defined by holographic lithography. The resist grating on the top left has a period of $a = 250$ nm. The resist grating on the top right with period $a = 650$ nm has been covered with aluminium stripes. The resist squares of the grid on the bottom left have widths and lengths of 170 nm and 420 nm, respectively. The period of the dot matrix on the bottom right is $a = 250$ nm.

of the grating period. Spectroscopic transmission experiments with radiation in the far-infrared regime (wavelength $\lambda \leq 1$ mm), as described in Section IV, require samples with active areas larger than 1 mm^2. A grating on an area of 5×5 mm^2 with stripes or dots of period $a = 250$ nm consists of 20.000 stripes or $4 \cdot 10^8$ dots, respectively! Patterning a corresponding area with focused particle beams consumes a considerable amount of time and requires high long-term stability of the exposure system. Holographic lithography thus is especially suited for the preparation of samples for spectroscopic experiments.

With the mask pattern on the crystal surface, various fabrication processes can be employed to transfer the geometry of the resist mask to the electron system. Commonly used techniques involve metal evaporation with lift-off, etching processes, ion bombardment,[40] or material regrowth.[41,42] Figure 6b depicts the so-called lift-off processes in which a metal pattern representing the inverse of the resist mask is fabricated. After fabrication of the resist mask, a metal layer is deposited on the masked surface. Afterwards, the resist is dissolved, removing those parts of the metal layer that cover the resist. The metal pattern thus created may be used as, e.g., a grating coupler for optical experiments, an etch mask for various etching processes, or a gate electrode. Holographically defined metal gratings on Si MOS structures or Al$_x$Ga$_{1-x}$As/GaAs heterojunctions have been used as passive grating couplers to couple far-infrared radiation to plasmon or intersubband excitations in 2D electron systems.[43–45] When they are used as gate electrodes one can induce a spatial variation of the electron density in 2D systems.

A grating-type gate in a Si MOS device at gate voltages close to the inversion threshold will induce only electrons beneath the metal stripes. Thus, an array of isolated electron channels is created. Spatially modulated electron densities also are induced in MOS devices with periodically modulated gate oxide thicknesses as shown in Fig. 3. To fabricate such devices, either a resist pattern or a metal grating is used as an etch mask. After an etching process that partially reduces the insulator thickness, the active area of the device is covered with a continuous metal gate.

In so-called dual-stacked gate devices, a grating-type gate is combined with a homogeneous top gate as illustrated in Fig. 8.[6,46,47] First, a grating-type electrode is fabricated on top of a relatively thin (~ 20 nm) oxide on the Si substrate as described earlier. This grating electrode is buried in a second oxide on top of which the continuous gate is deposited. The advantage of such a structure is that by application of different voltages to the lower and upper gates, the average charge density and the potential modulation may be controlled independently. Two different modes are possible to create an array of isolated electron stripes in this dual-stacked gate device. The first, the *subgrating inversion mode*, is illustrated in Fig. 8b. In this mode, the grating-type electrode is biased above the threshold voltage, thus

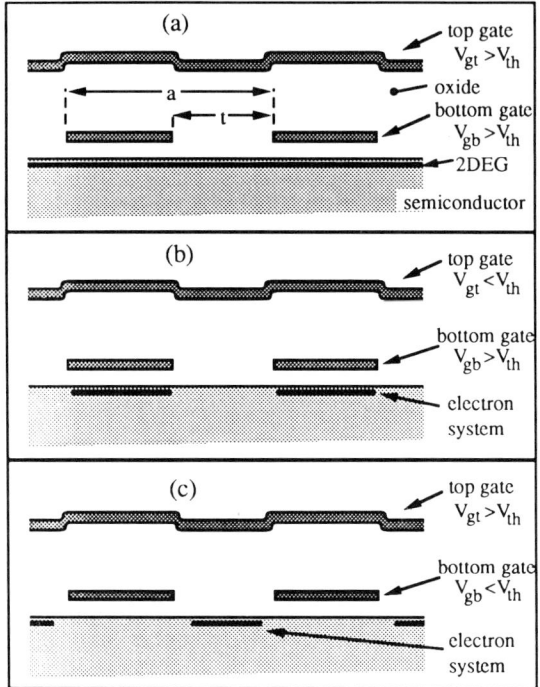

FIG. 8. Sketch of a Si MOS device with dual-stacked gate configuration. (a) Both gates are biased above threshold so that a 2D inversion layer is created. (b) Subgrating inversion mode. (c) Subgap inversion mode.

inducing electrons underneath the bottom gate ($V_{gb} > V_{th}$). The continuous top gate is biased below threshold, so that the regions between the stripes remain isolating ($V_{gt} < V_{th}$). In the second mode, the *gap inversion mode* depicted in Fig. 8c, the grating-type electrode is biased below threshold ($V_{gb} < V_{th}$) and inversion channels are induced by the top electrode between the grating stripes ($V_{gt} > V_{th}$).

Quite similar field-effect devices can be prepared on heterojunctions as sketched in Fig. 9. Modulation-doped heterojunctions generally consist of several crystalline layers of different composition or doping grown epitaxially on a substrate crystal. In an $Al_xGa_{1-x}As/GaAs$ heterojunction, a typical sequence of layers starting from the substrate consists of an undoped GaAs buffer layer (~ 1 μm), an undoped $Al_xGa_{1-x}As$ spacer layer (~ 20 nm), an n-doped $Al_xGa_{1-x}As$ layer (~ 30 nm), and a GaAs cap layer (~ 10 nm). Electrons from the donors in the doped $Al_xGa_{1-x}As$ layer occupy the 2D electron states at the $GaAs/Al_xGa_{1-x}As$ interface. Thus, in contrast to MOS devices, a 2D electron system exists even without a biased gate electrode. The

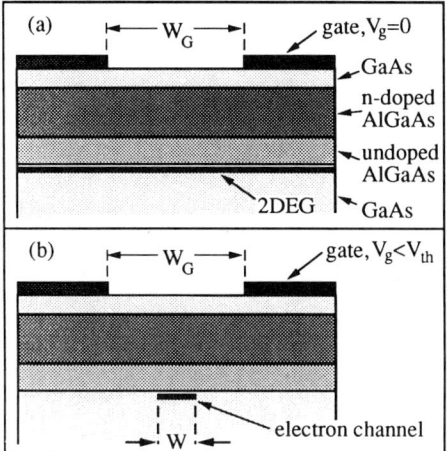

FIG. 9. (a) $Al_xGa_{1-x}As/GaAs$ heterojunction with a split front gate at zero bias and with gate opening W_G. The layer sequence and the position of the 2D electron system are indicated. (b) Same structure at negative gate bias below threshold voltage.

carrier density in the 2D electron system depends on various design parameters such as the doping of the active $Al_xGa_{1-x}As$ layer, the thickness of the $Al_xGa_{1-x}As$ spacer layer, and the position of the Fermi energy with respect to the GaAs gap at the heterojunction surface. These parameters are fixed once the specimen is grown. A Schottky gate deposited on the heterojunction surface allows one to continuously alter the carrier density during a measurement. In particular, at sufficiently low negative voltages between the electron system and the Schottky gate, the areas below the Schottky gate will be completely depleted of carriers. In the following, the gate voltage at which the carrier density underneath the Schottky gate drops to zero will be called *threshold voltage* in analogy to the inversion threshold voltage in MOS devices.

With the Schottky gate, small electron systems can be created if the gate is patterned so that it has small openings. At gate voltages below threshold voltage, the electron system will be confined to the areas that are not covered by the gate. This *depletion gate technique* was first applied by Thornton et al.[48] in a split-gate device as shown in Fig. 9, where a narrow electron channel of effective width W is created beneath the narrow slit of width W_G in the Schottky gate. Note that due to fringing fields, the effective width W of the remaining electron channel can be considerably smaller than the width W_G of the gate opening.

Slightly modified versions of the depletion gate technique were used later to induce laterally periodic density modulations or isolated arrays of electron channels for transport,[49–52] optical,[53] and capacitance measurements.[54] In devices used by Hansen et al.,[53] the distance between a continuous gate and the electron system was modulated laterally by a holographically defined insulator grating as sketched in Fig. 10a. Thus, a periodic carrier density modulation is induced at gate voltages above the threshold voltage. Note that since the Fermi level pinning at the heterojunction surface may be modified by the Schottky gate or the insulator grating on the heterojunction surface, a slight modulation of the carrier density may be expected even without the application of a gate voltage. At gate voltages below the threshold voltage, isolated electron channels are formed below the insulator stripes. As shown in Fig. 10b, in the devices fabricated by Smith et al.,[54] an undoped GaAs cap layer about 30 nm thick was removed partially by a reactive ion etching process through an etch mask consisting of lithographically defined metal stripes.[55] After the etching process, the surface was covered with a continuous gate, so that the distance between the electron system and the Schottky gate is modulated by the remaining parts of the cap layer. The field effect-induced carrier depletion underneath the gate parts that are close to the electron system is augmented by the band bending induced by the etching process and the Schottky gate.

In fact, it was found that the band bending induced by so-called *shallow mesa etching* procedures may be sufficient to confine the electron system. As

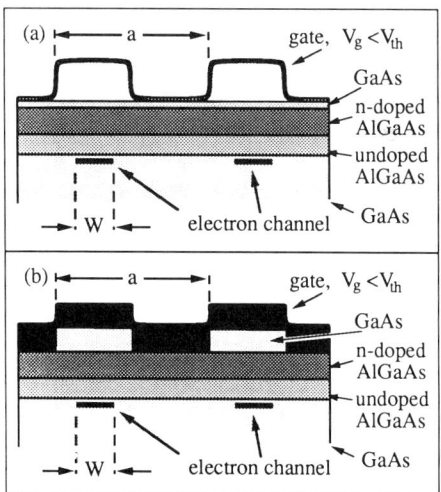

FIG. 10. $Al_xGa_{1-x}As$/GaAs heterojunctions with corrugated continuous gates. The positions of the narrow electron channels at gate bias below threshold are indicated. (a) Corrugation by an insulator grating, (b) by an etched grating in the GaAs cap layer.

shown in Fig. 11a, in shallow etched devices, only a part of the layer sequence above the interface where the 2D electron system resides is removed. The removal of material as well as surface states and damage introduced by the fabrication cause sufficient band bending to provide the confinement potential in these structures.[56,57]

In *deep mesa etched* devices, the material is removed down to the GaAs buffer layer as illustrated in Fig. 11b. Thus, heterojunction patterns are created, in which electron systems are confined by the surface potentials of the etched sidewalls. Carrier trapping in surface states of large density in the exposed sidewalls of the etched geometries can cause large surface potential gradients, so that the extension of the electron system can be much smaller than the etched heterojunction patterns. In deep mesa etched $Al_xGa_{1-x}As/GaAs$ heterojunction wires, depletion widths W_d of the order of several 100 nm were found.[58] Since the position of the Fermi level pinning in the band gap and the surface state density are sensitive to details of the fabrication process as well as to the type of the semiconductor material, the depletion widths of deep mesa etched devices can vary in a large range. With deep mesa etching techniques, a multiple quantum well wire as depicted in Fig. 11c can be fabricated, starting from a doped multiple quantum well heterostructure.[59] In

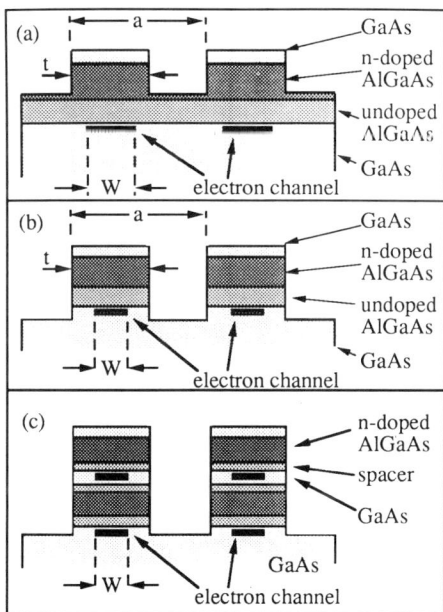

FIG. 11. Cross sections of $Al_xGa_{1-x}As/GaAs$ heterojunction wires. (a) Shallow mesa etched, (b) deep mesa etched single well, and (c) deep mesa etched multiple quantum well wire.

such a structure, vertically stacked narrow electron channels are created, which can interact via Coulomb interaction or tunneling processes.

In InSb MOS devices, lateral confinement is provided by metal patterns directly deposited onto the InSb crystal surface as depicted in Fig. 12. In the areas of the InSb surface that are covered by metal, the Fermi level is pinned within the InSb band gap. The surface is covered after deposition of the metal pattern with an oxide and a continuous gate on top. With positive voltages applied at the continuous gate, confined electron systems are induced between the metal patterns of the Schottky gate. Arrays of very narrow electron channels and electron dots have been created in this fashion between metal gratings and grids, respectively.[53,60]

Previously described examples of lithographic preparation techniques represent only a very limited selection. Other techniques involve, e.g., epitaxial growth on corrugated substrates,[41] regrowth on etched heterostructures,[61,62]

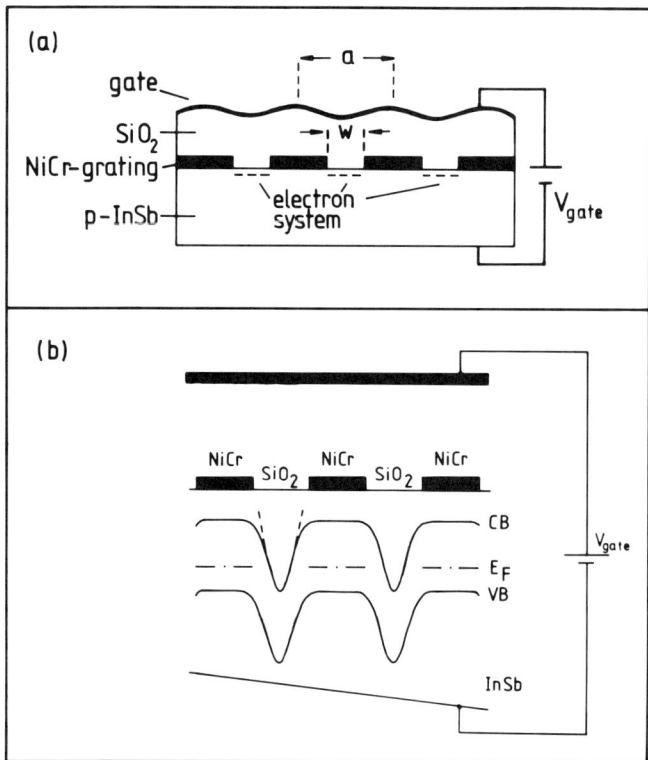

FIG. 12. (a) Cross section of an InSb MOS device with a grating-type Schottky gate. (b) Positions of the conduction and valence bands with respect to the Fermi level at positive gate bias.

ion-implantation-enhanced compositional interdiffusion,[63,64] or strain modulation imposed by selective etching of initially uniformly strained heterojunction layers.[18] Among other things, these techniques offer the possibility to generate a lateral band-gap modulation and thus are of enormous importance for optical investigations in the band-gap energy regime. Here, we have focused on fabrication techniques used for devices for which experimental results are described in the following sections.

In lithographic processing steps, it is inevitable that extrinsic mask material, which may affect purity, is deposited onto the crystal surface. Most desirable therefore, are, maskless techniques, in which the patterns are imposed directly on the electron system. One example of such methods, applied relatively early, is the preparation of electron systems on vicinal surfaces, as described in the introduction. Recently, this method has been revived for heterostructures grown by MBE. Alternating the composition of submonolayers deposited on vicinal planes, a compositional superlattice can be realized, since layer nucleation preferentially is initiated at the step edges of the misoriented surfaces.[19–22]

A large potential for further improvement also may be offered by focused-ion-beam techniques using exposure systems akin to electron beam writers. Ion-implantation-induced local damage, changes of composition, or doping can be used to define microstructures directly with focused ion beams.[62,65–67] Since the lithographic step thus can be avoided, further technological advances will make it feasible to combine focused-ion-beam systems with *in situ* MBE.

Extrapolating the rapid technological advances, we may anticipate the development of sophisticated fabrication techniques allowing the creation of 2D or even 3D superlattices with periods comparable to those of epitaxially grown superlattices. Besides the enormous technological implications, this will have considerable impact on the feasibility of experimental investigations. In this sense, the investigations described in the following sections can be viewed as pioneering work in a new field of scientific research.

III. Transport in Periodic Nanostructures

In this section, static or quasi-static transport of electron systems with different lateral periodic potential modulation will be discussed. The division of the section into density-modulated 2D systems and isolated systems with 1D or 0D behavior suggests itself, since such systems represent the weak and tight-binding limits of electron systems in superlattice potentials. Our strategy to give a concise overview on the subject is to discuss a few experiments in detail and refer the interested reader to original publications, otherwise.

Selection is somewhat arbitrary, but intends to demonstrate the particularities and broad possibilities of laterally periodic systems.

1. Density-Modulated Systems

Bloch's theory of electron motion in crystalline lattices, the basis for understanding the electronic properties of solids, has constantly inspired physicists to generate artificial band structures for new electronic properties. According to Bloch's theorem, the interaction of electrons with the lattice potential results in energy bands with dispersions dependent on the electron potential in the lattice. Electrons within the Bloch bands are described as quasi-free particles with effective mass determined by the dispersion of the band energy with respect to the quasi-momentum in the Brillouin zone,

$$m^* = \hbar^2 \left(\frac{d^2 E}{dk^2}\right)^{-1}. \tag{1}$$

The Bloch bands will be modified when a superlattice potential $V(x)$ with period a is added to the crystal potential. The momentum space of the Brillouin zone is divided up into minizones of size $2\pi/a$ and the electronic energies now form minibands separated from each other by minigaps. For example, in a sinusoidal potential, $V(x) = V_1 \cos(\mathbf{Kr})$ with characteristic wave vector $\mathbf{K} = 2\pi/a$, the energy dispersion forming the first minigap is

$$E^{\pm}(\mathbf{k}) = \frac{1}{2}[E(\mathbf{k}) + E(\mathbf{k} - \mathbf{K})]^2 \pm \sqrt{\frac{1}{2}[E(\mathbf{k}) - E(\mathbf{k} - \mathbf{K})]^2 + V_1^2}. \tag{2}$$

Generally, within first-order perturbation theory, the width of the nth gap is equal to the amplitude V_n of the nth Fourier coefficient of the superlattice potential. The effective mass of the electron in the miniband, defined by Eq. (1), is strongly dependent on its quasi-momentum in the minizone.

By tuning the superpotential amplitude and periodicity, it is possible to control parameters of the electron bands and thus generate and study novel electronic properties. Negative differential resistance (NDR), the formation of a Stark ladder, and high-frequency oscillations of the electron motion, so-called Bloch oscillations, are examples of anticipated properties. Electrons in the upper part of the miniband experience an energy dispersion corresponding to a negative mass ($d^2E/dk^2 < 0$) and thus exhibit similar properties as electrons in valence bands. An electron accelerated by an electric field $\mathbf{E}$ from the bottom of a miniband will change its mass and, in the absence of scattering processes, decelerate when it reaches the top part of the miniband. This behavior would continue periodically with frequency $v = eEa/h$ and thus localize the electron to a spatial region of $x = \Delta/eE$, where Δ is the energy width of the miniband. Consequently, in sufficiently high fields, the miniband could be viewed as a ladder of spatially separated discrete energy levels, the so-called Stark ladder.

The importance of Zener tunneling between the minibands remains unclear, and the existence of the Bloch oscillation and a Stark ladder has been disputed intensively.[68] The realization of a Bloch oscillator in intrinsic material seems to be unfeasible, since in semiconductors studied thus far, optical phonon scattering and avalanche ionization will prevent carriers from reaching the negative mass range of the Bloch band.[69] Esaki and Tsu[1] were the first to point out that a 1D superpotential can be realized in synthesized semiconductor superlattices consisting of alternating thin semiconductor layers of different alloy composition or doping density controlled by epitaxial growth. The relatively large period of the artificial superlattice results in small Brillouin zones in reciprocal space and correspondingly small widths of the Bloch bands. The period of Bloch oscillations in these so-called minibands thus would be reduced and the restrictions imposed by scattering processes may be less severe. The publication by Esaki and Tsu initiated a new field of semiconductor research[4] in whose course negative differential resistance[70] and, more recently, evidence for a Stark ladder[71] have been found. Until now, however, Bloch oscillations have eluded experimental observation.

In a 3D electron system with a 1D superlattice in the x direction, only electrons with momentum at the minizone boundary in the k_x direction experience a true minigap.[72-76] The energy gap becomes smaller and eventually vanishes for electrons of the same energy but with finite momentum $k_{y,z}$ in the remaining two directions of free dispersion perpendicular to the reciprocal superlattice vector **K**. Thus, electrons may enter upper unoccupied minibands by scattering processes from intermediate states with finite momentum perpendicular to **K**. It thus is anticipated that superlattice effects are enhanced if the superstructure is incorporated in an electron system of reduced dimensionality. True minigaps in the density of states will occur if the dimensionality of the potential modulation $V(\mathbf{r})$ coincides with the dimensions of free dispersion.[75,76]

Such advantageous properties as high mobility and the possibility to adjust the carrier density make 2D electron systems favorable for investigations of superlattice effects. The first experiments were performed with MOS devices prepared on vicinal planes, where a 1D periodic potential is created by the misoriented crystal surface.[26,33] The effect of the minigaps on the transport coefficients was investigated as a function of carrier density and magnetic field strength. Magnetic breakdown[77] of the energy gaps and additional Shubnikov–de Haas oscillation periods corresponding to different orbits on the Fermi surface are observed at magnetic fields larger than the breakdown field, $B_T \approx m^* E_g^2/e\hbar E_F$, where E_g is the size of the minigap.[32]

Realization of artificial superlattice potentials by surface corrugation,[5] arrangements of external electrodes,[6,7,24,78] and lateral variation of epitaxially grown alloy composition[76] have been proposed. Although present lithographic techniques approach the 10 nm regime, fabricated structures employing these concepts still have significantly larger periods than those realized on vicinal planes. However, the important advantage of these devices is that it is possible to control the strength of the potential modulation. So far, experiments have been performed on devices where such potential modulation is induced by external electrodes.[23,47,49,79,80,81]

Analytical estimates of the potential modulation, and thus the size of the minigaps, have been made for an idealized gate configuration. The gate is assumed to consist of a metal grating of period a. The separation of the metal stripes is t. The voltage on the metal stripes is $V_g \neq 0$ and zero otherwise.[7,24,82] If the potential screening is approximated by the Thomas Fermi model and the finite thickness of the inversion layer is assumed to be negligible compared to the distance d between gate and inversion layer, the nth Fourier component of the effective potential modulation in the plane of the electron system can be approximated by[24]

$$V_{\text{eff}}^{(n)} = \frac{2V_g}{n\pi} \frac{\sin(n\mathbf{K}t/2)}{[\varepsilon_s/\varepsilon_i + \coth(n\mathbf{K}d)]\sinh(n\mathbf{K}d)(1 + \mathbf{q}_s/n\mathbf{K})}. \tag{3}$$

Again, t is the width of the gap between the metal stripes of the gate, V_g the gate voltage, $\mathbf{K} = 2\pi/a$ is the reciprocal lattice vector of the superlattice, ε_s and ε_i are the dielectric constants of the semiconductor and of the insulator between gate and semiconductor, respectively, and $\mathbf{q}_s$ is the screening wave vector.[24] If the average electron density is adjusted solely with the grating gate of typical period $a = 200$ nm, the electron density, and thus the gate voltage at which the Fermi energy lies in the center of the first minigap, is quite small. This is confirmed in a self-consistent calculation based on a mean-field approximation, which shows, furthermore, that at very low density, the density modulation can result in isolated electron channels beneath the gate stripes.[83] The potential modulation can be considerably stronger if the average electron density is adjusted independently. This can be achieved with a separate homogeneous gate as in the dual-stacked gate configuration that is depicted in Fig. 8 or, in heterojunctions as in Fig. 9, with the carrier density at zero gate voltage controlled by the doping of the active $Al_xGa_{1-x}As$ layer.

Transconductance oscillations have been observed by Warren et al.[47] in Si MOS devices with a dual-stacked gate configuration. (See Fig. 8.) In these structures, the electron density and potential modulation can be controlled independently by applying different gate voltages to the homogeneous top gate and the buried grid gate. Reproducible oscillations of the transconductance as a function of the voltage applied at the grid gate are reported to occur at onset temperatures between 4K and 8K. They were not observed on homogeneous reference samples. The oscillations are interpreted as a consequence of the formation of minibands in the superlattice potential, although the elastic mean free path in Si MOS devices is expected to be significantly smaller than the period $a = 0.2$ μm of the grating gate. In explanation, it was suggested that an observable effect in transport measurements already could arise if the inelastic mean free path of the electrons is larger than the superlattice period.[47] Similar experiments have been performed on high-mobility heterostructures.[79-81] Grating gates or grid gates on top of the heterojunction devices are used to induce 1D or 2D potential modulations, respectively. The period of the superstructure was $a = 0.2$ μm, as in the Si MOS devices[80,81] or even larger ($a = 0.5$ μm in Ref. 79). However, the elastic mean free path is significantly longer (in the order of 1 μm) due to the high mobility in the $Al_xGa_{1-x}As/GaAs$ heterojunctions (about 25 m^2/Vs at carrier density $n_s = 3 \cdot 10^{11}$ cm^{-2}). As in the Si MOS devices, low temperature oscillations of the transconductance as function of gate voltage were observed that did not occur in samples prepared with homogeneous gates. Additionally, negative differential resistance was observed at high source-drain biases in grid gate devices.[49,81] The origin of the negative differential resistance is not understood, but explanations in terms of onset of Bloch oscillations or sequential resonant tunneling between spatially separated electron dots have been suggested.[49,81] To date, there have been no reports on the magnetic field behavior of the observed conductance oscillations or the negative differential resistance, although one may anticipate that application of a magnetic field will aid a quantitative understanding of the observed conductance oscillations.

Corresponding to the larger superlattice period, experiments on these artificial superlattices are much more difficult than those performed on the natural superlattices of high-index surfaces.[26,32] In present artificial lateral superlattices, the densities at which the first minigap is expected typically are about an order of magnitude smaller. Furthermore, at high source-drain voltages, the effective voltage inducing carriers in the electron channel varies beneath the gate, and the actual potential distribution underneath the gate is unclear. Although the observations reported so far are promising, a careful quantitative analysis of the results with potential modulations and density profiles calculated self-consistently from the experimental geometries still is missing.

The zero-field negative differential resistances observed on superlattice devices on $Al_xGa_{1-x}As/GaAs$ heterojunctions at high source-drain fields demonstrate that we might be just on the verge of realization of the Bloch oscillator.[81,84] It is anticipated that the oscillatory motion of miniband electrons subject to an electric field generates microwave radiation of frequency $v = eEa/h$ (e.g., $v = 60$ GHz for $E = 10^3$ V/m and $a = 250$ nm).

More recently, an interesting and unexpected effect of a weak superlattice potential on the magneto-transport coefficients has been observed.[52,85] In a magnetic field, the diagonal components of the resistivity tensor were found to exhibit oscillations periodic in $1/B$ akin to Shubnikov–de Haas oscillations. However, they appear at lower magnetic fields than the Shubnikov–de Haas oscillations and have a much weaker temperature dependence. Although the observations were not anticipated, they found a satisfying quantitative explanation afterwards.[52,86,87] They basically arise because two different length scales compete, the superlattice period a and the cyclotron radius R_c, which is the radius of the classical orbital motion of a free electron at the Fermi energy E_F in a magnetic field B,

$$R_c = \frac{\sqrt{2m^*E_F}}{eB}. \tag{4}$$

Let us first review briefly the basic properties of a homogeneous electron system in a magnetic field as far as it is important to understand the oscillations in a weak superlattice potential. A magnetic field applied perpendicular to the plane of a 2D electron system will localize the electron wave functions spatially on a length scale defined by the magnetic length $l_B = (\hbar/eB)^{1/2}$. In a homogeneous system, the dispersion of free electrons transforms into the discrete spectrum of Landau levels separated by the cyclotron energy $\hbar\omega_c = \hbar eB/m^*$,

$$E_N = \hbar\omega_c\left(N + \frac{1}{2}\right). \tag{5}$$

Here, the quantum number $N = 0, 1, 2, \ldots$ is the Landau level index. In an unperturbed electron system, the degeneracy of the Landau levels is high. Each Landau level contains an areal density of electron states n_L equal to the number of flux quanta, $\Phi_0 = h/e$, penetrating the unit area multiplied by the spin and valley degeneracies g_s and g_v,

$$n_L = g_s g_v \frac{e}{h} B = g_s g_v \frac{1}{2\pi l_B^2}. \tag{6}$$

The other important length scale of the homogeneous electron system is the Fermi wavelength λ_F determined by the areal density n_s of the electrons ($n_s = g_s g_v \pi/\lambda_F^2$). The ratio n_s/n_L determines how many Landau levels are occupied and the filling fraction of the uppermost, partly filled level. The well-known Shubnikov–de Haas oscillations of the transport coefficients arise because the elastic scattering processes are dependent on this filling fraction.[10,88] If the symmetric gauge is used to calculate the electron wave functions, the probability amplitude to find the electron peaks on a circle with a radius that corresponds to the classical Landau radius R_c of Eq. (4) at high Landau quantum numbers N

$$R_c = l_B(2N + 1)^{1/2}. \tag{7}$$

In the experiment by Weiss et al.,[85] a superlattice period was imposed onto an originally homogeneous $Al_xGa_{1-x}As/GaAs$ heterojunction taking advantage of the persistent photoeffect.[89,90] Illuminating the surface of an $Al_xGa_{1-x}As/GaAs$ heterojunction with an interference pattern of visible light at low temperature ($T = 4.2$ K) creates a spatially modulated electron concentration in its 2D electron system.[91] The experimental setup is sketched in Fig. 13. A fixed angle of 56° and one of two different laser wavelengths, $\lambda = 488$ nm or $\lambda = 633$ nm, for the interfering beams were selected. The transport coefficients $\rho_\parallel$ and $\rho_\perp$, as well as the Hall components, were measured on two lithographically defined Hall bars oriented parallel and perpendicular to the interference fringes, as shown in Fig. 13 and the insert of Fig. 13a.

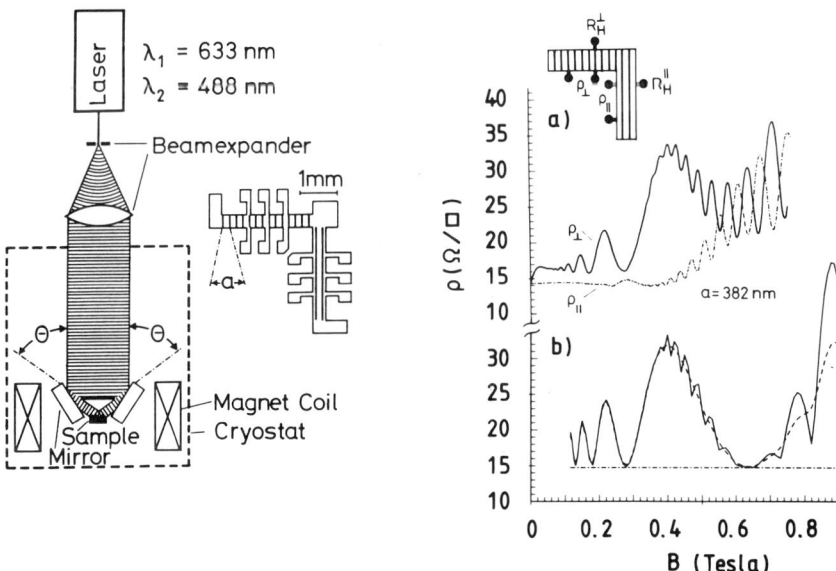

FIG. 13. Magneto-resistance oscillations in a weak superlattice potential. (a) Experimental traces, (b) calculated traces. (From Ref. 86.)

The experimental traces of the magneto-resistivities depicted in Fig. 13b were recorded at $T = 2.2$ K after illumination with an interference pattern of period $a = 382$ nm. The short period oscillations that become apparent in both resistivity components at magnetic fields above $B = 0.4$ T are Shubnikov–de Haas oscillations. At lower magnetic fields, additional pronounced oscillations are observed, with different strength in the $\rho_\parallel$ and $\rho_\perp$ components. The oscillations in $\rho_\parallel$ are phase-shifted by 180° with respect to those in $\rho_\perp$. Furthermore, a large positive magneto-resistance at magnetic fields $B < 0.5$ T is accompanied with the oscillations in the $\rho_\perp$ component. The Hall resistances do not deviate from linear behavior up to those fields at which Shubnikov–de Haas oscillations are observed. From the zero-field resistance and the Hall resistance, an average carrier density of $n_s = 3.16 \cdot 10^{11}$ cm^{-2}—in agreement with the Shubnikov–de Haas periodicity—and a mobility of $\mu = 1.3 \cdot 10^6$ cm^2/Vs where derived. The additional low field oscillations were studied as a function of carrier density, mobility, and temperature. The temperature dependence of the oscillations was found to be much weaker compared to the temperature dependence of the Shubnikov–de Haas oscillations, indicating that quantization into Landau levels does not play an important role. For this reason, it can be ruled out that the new oscillations arise from the occupation of a higher 2D subband. Comparing samples with different mobility, it was found that the low field oscillations start at fields where the Landau orbit $2\pi R_c$ becomes smaller than the elastic mean free path.

Similar experiments were performed by Winkler et al.[52] on Al$_x$Ga$_{1-x}$As/ GaAs heterojunctions with a photoresist-patterned front gate. The period of the photoresist grating was $a = 500$ nm. Application of a gate voltage results in a variation of the potential modulation as well as of the carrier density. Furthermore, the amplitude of the effective potential modulation can be estimated from the gate geometry, as described earlier in Eq. (3). In capacitance measurements, two different carrier densities, n_{s1} underneath the closer and n_{s2} beneath the more distant parts of the gate, could be revealed, so that in addition, the amplitude V_0 of the potential modulation could be estimated from experiment using the simple Thomas Fermi approach with the 2D density of states at zero magnetic field:[52] $V_0 \approx (n_{s2} - n_{s1})/D(E_F)$. In the fan-charts (Fig. 14), an arbitrary running index i is plotted as function of the $1/B$ positions of the extrema in the low-field magneto-resistance.

The experimental results for the dependence of the oscillations on the average carrier density and the period can be summarized as follows: Extrema in the component $\rho_\perp$ arise whenever the classical cyclotron diameter $2R_c$ at the Fermi energy is a multiple of the period a, such that

$$\frac{2R_c}{a} = m + \varphi, \tag{8}$$

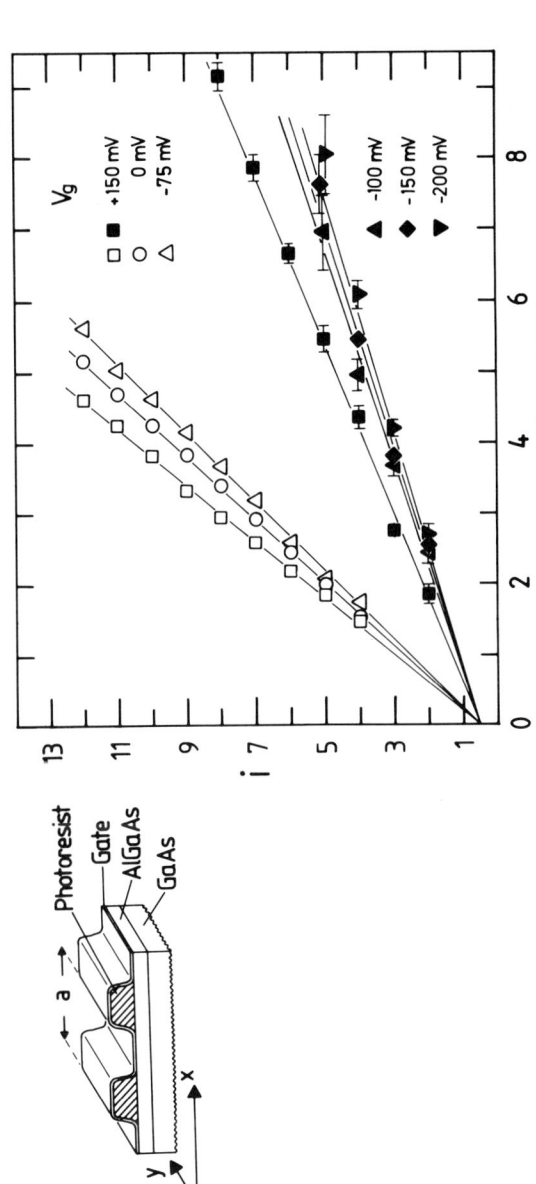

FIG. 14. Fan-charts of the extrema in the low-field magneto-conductance oscillations recorded on an $Al_xGa_{1-x}As/GaAs$ heterojunction with modulated gate of period $a = 500$ nm. The device geometry is sketched on the left. (From Ref. 52.)

with the index $m = 1, 2, 3,\ldots$ and a phase factor $\varphi = 0.18 \pm 0.06$ for maxima or $\varphi = -0.26 \pm 0.06$ for minima of the magneto-conductance.

The straight lines through filled symbols in Fig. 14 are calculated with Eq. (8) using a phase $|\varphi| = 0.25$ and a cyclotron radius R_c according to Eq. (4), with the Fermi energy E_F determined from the Shubnikov–de Haas oscillations at higher magnetic fields and a period $a = 500$ nm. In the gated structure of Winkler et al., a second oscillation period was observed (open symbols in Fig. 14), corresponding to a period $a = 168$ nm ≈ 500 nm/3.

As pointed out by Beenakker,[87] the weak temperature dependence of the resistivity oscillations indicates that the additional low field oscillations should be amenable to a classical description. Furthermore, the Landau radius and the superlattice period are considerably larger than the Fermi wavelength at the densities ($n_s \approx 2 \cdot 10^{11}$ cm^{-2}) and magnetic field strengths ($B < 0.5$ T) investigated in the experiments: $R_c \gg a \gg \lambda_F$. Classically, the guiding center (x_0, y_0) around which the electrons orbit in a magnetic field **B** along z direction performs a drift motion in an electric field **E**, the so-called $E \times B$ drift. If, for instance, the electric field is applied in the x direction, the guiding center drifts according to $x_0 = 0$ and $y_0 = E_x/B$. Here, E_x is the field component at the location of the electron.

In an x-dependent superlattice potential $V(x)$ with period $a \ll R_c$, the guiding center drift oscillates rapidly. Thus, only the drift motion at the extremal points $(x_0 \pm R_c, y_0)$ of the electron orbit will not average out and contribute to a net drift motion of the center coordinate, evaluated by averaging the drift motion along one cyclotron orbit. As shown in Fig. 15, the net drift motion of the center coordinate depends on the radius of the orbit R_c as well as the location of the center coordinate x_0 within the superlattice

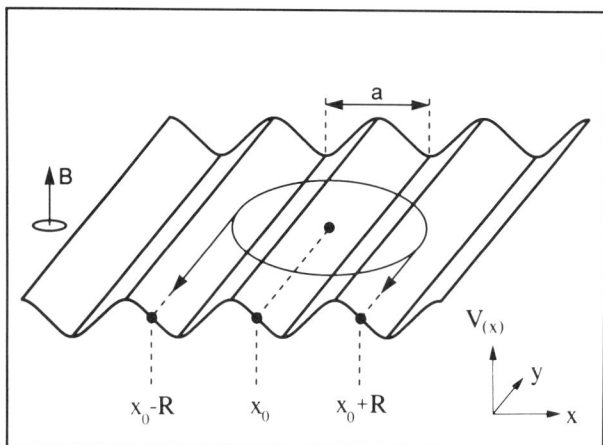

FIG. 15. Sketch of a cyclotron orbit in a periodic electrostatic potential. (From Ref. 87.)

period. A mean-square drift velocity can be evaluated by averaging the square of the net drift velocities for different center coordinates x_0. Assuming a sinusoidal superlattice potential, $V(x) = V_0 \sin(2\pi x/a)$, the result is

$$\langle v_{\text{drift}}^2 \rangle = \frac{1}{2} v_F^2 \left(\frac{eV_0}{E_F}\right)^2 \frac{R_c}{a} \cos^2\left(2\pi \frac{R_c}{a} - \frac{\pi}{4}\right). \tag{9}$$

The same result can also be derived from a purely quantum mechanical approach.[52,86] The degeneracy of the discrete Landau levels of the unperturbed Hamiltonian is lifted in a 1D superlattice potential so that Landau bands are created. The dependence of the energy on center coordinate and Landau radius arises because the wave functions sense the average superlattice potential over an interval of $2R_c$ around the center coordinate x_0. The energy dispersion is calculated most easily in first-order perturbation theory with the wave functions of the asymmetric Landau gauge,[82,92]

$$E_N^{(1)}(x_0) = \hbar\omega_c\left(N + \frac{1}{2}\right) + V_0 \cos\left(2\pi \frac{x_0}{a}\right) \exp\left(-\pi^2 \frac{l_B^2}{a^2}\right) L_N\left(2\pi^2 \frac{l_B^2}{a^2}\right). \tag{10}$$

Here, we have the Laguerre polynomials L_N and the wave vector $k_y = x_0/l_B^2$ of the plane wave ansatz in y direction. If the condition $2\pi R_c/a > 1$ is fulfilled, the perturbed energies can be approximated by

$$E_N^{(1)}(x_0) = \hbar\omega_c\left(N + \frac{1}{2}\right) + V_0 \sqrt{\frac{a}{\pi^2 R_c}} \cos\left(2\pi \frac{x_0}{a}\right) \cos\left(2\pi \frac{R_c}{a} - \frac{\pi}{4}\right). \tag{11}$$

The dispersion given by Eq. (10) and the corresponding density of states $D(E)$ are plotted in Fig. 16 and demonstrate how the band energies depend on the center coordinate x_0 and the radius R_c via the Landau index N. A comparison of the first-order perturbation result, Eq. (10), with a calculation performed by numerical diagonalization has been presented by Gerhardts et al..[86]

From the energy dispersion given by Eqs. (10) or (11), a group velocity $v_g = \hbar^{-1} \partial E_N^{(1)}/\partial k_y$ can be defined that reproduces the classical mean-square

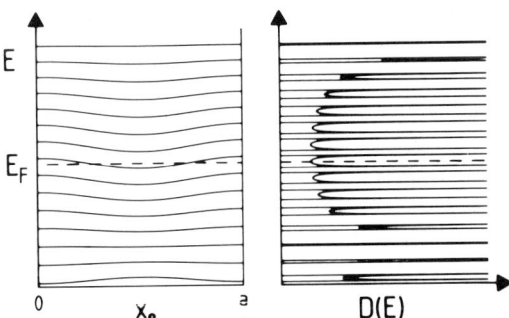

FIG. 16. Electron energies versus center coordinate x_0 and density of states in a periodic potential modulation in the presence of a magnetic field. (From Ref. 52.)

drift velocity, Eq. (9), in the limit of a large Landau radius $R_c \gg a$. The Boltzmann[52,87] and Kubo formalism[86] in constant relaxation time approximation both reproduce the Drude conductivity tensor except for the σ_{yy} component, which gets an additional contribution from the finite drift velocities in y direction parallel to the superlattice stripes. This component predominantly influences the ρ_{xx} component of the resistivity tensor, because the non-diagonal components of the conductivity show no noticeable oscillations and are much larger than the diagonal components. Within the approximation of Eq. (10) and with $\rho_{xx} = \sigma_{yy}/(\sigma_{xx}\sigma_{yy} - \sigma_{xy}\sigma_{yx})$ and $\sigma_{xx} \approx \sigma_{yy} \ll |\sigma_{xy}|$, the longitudinal resistivity for currents along the superlattice wave vector becomes

$$\rho_{xx} = \rho_0 \left[1 + \left(\frac{eV_0}{E_F}\right)^2 \frac{l_e^2}{aR_c} \cos^2\left(2\pi \frac{a}{R_c} - \frac{\pi}{4}\right) \right]. \quad (12)$$

Here, we have the zero-field resistivity, $\rho_0 = h/e^2 k_F l_e$, and the elastic mean free path, $l_e = v_F \tau_e$ determined by the momentum relaxation time τ_e. Equation (12) explains the periodicity of the oscillations and predicts the phase shift to be $\varphi = \pm \pi/4$ for the maxima and minima, respectively. Furthermore, considering the simplicity of the model, it describes well the amplitude of the oscillations with relaxation times extracted from the zero-field resistivity and a potential modulation amplitude V_0 determined from the measured density modulation as described before.[52] Even a very small amplitude compared to the Fermi energy will result in observable oscillations, provided the mean free path is large.

In Fig. 13b, the result of a calculation using the Kubo formalism starting from the first-order perturbation result, Eq. (9), is depicted for two different temperatures and a scattering time, $\tau_e = 5 \cdot 10^{-11}$s, corresponding to the $B = 0$ mobility of the sample in Fig. 13a. The modulation amplitude was chosen to be $V_0 = 0.3$ meV. This calculation reflects the weak temperature dependence of the low-field oscillations and even reproduces the experimental observation that the maxima in ρ_{xx} occur with a phase φ slightly smaller than $\pi/4$. However, the weaker and 180° phase-shifted oscillations of the parallel resistivity component ρ_{yy} can not be obtained in a constant relaxation time approximation. However, a consistent theory of magnetotransport and collision broadening based on the self-consistent Born approximation with a magnetic field-dependent relaxation time reproduces all different types of oscillations in ρ_{xx} and ρ_{yy}, as well as the Shubnikov–de Haas oscillations in higher magnetic fields.[93,94] Such a theory also explains the modulation of the envelope of magneto-capacitance oscillations observed by Weiss et al. in gated heterojunctions.[95]

The origin of the strong positive magneto-resistance at fields below $B = 0.05$ T in Fig. 13 is still unclear. A large positive magneto-resistance below

$B = 0.1$ T has been observed by Winkler et al. as well.[52] Presently, the low-field magneto-resistance is the subject of experimental investigations on grating-type 1D as well as 2D grid-type lateral superlattices.[96–98]

The energy spectrum of Bloch electrons in a magnetic field has been discussed intensively, particularly because it involves an intriguing incommensurability problem.[99–106] The competing length scales are the magnetic length l_B and the lattice constant a. 2D electron systems with a 2D superlattice potential represent an excellent experimental system to probe existing theoretical predictions. Whereas the Landau levels in a 1D superlattice are subject to broadening into bands, a 2D potential will split up such Landau bands into subbands with energy gaps in between. It can be shown[99,102,104] that the number of subbands depends on the number α of flux quanta Φ_0 penetrating a unit cell a^2. The situation is especially peculiar in the case of irrational ratios $\alpha = p/q$: Then, for instance, the number of subbands in the lowest Landau band is equal to the numerator p of the ratio p/q. The commensurability problem first was discussed by Hofstadter for the magnetic field splitting of tight-binding Bloch bands.[101] He found a periodic energy spectrum as a function of the ratio $\alpha = p/q$ with period $\alpha = 1$. The spectrum within one period, often referred to as the *Hofstadter butterfly*, exhibits fractal structure. The Bloch band splits into q subbands, each containing an equal number of states. These subbands again are formed by self-similar clusters of subbands in a recursive way. The periodicity of the spectra allows for a correspondence of the spectra obtained in the Landau band theory and in the Bloch band theory, although the approaches are completely different.[99,101,102,104]

It is virtually impossible to observe this intriguing magnetic-level structure with the crystal lattice potential as the origin of the splitting. In a square potential of period $a = 0.2$ nm, a typical lattice constant in crystalline solids, fields of 10^5 T would be required to have one flux quantum per unit cell. However, in a 2D electron gas with a superlattice of period $a = 250$ nm, this field would be only 0.07 T. Thus, $1/B$ periodicity of fine structure in splitted Landau bands could be observable, provided refined lithography and low temperatures make an experimental resolution of such fine structure feasible. On 2D superconducting networks, closely related flux quantization effects already have been observed as fine structure of the magnetic field dependence of the critical temperature.[107]

2. ARRAYS OF QUASI-ONE-DIMENSIONAL SYSTEMS

Large potential gradients normal to the interface of MOS or heterojunction devices provide the confinement potential of 2D electron systems. With sufficiently large lateral gradients, the electron system can be confined laterally such that narrow electron stripes or small electron dots are created. Thus, it is possible to investigate the physics of 1D and even 0D electron systems artificially created in semiconductor crystals. A wide range of so-called quasi-one-dimensional conductors, i.e., organic as well as inorganic linear chain compounds with highly anisotropic band structure, has been investigated for more than a decade, mainly in view of the physics of Peierls transitions, Wigner crystallization, and charge density waves.[108–110] The systems described in the following may allow valuable extensions of these investigations.

The physical behavior of an electron system depends on the extension of the system compared to system-inherent length scales such as elastic or inelastic mean free paths, phase coherence lengths, cyclotron radius, quasi-particle wavelengths, or the Fermi wavelength. The system changes dimensionality with respect to a physical phenomenon if the spatial extension of the electron system drops below the length scale of the corresponding physical process.

Electron–electron interaction and weak localization corrections to the Boltzmann conductivity are governed by the thermal diffusion length $L_T = \sqrt{\hbar D/k_B T}$ and the phase coherence length $L_\phi = \sqrt{D\tau_\phi}$, respectively. Here, we have the diffusion constant $D = v_F^2 \tau_e/d$, which depends on dimensionality d and the phase coherence time τ_ϕ. The conductivity corrections become 1D in character once the channel width becomes smaller than these lengths, which at low temperatures (typically 1 K) can be of the order of microns in high-mobility devices.[48,111–114] The specific nature of the boundary becomes important when the channel width is smaller than the elastic mean free path $l_e = v_F \tau_e$. Then the electron motion across the channel is quasi-ballistic.[115]

The corrections will change character again if not only the sample width but also the sample length L becomes comparable to the phase coherence length.[116] So-called universal conductance fluctuations arise because with decreasing dimension, transport measurements no longer probe a statistical ensemble but rather a particular sample-specific configuration of scattering centers.[117,118] Furthermore, in samples with probe distances comparable to the coherence length, transport coefficients become non-local and critically dependent on the contact geometry.[116,117,119,120]

Weak localization and universal conductance fluctuations are quantum interference phenomena demonstrating wave function coherence over distances much longer than the elastic mean free path.

Here, we would like to discuss static transport experiments performed on arrays of 1D conductors in the regime, where the width of the electron channels becomes comparable to the Fermi wavelength. While the modes propagating along the wire still are quasi-continuous, the modes for perpendicular propagation in a channel of width W comparable to the Fermi wavelength λ_F are restricted to a discrete number of the order W/λ_F. Suppose the confinement is in x and z direction, the Hamiltonian is of the form,

$$\mathbf{H} = -\frac{\hbar^2}{2m^*}\nabla^2 + V(x,z), \quad (13)$$

and the eigenergies of the 1D system form 1D subbands,

$$E_\alpha(k) = E_\alpha + \frac{\hbar^2 k_y^2}{2m^*}, \quad (14)$$

in which the quasi-continuous dispersion with momentum $\hbar k_y$ in the y direction adds to the quantized subband energies E_α from confinement in the remaining two directions. Single-mode electron systems, i.e., systems with only one subband occupied, are strictly one-dimensional, whereas multimode systems often are referred to as quasi-one-dimensional. In this sense, a narrow electron channel often is addressed as an electron waveguide in single-mode operation in the quantum limit and in multimode otherwise. In our discussion, we summarize both cases under 1D phenomena and will make a distinction only where it seems especially important. In typical state-of-the-art devices, the energetic separation ΔE of subsequent modes is of the order of $\Delta E/k_B = 10$ K, so that measurements at low temperatures may reflect the quantization of the system if the elastic mean free path l_e is much larger than the channel width W. In this respect, the low temperature conductance of an array of 1D channels does not differ from the conductance of a single-channel device. The impact of universal conductance fluctuations, however, is reduced significantly, since random contributions of individual wires average out when they are measured in parallel.

The dependence of the static conductivity on the subband occupation in 1D systems has been studied theoretically by several authors.[121-126] For strictly 1D systems, low temperature mobilities far beyond those in 2D systems are predicted by Sakaki.[121] This prediction is based on the fact that the low temperature mobility is dominated by elastic scattering processes. In a 1D system with only one subband occupied, elastic scattering processes must have momentum transfer $\Delta k = 2k_F$ at low temperature ($k_B T \ll E_F$). Thus, the phase space volume of final states for elastic scattering is reduced drastically and the mobility will be affected only by higher Fourier components of the impurity potential. Those could be reduced by a spacer layer separating the impurities spatially from the channel as in modulation-doped heterojunctions. Motivated by Sakaki's argument, the influence of phonon scattering on the conductivity at higher temperatures was calculated.[122] Calculations of the conductivity in 1D systems with several subbands occupied and including level broadening by impurity scattering were performed by Kearney and

Butcher[125] and Das Sarma and Xie.[126] When the Fermi energy is swept across the bottom of a higher subband, elastic scattering processes into this subband become possible. These scattering processes will be most effective if the Fermi energy coincides with the bottom of a subband ($E_F = E_\alpha$) because the density of final states for scattering into the subband decreases like $1/(E_F - E_\alpha)^{1/2}$. Thus, with increasing Fermi energy, minima in the conductance are expected, provided that level broadening by impurities and inhomogeneities of the effective channel width are sufficiently small. Numerical calculations by DasSarma and Xie[126] indicate that the level broadening by impurities and temperature should be less than half of the subband spacing to make quantum structure observable in the conductance. Fluctuations of the effective channel width that cause additional broadening depend on the particular experimental system and are hard to estimate. Short wavelength fluctuations of gate electrodes or etched sidewalls may be damped appreciably in experimental systems with large depletion widths between electrode or sidewall and electron channel.[127] Since long-range fluctuations of the channel width will result in a spatially varying position of the Fermi energy with respect to the subband levels, however, they seriously can deteriorate experimental results.

Quite surprisingly, clear evidence of lateral quantization also was found in very short ballistic constrictions with length L much smaller than the elastic mean free path.[128,129] Resistance steps quantized in h/e^2 for each mode occupied within the constriction were observed that can be viewed as *contact resistances* between the electron states in the 2D contact areas and the discrete modes of the constriction. Here, we would like to discuss experiments performed on long, narrow channels and refer the reader interested in short ballistic devices to Chapter 2 by van Houten *et al.*

Initial experimental results on a Si device with an array of 250 parallel long inversion lines induced by a dual-stacked gate configuration showed an oscillatory modulation of the conductance as function of the electron density. (See Fig. 17.)[130] As described in Section II, the dual-gate configuration with a grating gate between the homogeneous top gate and the semiconductor–insulator interface allows operation in two different modes. (See Fig. 8.) Oscillations of the conductance were observed only in the gap inversion mode, where the lower, grating-type electrode is biased to deplete the regions underneath the stripes and thus provides confinement for the channels that are induced in the gap between the grating lines when the top electrode is positively biased. The modulation of the conductance as a function of the voltage on the top gate that controls the inversion electron density is weak. However, the transconductance, i.e., the gate voltage derivative of the conductance, exhibits clear structure with gate voltage spacings between 1 V and 1.5 V, as can be seen in Fig. 17, where the transconductance is plotted versus the voltage

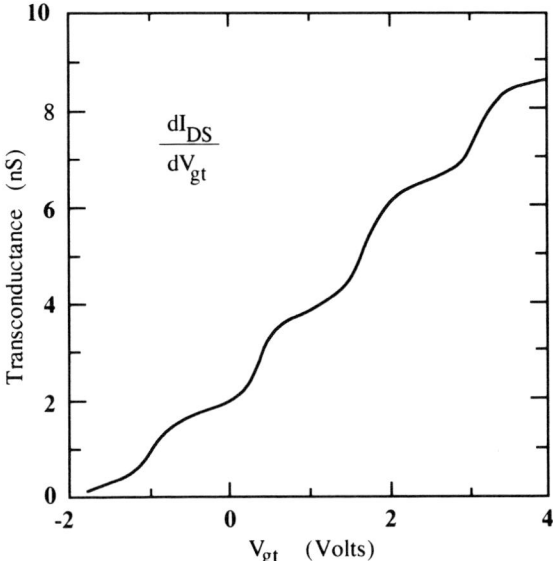

FIG. 17. Derivative with respect to the gate voltage of the source-drain current I_{DS} plotted versus the gate voltage applied at the top gate of a dual-stacked gate Si MOS device similar to the one sketched in Fig. 8 (From Ref. 130.)

V_{gt} on the top gate. From their dependence on temperature and mobility, it was inferred that the oscillations reflect the lateral subband quantization of the inversion channels.

The results of a numerical model calculation of the classical charge distribution induced by the gate electrodes supports this interpretation.[131] It was found that in the subgrating inversion mode, the channel width is almost fixed and approximately equal to the width $W = 100$ nm of the grating gate stripes. In contrast, calculations of the effective width in the gap inversion mode at low electron densities yield widths that are significantly smaller ($W \approx 50-60$ nm) than the gap width $W_G = 100$ nm. This explains why transconductance oscillations were observed only in the gap inversion mode but not in the subgrating inversion mode. Furthermore, the model calculation reflects the experimental finding that the bias of the grating gate shifts the threshold voltage but has little influence on the effective channel width as function of the electron density.

In Ref. 131, the renormalization of the external potential was calculated with a classical quasi-continuum charge distribution. In a more refined model, the charge distribution is calculated self-consistently solving the Schrödinger equation in the screened electrostatic potential.[132,133] The shape of the potential well and the electron density calculated with the self-consistent model hardly differ from results of a classical approach. The zero point energy slightly shifts the threshold voltage and the calculated capacitance is slightly smaller than in the quasi-continuum approach as a result of the larger distance of the quantum mechanical electron density distribution from the semiconductor–insulator interface. However, the self-consistent model allows one to determine the quantum levels in the device and the gate voltages V_α at which the Fermi energy intersects the bottom E_α of a 1D subband. Although the results of the quantum mechanical model are in qualitative agreement with the experimental results by Warren et al.,[130] a precise quantitative agreement with experimental gate voltages V_α could not be obtained. The calculated differences ΔV_α of gate voltages at which the Fermi energy crosses adjacent quantum levels turned out to be much larger than the gate voltage separations of the experimental transconductance minima. Furthermore, the calculated gate voltage separations increase with increasing quantum level index, suggesting the effective confinement potential to be square-well-like, whereas the experimental gate voltage separations are almost constant.

Self-consistent quantum mechanical calculations also have been performed for electron channels in $Al_xGa_{1-x}As/GaAs$ heterojunctions with a split front gate as depicted in Fig. 11.[134] As discussed in Section II, the regions beneath the front gate will be depleted from mobile electrons at gate voltages below a threshold voltage, $V_g = V_d$, thus leaving a narrow channel underneath the opening of the gate. Such a device was used by Berggren et al.[135] to investigate the magneto-conductance in 1D channels, as will be discussed next.

In Fig. 18, the numerically calculated energy levels and electron densities in a device with a 400 nm gate opening are depicted as a function of the bias on the split gate. The calculation is carried out for a heterojunction consisting of a 24 nm-thick undoped GaAs cap layer, 36 nm n-doped $Al_xGa_{1-x}As$ with a donor concentration $N_D = 6 \cdot 10^{17}$ cm^{-3}, a 10 nm-thick undoped $Al_xGa_{1-x}As$ spacer on a slightly p-doped ($N_A = 1 \cdot 10^{14}$ cm^{-3}) GaAs substrate. Furthermore, it is assumed that the exposed surface of the GaAs cap has a constant negative surface charge density of $1.6 \cdot 10^{12}$ cm^{-2} in the region of the gate opening. The energies of the first five 1D subbands are denoted E_{00} to E_{04}, with indices corresponding to the modes of propagation in z direction and lateral direction, respectively. Below a gate voltage of $V_g = -1.55$ V, no electrons are in the channel and the energy separation of adjacent subbands is about 5 meV. The fact that at low densities the separation is calculated to be nearly independent of the subband index demonstrates that the self-consistent confinement potential is of nearly parabolic shape. In the limiting case of a vanishing electron density, the confinement

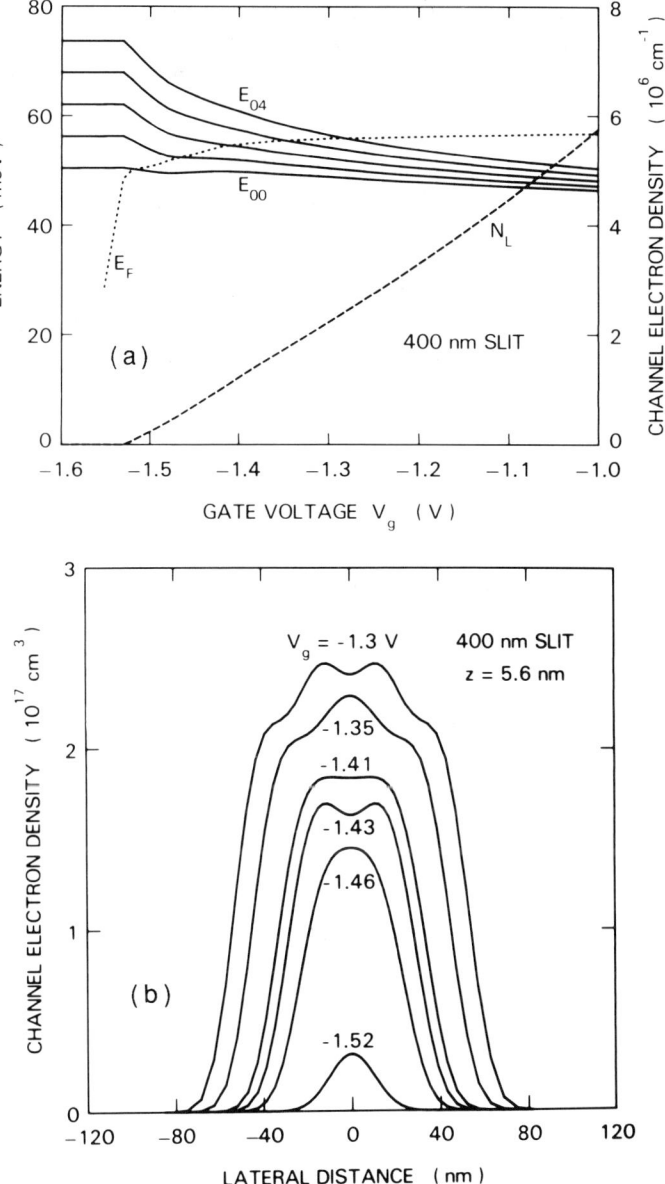

FIG. 18. Calculated energy spectrum and electron density for a split-gate device on GaAs. (a) Gate voltage dependence of subband edges and Fermi energy. The gate voltage dependence of the 1D linear electron density (dashed line) is depicted with its scale on the right-hand side. (b) Electron density distribution at a distance $z = 5.6$ nm from the $Al_xGa_{1-x}As/GaAs$ interface. (From Ref. 134.)

potential in the split-gate device also can be calculated analytically as demonstrated by Davies.[136] Similar values for the subband separations are obtained, although the heterojunction surface in the gate opening is treated differently. In Fig. 18a, the subband separation decreases with increasing electron density, indicating that the effective confinement potential becomes wider. The subband separation is about 1 meV if five subbands are occupied and slightly increases with increasing subband index, demonstrating that the Hartree potential gradually transforms from parabolic to square-well shape.

Figure 18b demonstrates that the total charge density as well as the lateral extension of the electron system decrease with decreasing gate voltage. The electron densities at six different gate voltages are plotted as a function of the coordinate perpendicular to the wire in a distance of $z = 5.6$ nm from the $Al_xGa_{1-x}As/GaAs$ interface. This distance corresponds to the average distance of the charge density distribution in z direction. The electron channel is well-confined in the lateral direction to widths below 160 nm at all gate voltages depicted. Thus, the lateral extension of the electron density distribution is significantly smaller than the gate opening. The modulation of the lateral density distributions at higher gate voltages arises from the contributions of the probability densities in higher occupied subbands. However, in spite of its origin as a summation of discrete subband contributions, the total electron density remains quite smooth.

Good agreement between self-consistent calculations and experiments was obtained by Smith et al.[54] in capacitance measurements on $Al_xGa_{1-x}As/$ GaAs heterojunction devices with a corrugated gate as illustrated in Fig. 10b. As described in Section II, the devices used by Smith et al. have a continuous gate covering a grating-type cap layer on the heterojunction. The GaAs stripes had widths between 200 nm $\leq W_{geom} \leq$ 400 nm. The electron density in the channels is varied by injecting carriers from the doped substrate through a thin tunnel barrier of undoped GaAs. This technique of providing a back contact to the electron system proved useful in previous magneto-capacitance measurements on 2D systems, in which a magnetic-field-independent low-resistance back contact is important.[137,138]

The capacitance measured between the gate and the back contact essentially consists of two parts that can be viewed as series capacitances. The first one, C_{geom}, describes the geometry of the electrodes, namely the gate, the electron system, and the back contact. The second part describes the rise of the Fermi energy with increasing channel occupation and essentially is proportionally to the density of states at the Fermi energy $D(E_F)$ if many-particle contributions and band non-parabolicity are neglected,[137]

$$\frac{1}{C_{meas}} = \frac{1}{C_{geom}} + \frac{1}{e^2 D(E_F)}.$$

Thus, the density of states in the electron system contributes directly to the capacitance measured between the gate and the back contact. Smith et al.[54] observed oscillations in the low temperature capacitance measured as function of the gate voltage. The variations of the capacitance have been attributed to the rapid variation of the density of states at the Fermi energy if a 1D subband starts to be occupied. In the model calculations for a device of geometrical width $W_{geom} = 200$ nm, the energy separation of 5 meV between the first and the second subbands, is obtained at a gate voltage where the second subband starts to be occupied. The width of the confinement potential at this gate voltage was determined to be 90 nm. Similar measurements on devices with 0D electron systems will be discussed in the following section.

The complexity of the numerical models to analyze the gate voltage dependence of conductance or capacitance in present devices originates partly from the fact that the gate bias in the experimental systems changes both the channel density and the effective confinement potential simultaneously. Presently, it is not clear whether there exists an analytical approach for the description of the interplay between electron density and channel geometry sufficiently accurate to analyze experimental data. It thus is desirable to have a method in which the response is investigated as a function of an independent variable keeping the electron density and the channel geometry constant. Moreover, an independent variable also provides the possibility to investigate systems in which the electron density can not be changed during the experiment, as for instance in etched heterojunction wires without gates. Such a variable is provided by a magnetic field applied perpendicularly to the 1D channels. As will be shown, the magnetic field increases the density of states in the 1D subbands, so that the subband occupation is changed while the electron density and—to a first approximation—the potential shape are kept constant. Magnetic-field-dependent screening may alter the effective potential at high field strengths.[139] Since the magnetic field localizes the electron wave functions in the plane perpendicular to the field to a length scale of the cyclotron radius R_c, the electronic properties will become virtually 2D once the cyclotron radius is much smaller than the channel width perpendicular to the field. Thus, at sufficiently high fields, the magneto-conductance exhibits Shubnikov–de Haas oscillations. These, however, deviate from 2D behavior at low fields, where the electric confinement starts to dominate the system properties.

Such deviations from 2D behavior first were observed by Berggren et al.[135] in magneto-transport experiments performed on $Al_xGa_{1-x}As/GaAs$ heterojunctions in which a narrow electron channel was created by depleting the 2D electron gas between two contact areas with a biased split front gate. The width of the slit in the gate was 1 µm, the length 5 µm. Effective widths of the electron channel were determined from the Boltzmann conductance $g_B =$

$N(E_F)De^2W/L$ with diffusion constant D and with the conductance g_B extracted from the temperature-dependent interaction corrections to the channel conductance. At gate voltages $-3.0\text{ V} \geq V_g \geq -3.2\text{ V}$, channel widths in the range $250\text{ nm} \geq W \geq (150 \pm 20)\text{ nm}$ were found. Thus, at low gate voltages, the effective channel width is found to be smaller by far than the gate opening.

In Fig. 19, a running index is plotted versus the reciprocal magnetic field positions of the magneto-conductance maxima measured at different gate biases.[135] Usually, a fan-chart of Shubnikov–de Haas oscillations in 2D systems, in which the Landau level indices are plotted versus the reciprocal field positions of the magneto-conductance maxima, shows a linear dependence of the data points with the slope $n_s h/g_s g_v e$ depending on the carrier density n_s. (See Eq. 6.) In Fig. 19, the data points depart from such linear $1/B$ behavior at low fields ($B < 1$ T) and gate biases smaller than $V_g = -3.0$ V.

To explain the deviations from linear $1/B$ behavior, we briefly discuss in the following the effect of a perpendicular magnetic field on the electron states in a narrow channel. To simplify the Hamiltonian of Eq. (13), the confinement

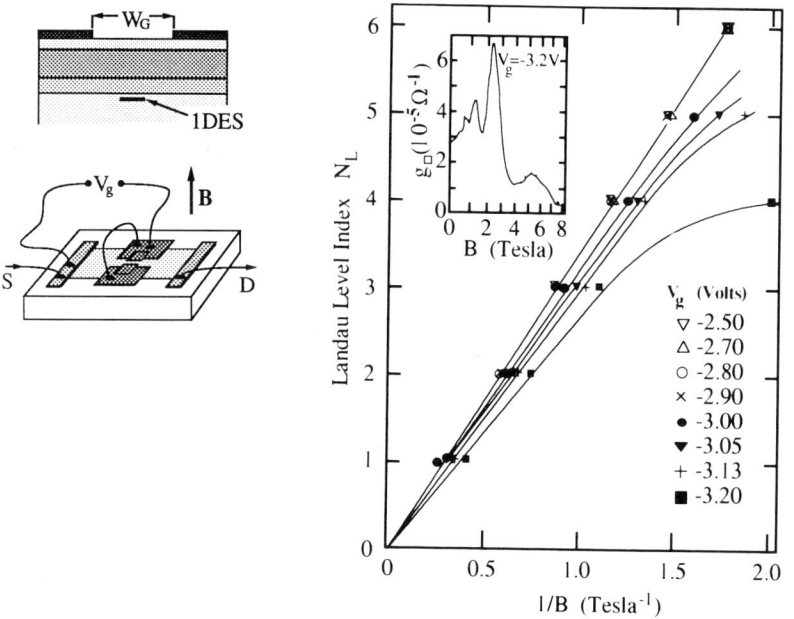

FIG. 19. Fan-chart of conductance maxima in an $Al_xGa_{1-x}As/GaAs$ heterojunction with split gate. The inset shows the conductance of the device at gate voltage $V_g = -3.2$ V. The sketch on the left indicates the device geometry. (From Ref. 135.)

potential in the lateral dimension (say, in the x direction) is assumed to be separable from the strong confinement in z direction normal to the interface: $V(x, z) = V(x) + V(z)$. Numerical calculations show that this is a reasonable approximation in experimental systems investigated so far.[133,134,140,141] The magnetic field is applied in z direction, so that the Hamiltonian reads:

$$\mathbf{H} = -\frac{\hbar^2}{2m^*}\nabla^2 + \frac{1}{2}m^*\omega_c(x - x_0)^2 + V(x) + V(z). \quad (15)$$

Here, the center coordinate $x_0 = k_y l_B^2$ is introduced and the asymmetric Landau gauge $A = B(0, x, 0)$ is chosen to take advantage of the symmetry. Furthermore, spin-splitting is neglected. Formally, the effect of the magnetic field in the Hamiltonian of Eq. (15) is to add to the electric confinement $V(x)$ a parabolic potential $m^*\omega_c^2(x - x_0)^2/2$ with center coordinate $x_0 = k_y l_B^2$. Accordingly, the additional magnetic confinement results in the formation of hybrid subbands with the same sublevel structure as in the pure electrostatic case, but with magnetic-field-dependent dispersion and band separation. Both the energy separation and the density of states increase with magnetic field approaching the Landau level structure in the high-field limit.

It is instructive to use a harmonic potential ansatz $V(x) = \frac{1}{2}m^*\Omega_0^2 x^2$ for the electric confinement because it allows analytical description of the transformation from 1D subbands into hybrid bands.[135,143] With this ansatz, the eigenenergies are

$$E_{N,i}(\tilde{\omega}, k_y) = E_i + \hbar\tilde{\omega}\left(N + \frac{1}{2}\right) + \left(\frac{\Omega_0}{\tilde{\omega}}\right)^2 \frac{\hbar^2 k_y^2}{2m^*}, \quad (16)$$

with the squared hybrid frequency $\tilde{\omega}^2 = \omega_c^2 + \Omega_0^2$. The magnetic-field dependence of the hybrid band edges is depicted in Fig. 20 together with the Fermi energy calculated from the corresponding density of states with parameters chosen for an electron channel in GaAs,

$$D(E) = \frac{2}{\pi\hbar}\sqrt{\frac{m^*}{2}}\frac{\tilde{\omega}}{\Omega_0}\sum_{N \leq N_{max}}\left[E - \hbar\tilde{\omega}\left(N + \frac{1}{2}\right)\right]^{-1/2}. \quad (17)$$

N_{max} denotes the highest occupied hybrid level. The Fermi energy oscillates with maxima whenever it crosses a new hybrid subband. This behavior is similar to the oscillations of the Fermi energy between the highest occupied Landau levels in 2D systems. Since scattering effects depend on the density

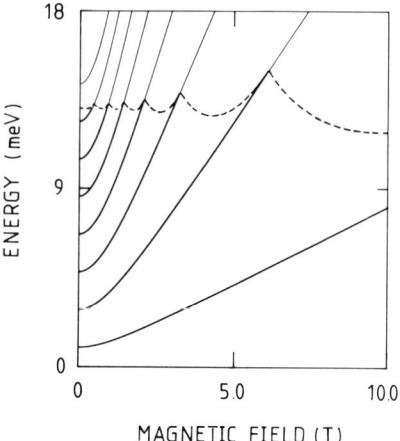

FIG. 20. Energies of the subband edges in a magnetic field applied perpendicularly to a 1D channel in the harmonic confinement model with quantum-level separation $\hbar\Omega_0 = 1.8$ meV. The position of the Fermi level (dashed line) is depicted assuming a constant 1D electron density $n_l = 4.8 \cdot 10^6$ cm^{-1}. (From Ref. 142.)

of states, which changes rapidly at the subband edges (as seen in Fig. 1), the low-temperature ($k_B T \ll \hbar\tilde{\omega}$) channel resistance is expected to oscillate with the same period. This behavior gradually transform into Shubnikov–de Haas oscillations at high magnetic fields ($\hbar\Omega_0 \ll \hbar\omega_c$), with maxima of the longitudinal resistance when the Fermi energy is within a Landau level.[144] At low magnetic fields, the aforementioned deviations from linear $1/B$ behavior occur. In particular, the number of oscillations is limited to the number of subbands occupied in the 1D system at $B = 0$, whereas in a 2D system the number of Landau levels under the Fermi level approaches infinity with vanishing field strength.

The numerical calculations by Laux et al.[133,134] demonstrate that a harmonic confinement potential is a good approximation only as long as the electron density in the channel is low. Coulomb interaction of the electrons within the channel results in a Hartree potential with a flat bottom, so that the effective potential is intermediate between a parabolic well and a square-well potential. This fact is taken into account approximately in the potential ansatz,

$$V(x) = \begin{cases} \dfrac{m^*\Omega_0^2}{2}\left(|x| - \dfrac{w_b}{2}\right)^2, & |x| > \dfrac{w_b}{2} \\ 0, & |x| < \dfrac{w_b}{2} \end{cases},$$

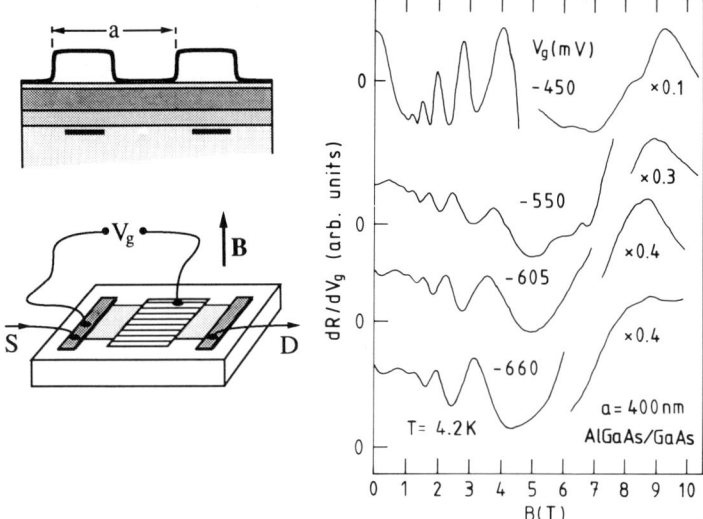

FIG. 21. Experimental traces of magneto-conductance oscillations. Here, the gate voltage derivative is measured versus magnetic field strength. The device geometry is sketched on the left. (From Ref. 51.)

which also has been used to analyze the magneto-conductance oscillations in narrow channels.[135,145,146] The model potential consists of a flat bottom of width w_b and parabolic walls described by the parameter Ω_0. It describes channels with high electron densities more realistically, but has the disadvantage that calculations are performed numerically and—without knowledge of the real potential—choices of the parameters Ω and w_b are somewhat arbitrary. Furthermore, it was found that within the accuracy of present experimental results, the different potential models produce values for parameters such as the effective channel widths or linear densities that are comparable within the experimental accuracy.[147–149]

In the following, results of transport experiments on various devices with periodic arrays of 1D channels are described. Magneto-transport experiments are performed to prove lateral quantization and to determine relevant parameters such as the 1D electron density n_l, the quantization energy $\hbar\Omega_0$, and the effective channel width W. Characterization usually is based on the harmonic confinement model because it can be treated analytically.

In Fig. 21, results of magneto-transport experiments on an $Al_xGa_{1-x}As/GaAs$ heterojunction with a microstructured gate by Brinkop et al.[51] are shown. As sketched in the left-hand side of Fig. 21, the distance between the gate and the electron system is modulated periodically by a holographically defined insulator grating of period $a = 400$ nm, which lies between the heterojunction surface and the front gate. As described in Section II, application of a negative bias preferentially depletes the areas between the insulator stripes, where the distance between the gate and the electron system is small. Thus at gate voltages V_g below a threshold voltage V_d, narrow electron channels remain beneath the insulator stripes. From capacitance measurements, the threshold voltage of the device in Fig. 21 is determined as $V_d = -0.5$ V. The sample consists of two contact areas serving as source and drain and the gated area in between with stripes oriented so that a current can be passed along their direction. About 7,000 channels are measured in parallel. To eliminate resistance oscillations arising from the 2D contact areas, the gate voltage derivative of the source-drain resistance R at constant current $I = 1$ μA is recorded as function of the magnetic field. Note the complete absence of random structure in the traces of Fig. 21, which demonstrates successful averaging of universal conductance fluctuations in multi-wire arrays.

In the fan-chart in Fig. 22, the running index n of the oscillation maxima is plotted versus their reciprocal field positions. Since the transconductance is recorded, positions of the maxima are not expected to coincide with the magnetic field values, at which the Fermi energy crosses a hybrid subband edge. Thus, the index is phase-shifted so that the data extrapolate to $n = 0$ at $B = 0$. Within the harmonic confinement model, calculated reciprocal field values, at which sublevels with index n are depopulated, are denoted by open symbols. The best fits in the gate voltage range between -0.5 V $\leq V_g \leq -0.66$ V are

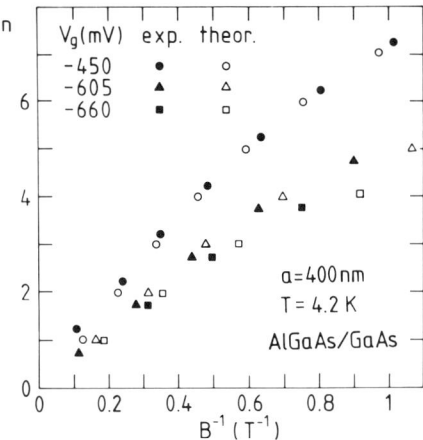

FIG. 22. Fan-charts of the transconductance maxima of the device in Fig. 21 (From Ref. 51.)

obtained with quantization energies $\hbar\Omega_0$ ranging from 1.5 meV to 2.0 meV and Fermi energies between 14.5 meV and 11.7 meV. The errors are about 20% and 10%, respectively. Thus, without magnetic field, between nine and five subbands are occupied.

Because of the soft potential walls in the harmonic confinement model, a definition of the effective channel width is not straightforward. Generally, two different widths can be evaluated, namely the width W_F of the parabolic potential at the Fermi energy and a mode average $\langle W \rangle$ of the wave functions between the classical turning points. The wave function of the Nth sublevel extends in the confinement direction on a width,

$$W_N = 2\sqrt{\frac{\hbar}{m^*\Omega_0}}\sqrt{2N+1}.$$

The mode average $\langle W \rangle$ thus may be defined by

$$\langle W \rangle = \frac{1}{N_{max}} \sum_{N=0}^{N_{max}} W_N. \tag{18}$$

At high-level index ($N \gg 1$), the width $W_{N_{max}}$ of the highest mode is akin to the classical turning point separation and close to the width W_F of the potential at the Fermi energy,

$$W_F = \sqrt{\frac{8E_F}{m^*\Omega_0^2}}. \tag{19}$$

Furthermore, at high-index N_{max}, the mode average $\langle W \rangle$ is related to the width at the Fermi energy by $\langle W \rangle = 2W_F/3$.

The widths W_F at the Fermi energy determined for the data in Fig. 22 decrease from $W_F = 240$ nm to 160 nm in the gate voltage regime, -0.5 V $\leq V_g \leq -0.66$ V. Surprisingly, deviations from linear behavior in the fan-chart start to be observable at the gate voltage $V_g = -450$ mV, where according to the threshold voltage of $V_d = -0.5$ V, the electron system is expected still to be a density-modulated 2D system. This indicates that the Landau level structure is affected by the strong density modulation. However, the tight-binding model described here may not be appropriate to analyze the deviations in the gate voltage regime above $V_g \geq V_d$, where the electron channels are not isolated yet.

Strong deviations from linear behavior as demonstrated in Fig. 23 also are observed on 1D electron channels in InSb MOS devices.[150] These devices consist of a grating-type Schottky gate of period $a = 250$ nm that is

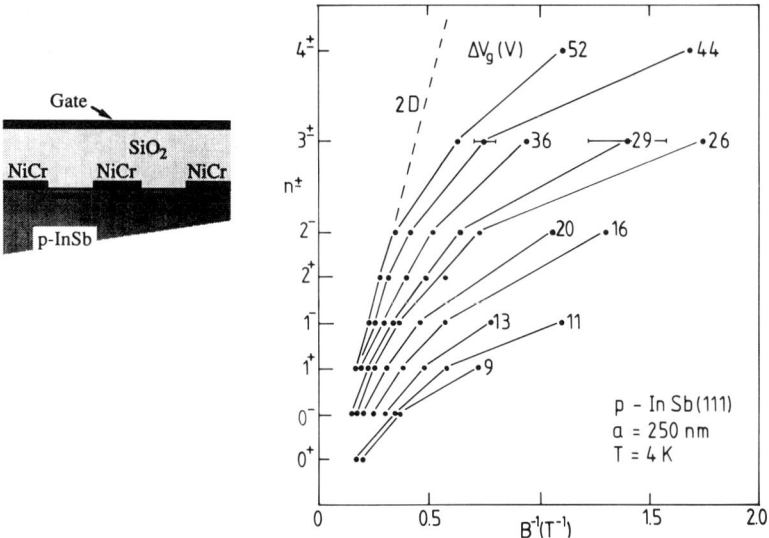

FIG. 23. Fan-charts of transconductance maxima measured in an InSb MOS device. The voltage ΔV_g denotes the gate voltage above inversion threshold. The device geometry is sketched on the left. (From Ref. 150.)

evaporated directly onto the InSb substrate and an insulator separating the homogeneous top gate from the NiCr Schottky gate (see Fig. 12). Four contacts allow one to measure the conductance parallel as well as normal to the grating. Inversion channels are formed between the stripes of the Schottky gate, with density control provided by the voltage V_g on the homogeneous top gate. In the gate voltage range studied on these devices, virtually no quasi-static conductivity was measured perpendicularly to the channels, indicating that the Schottky gate always keeps the channels isolated.

In the InSb inversion systems, spin-splitting and band non-parabolicity are not negligible. Different spin levels are indicated by $\pm$ signs attached to the quantum index as in the fan-chart of Fig. 23. The low effective electron mass in InSb results in large quantization energies. Because of the low effective mass, on the other hand, band non-parabolicity influences the magnetoconductance oscillations.[151] With the harmonic confinement model including band non-parabolicity as well as spin-splitting, 1D subband separations of $\hbar\Omega_0 = 9$ meV that are nearly gate voltage-independent were derived from the date of Fig. 23. The channel width at the Fermi energy increases with the gate voltage, and channel widths W_F in the range 70 nm to 90 nm were found in the gate voltage regime, $9 \text{ V} \leq \Delta V_g \leq 52 \text{ V}$.

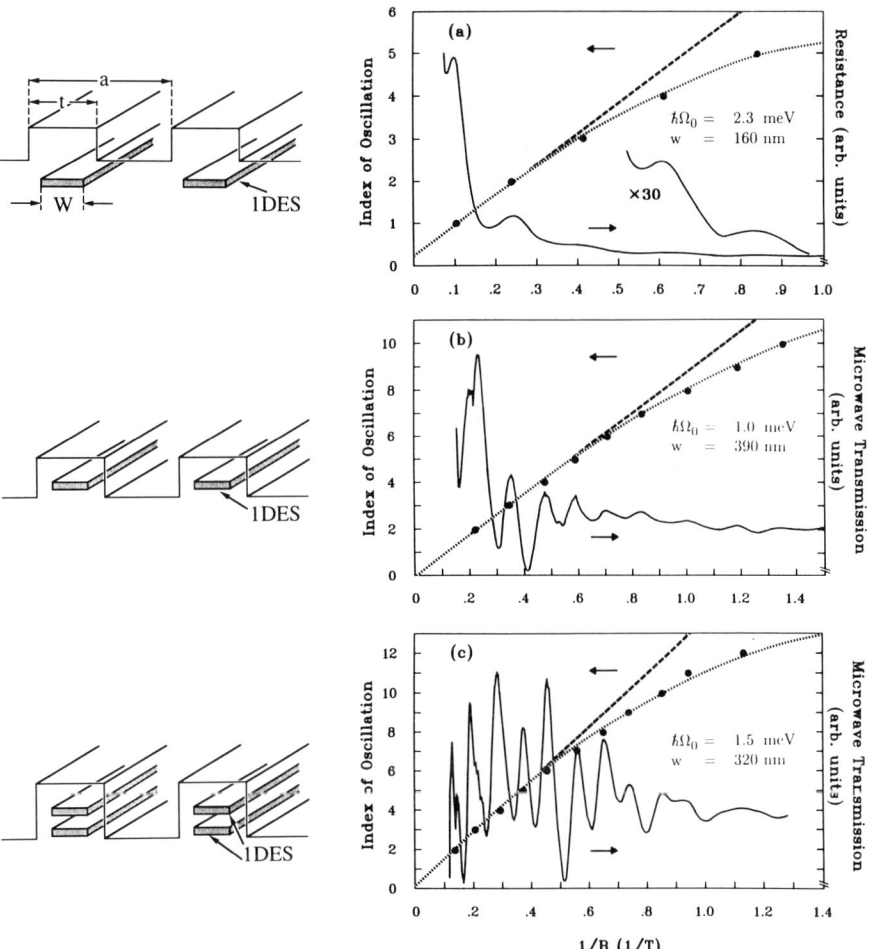

FIG. 24. Magneto-conductance oscillations and fan-charts of etched $Al_xGa_{1-x}As/GaAs$ heterojunction wires. The depths up to which the epitaxial layers are removed are indicated on the left. (From Ref. 59.)

Quantization of electron states in etched heterojunction wires was demonstrated via magneto-transport measurements by Demel et al.[59] shown in Fig. 24. As described in Section II, etched heterojunction wires can be fabricated by various techniques, depending on the particular etching process and on the depth up to which the epitaxial layers are removed. The data in Fig. 24a are taken on a device fabricated by a shallow mesa etching technique resulting in a structure that is sketched on the left of the data and in more detail in Fig. 11a. Here, the n-doped $Al_xGa_{1-x}As$ layer is removed completely down to the $Al_xGa_{1-x}As$ spacer layer so that the electron channels reside beneath stripes of n-doped $Al_xGa_{1-x}As$ of width $t = 250$ nm. The deviations from linear $1/B$ dependence could be fitted best with parameters $\hbar\Omega_0 = 2.3$ meV and $E_F = 15$ meV. Thus, about six subbands are occupied, the 1D electron density is $n_l = 4.5 \cdot 10^6$ cm^{-1}, and the effective channel width is $W_F = 160$ nm.

Apparently, the effective width of the channel is smaller than the geometrical width of the etched $Al_xGa_{1-x}As$ stripes by a depletion width of about $W_{depl} = 45$ nm on each side of the channel. Somewhat larger depletion widths were obtained on deep mesa etched devices like the one of Fig. 24b. Here, the geometrical width of the etched heterojunction wire is $t = 550$ nm. From analysis of the magneto-transport oscillations with the parabolic confinement model, parameters $\hbar\Omega_0 = 1.0$ meV and $E_F = 17$ meV are found. Then 16 subbands are occupied, the 1D density is $n_l = 13 \cdot 10^6$ cm^{-1}, and the potential width at the Fermi energy is $W_F = 390$ nm. From comparison of the geometrical wire width $t = 550$ nm with $W_F = 390$ nm, a depletion width of 80 nm can be deduced. However, in deep mesa etched wires of widths $t \leq 400$ nm, no free electrons were found, indicating the enhanced influence of the etched sidewalls on the electron channels in these GaAs structures.

The data of Fig. 24c demonstrate that 1D confinement can be achieved in deep mesa etched multiple quantum well wires, too. In the device of Fig. 24c, two electron channels are created in a two-quantum-well system. The quantum wells are 50 nm wide and separated from each other by about 130 nm of higher band gap material. Thus, the vertically stacked channels can be regarded as being electronically decoupled. However, they may still interact via Coulomb interaction, as will be seen in Section 5 on far-infrared excitations of confined electron systems. The data of Fig. 24c can be described with the harmonic oscillator parameters $\hbar\Omega_0 = 1.5$ meV and $E_F = 26.1$ meV. Then about 17 subbands are occupied and the effective width $W_F = 320$ nm and the 1D electron density $n_l = 16 \cdot 10^6$ cm^{-1} are comparable to the values found in the single-channel heterojunction wire.

The data of Fig. 24b and c also demonstrate that quasi-static quantum transport can be observed in microwave transmission experiments. Instead of the DC resistance along the wires, the microwave transmission through the device is measured. In the small-signal limit, changes of the transmission are directly proportional to variations of the device conductivity, as will be discussed at the beginning of Section IV on far-infrared experiments. The microwave technique is especially advantageous if devices for optical measurements are tested, since no ohmic contacts are needed.

Microwave transmission data obtained similarly on a wire array etched in an InGaAs/InAlAs/InP heterostructure are shown in Fig. 25.[152] Again, the wires of width $t = 300$ nm are fabricated by a deep mesa etch. The elec-

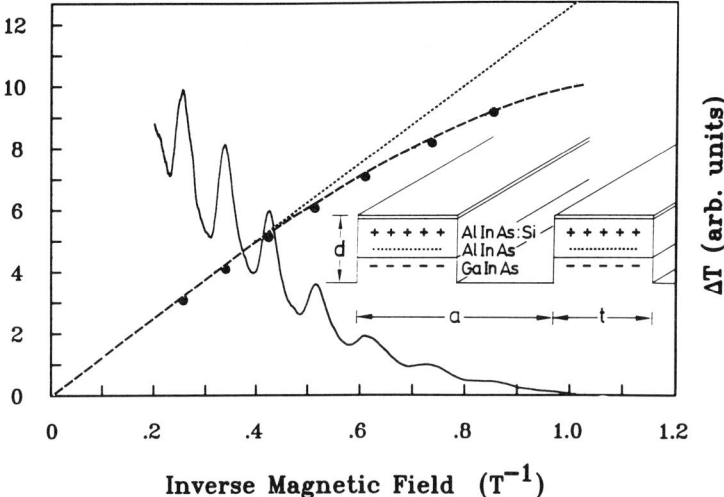

FIG. 25. Fan-chart with running indices 0 to 12 and magneto-conductance oscillations of an etched InGaAs/InAlAs/InP heterostructure wire grating plotted versus inverse magnetic field. The conductance is measured with microwaves. The inset depicts the device geometry. (From Ref. 152.)

tron system resides in InGaAs with effective electron mass $m^* = 0.04$ m_e. Because of the smaller electron mass, higher subband separations than in GaAs may be expected. Within the harmonic oscillator model, parameters $\hbar\Omega_0 = 2.5$ meV and $W_F = 300$ nm are deduced. About 15 subbands are occupied and a 1D electron density $n_l = 14 \cdot 10^6$ cm^{-1} is evaluated for the experiment of Fig. 25. Thus, within the harmonic oscillator model, the effective channel width and the geometrical width are approximately the same. This demonstrates that the depletion widths in etched heterostructure devices are material-dependent. Smaller surface state densities and pinning of the Fermi level at energies closer to the conduction band edge can result in significantly smaller depletion widths in InGaAs/InAlAs devices as compared to $Al_xGa_{1-x}As$/GaAs wires. Furthermore, the lower effective mass results in larger subband spacings compared to devices of similar effective widths and electron densities.

Previously described arrays of 1D electron systems are the basis for measurements of optical excitations in the far-infrared (FIR) regime. Strong resonances of the FIR conductivity are observed in wire arrays, which will be discussed in Section IV. In the present section, we mainly discussed experiments performed to obtain information on the geometry of the narrow channels and the associated energy quantization of the electron states. We have seen that conductance as well as capacitance oscillations can be observed as functions of a gate-bias-controlled electron density or a magnetic field applied perpendicularly to the channels. Analysis of the experimental data with theoretical models turned out to be complex in the case of varying electron density, since the channel geometry and thus also the quantization energies change along with the density. In contrast, it may be assumed that a magnetic field in first-order approximation has no influence on the electrostatic potential. With an analytical model based on a simplified confinement potential of parabolic form, the channel geometry and quantization may be parameterized from magneto-conductance data. Thus, magnetic fields provide a useful tool to characterize 1D electron systems in different experimental devices. The price paid for the relative ease of this tool is that the parameters do not describe accurately the actual geometry in systems with high electron densities, where the actual potential form differs from the harmonic potential ansatz.

One goal of present experimental efforts is to realize 1D systems with only one subband occupied at electron densities sufficiently high and homogeneous for pronounced signals and well-defined experimental conditions. Very recent publications report already about investigations in the regime of the 1D quantum limit.[153,154] In single, narrow channels on Si MOS structures[153] as well as $Al_xGa_{1-x}As$/GaAs heterojunctions,[154] strong, quasi-periodic conductance oscillations were observed as a function of the electron density in the quantum limit. This intriguing phenomenon was interpreted in terms of 1D-charge density waves pinned to potential fluctuations in the channel. Others have suggested a different explanation that involves the Coulomb gap associated with the charging of electron states between two impurity-induced tunneling barriers in the channel.[155] Further investigation should clarify the origin of the observed oscillations.

3. Quasi-Zero-Dimensional Systems

In analogy to the transformation of a 2D electron system to an array of 1D electron channels by a sufficiently strong lateral periodic potential $V(x)$, a potential $V(x, y)$ in both lateral directions can generate either a grid-type electron system or an array of isolated 0D electron systems. In the latter case, the electron system is bound in all spatial dimensions and the energy spectrum of the electron states consists of discrete energy levels like those of semiconductor impurities or atoms. However, the confinement potential of such artificially generated so-called electron dots differs significantly from the Coulomb potential of semiconductor impurities. Also, in such dots, it is possible to control the number of electrons bound in the dot and it may be much larger than the number of electrons bound to an impurity atom. Moreover, the coupling between different dots may be controlled by variation of the distance and the potential barrier between adjacent dots. These properties make electron dots an interesting experimental system.

Experiments on single 0D electron systems are discussed in detail in Chapter 1 by Reed. Here, we would like to discuss briefly capacitance measurements on matrices of electron dots,[156,157] which in analogy to the magneto-conductance measurements on 1D channels described in Section 2 are subsidiary to the spectroscopic measurements discussed in Section IV of this chapter.

The way contacts to quantum dots are provided becomes a crucial part of the experiment, since with decreasing size of the few electron systems, the properties of the contact increasingly contributes to the experimental results. To reduce the influence of a contact onto the electron states of the quantum dot, carrier injection through a tunnel barrier is preferable to direct contact to the metallic electron system of a probe. For the capacitance measurements on 0D electron systems, $Al_xGa_{1-x}As/GaAs$ heterojunctions grown on a doped substrate have been used, as in the capacitance measurements on quantum wires. The doped GaAs substrate serves as a back contact to the 0D electron systems. Carrier exchange between dot and back contact takes place through a small tunnel barrier formed by about 80 nm of undoped GaAs separating the doped GaAs substrate from the interface at which the electron system resides. Like in the case of capacitance devices for 1D quantum wires, the confinement of the 0D electron systems is created by the modulation of the conduction band generated with an etched GaAs cap layer in connection with a homogeneous Schottky gate on top. (See Fig. 10b.) The heterojunction cap layer consists of 30 nm-thick GaAs squares with edges of 200 nm, 300 nm, and 400 nm widths in different devices. Like in the quantum wires, the extension of the electron systems below the GaAs cap squares is expected to be significantly smaller than the geometrical extension of the etched structures.

To get sufficiently large capacitance signals in the experiments, very many (about 10^5) electron dots are measured in parallel. The gate voltage derivative of the sample capacitance is recorded as function of the gate voltage at different magnetic fields applied perpendicularly to the sample surface, i.e., in the direction of the strongest confinement of the 0D electron systems. Figure 26 shows representative experimental traces recorded on 300 nm dots at four different magnetic fields. The abrupt onset of the derivative signal at a gate voltage of $V_g = -0.27$ V indicates that at this voltage, electrons start to occupy the potential minima of the 0D systems. At higher gate voltages, pronounced oscillations are observed in the capacitance derivative. The

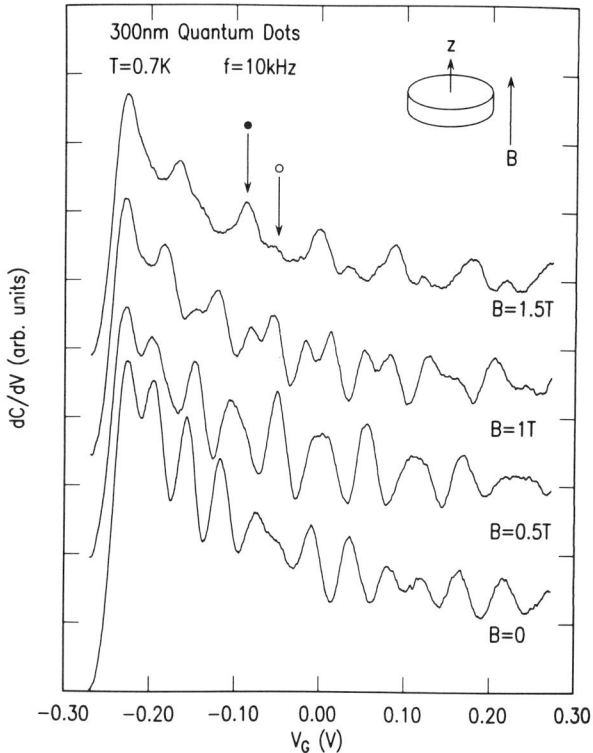

FIG. 26. Gate voltage derivative of the capacitance for an $Al_xGa_{1-x}As/GaAs$ device with dots of 300 nm diameter. The traces are taken at temperature $T = 700$ mK and at four different magnetic field strengths applied in the direction indicated in the inset. (From Ref. 157.)

capacitance derivative is recorded to enhance these oscillations against the background capacitance. Throughout the whole gate voltage range recorded the electron system consists of electron dots that are connected only by the common back contact. The capacitance oscillations reflect the fact that carrier injection from the back contact into the electron systems is possible only at discrete gate voltages. Whenever the Fermi energy in the back contact is sufficiently high to enable carrier exchange with the quantum levels of the 0D system, the capacitance exhibits a maximum. Correspondingly, it is found that with decreasing dot size, the gate voltage separation of capacitance maxima increases.

The different traces in Fig. 26 demonstrate that the positions of the capacitance maxima depend on the magnetic field applied perpendicularly to the sample surface. The positions of capacitance maxima observed at high voltages above the threshold voltage already change rapidly at very low magnetic fields ($B < 0.2$ T) and exhibit a fairly complex behavior. At a magnetic field of about 1 T, strong splittings are observed. The behavior of samples with the same dot size is sample-independent. In samples with smaller dot sizes, splitting occurs at higher magnetic fields (1.5 T in 200 nm dots) and in larger dots at smaller fields (0.6 T in 400 nm dots). Figure 27 shows the magnetic field-dependent gate voltage positions of the oscillation maxima in detail. The strong maxima at fields above $B = 1.2$ T, indicated with filled symbols, gain in strength with increasing magnetic field and persist to higher fields ($B > 2$ T), whereas weak maxima vanish. Furthermore, the gate voltage separation increases nearly linearly with the magnetic field at $B > 2$ T. At magnetic fields above 4 T, spin-splitting similar to the one of Landau levels is observed. At extremely high fields ($B > 15$ T), additional structures in the capacitance signal arise at gate voltages that correspond approximately to fractional filling factors of Landau levels, as will be discussed later.

The low magnetic field behavior of the capacitance oscillations may be understood qualitatively by considering the energy degeneracy of single-particle levels in a quantum dot in the two limiting cases of zero field and very strong magnetic fields, respectively. Without magnetic field, the energy levels in a quantum dot are determined solely by the electrostatic confinement. The degeneracy is low and depends only on the symmetry of the confinement potential. In the strong magnetic field case, the energy levels are determined dominantly by the magnetic confinement. When the magnetic length is much smaller than the spatial extension of the electron system, it is possible to distinguish between edge states at the boundaries of the dot and bulk-like states that have little probability to sense the electrostatic confinement. Thus, the bulk-like states are similar to Landau states in homogeneous 2D systems. Their degeneracy is very high and increases with increasing magnetic field. Simultaneously, the fraction of edge states in each dot decreases, since the magnetic length decreases. Thus, at high magnetic fields, the majority of the electron states are Landau-level-like with high degeneracy of the energy levels. Accordingly, the 0D energy levels must be very sensitive to a magnetic field showing a complex rearrangement at intermediate magnetic fields to transform into highly degenerate Landau levels.

The preceding behavior is demonstrated, for instance, by model calculations of single-electron energy levels of quantum dots in a magnetic field. Figure 28 shows the result of a calculation performed for an electron in a square-shaped box with square-well confinement perpendicular to the surface. The width of the box is assumed to be 120 nm, its depth 600 meV. The motion in the plane perpendicular to the magnetic field is assumed to be separable from the motion in the direction of the magnetic field. Furthermore, the confinement in the direction of the magnetic field is assumed to be so strong that at all energies considered, the system is in the quantum limit with respect

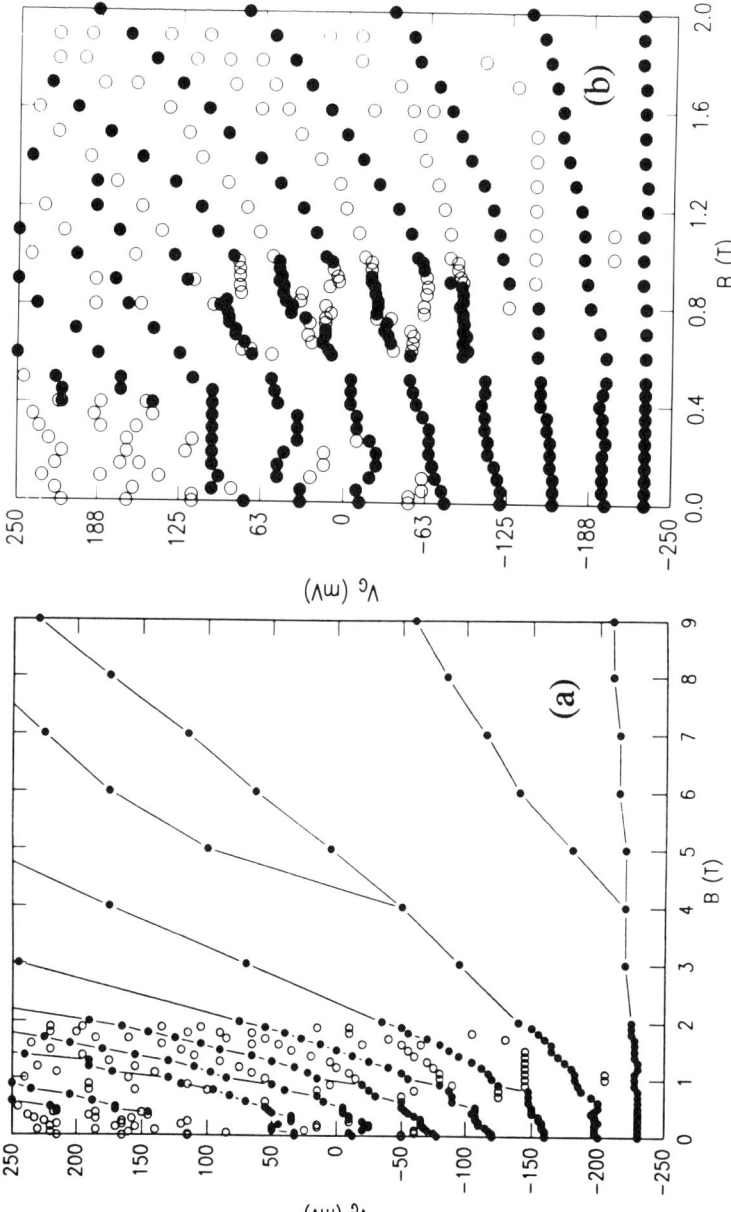

FIG. 27. (a) Magnetic field dependence of the maxima in the capacitance derivative measured as function of the gate voltage on an $Al_xGa_{1-x}As/GaAs$ device with 300 nm dot size. Filled and open circles indicate strong and weak maxima, respectively. (b) Extended scale of the low-field range in (a). (From Ref. 157.)

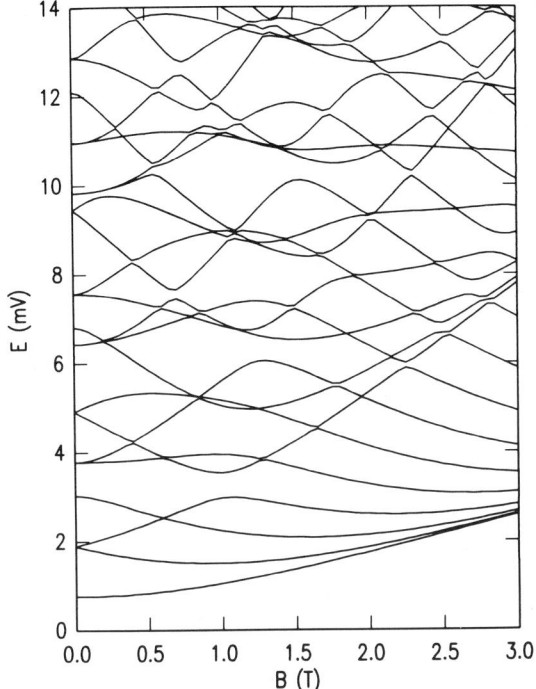

FIG. 28. Electron energies as function of a magnetic field in a 120 nm square-shaped square-well quantum dot calculated numerically. (From Ref. 157.)

to motion in this direction. The zero-field degeneracy of the energy levels is lifted by the magnetic field so that at high level indices, a fairly complex behavior arises. After crossing or anti-crossing at intermediate fields, the energy levels merge to form Landau-level-like states at higher magnetic fields. The formation of highly degenerate levels is clearly observable at $B = 3$ T at energies of about 3 meV and 8 meV and becomes more and more pronounced at higher fields. (See, e.g., Fig. 2 in Ref. 158 or Figs. 1 and 2 in Ref. 159.) Selection of a different dot size simply changes the scale of the magnetic field, since the Hamiltonian is invariant to a change of scale in real space. Although there are pronounced differences in details, the qualitative behavior of the energy levels as described in the preceding does not change if a different ansatz for the confinement potential is chosen.[158-162]

For a quantitative comparison of the measured capacitance oscillations with calculated quantum levels, preceding models are too simple mainly for the following two reasons. First, the number of electrons changes with the gate voltage and the actual quantum levels are many-particle states that may differ significantly from energy levels calculated in a single-particle model. Second, the experimental capacitance spectra represent a statistical average of very many 0D electron systems measured in parallel.

The fact that electron–electron interaction and correlation influence the energy spectra in 0D systems renders the calculation of quantum levels, as well as the connection between gate voltage and Fermi energy within the 0D system, a non-trivial problem. The confinement potential changes self-consistently with the electron occupation of the dot, so that, for instance, the effective dot size increases with increasing gate voltage. When injecting an additional electron from the back contact into the low-capacity electron system, the energy levels of the electron dot are lifted with respect to the Fermi energy of the back contact by the Coulomb gap energy e^2/C. Furthermore, calculations by Bryant[163] demonstrate that in 0D electron systems of present size, the electron states of the few-electron systems become strongly correlated and the energy levels may depart significantly from those calculated for non-interacting single particles.

Averaging over about 10^5 different electron dots also influences the experimental data, since slight differences in the level structure of different dots caused by lithographic imperfections or impurities located in the vicinity of the electron systems may deteriorate the experimental resolution or even change the form of the spectra. Presently, it is unclear whether this can be modeled with inhomogeneous broadening of calculated energy levels or whether a more sophisticated theory of spectra in disordered and possibly correlated systems has to be applied.

Since the few-electron system of a quantum dot is expected to be strongly correlated, the question arises whether it still is possible to observe fractional states at very high magnetic fields. In 2D systems, Hall plateaus have been observed at fractional filling factors p/q in addition to the plateaus of the integer quantized Hall effect.[164] Plateaus at fractional filling factors are explained in terms of a highly correlated ground state of quasi-particles that is separated from its excitations by an energy gap.[165] Since at high magnetic fields the magnetic length becomes much smaller than the extension of presently realized electron dots, it may be anticipated that at sufficiently high fields, fractional states should be observable. However, if the occupation of the electron system is low, the charge density no longer is homogeneous and all electrons sense the boundary of the system. Investigations on 2D[138] as well as 1D electron systems[166] have demonstrated that fractional states can be observed with the capacitance technique.

Figure 29 shows traces of the capacitance derivative recorded at high magnetic fields on 300 nm dots.[167] Only the lowest spin-split Landau level is occupied at the gate voltages depicted. The next spin level starts to be occupied at the strong minimum observed at high gate voltages on each trace.

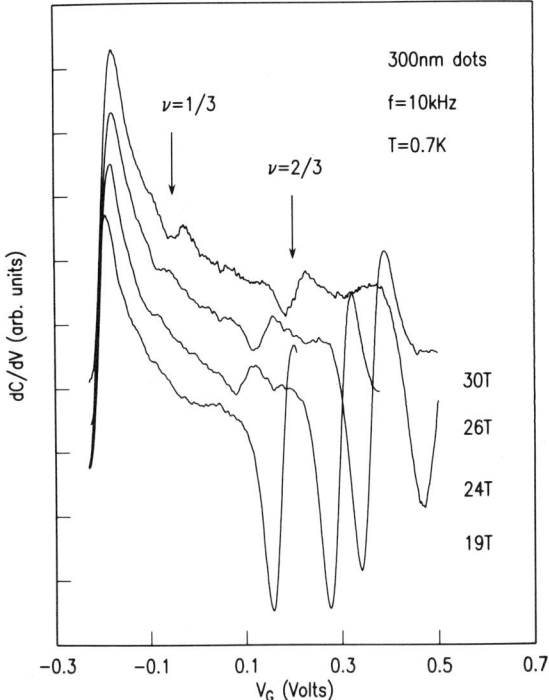

FIG. 29. Gate voltage derivative of the capacitance recorded on 300 nm dots in an $Al_xGa_{1-x}As/GaAs$ heterojunction at high magnetic fields. The structures at fractional filling factors $v = 1/3$ and $2/3$ are marked by arrows at the $B = 30$ T trace. (From Ref. 167.)

With increasing magnetic field, two additional structures arise and become more and more pronounced. Interpolating between the gate voltages at filling factors $n = 0$ and $n = 1$ and assuming a constant capacitance, the gate voltage positions of these structures correspond approximately to filling factors $v = 1/3$ and $2/3$. That the additional structures in the high-field capacitance are caused by the existence of fractional gaps also is augmented by the temperature dependence. The strength of the structure decreases monotonically with increasing temperature and vanishes at about $T = 3$ K. So far, fractional states have been observed in 300 nm and 400 nm but not in 200 nm dots. This may be caused by the low number of electrons contained in the smallest dots, the correspondingly increased influence of the boundary, or reduced screening of imperfections. The number of electrons contained within the 200 nm dots is estimated to be not larger than 13 at the gate voltages applied. Presently, it is not clear whether boundaries in small electron systems principally destroy fractional quantization. Assuming an effective width of 120 nm in the 300 nm dots at a magnetic field of $B = 24$ T and filling factor $v = 1/3$, only about 22 electrons occupy each dot. Thus, high-field measurements on quantum dots may allow for the first time direct comparison of experimental results with numerical few-particle calculations of fractional states.[168]

Mesoscopic electron dots realized in semiconductors represent the link between atomic-like impurities in semiconductors and macroscopic electronic devices.[169] Various publications have pointed out the potential technological applications of 0D systems. (See, e.g., Refs. 170 and 171.) In this section, we have focused on the magnetic field behavior of the quantum levels in such systems. In contrast to the one-to-one correspondence in 1D quantum wires between the electric subband structure at zero magnetic field and the Landau-level-like hybrid bands at high magnetic fields, the quantum levels in 0D systems exhibit a far more complex behavior when they transform into Landau-level-like states. In the following sections, the far-infrared excitations in such systems will be discussed. Again, it will be found that the excitations in 1D electron channels differ from those in spatially completely bound 0D systems when a magnetic field is applied.

IV. Far-Infrared Spectroscopy of Electronic Excitations

The characteristic energies of intraband electronic excitations of laterally confined electron systems at semiconductor interfaces typically are of order $\hbar\omega = 10$ meV. The corresponding photon frequency thus lies in the far-infrared regime and is represented by a wave number $\bar{\nu} = \omega/2\pi c$ of about 100 cm^{-1}. In this section, we describe some of the experimental methods and possibilities of far-infrared spectroscopy in the study of 1D and 0D electron systems in quantum wires and quantum dots. The experimental equipment essentially is the same as that used for quite a time in the study of 2D electron systems, namely Fourier transform spectrometers[172,173] and optically pumped or molecular far-infrared lasers.[174] To detect 2D intersubband resonances, sophisticated methods had to be applied to obtain light polarization perpendicular to the electron layers, such as strip-line arrangements,[174,175] grating couplers,[176] or tilted magnetic field experiments.[177,178] However, in contrast to 2D intersubband resonances, transitions between 1D subbands and discrete 0D states can be excited with light polarization parallel to the surfaces, i.e., with light impinging perpendicularly onto the samples. This simplifies the experimental setup considerably. In many cases, infrared spectroscopy is being complemented by Raman spectroscopy[179,180] and interband optical studies.[181,182]

Infrared spectra are taken mostly in a simple transmission geometry. Radiation either is generated at discrete laser frequencies or a Hg lamp is used in conjunction with a Fourier transform spectrometer to measure spectra in the frequency domain. Figure 30 schematically pictures a typical setup for Fourier transform spectroscopy. The infrared radiation from the lamp is transmitted through a Michelson interferometer, focused onto the sample, and finally detected with a cryogenic bolometer. To achieve sufficiently large signals, the active sample area must be significantly larger than the infrared wavelengths and typically is several mm in diameter. For lateral nanostructures, this means that a device must contain homogeneous arrays of order 10^4 quantum wires or 10^8 quantum dots, respectively. The samples are cooled to liquid helium temperatures and often are placed in the center of a superconducting solenoid. With an appropriately chosen beamsplitter and suitably selected velocity of the moving mirror in the Michelson interferometer, the detector records an interferogram whose Fourier transform is the far-infrared spectrum of the setup.

To determine the transmission coefficient T of a given sample, it is necessary to take interferograms with and without the sample. Then one obtains the transmission coefficient by division of sample and reference spectra. In the case of the electron systems discussed here, the reference spectrum can be measured without removing the device. This is done either by a biased gate to switch the number of electrons between zero and a desired value or by using

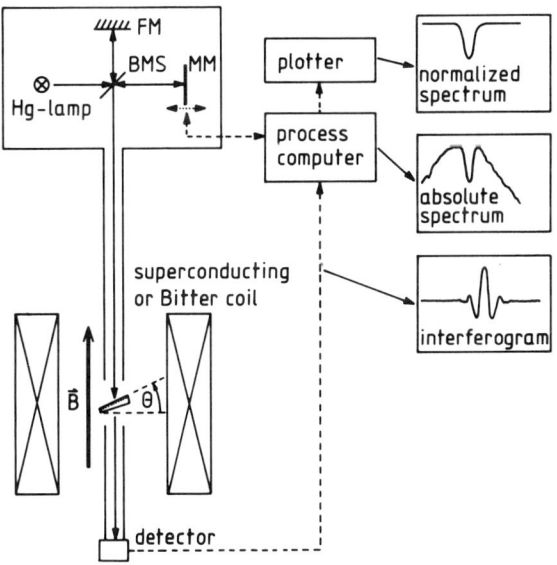

FIG. 30. Experimental setup for far-infrared transmission measurements in a magnetic field. Usually, the sample is cooled to 1K to 4.2 K by the liquid helium bath of the superconducting magnet. The sample may be titled by an angle θ with respect to the magnetic field direction. (From Ref. 183.)

a magnetic field to shift the frequency of the electronic excitations into a remote frequency domain in order to record the signal without low-dimensional electron gas. Thus, one can measure the ratio $T(n_s)/T(n_s = 0)$ or $T(B)/T(B = 0)$ with high sensitivity.

The measured transmission coefficient T for light polarization parallel to the sample surface can be related to the complex dynamical sheet conductivity $\sigma(\omega) = \sigma_1 + i\sigma_2$. This can be done in simple terms provided the semiconductor substrate is nonabsorbing, does not give rise to interferences, and has the low-dimensional electron system in close vicinity to its surface. These conditions are fulfilled in most cases. Interferences are avoided by wedging the back of the sample. One then obtains the relations,[184]

$$\frac{T(n_s)}{T(0)} = \frac{(1 + \sqrt{\varepsilon})^2}{(1 + \sqrt{\varepsilon} + \sigma_1/Y_0)^2 + (\sigma_2/Y_0)^2} \qquad (20a)$$

and

$$\frac{\Delta T}{T} = 1 - \frac{T(n_s)}{T(0)} \cong 2\frac{\text{Real } \sigma(\omega)}{Y_0(1 + \sqrt{\varepsilon})}, \qquad (20b)$$

with the vacuum admittance $Y_0 = (\varepsilon_0/\mu_0)^{1/2}$. The latter equation of Eq. (20b) applies in the small signal limit $\Delta T/T \ll 1$.

In the case of quantum wires, we no longer have an isotropic sheet conductivity. As long as the light is linearly polarized and there is no magnetic field, we can define effective sheet conductivities σ_l/a from the corresponding 1D conductivity components σ_l along and perpendicular to the channels, since the wavelength of the far-infrared radiation is much longer than the grating constant of the wire array ($\lambda \gg a$). These conductivity components include possible interactions between adjacent channels and with a grating coupler. Similarly, one can define an effective sheet conductivity for a dot array. In the following, we will discuss infrared transmission experiments starting with results obtained on moderately density-modulated systems. We then will proceed with systems confined to 1D electron channels and finally will discuss spatially completely confined 0D systems.

4. Plasmons in Density-Modulated Systems

Far-infrared studies of 2D electron systems involving laterally periodic structures started with spectroscopy of 2D plasmon excitations.[43,185–187] In such experiments, grating couplers of period a placed sufficiently close to a 2D electron system are employed to spatially modulate the incident field at the location of the inversion layer and thus couple infrared radiation of wavelength $\lambda \gg a$ to 2D plasmons of discrete wave vectors $q = 2\pi m/a$ ($m = 1, 2, \ldots$). The initial experiments served to verify experimentally the long-wavelength dispersion relation of 2D plasmons and to study the effects of magnetic fields on plasmon excitations. In the absence of magnetic fields and for isotropic and parabolic subbands, the 2D plasmon dispersion in the regime of wave vectors $\omega/c \ll q \ll k_F$ may be written as

$$\omega_p^2 = \frac{e^2 n_s q}{2\bar{\varepsilon}\varepsilon_0 m^*}. \tag{21}$$

Here, $\bar{\varepsilon}$ is an effective dielectric constant at the interface that also accounts for screening effects by metallic layers placed in the vicinity of the 2D electron system, and m^* is the effective optical mass for motion in the plane of the 2D system. Consecutive experiments concentrated on plasmon spectroscopy as a tool for studying the optical mass in various 2D systems[188-190] and investigated deviations from the long-wavelength plasmon dispersion Eq. (21) that become visible at sufficiently large wave vectors corresponding to submicron wavelengths.[191]

In 2D plasmon as well as in 2D intersubband resonance experiments with grating couplers,[176,192] the gratings served as passive couplers only. From the very beginning of such experiments, however, it had been envisioned that similar grating structures can be used to modulate laterally the electronic properties of 2D electron systems.[7,24] In MOS devices, such periodic modulation of the 2D electron density at submicron length scales was achieved eventually either by a periodic modulation of the thickness of the gate oxide[23] or with grating-type gates.[46,193] On $Al_xGa_{1-x}As$/GaAs heterojunctions, a laterally periodic modulation of the 2D electron density first was realized by utlizing the persistent photoeffect via illumination through a grating structure.[91,191]

Anisotropic infrared excitation spectra reflecting a strong laterally periodic modulation of the inversion density were reported first by Batke et al.[46] in studies of Si MOS devices employing a floating grid gate similar as the one illustrated in Fig. 8. Lateral superlattice effects on the dispersion of 2D plasmons reflecting the lateral modulation of the 2D electron density were discovered by Mackens et al.[23]. In these experiments, MOS devices with periodic modulation of the oxide thickness were used to generate the density modulation. Infrared transmission spectra of such samples exhibit a strong plasmon mode at wave vector $q = 2\pi/a$ on top of a Drude-type $q = 0$ background signal as shown in Fig. 31.

The characteristic signature of the lateral density modulation is a splitting of the plasmon excitation into two modes $\omega_\pm$. The amount of the splitting

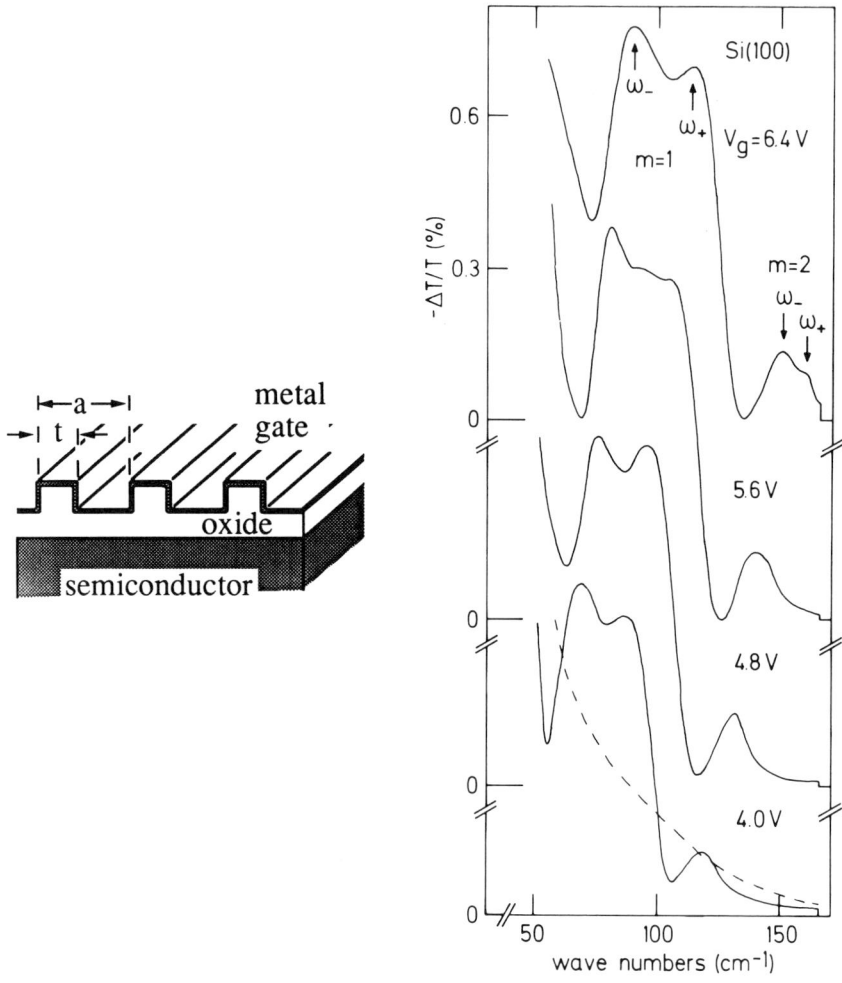

FIG. 31. Far-infrared spectra of 2D plasmon excitations in a device with a thickness-modulated gate oxide as sketched in the left part of the figure. The resonances are split due to the superlattice effect of a charge density modulation of period $a = 650$ nm. The spectra are measured with light polarized perpendicularly to the grating. The dashed line indicates the Drude background measured with parallel light polarization at $V_g - V_t = 4.0$ V. (From Ref. 23.)

easily can be calculated if the density modulation is treated within perturbation theory as shown by Krasheninnikov and Chaplik.[194] This model predicts the formation of minigaps in the 2D plasmon dispersion at the artificial Brillouin zone boundaries, $q_n = n\pi/a$ $(n = \pm 1, 2, \ldots)$, as shown schematically in Fig. 32. The size of these minigaps is determined by the nth Fourier

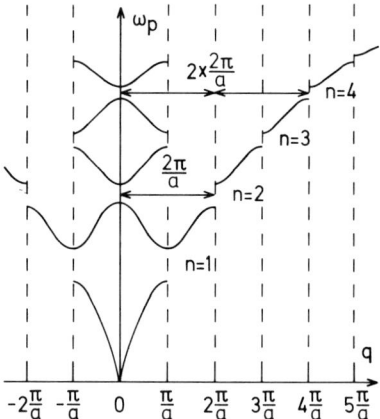

FIG. 32. Energy dispersion of a plasmon in a 2D system with a density modulation of period a in one of the lateral directions. It is assumed that the plasmon propagates along the direction of the modulation. (From Ref. 23.)

expansion coefficient n_n of the periodic charge density $n_s(x)$ and is given by

$$\frac{\omega^2_{+(q_n)} - \omega^2_{-(q_n)}}{\omega^2_p(q_n)} = 2\frac{n_n}{\bar{n}_s}, \tag{22}$$

where $\bar{n}_s$ denotes the average charge density that determines the unperturbed plasmon frequency ω_p. Since the grating-coupler effect of the density-modulated electron system only acts to excite plasmons at wave vector $q_{2m} = 2m\pi/a$, the splitting of the plasmon resonances reflects the corresponding even Fourier coefficient n_n of the electron density with index $n = 2m$. An observable splitting of the plasmon at $q = 2\pi/a$ hence requires a sufficiently strong modulation of the electron density with period $a/2$ and is expected for a system with fundamental period a only when the density modulation has sufficient harmonic content. For a MOS device with modulation of the gate oxide as sketched in Fig. 31, this implies that the ratio of the width $(a - t)$ of the thin oxide to the period a must be sufficiently different from $(a - t)/a = 0.5$, since otherwise, the even Fourier components vanish. Correspondingly, no splitting of the 2D plasmon dispersion is discernible in devices with $(a - t)/a = 0.5$.

Superlattice effects on the 2D plasmon dispersion already are observable at length scales well above the elastic mean free path of the inversion electrons $l_e = v_F \tau_e$, since the mean free path of the collective plasmon excitation, $l_p = (\partial \omega/\partial q)\tau_e$, usually is much longer than l_e. This reflects the fact that in 2D systems, the group velocity of a collective mode usually is much larger than the Fermi velocity v_F. To observe superlattice effects of the periodic density modulation on the single-electron energies, one naively expects to require devices in which the period a is significantly shorter than the elastic electronic mean free path l_e. For typical inversion layers in Si MOS devices with a low temperature mobility of order 10^4 cm^2/Vs, this would require superlattice periods below 100 nm, which are difficult to fabricate. In $Al_xGa_{1-x}As/GaAs$ heterojunctions with low-temperature mobilities exceeding 10^6 cm^2/Vs, superlattice effects can be expected to become observable with modulation periods of several 100 nm. Therefore, efforts concentrated on fabricating devices with periodic modulation of the inversion electron density on the basis of high-mobility $Al_xGa_{1-x}As/GaAs$ heterojunctions.

As a first success, Hansen et al.[53] introduced a comparatively simple field-effect-tunable heterojunction in which a 2D electron layer at the $Al_xGa_{1-x}As/GaAs$ interface could be density-modulated periodically via a gate electrode on top of a thickness-modulated insulator. The operation and the static transport properties of such devices already have been discussed in detail in Sections II and III. Here, we want to concentrate on the infrared excitation spectra observed at low temperatures.[53,142] Figure 33 displays the dependence of the dominant infrared excitations on gate bias. At gate voltage $V_g = 0$, the electron system essentially is an unmodulated 2D electron layer with density $n_s = 6 \cdot 10^{11}$ cm^{-2}. Correspondingly, the most prominent excitation is a 2D plasmon with wave vector $q = 2\pi/a$ excited by the grating-coupler effect of the metal-covered insulator grating. With increasing negative gate bias inversion, electrons get depleted predominantly below those parts of the gate that are closest to the heterojunction interface. Hence, the average electron density $\bar{n}_s$ in the inversion layer decreases as mirrored by the lowering of the plasmon frequency. Simultaneously, the amplitude of the periodic density modulation of the inversion layer increases. The fact that there is no discernible splitting of the plasmon frequency at negative bias indicates that the width of the grating stripes of the photoresist grating is close to half the period and, therefore, the second Fourier component of the density modulation is rather small.

At a gate bias of $V_g = V_d = -0.5$ V, the areas closest to the gate are fully depleted and the electron system transforms into an array of narrow inversion wires that are separated by insulating channels. In the infrared spectra, this is reflected by two prominent changes. One is that the resonance frequency of the strong infrared excitation suddenly increases with decreasing gate voltage. The other is that the oscillator strength of the resonance rapidly gains strength in a narrow regime of gate voltages around $V_g = -0.5$ V. Both phenomena reflect the confinement of the inversion electrons to narrow inversion channels. The sudden gain in oscillator strength can be understood as resulting from a transformation of the character of the infrared resonance.

5. ELECTRONS IN LATERALLY PERIODIC NANOSTRUCTURES 341

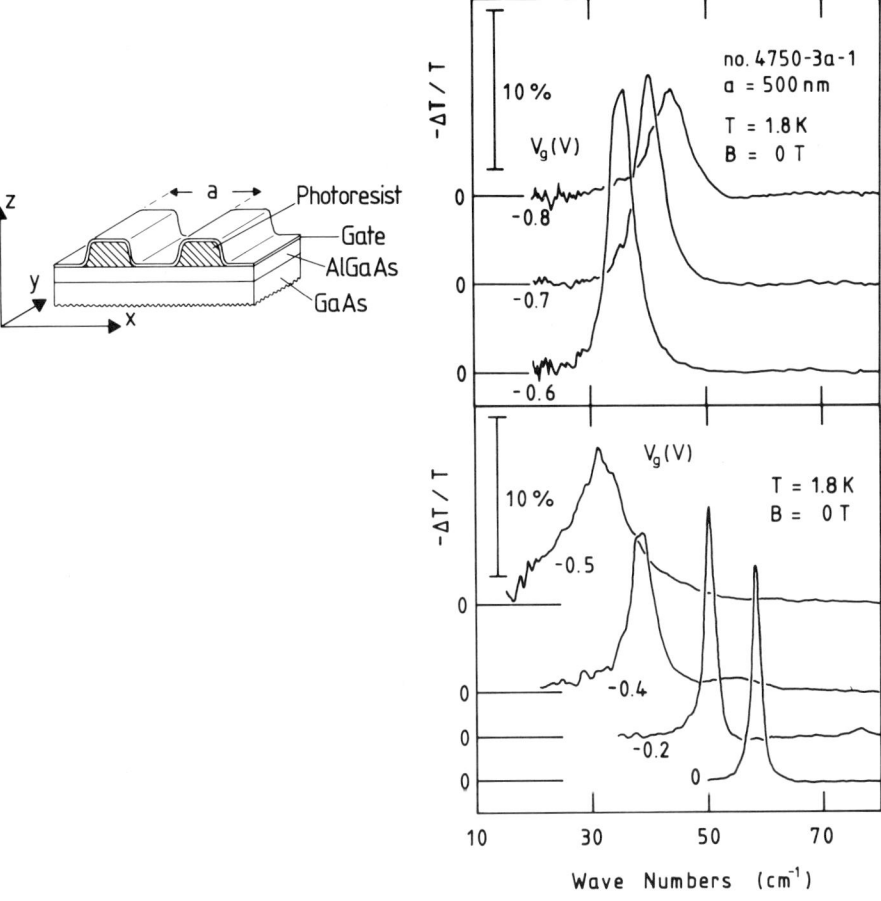

FIG. 33. Infrared spectra of an $Al_xGa_{1-x}As/GaAs$ heterojunction with microstructured gate of period a. The radiation is polarized perpendicularly to the grating. A cross section of the sample is sketched on the left. (From Ref. 195.)

At $V_g > -0.5$ V, the infrared resonance is described adequately as a plasmon of a density-modulated 2D system with wave vector $q = 2\pi/a$ that is excited predominantly by the grating-coupler effect. At $V_g < -0.5$ V, zone folding of the inversion electron system makes the distinction between $q = 2\pi/a$ and $q = 0$ modes oblivious. Hence, the infrared resonance becomes essentially a $q = 0$ resonance of an array of inversion wires with its strength no longer controlled by the efficiency of the grating coupler but mainly determined by the average density of inversion electrons per unit area n_l/a, where n_l denotes the linear density of electrons per wire. As will be discussed in detail in the following section, the resonance frequency then reflects the confinement of carriers to narrow wires. For dominantly classical confinement, it becomes the depolarization frequency of electrons in a narrow wire. For dominantly quantum confinement, it becomes the intersubband resonance frequency of transitions between adjacent 1D subbands.

Another feature of the transition from 2D to 1D is a marked broadening of the resonance at the threshold gate voltage V_d. Besides broadening caused by inhomogeneities in the threshold regime, this broadening may be a consequence of Landau damping: Not only the plasmon branch but also the single-particle excitations, which in the homogeneous electron gas arc well-separated from the plasmon, are folded back. It has been shown that damping increases with increasing modulation until, in the transition regime, the plasmon ceases to be a well-defined resonance. With further increased modulation amplitude, damping then diminishes as the system begins to exhibit 1D intersubband transitions between bands with flat dispersion in the confinement direction.[196]

The transition from a density-modulated 2D-like inversion layer to an array of inversion channels is directly visible as a sudden change of the slope of the resonance frequency with gate voltage in Fig. 34. In the density-modulated regime, the frequenccy is well-described as that of a 2D plasmon reflecting the average areal electron density $\bar{n}_s$.[53] Correspondingly, the squared resonance frequency decreases approximately linearly with gate bias that is proportional to the average density.

The description of the infrared resonance at voltages $V_g > -0.5$ V as a 2D-like plasmon is asserted further by the infrared spectra in a magnetic field applied perpendicularly to the sample surface. Figure 35 shows infrared spectra in a moderate magnetic field for closely spaced values of the gate bias between -40 mV and -480 mV. For a low gate bias, two infrared modes are

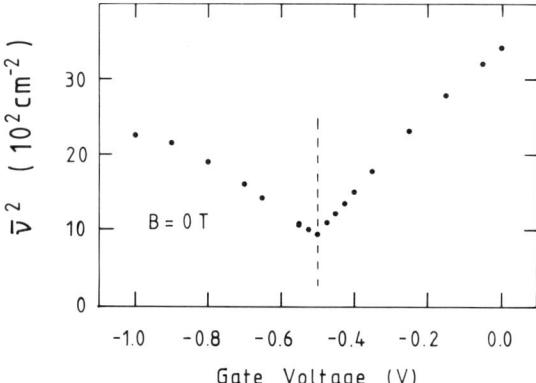

FIG. 34. Square of the resonance frequency pf Fig. 33 plotted versus the gate voltage.

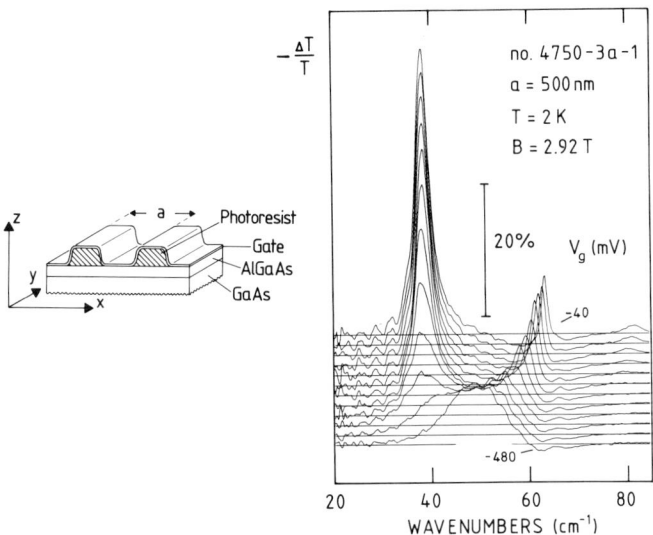

FIG. 35. Infrared spectra of an $Al_xGa_{1-x}As/GaAs$ heterojunction with microsctructured gate at a magnetic field $B = 2.9$ T applied perpendicularly to the sample surface. (From Ref. 142.)

discernible. The strong mode at lower frequency is the cyclotron resonance of the density-modulated 2D system and thus the $q = 0$ mode. With the gate bias approaching the transition point $V_g = V_d$ to arrays of narrow channels, this resonance decreases in strength but hardly change its frequency except in the very vicinity of the transition. The weaker resonance at higher frequency is the grating-coupled magneto-plasmon at wave vector $q = 2\pi/a$. As expected, its resonance frequency decreases with decreasing gate voltage and hence decreasing average density. As the gate bias approaches the value $V_g = -0.5$ V, both the cyclotron resonance and the magneto-plasmon resonance merge into a single infrared excitation.

A similar behavior is observed at all magnetic fields studied. The gate voltage dependence of the observed resonance frequencies for two other magnetic field values is summarized in Fig. 36. In the 2D regime, one observes magneto-plasmon modes at wave vector $q = 2\pi/a$ as well as the cyclotron resonance at $q = 0$. At certain magnetic field values, such as $B = 2.88$ T, the magneto-plasmon modes are split because of non-local interaction with harmonics of the cyclotron resonance,[142] a phenomenon well-known from studies of 2D systems.[191] As one approaches the transition to the 1D regime, all resonances collapse into a single $q = 0$ mode. At low magnetic fields, this happens rather abruptly at $V_g = -0.5$ V. At high magnetic fields, the transition becomes more smooth and the resonance frequency of the cyclotron resonance slowly increases as one approaches the transition voltage. This behavior may reflect

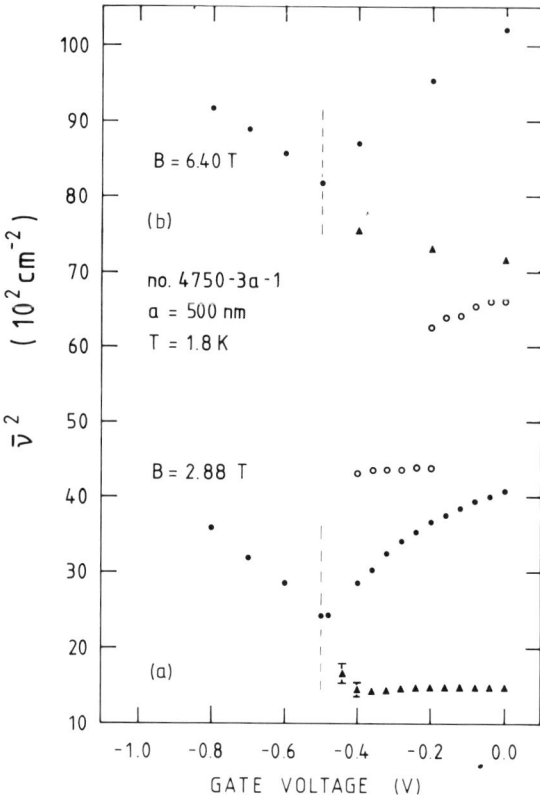

FIG. 36. Square of the resonance position versus gate voltage at two different magnetic fields. The dashed line indicates the gate voltage V_d at which the density-modulated 2D system transforms into an array of isolated quasi-1D channels. Cyclotron resonances in the 2D regime are marked by triangles, strong plasmon resonances by full circles, and weak ones by open circles. (From Ref. 142.)

modified screening in strong magnetic fields.[139] The smooth transformation of both the collective magneto-plasmon resonance and the single-particle cyclotron resonance into a single $q = 0$ resonance that happens as one tunes to the transition from a periodically modulated 2D system to an array of isolated inversion wires is not quantitatively understood yet. Present experimental results as well as various theoretical models[197–199] indicate that in a strongly modulated system close to and beyond the transition, the infrared excitation has both a collective contribution reflecting the dynamic polarization of the inversion charge as well as a single-particle contribution reflecting the confinement potential. Both aspects will be discussed in further detail in the following section.

5. Excitations in Confined Systems

Far-infrared spectroscopy of laterally confined electron systems on semiconductors has begun with the study of periodic arrays of mesoscopic electron discs[15] and stripes[16,17] on $Al_xGa_{1-x}As/GaAs$ heterojunctions. The confined systems were prepared using conventional mask photolithography or holographic lithography and wet etching. This way, the originally 2D electron systems of the heterojunctions are converted into free-standing individual electron systems with dimensions of some micrometers down to values of some hundred nanometers. Quantization of the electron states still is negligible at such widths and a classical description can account for the observed resonances. In the next section, we first briefly review the work on quasi-macroscopic stripes and then proceed to quantized 1D channels. In the subsequent section, we treat the corresponding approach from electron discs to 0D systems in quantum dots.

a. Arrays of Quasi-1D Channels

Electronic excitations of deep mesa etched narrow electron stripes on $Al_xGa_{1-x}As/GaAs$ heterojunctions have been studied by Allen et al.[16] and Hansen et al.,[17] both obtaining similar results. Figure 37 shows spectra in low magnetic fields for the two principal polarizations of the incident far-infrared radiation, namely parallel and perpendicular to the stripes. In zero magnetic field, a strong resonance at about 40 cm^{-1} is present in the perpendicular polarization but none is observed in the parallel one. In finite magnetic fields, a resonance also develops in parallel polarization and reaches about the same strength as the one in perpendicular polarization at field strengths $B \gtrsim 3$ T. The resonance frequencies plotted in Fig. 37 demonstrate that the resonance positions do not depend on polarization and closely follow the relation,

$$\omega^2 = \omega_0^2 + \left(\frac{eB}{m^*}\right)^2, \tag{23}$$

with an intercept $\hbar\omega_0 = 5$ meV at $B = 0$ and mass $m^* = 0.072\ m_e$, which compares well with the cyclotron mass of the unstructured sample. Fan-charts of Shubnikov–de Haas oscillations are linear and thus allow one to determine the 2D electron density $n_s = (4.8 \pm 0.2) \cdot 10^{11}$ cm^{-2}. The widths $W = (0.45 \pm 0.04)$ μm of the individual stripes are read from SEM micrographs.[17] In such relatively wide stripes, quantization of the electron system in the lateral potential is not expected to be important and classical theory may account for the observed resonance positions and oscillator strengths. The square of the classical depolarization frequency at $B = 0$, which,

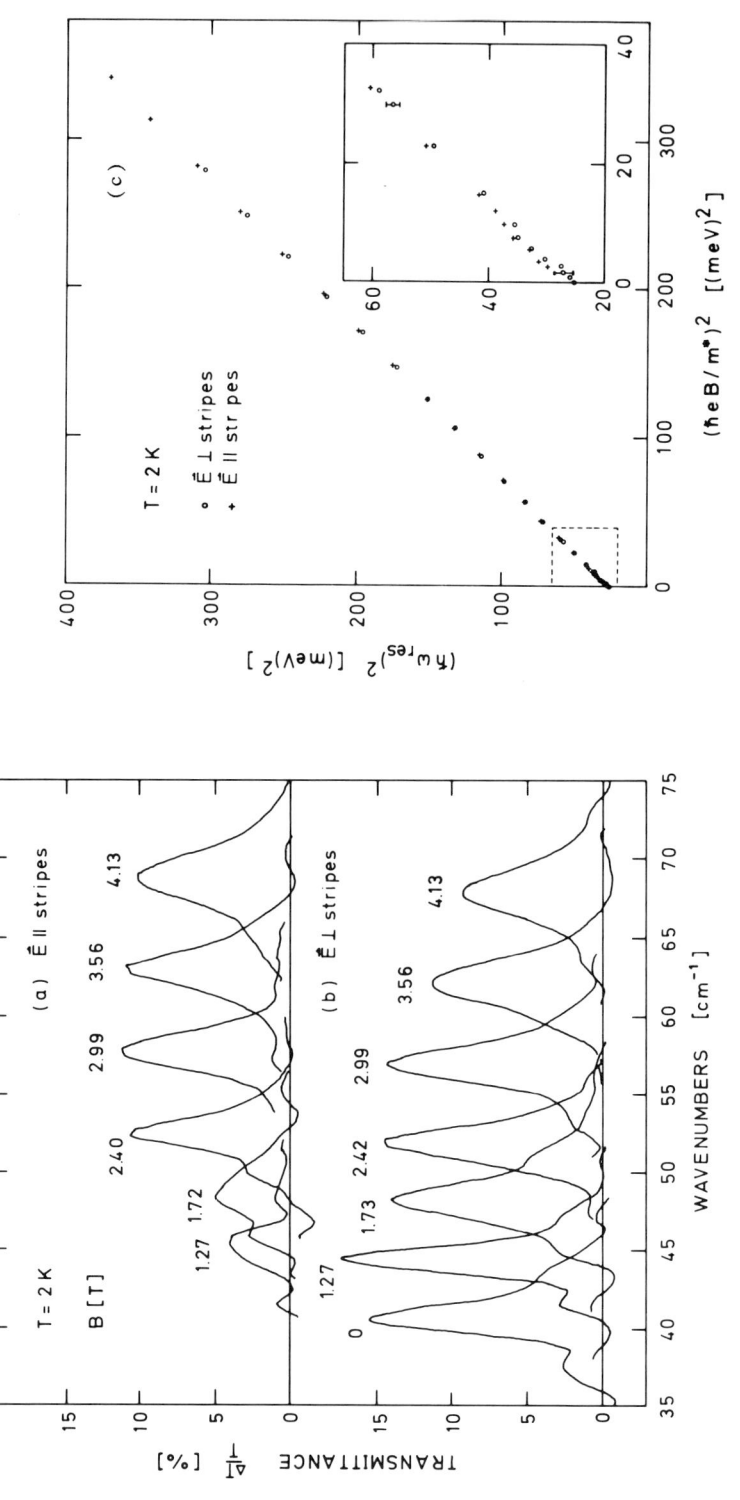

FIG. 37. Far-infrared transmittance spectra of a laterally microstructured $Al_xGa_{1-x}As/GaAs$ heterostructure at different magnetic fields for (a) parallel polarization and (b) perpendicular light polarization. (c) Squared resonance positions versus squared magnetic field strength. (From Ref. 17.)

figuratively speaking, describes the sloshing back and forth of the electrons between the sidewalls of the wire, is given by the expression,

$$\omega_d^2 = \alpha \frac{e^2 n_s}{\varepsilon_0 \bar{\varepsilon} m^* W}, \tag{24}$$

with a numerical factor α of order unity and an appropriately averaged dielectric constant $\bar{\varepsilon}$.[142] This dielectric constant takes into account the presence of semiconductor, insulator, and metallic gate and by no means is easy to specify. Also, the factor α renders difficult a more detailed comparison between experiments and existing theories. In the evaluation of their experiments, Hansen et al.[17] assume values $\bar{\varepsilon} = 12.5$ and $\alpha = 2$.

The real parts of the classical conductivities, which are related closely to the recorded relative changes of transmittances according to Eq. (20), are[17]

$$\sigma_{xx} = \sigma_0 \frac{\omega^2(1 + \omega^2 \tau_e^2 + \omega_c^2 \tau_e^2)}{\omega^2(1 + \omega_0^2 \tau_e^2 + \omega_c^2 \tau_e^2 - \omega^2 \tau_e^2) + (\omega_0^2 - 2\omega^2)^2 \tau_e^2} \tag{25}$$

for perpendicular polarization and

$$\sigma_{yy} = \sigma_0 \frac{\omega_0^2 \tau_e^2 (\omega_0^2 - 2\omega^2)}{\omega^2(1 + \omega_0^2 \tau_e^2 + \omega_c^2 \tau_e^2 - \omega^2 \tau_e^2) + (\omega_0^2 - 2\omega^2)^2 \tau_e^2} + \sigma_{xx} \tag{26}$$

for parallel light polarization. With constant scattering time τ_e and increasing cyclotron frequency ω_c, the conductivity at resonance for perpendicular polarization is expected to decrease monotonically from the static conductivity $\sigma_0 = e^2 n_s \tau_e / m^*$ to $\sigma_0/2$. At the same time, the corresponding conductivity for parallel polarization increases from zero to $\sigma_0/2$ provided the scattering time is sufficiently long ($\omega \tau_e \gg 1$). The experimental spectra in Fig. 37 show a more complex behavior than the previous prediction that describes the overall variation of the intensity only qualitatively. This may be caused by the etched heterojunction surface that modulates the radiation field similar to a grating coupler. Furthermore, the assumption of a constant and isotropic scattering time may be too simple. A quantitative evaluation of the resonance positions and amplitudes yields values $n_s = (4.0 \pm 0.2) \cdot 10^{11}$ cm^{-2} and $W = (0.49 \pm 0.2)$ μm. The fairly close agreement of the electron density with that of the unstructured sample as well as of the electronic width with the geometrical one is surprising, since it implies that the electrons are distributed over the full width of the stripes without significant depletion zones at their edges.

So far, we have considered only the lowest-lying depolarization mode. Higher-order modes of a confined 2D electron gas first were observed in laterally confined systems on liquid helium.[200,201] In isolated electron stripes on semiconductors, they have been addressed as dimensional resonances,[202] confined plasmons,[203] or waveguide-like modes.[204] Whereas in weakly modulated systems, the plasmon-like collective excitations are described best in a reduced minizone scheme, the resonances in arrays of individual stripes with no electrons between stripes essentially are standing-wave-type dimensional resonances of individual stripes. Such transitions from propagating collective-type excitations to dimensional resonances are well-known in magnetically ordered systems where propagating spin-wave modes become standing-wave-type magnetostatic modes when dimensions of the system are reduced and become comparable to the wavelength of the collective excitations.[205] In the case of weak density modulation, small energy gaps in the dispersion are induced at the zone edges whose sizes are proportional to the corresponding Fourier components of the density modulation. (See Eq. (22) in Section 4.) In the case of a single isolated stripe, one obtains discrete resonance frequencies governed by the width of the stripe, namely ($k_y = 0$),[204]

$$\omega_{dn}^2 \cong \frac{\pi}{2} \frac{e^2 n_s}{\varepsilon_0 \bar{\varepsilon} m^*} \left(\frac{n}{W}\right) \qquad (27)$$

Higher orders $n = 2, 3, \ldots$ of dimensional resonances in isolated electron stripes have been observed only recently in a dual-gate device on Si like the one illustrated in Fig. 8.[202] As described in Section III, the dual-stacked gate allows one to tune the channel width and the electron density by gate voltages V_{gt} and V_{gb} applied between the Si substrate and the top gate and the bottom gate embedded in the SiO_2 insulator, respectively. Furthermore, the device also allows one to squeeze the widths W by the application of a so-called substrate bias.[10] In 1D and 0D systems, this substrate-bias technique for the first time allows nearly independent control of channel width and electron density. It is discussed in the following section in connection with the transition to quantized 0D systems.

Figure 38a shows spectra recorded in the subgrating inversion mode where the electron channels are located at the $Si-SiO_2$ interface underneath the stripes of the bottom gate. With different bottom gate voltages ΔV_{gb} above inversion threshold, various electron densities are adjusted. The top gate voltage $\Delta V_{gt} = -8$ V creates a lateral confinement potential with depth exceeding the band gap of Si. It is interesting to note that, consequently, at this top gate voltage, free holes are induced at the $Si-SiO_2$ interface between the electron channels, thus forming a lateral nipi structure. In Fig. 38a, a series of well-defined dimensional resonances is observed. Classically, it is expected that only modes with anti-symmetric charge distribution at opposing channel boundaries are excited. In fact, only modes of odd order $n = 1, 3, 5, \ldots$ are observed. In Fig. 39, the squared resonance frequencies are displayed versus

5. ELECTRONS IN LATERALLY PERIODIC NANOSTRUCTURES

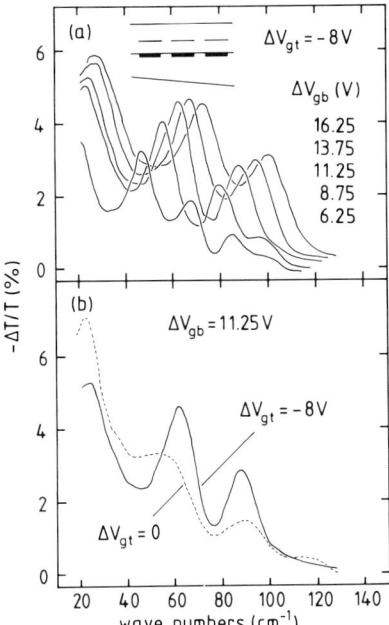

FIG. 38. Transmission spectra of a periodic array of electron channels of width $W \approx 1.5$ μm on Si. The channels are created in the subgrating inversion mode. The depletion charge is $N_{\text{depl}} = 0.9 \cdot 10^{11}$ cm^{-2}. In (a), the electron density n_s is varied via ΔV_{gb}; in (b), the confining potential is altered via ΔV_{gt}. The location of the electron channels beneath the grating stripes is indicated in the inset. (From Ref. 202.)

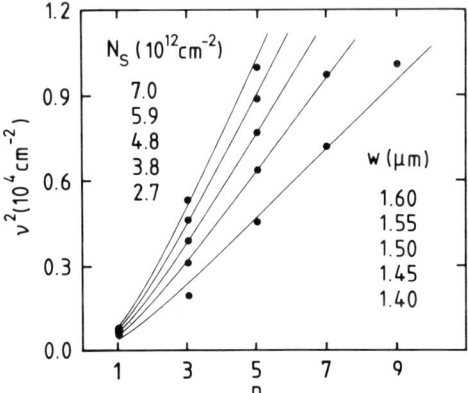

FIG. 39. Squared resonance frequencies of the dimensional resonances in Fig. 38a versus mode index n. The solid lines are calculated from Eq. (27) as described in the text. The channel widths W are adjusted for best fits for each electron density N_s. (From Ref. 202.)

index n. The solid lines represent the result calculated from Eq. (27) but with an additional prefactor $2/\pi$. This reproduces for $n = 1$ Eq. (24) with $\alpha = 1$. Furthermore, it is assumed that the effective dielectric constant $\bar{\varepsilon}$ depends on the order n of the resonances. The electron densities n_s are measured by Shubnikov–de Haas oscillations; the widths W are fitting parameters. The slight density dependence of the widths is thought to reflect fringing field effects.[202]

Generally, there is good agreement between experimental resonance positions and the dispersion law of Eq. (27). However, in less deep potentials, the oscillator strengths of higher modes ($n > 1$) decrease and deviations from the classical dispersion law are observed in Fig. 38b. From this, it has been concluded that square-well confinement is essential for observing higher-order resonances. Higher-order resonances also disappear when the channel width becomes smaller.[202]

Far-infrared excitations in current GaAs heterojunction wire gratings predominantly have collective character.[51,203] The plasmon character in free-standing multi-layered wire arrays is manifested by the observation of a number of n-layer-coupled local plasmon modes in an n-layered system. The essentially classical interpretation is consistent with single-particle quantum energies $\hbar\Omega_0 = 1$ to 2 meV revealed by Shubnikov–de Haas experiments.[59] This means that the far-infrared excitations are significantly higher in energy than the 1D subband spacings. Spectra for a five-layered sample are depicted in Fig. 40.[206] In small magnetic fields, the resonances are observed only

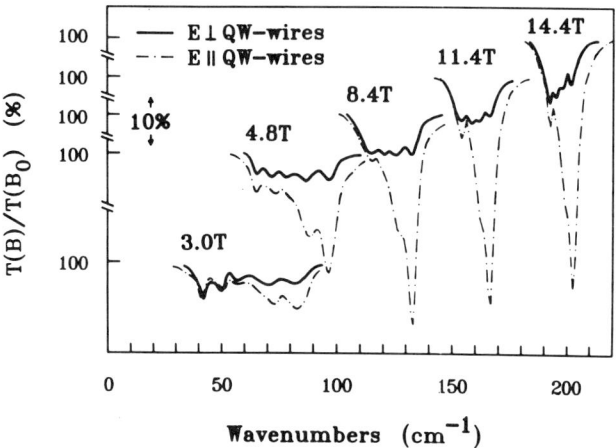

FIG. 40. Transmission spectra for a five-layered quantum wire system. Full and dashed-dotted lines indicate polarization of the incident FIR radiation perpendicular and parallel to the wires, respectively. (From Ref. 206.)

in perpendicular polarization quite similar to the case of a one-layered wire grating.[17] Resonances for parallel polarizations increase in their strengths as the magnetic field increases and the positions are identical in both polarizations. It remains to be understood why in high magnetic fields, parallel polarization excites much more effectively than perpendicular polarization, quite in contrast to the one-layered wire grating.

The observation of several modes at a fixed magnetic field strongly resembles layer-coupled plasmons in homogeneous multi-quantum wells that have been examined by Raman spectroscopy.[207–209] Figures 41a–c reveal the similarity of the far-infrared excitations in multi-layered wire arrays to magneto-plasmons. The magnetic field dependence of the resonance positions ω_i of each of the modes ($i = 1$ to 5) is described by the condition,

$$\omega_i^2 = \omega_i^2(B = 0) + \omega_c^2.$$

The predominant collective contributions to the resonance frequencies in their systems have been described by Demel et al.[203] as local collective excitations in a *plasmon in a box* model. This approach starts from the dispersions of a homogeneous 2D *n*-layered system that gives rise to *n* continuous plasmon branches separated by the Coulomb interaction. The mode frequencies $\omega_i(k_x)$ in the microstructure then are obtained at the wave vector $k_x = \pi/W_i$, as is shown for the two-layered system in Fig. 42. The channel widths effective for each of the local plasmon modes *i* may be slightly different, since the plasmons may be reflected at the walls with different phases. In fact, values $W \simeq 500$ nm that are comparable to the geometrical width are deduced from a best fit to the experimental frequencies. A quantitative evaluation within this simple model is not always satisfactory; however, it provides a vivid picture of the basic physical situation.[203]

The far-infrared excitations in the aforementioned nanostructures can be described as classical depolarization resonances to a very good approximation. Quantization into 1D subbands is revealed convincingly only by the aperiodic nature of the Shubnikov–de Haas oscillations. (See Section 2.) In other words, the single-particle subband spacings are less than the energies of the far-infrared resonances by at least half an order of magnitude. The transition from classical excitations to transitions between quantized subbands, i.e., from classical to quantum confinement, has been studied recently by Alsmeier et al. in dual-stacked gate structures on Si.[202]

A reduction of the channel width can be achieved readily by inducing the electrons at the SiO_2–Si interface underneath the bottom gate openings via the gate voltage ΔV_{gt}, as sketched in the inset of Fig. 43a. The excitations observed in this configuration are displayed in Fig. 43b. By permanently exposing the sample to band-gap radiation, a quasi-accumulation condition[10] is realized and dimensional resonances represented by the solid traces are

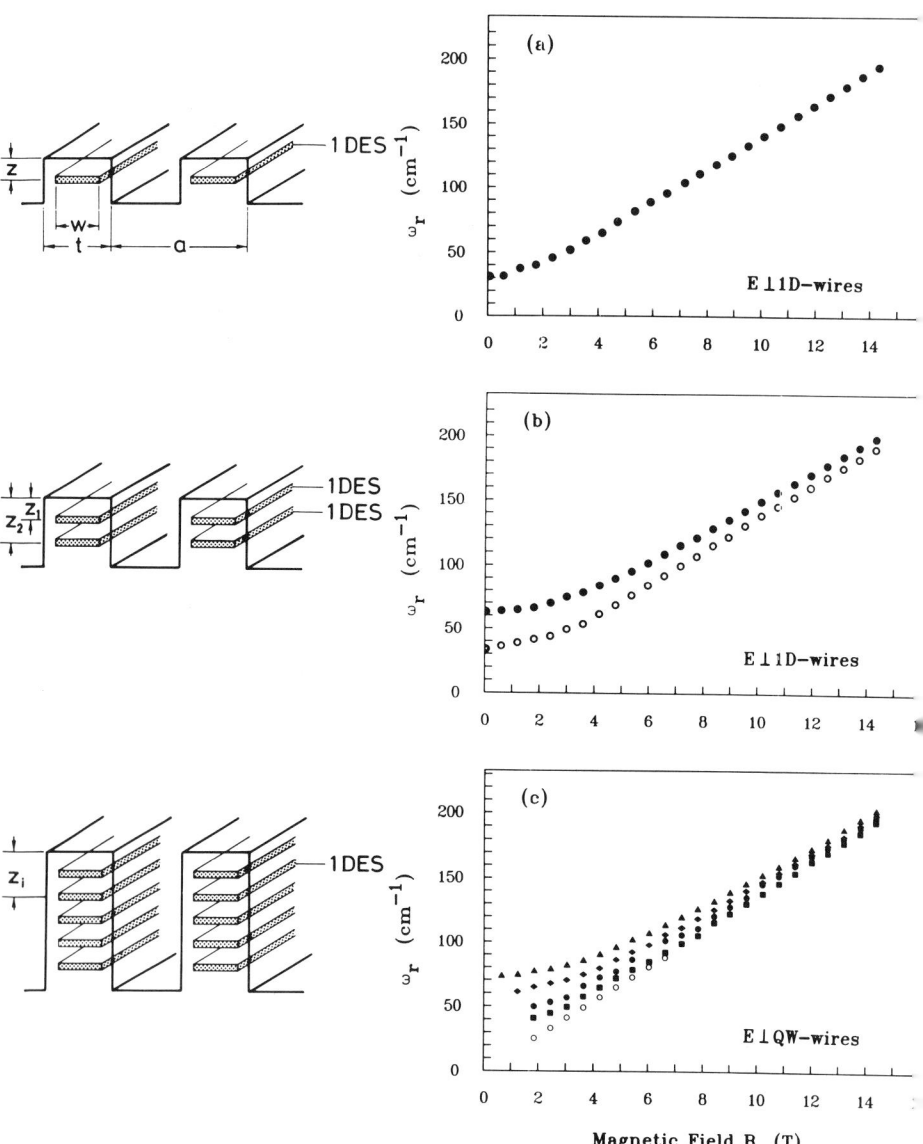

FIG. 41. Resonance positions measured on (a) a one-layered, (b) two-layered, and (c) five-layered quantum wire systems. (From Ref. 206.)

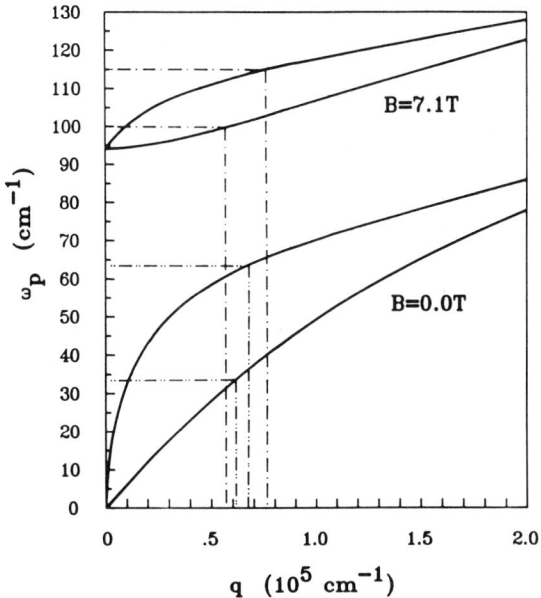

FIG. 42. Calculated plasmon dispersions for a two-layered homogeneous system. The dash-double-dotted (dash-dotted) lines indicate the experimentally observed resonance energies for $B = 0$ ($B = 7.1$ T). The intersections with the dispersions are projected onto the q axis to determine quantized wave vectors $q = \pi/W_i$ of the local plasmons. (From Ref. 203.)

observed. Only the fundamental $n = 1$ resonance is observed. The dashed traces are obtained for a substrate bias $V_{SB} = 18$ V, i.e., with the gate potential at the top gate increased in the dark after charging the channels at the indicated gate voltages ΔV_{gt}. This further squeezes the channel width but does not change noticeably the oscillator strength and thus the linear electron density n_l.

In Fig. 43b, the squared resonance frequencies versus gate voltage are displayed for various depletion conditions. The quasi-accumulation data ($N_{depl} \simeq 0$) are well-described by the lowest dimensional resonance, as is evident from the agreement with the solid line calculated according to Eq. (27). We note that for quasi-accumulation, the resonance frequency extrapolates to zero for vanishing gate voltage ΔV_{gt} or electron density n_l, as is expected indeed for classical confinement according to Eq. (27). For finite depletion charge densities, intercepts as high as 5 meV are obtained. This provides convincing proof of quantum confinement on Si and it means that subband spacings of 5 meV could be realized at the smallest electron densities. The increase of the resonance frequencies at higher densities reflects the collective contribution to the excitations that again can be described classically.[202]

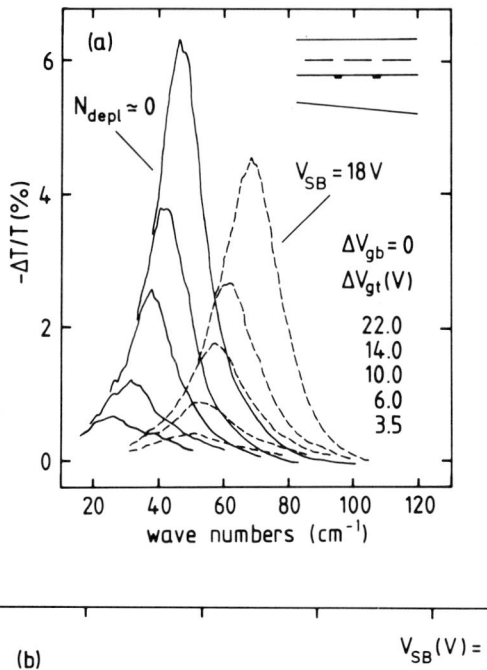
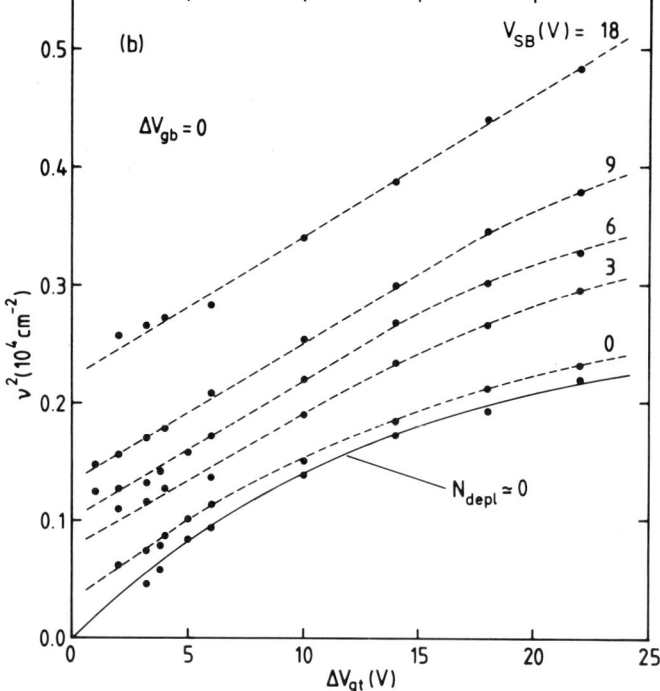

FIG. 43. (a) Transmission spectra for periodic arrays of narrow electron channels on Si at different voltages ΔV_{gt} or linear electron densities n_l. Solid and dashed lines are measured in quasi-accumulation ($N_{depl} \simeq 0$) and at substrate bias $V_{SB} = 18$ V ($N_{depl} = 3.8 \cdot 10^{11}$ cm^{-2}). The inset shows the location of the channels. (b) Squared resonance frequencies. The solid line is calculated from Eq. (27); the dashed lines are guides to the eye. (From Ref. 202.)

The quantization energies on Si are astonishingly high, particularly in view of the relatively high effective electron mass $m^* = 0.19\, m_e$. In this respect, the narrow-gap semiconductor InSb with its much lower effective mass at the conduction band edge, $m_0^* = 0.014\, m_e$, is more favorable to study quantum confinement.[53] Normalized spectra of a wire array on this material are depicted in Figs. 44a and b for light polarizations parallel and perpendicular to the inversion channels at various gate voltages ΔV_g above inversion threshold in the absence of a magnetic field.[210] In parallel polarization, Drude-type spectra corresponding to the free motion along the channels are observed; in perpendicular polarization, we detect depolarization-shifted intersubband resonances. The dashed line in Fig. 44b is a fit to the spectrum for gate voltage $\Delta V_g = 50$ V, assuming the real part of the classical conductivity to be

$$\sigma_l = \sigma_{l0}\left[1 + \left(\frac{\omega^2 - \omega_0^2}{\omega}\right)^2 \tau_e^2\right]^{-1}, \quad (28)$$

with the static 1D conductivity $\sigma_{l0} = e^2 n_l \tau_e / m^*$. It yields a mobility $\mu = e\tau_e/m^* = 16{,}000$ cm^2/Vs and an areal density $\bar{n}_s = n_l/a = 5 \cdot 10^{11}$ cm^{-2} in the channels. Spectra in a magnetic field are shown in Figs. 44c and d. Note that for both light polarizations, the same spectral shapes are observed. The resonances are similar to cyclotron resonances of a homogeneous 2D electron system, as is explained by the rather high magnetic field strength $B = 5.4$ T

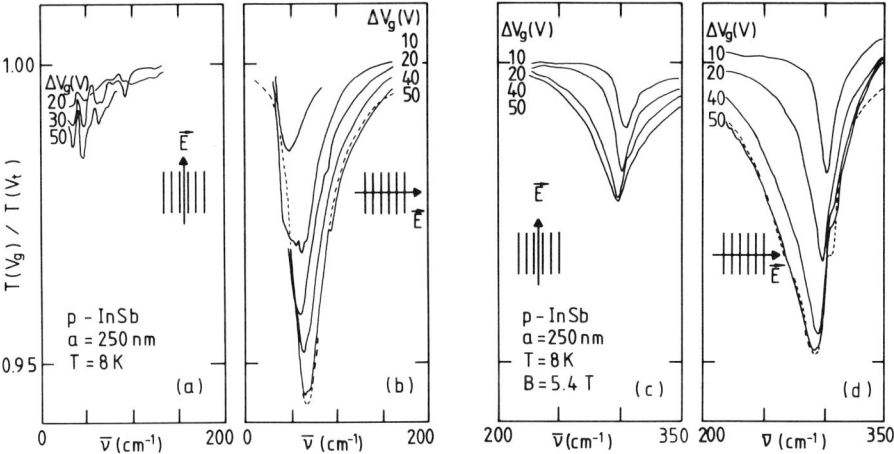

FIG. 44. Intersubband resonances of quasi-1D electron channels on InSb. Light polarizations parallel and perpendicular to the wires are indicated and magnetic field strengths are $B = 0$ (a,b) and $B = 5.4$ T (c,d). The dashed lines represent Drude fits to the $\Delta V_g = 50$ V spectra. (From Ref. 210.)

and the related magnetic length $l_B = 11$ nm, which is much less than the channel width $W \simeq 100$ nm. In fact, subband-shifted cyclotron resonances at hybrid frequencies $\tilde{\omega} = (\omega_c^2 + \Omega_0^2)^{1/2}$ are observed. A small collective contribution to the resonance frequency is ignored in this equation.[198] In strong magnetic fields, the cyclotron frequency clearly exceeds the frequency in the absence of a magnetic field. The dashed line in Fig. 44d is a cyclotron resonance fit, again for voltage $\Delta V_g = 50$ V, using the 2D Drude magneto-conductivity ($\omega_c \gg \omega_0$). Here, occupation of three 2D subbands, $i = 0, 1, 2$, was taken into account because of the high electron density at this gate voltage and the low 2D density of states on InSb.[211] Though the spectra are normalized with the transmittance at threshold voltage, the signals measured in parallel polarization in Fig. 44c are about a factor of two smaller than the ones in perpendicular polarization in Fig. 44d. This is explained by the polarizing effect of the metal stripes, which act as a wire grid polarizer.

Resonance frequencies versus magnetic field strength are summarized in Fig. 45a for three gate voltages ΔV_g. The inset once more presents the data in low magnetic fields on an expanded scale. Transmittance spectra could not be taken close to the reststrahlen band (183–194 cm^{-1}) of InSb, which is indicated by the horizontal dashed lines. In strong magnetic fields above the reststrahlen band, we have subband-shifted cyclotron resonances with frequencies that decrease slightly with increasing gate voltage. This decrease of the cyclotron frequency compares well with the one observed in a ho-

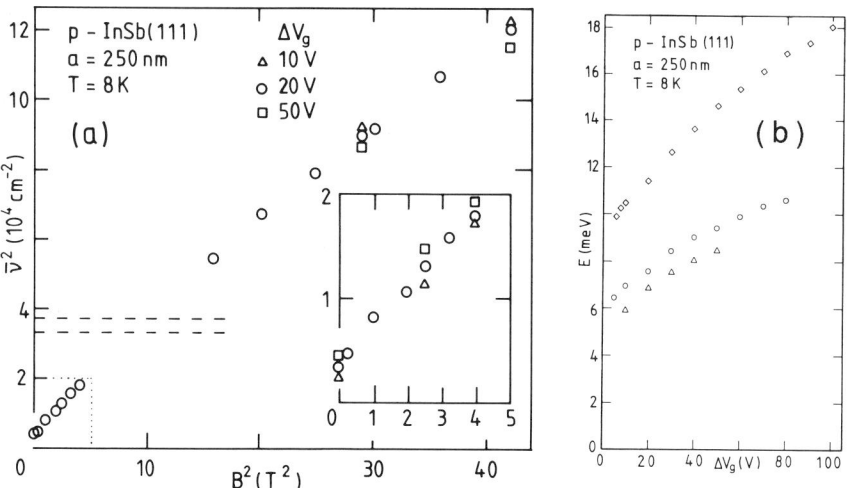

FIG. 45. (a) Resonance positions of 1D subband-shifted cyclotron resonances for three gate voltages. The inset shows the data in low magnetic fields on an enlarged scale in the same units. (b) Intersubband resonance positions ($B = 0$) for three different samples versus gate voltage. (From Ref. 210.)

mogeneous 2D electron gas on InSb and is a consequence of band nonparabolicity.[211] In the absence of magnetic fields, we observe resonance transitions between 1D subbands with frequencies that increase with gate voltage, i.e., number of free electrons induced into the inversion channels. This increase of resonance energies is depicted in Fig. 45b for three samples of different lateral quantization energies. Since collective contributions may shift intersubband resonance energies from the subband spacings, the latter can be determined safely only from the infrared experiments at vanishing density n_l via extrapolation to zero gate voltage $\Delta V_g = 0$. These subband spacings at vanishing n_l have energies between 5 and 10 meV. Since the intersubband resonance energies even at $\Delta V_g = 50$ V are less than twice as large, we expect depolarization not to be dominant over the whole range of gate voltages studied.

This is further asserted by comparison to single-particle subband spacings evaluated from magneto-transport data (as seen in Section 2), which allows one to determine the collective contributions to the 1D intersubband resonances. This approach has been employed by a number of authors,[51,206] but we first adhere to the InSb system.[212] In Fig. 46, intersubband energies

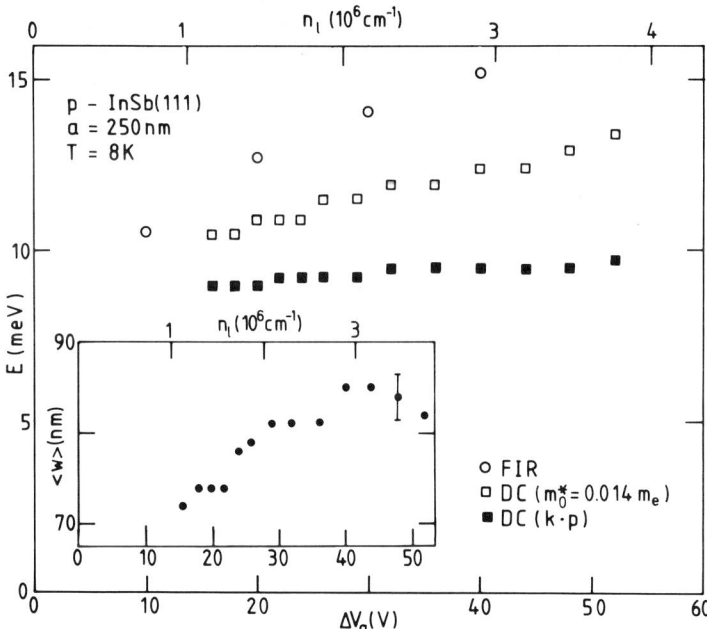

FIG. 46. Subband spacings of 1D inversion channels versus gate voltage ΔV_g or linear density n_l deduced from DC magneto-resistance oscillations using the EMA (open squares) and the $k \cdot p$ approximation (closed squares). For both approximations, the linear densities n_l are almost identical. The inset shows the channel width $\langle W \rangle$. Intersubband (FIR) energies (open circles) exceed the subband spacings due to the depolarization shift. (From Ref. 212.)

(circles) are compared to subband spacings (squares) deduced from Shubnikov–deHaas oscillations of the magneto-resistance assuming the effective mass approximation (EMA) as well as the $k \cdot p$ approximation. In the $k \cdot p$ approximation, the effective electron mass at the Fermi energy no longer is constant but is estimated to be ($B = 0$)

$$m^* \cong m_0^* \left[1 + \frac{2}{E_g} \left(\frac{\hbar^2 k_y^2}{2m_0^*} + \frac{1}{2} \hbar \omega_0 \left(n + \frac{1}{2} \right) + \frac{1}{3} E_i \right) \right] \qquad (29)$$

as a consequence of band non-parabolicity.[151] Since this mass increases as the Fermi energy increases, the subband spacings are less than the corresponding values calculated from the effective mass approximation. As a result of depolarization, the far-infrared energies exceed the subband spacings and increase pronouncedly with gate voltage or electron density. In the limit $n_l \to 0$, however, all three sets of data approach the same subband spacing $\hbar \Omega_0 = 9$ meV, since depolarization effects vanish in the limit of zero electron density and non-parabolicity becomes unimportant for low zero-point energies. For the gate voltage $\Delta V_g = 40$ V, a depolarization shift $\hbar \omega_d = 11.6$ meV is deduced from Fig. 46 if we compare the spectroscopic data to the $k \cdot p$ values. This is somewhat less than the value predicted from Eq. (24).

It is interesting to compare the magnitude thus extracted of collective contributions and single-particle subband spacings for InSb to their counterparts in $Al_xGa_{1-x}As/GaAs$ heterojunctions, which are shown in Fig. 47. As

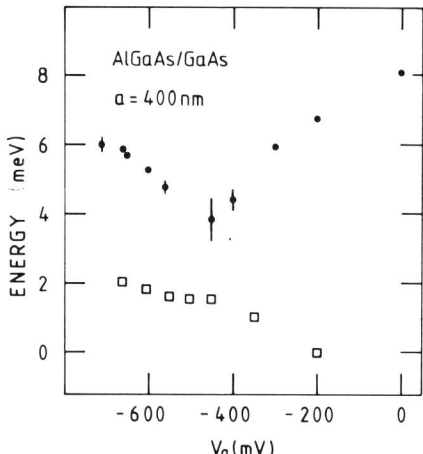

FIG. 47. Resonance energies of the far-infrared excitations (closed circles) at zero magnetic field and spacings of adjacent 1D subbands (open squares) as determined from quantum oscillations of the magneto-resistance. (From Ref. 51.)

a function of the gate voltage V_g, the energy of the far-infrared excitations runs through a minimum at voltage $V_g = V_d \approx -0.5$ V. At zero voltage, the excitation is the plasmon of the homogeneous 2D electron gas. It becomes the plasmon of a density-modulated system (as seen in Section 4) when the gate voltage is decreased. At voltages $V_g \le -0.5$ V, depolarization-shifted intersubband resonances are observed. As the resonance energy in this regime is significantly higher than the subband spacing, it was concluded that the resonance energy is dominated by collective contributions in the GaAs heterojunctions.

Recent theories on periodic multi-wire arrays consider the quantization that leads to 1D subbands and the Coulomb interaction between adjacent stripes that leads to a plasmon character of the excitations.[197,198,213] To account for this two-fold nature, the term *intersubband plasmons* has been introduced for excitations with polarizations perpendicular to the wires.[197] The energy dispersion of the intersubband plasmons is caused by the Coulomb coupling of virtual intersubband excitations in the individual wires, which acts independent of whether electrons can tunnel between wires or not. In this respect, their nature is similar to plasmons in multi-layer superlattices, which have been studied previously by Raman spectroscopy.[207-209] It should be noted here that in the far-infrared experiments, the wave vector transferred to the system is zero or, equivalently, a multiple of the reciprocal lattice vector $2\pi/a$.

Earlier theories on lateral multi-wire superlattices have considered a two-level approximation.[197,198] More recent work[213] has extended this to the general case of an arbitrary number of subbands. However, if the confining potential of a single wire is parabolic, a system of a large number of occupied subbands is approximately equivalent to a two-level scheme.[213] This means that all the virtual intersubband resonances of n occupied subbands interfere constructively to produce a single intersubband plasmon with dominant oscillator strength. The energy of this excitation can have a large shift from the single-particle subband separation. The other excitations, have energies close to the subband separation and couple only very weakly to the far-infrared field.

The Coulomb interaction also should result in higher harmonics of the fundamental collective resonance.[198,213] Such oscillator strengths have been calculated recently.[196] However, no higher collective excitations have been observed so far in quantum-confined systems. Higher harmonics have been observed only in the classical regime of the Si MOS system as discussed before (See Fig. 38.) It thus may suffice to restrict discussion to a two-level scheme when we compare experiments in the quantum-confined regime to theory. Analytical expressions for the resonance frequency ω have been derived by Chaplik[198] as well as Que and Kirczenow[213] for the case that the Coulomb-coupled transitions between many different 1D subbands are equivalent to the fundamental excitation in a two level scheme,

$$\omega^2 = \Omega_0^2 + \frac{e^2 n_l}{\pi \varepsilon_0 \bar{\varepsilon} m^* l_0^2} \cdot J(N). \tag{30}$$

In this equation, we have the linear electron density n_l and the characteristic oscillator length $l_0 = (\hbar/m^*\Omega_0)^{1/2}$ of the lateral potential. The factor,

$$J(N) = \frac{\lambda}{N} \sum_{n=1}^{\infty} [L_{N-1}^1(\lambda n^2)]^2 e^{-\lambda n^2} \cdot n \tag{31}$$

with abbreviation $\lambda = 2\pi^2 l_0^2/a^2$ and associated Laguerre polynomials L_{N-1}^1 describes the coupling of the depolarization fields of the individual wires. In Eq. (30), the electron density is assumed to be a δ function in the z direction. If the finite size of the wave function in the z direction is taken into account, an additional factor appears in the second term of Eq. (30) as given in the literature.[197,213] The Fermi energy is located between levels $N - 1$ and N. In the limit of an isolated stripe ($l_0 \ll a$), the factor $J(N) = 1/2$ is independent of index N and we recover the shift of Eq. (24) up to a numerical factor of order unity. Theories are conflicting as regards the electron density n_l that enters Eq. (30). In the two-level approximation,[198] the difference Δn_l of the two subbands adjacent to the Fermi energy appears, whereas a sum rule derived by Que and Kirczenow[213] results in the density of the lowest 1D subband. Experiments indicate that taking into account the total electron density provides the best description.[151]

In the limit of closely packed channels, i.e., $W \approx a$, the collective contribution becomes smaller, since the superposition of the depolarization fields of all channels becomes more uniform. Comparison of experimental and theoretical collective shifts of the resonance frequencies for GaAs and InSb has been discussed by Que and Kirczenow[213] and by Merkt et al..[212]

In a recent report, Brey et al.[214] have demonstrated that for strictly parabolic confinement, the infrared resonance frequency ω is identical to the characteristic frequency Ω_0 of the bare confining potential $m^*\Omega_0^2 x^2/2$, in close analogy to Kohn's theorem for cyclotron resonance excitations in an interacting electron gas.[215] This attaches a somewhat different interpretation to the intersubband resonance energies. Whereas the subband spacings determined from transport data reflect the self-consistent confining potential, the excitation energies reflect the bare potential, and the difference between the two reflects collective contributions. It appears that independent of the linear electron density within a parabolic bare potential, the many-body contributions to the self-consistent, static confinement potential on the one hand and the depolarization and exciton corrections to the resonance energies on the other hand cancel each other out. Since higher polynomial components of the bare potential will be small when the electrostatic confinement is defined by remote electrodes, as is the case in most of the present devices, the parabolic approximation of the bare potential should describe rather well the experimental situation at least for isolated channels. This also would explain readily the previously described observation that higher harmonics are absent in very narrow, quantum-confined channels. To summarize, the present theories qualitatively account for the experimental observations; a more quantitative evaluation is difficult mainly as a consequence of the unknown value of the effective dielectric constant $\bar{\varepsilon}$ in the complicated sample geometries.

b. *Matrices of Quasi-Zero-Dimensional Dots*

Two-dimensional electron discs first were studied by Allen et al.[15] in periodic arrays of mesa-etched $Al_xGa_{1-x}As/GaAs$ heterojunctions. Somewhat later, confined 2D electron systems restricted to macroscopic discs were examined by Glattli et al.[201] and by Mast et al.[200] on the surface of liquid helium. The excitations in the semiconductor and helium systems resemble each other and are described as classical depolarization modes. In magnetic fields applied perpendicularly to the discs, two sets of modes are identified: bulk modes whose squared frequency increases linearly with the squared cyclotron frequency and edge modes whose frequency decreases inversely with magnetic field strength in the strong-field limit.[216] On semiconductor systems, the latter modes are studied in the far-infrared regime and at radio frequencies. At radio frequencies, single-disc samples are investigated and the collective current modes observed only in high magnetic fields commonly are addressed as edge magneto-plasmons.[217–219] In far-infrared studies, periodic arrays are employed because of intensity reasons.[15,60,220] In these studies, emphasis first was on excitations of isolated dots. Most recently, it became possible to study electrical coupling between dots in a gate voltage-controlled device that allows one to induce a transition from a homogeneous 2D system over a system that is density-modulated in both lateral directions into an array of isolated 0D electron dots.[221] Presently, the far-infrared work focuses on single-particle excitations in few-electron systems and on the intriguing coupling phenomena when the wave functions of adjacent dots start to overlap.

Experimental traces of the high-frequency resistance extracted from transmission experiments and resonance positions of macroscopic electron discs on $Al_xGa_{1-x}As/GaAs$ heterojunctions with radii $a = 1.5$ μm are shown in Fig. 48. Unlike in electron systems on liquid helium, just two modes are observed. In the limit of strong magnetic fields, the position of the higher mode

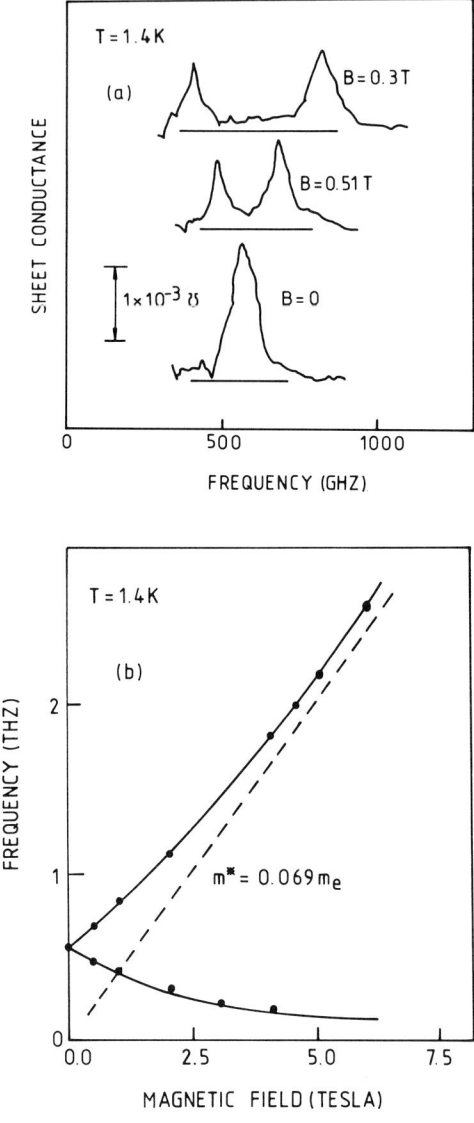

FIG. 48. (a) Sheet conductance of an array of electron discs on $Al_xGa_{1-x}As/GaAs$ heterojunctions. The magnetic field is applied perpendicularly to the surface. (b) Resonance positions versus magnetic field strength. The solid lines are calculated from Eq. (33); the dashed line is the cyclotron frequency calculated with the indicated effective mass. (From Ref. 15.)

approaches the cyclotron frequency, whereas that of the lower-lying one tends to zero. Simple classical theory yields a high-frequency sheet conductivity,[222]

$$\sigma_{\pm}(\omega) = \frac{e n_s \mu}{1 + \left(\frac{\omega_0^2}{\omega} - \omega \pm \omega_c\right)^2 \tau_e^2}, \tag{32}$$

for the two circular light polarizations. In this equation, we have the areal density of electrons n_s in the discs and a mobility $\mu = e\tau_e/m^*$ defined with a phenomenological relaxation time τ_e. For linearly polarized light and in a magnetic field, one obtains from the preceding equation two resonances with frequencies,

$$\omega_{\pm} = \sqrt{\omega_0^2 + \left(\frac{\omega_c}{2}\right)^2} \pm \frac{\omega_c}{2}. \tag{33}$$

In a purely classical picture, the zero-field resonance frequency ω_0 may be interpreted as depolarization frequency ω_d,[223]

$$\omega_d^2 = \frac{3\pi}{16} \frac{e^2 n_s}{\varepsilon_0 \bar{\varepsilon} m^* a}. \tag{34}$$

Actually, this frequency ω_d is the one of the lowest symmetric dipole mode. It differs by a factor of 3/4 from the earlier result of Allen et al.[15]. Equations (33) and (34) account very well for the observed resonance positions and Eq. (32) provides a qualitative description of their intensities.

Similar to the case of wire structures, the transition from classical discs to quantum dots has been examined recently with stacked-gate devices on Si.[220] For a particular set of devices discussed here, the bottom gate (Fig. 8) is a NiCr mesh with periodicity $a = 400$ nm in both directions and circularly shaped openings of diameter $t = 150$ nm. Far-infrared transmission spectra are depicted in Fig. 49. Again, different voltages V_{gt} and V_{gb} can be applied between the Si substrate and the top and bottom gate, respectively. In the presence of band-gap radiation, the electron dots are induced underneath the openings by a sufficiently positive gate voltage V_{gt}. This voltage essentially rules the electron number per dot n_0, whereas the negative voltage V_{gb} of the bottom gate serves to isolate the dots and to vary continuously the depth of the lateral potential from zero to a value even exceeding the band gap of Si. The dot diameter W can be reduced further via substrate bias, again in close analogy to the situation in the corresponding wire structures.

Figure 49a shows the relative change of transmittance at zero magnetic field for various voltages V_{gt}, i.e., electron numbers and for two substrate biases V_{SB}.

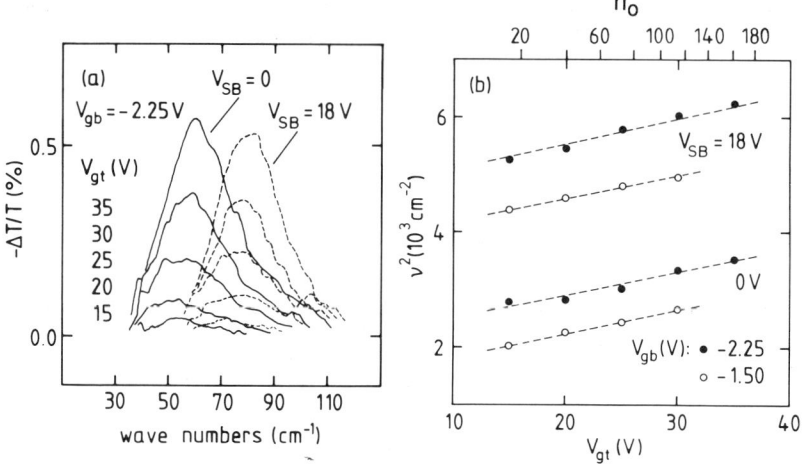

FIG. 49. (a) Dimensional resonances of a periodic array of electron dots on Si. Solid and dashed lines are taken for substrate bias $V_{SB} = 0$ and $V_{SB} = 18$ V, respectively. (b) Squared resonance positions. The dashed lines are guides to the eye. The electron numbers n_0 are valid only for $V_{gb} = -2.25$ V (closed circles). (From Ref. 220.)

For decreasing voltages V_{gt}, i.e., decreasing electron number, the resonances shift to lower frequencies. With increasing substrate bias, the resonances occur at higher values, indicating increased confinement. It is judged readily from the oscillator strengths that the substrate bias does change only marginally the number of confined electrons. Note that at the bottom voltage $V_{gb} = -2.25$ V, free holes are induced at the Si–SiO$_2$ interface under the bottom gate surrounding the electron dots and thus a lateral nipi structure forms.

The squared resonance frequencies are displayed versus gate voltage V_{gt} for two different bottom gate and substrate-bias voltages in Fig. 49b. The upper nonlinear scale is valid only for the data taken at voltage $V_{gb} = -2.25$ V (solid circles) and presents electron numbers n_0 derived from oscillator strengths. Numbers for the voltage $V_{gb} = -1.5$ V typically are a factor of three higher. The dashed lines in this figure are guides to the eye and indicate a linear dependence of the squared resonance energies on the top gate voltage V_{gt}. Extrapolating the squared frequencies to the conductivity threshold, which is reached at $V_{gt} \simeq 10$ V, one obtains finite intercepts. This is in clear contrast to the behavior expected for classical discs in which, according to Eqs. (34) or (24), the depolarization frequency vanishes in this limit. From the intercepts, discrete quantum-state separations $\hbar\Omega_0 = 5$ to 9 meV are extrapolated at vanishing electron numbers. Assuming a harmonic potential to be discussed next, and taking into account the spin and orbital degeneracies, one can estimate that for $n_0 = 20$ electrons per dot, the oscillator state

$N = 3$ is the highest one populated. Hence, electronic dot diameters $W = 2(2\hbar(N + 1)/m^*\Omega_0)^{1/2}$ between 37 nm and 50 nm are realized. This estimate, however, does not include screening, which decreases the quantization energies with increasing electron density, resulting in larger dot sizes. Because of screening, the resonances at high gate voltages in Fig. 49b are expected to be dominated by classical depolarization.[224]

Figure 50 shows spectra for dots with about 140 electrons and strong confinement in magnetic fields applied perpendicularly to the sample surface. The dashed lines have been calculated from the classical high-frequency conductivity in Eq. (32) and a phenomenological scattering time $\tau = 0.2$ ps. Significant deviations between experimental and calculated spectra are found at intermediate magnetic field strengths. (See $B = 7.5$ T.) Only at higher magnetic fields where the dot radius R becomes larger than the cyclotron radius R_c the classical prediction Eq. (32) of equal strengths of the two splitted modes $\omega_\pm$ is restored. From comparison of the observed line-shapes with theoretical predictions for the classical and quantum-confined cases, respectively, it was concluded that quantum confinement is reflected in reduced ratios of oscillator strengths and resonance amplitudes.[212,220]

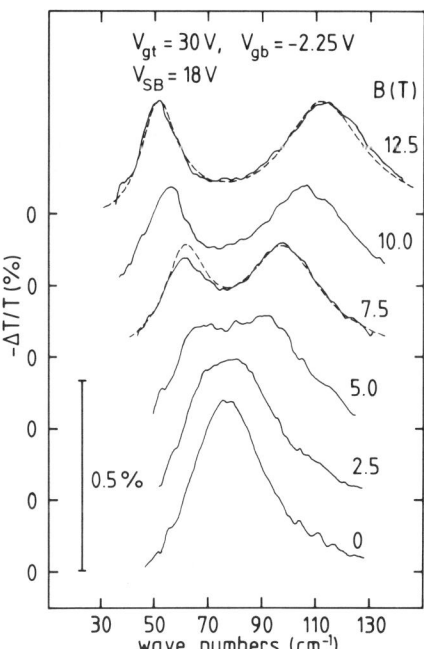

FIG. 50. Transmittance spectra of an array of electron dots on Si in magnetic fields applied perpendicularly to the surface. The dotted lines are calculated from Eq. (32). (From Ref. 220.)

Energy spectra of totally confined systems in a normal magnetic field have been discussed in Section 3 in connection with capacitance measurements on arrays of quantum dots. As for 1D electron systems, a confinement model based on the harmonic oscillator potential $m^*\Omega_0^2 r^2/2$ is most convenient because it provides a simple analytical solution for the single-particle eigenenergies in magnetic fields,[160]

$$E_{n,m} = (2n + |m| + 1)\hbar\sqrt{\Omega_0^2 + \left(\frac{\omega_c}{2}\right)^2} + \frac{\hbar\omega_c}{2}m, \qquad (35)$$

with radial $n = 0, 1,\ldots$ and azimuthal $m = 0, 1,\ldots$ quantum numbers. At low electron numbers, we can assume that only the lowest 2D subband $i = 0$ corresponding to motion in z direction is occupied. The quantum mechanical selection rules[212] allow for just two transitions,

$$\omega_\pm = \sqrt{\Omega_0^2 + \left(\frac{\omega_c}{2}\right)^2} \pm \frac{\omega_c}{2}, \qquad (36)$$

excited by the two circular light polarizations. Note that this result is formally identical to the classical Eq. (33), but the physical origin of the resonance frequency at zero magnetic field is different. Classically, this frequency denotes the depolarization frequency ω_d that tends to zero as the electron number vanishes; quantum mechanically, it is the single-particle frequency Ω_0 of the confining potential $m^*\Omega_0^2 r^2/2$ and is independent of electron number as long as the external potential is kept fixed.

The dipole matrix elements also can be calculated analytically and are given in Table I, in which absorption and emission processes are indicated

TABLE I

SELECTION RULES, DIPOLE MATRIX ELEMENTS, AND TRANSITION FREQUENCIES FOR CIRCULAR LIGHT POLARIZATIONS ± IN THE PARABOLIC WELL WITH THE SQUARED CHARACTERISTIC LENGTH $L^2 = \hbar/m^*(\omega_c^2/4 + \omega_0^2)^{1/2}$. PLUS AND MINUS SIGNS IN FRONT OF THE TRANSITION FREQUENCIES DENOTE ABSORPTION AND EMISSION PROCESSES $(n, m) \to (n', m')$, RESPECTIVELY.

		frequency	
$(n', \|m'\|)$	squared dipole matrix element	$m' \geq 0$	$m' < 0$
$n-1, \|m\|+1$	$n \cdot L^2$	$-\omega_-$	$-\omega_+$
$n, \|m\|+1$	$(n+\|m\|+1) \cdot L^2$	$+\omega_+$	$+\omega_-$
$n+1, \|m\|-1$	$(n+1) \cdot L^2$	$+\omega_-$	$+\omega_+$
$n, \|m\|-1$	$(n+\|m\|) \cdot L^2$	$-\omega_+$	$-\omega_-$

by plus and minus signs in front of the corresponding transition frequencies, respectively. The relation $\Sigma f_-/\Sigma f_+ = \omega_-/\omega_+$ for the sums of oscillator strengths for all possible transitions at any Fermi energy is verified readily with these matrix elements. This ratio of oscillator strengths is identical to the classical one, which is obtained by integrating the conductivities $\sigma_\pm$ of Eq. (32) in the limit $\omega_+\tau \gg 1$.[222]

The energies of the lowest levels and allowed dipole transitions are depicted in Fig. 51. For $B = 0$, we have equally spaced oscillator levels $(2n + |m| + 1)\hbar\omega_0$. In high magnetic fields ($\omega_c \gg \Omega_0$), all levels with quantum numbers $n = 0$, $m \leq 0$, merge and form a highly degenerate ground state of energy $\hbar\omega_c/2$. Levels $n = 1$, $m \leq 0$, merge to a common excited state of energy $3\hbar\omega_c/2$. Levels with angular momentum $\hbar m > 0$ have much higher energies and do not contribute to the signals, since they no longer are occupied. In this high magnetic field limit, transitions become cyclotron resonances between Landau levels and the electron gas exhibits 2D behavior. Simultaneously, the oscillator strength of the ω_- transitions vanishes. These ω_- resonances indeed are characteristic of a confined system with a radius less than or comparable to the magnetic length l_B. However, since the classical edge

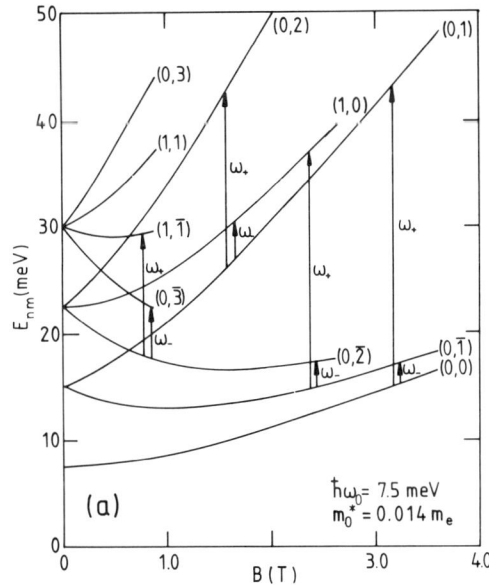

FIG.51. Eigenenergies of the harmonic oscillator potential $m^*\Omega_0^2 r^2/2$ versus the strength of a magnetic field $B \parallel z$. (n, m) indicate radial and azimuthal quantum numbers, respectively, with $\bar{m} = -m$. The quantization energy $\hbar\omega_0$ and the effective mass m_0^* apply to typical values in InSb devices. (From Ref. 60.)

magneto-plasmon has the same dispersion in a magnetic field, the observation of the ω_- resonance indicates lateral quantization only if the electron number is low enough to yield a negligible depolarization frequency at $B = 0$.[214]

Dots with very few electrons have been realized on InSb samples in which one can tune the number of electrons in the range $n_0 = 1$ to about 20 by the gate voltage.[226] The electron number is determined from spectra obtained in strong magnetic fields when the magnetic length is much less than the characteristic oscillator length of the confining lateral potential ($l_B \ll l_0$). In this limit, the signal is well-described by classical cyclotron resonance absorption, as is exemplified in Fig. 52. For the quantitative evaluation of the electron number, an extended form of Eq. (20) was employed, including an effective sheet conductivity of the gate configuration.[210] In this analysis, an effective sheet conductivity $\sigma_s = en_0\mu/a^2$ of the dot electrons is introduced, which contains an average areal electron density $\bar{n}_s = n_0/a^2$. This is justified as long as the wavelength of the incident radiation is much longer than the grating constant ($\lambda \gg a$).

Even the limit of a single electron per dot already could be realized in the InSb device.[226] To be more precise, it can be concluded from the change of transmittance that on the average, one electron per dot contributes to the signal at a gate voltage of about $\Delta V_g = 1$ V above inversion threshold. However, because of the geometrical homogeneity of the dot array (as seen in the micrograph in Fig. 7), the occupation of individual dots is thought to be quite homogeneous. The observation of one electron per dot at voltage $\Delta V_g = 1$ V is fully consistent with the analysis of the corresponding 1D channels depicted in Fig. 46, where a linear density $n_l \simeq 1 \cdot 10^6$ cm^{-1} was deduced for gate voltage $\Delta V_g = 10$ V. An approximately linear relationship between electron number and gate voltage is appropriate and yields at $\Delta V_g = 1$ V one electron per 100 nm, a length that corresponds to a typical dot diameter.

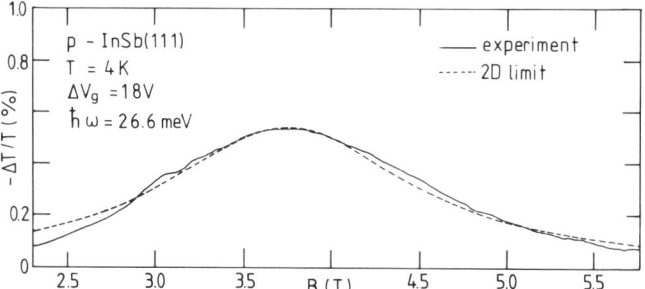

FIG. 52. Laser spectrum (solid line) for a laser energy $\hbar\omega$ at which the resonance occurs in sufficiently high magnetic fields such that the magnetic length is much less than the dot diameter. The dashed line is calculated from Eq. (32) and provides a good description with an areal electron density $\bar{n}_s = n_0/a^2$ with $n_0 = 20$ and $a = 250$ nm. (From Ref. 226.)

The resonance energies of ω_+ (upper branch) and ω_- (lower branch) resonances versus magnetic field strength are depicted in Fig. 53. The solid lines are calculated for a harmonic oscillator well ($\hbar\omega_0 = 7.5$ meV) using the effective mass approximation (EMA) result of Eq. (36). The dotted lines take into account corrections due to band non-parabolicity in a simplified $k \cdot p$ scheme,[227]

$$\Delta E^{k \cdot p} = -\frac{1}{E_g} \langle inm | (E_{inm}^{EMA} - U)^2 | inm \rangle. \tag{37}$$

At lower energies, the corrections due to band non-parabolicity are minor; hence, the dotted and solid lines of the lower branch almost coincide. At the higher energies of the upper branch, the apparent electron mass increases and the corresponding energies $\hbar\omega_+$ bend downwards. In magnetic fields $B \geq 4$ T, when the cyclotron radius is much less than the dot diameter, the resonance no longer is discernible from cyclotron resonance of a homogeneous 2D system for which such non-parabolic effects are well-understood.[228-230]

The inset of Fig. 53 shows the resonance energy $\hbar\omega_0$ as a function of gate voltage or electron number in the absence of a magnetic field. This energy is extrapolated from its Zeeman splitting in finite magnetic fields and does not depend on electron number in the range $n_0 = 1$ to 20 within experimental error. The independence of the excitation energy on electron number is clearly distinct from the situation in real atoms if one compares, e.g., hydrogen and helium. In the quasi-atomic quantum dots, however, the electrons are not

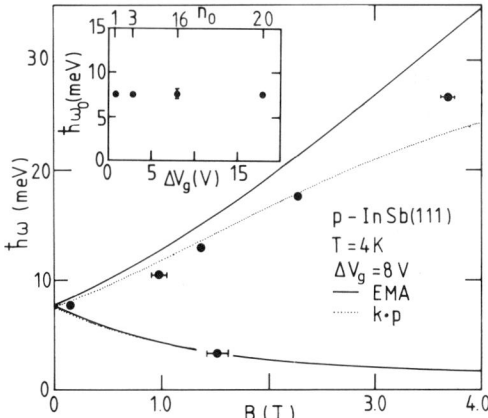

FIG. 53. Zeeman splitting of the resonance position of quantum dots. The inset gives the independence of the quantization energy on gate voltage and electron number n_0. The solid lines are calculated from the EMA result in Eq. (36), the dotted lines from a simple estimate within $k \cdot p$ approximation. (From Ref. 226.)

bound in a Coulomb potential, but in a more or less parabolic well. This fact, together with the low effective electron mass and the large dielectric constant of the semiconductor, strongly modifies the electron–electron interaction. Our observation is explained most readily assuming a harmonic confining potential whose curvature does not change significantly with the gate voltage. In the situation of just one electron per dot, the characteristic energy $\hbar\Omega_0$ is observed *a priori*. Recently, as mentioned at the end of the previous section, it has been shown generally that in a single dot with parabolic confinement, infrared transition energies measure the bare potential independently of the electron–electron interaction and the number of electrons involved.[214] In the case of more electrons, we can rely on the theoretical result of Shikin *et al.*[199] and Brey *et al.*[214] that the frequency of the dipole mode is the frequency of the center of mass motion of the electron plasma that coincides with the one of the individual motion of one electron in the bare potential. Using this explanation, we take for granted that there is neither coupling between the excitations of individual dots via Coulomb interaction[224] nor a lateral modulation of the incident far-infrared field.

Relating the infrared frequencies to the bare potential via the solution of Brey *et al.*[214] does not answer the question for the quantum level spacings determined by the self-consistent Hartree potential. Theoretically, two electrons in a quantum dot have been treated by Bryant[163] for rectangular and square shapes and infinite barrier heights and by Chaplik[231] for harmonic wells. In particular, it has been pointed out by Bryant that the single-particle spacings scale as $1/L^2$ with the dot dimension L, whereas the Coulomb interaction matrix elements scale as $1/L$. Hence, for small dots, the electrons exhibit independent particle behavior. This is not true in larger boxes, where strong exchange splittings and eventually complete reordering of levels occur. For a harmonic well, the Coulomb and the exchange integral can be calculated analytically.[231] In this perturbational approach, which is valid as long as the effective Rydberg energy R^* is small compared to the oscillator energy $\hbar\Omega_0$, the ground state and the first excited singlet ($S = 0$) state are shifted by the same energy $(\pi R^* \hbar\Omega_0)^{1/2}$. This means that the same excitation energy is observed in spectroscopic experiments ($\Delta S = 0$) for one and two electrons, namely the bare oscillator energy $\hbar\Omega_0$. This result remains valid for any form of the interaction between two electrons in a parabolic well, as, for instance, a Coulomb interaction modified by screening in adjacent metals. For higher electron numbers n_0, the perturbational approach predicts shifts of the excitation energy proportional to $n_0^{1/8}$. Again, as for electrons in quantum wires, the increase of collective contributions to the resonance energy goes along with a corresponding decrease of the single-particle level spacings so that the general result of Brey *et al.*[214] remains valid.

First experiments on far-infrared spectroscopy of coupled quantum dots have been carried out using $Al_xGa_{1-x}As/GaAs$ heterojunctions in which the coupling of dots can be tuned with a gate voltage.[221] In the regime of strong coupling, the electron layer forms a mesh and thus is similar to the system that was envisaged by Hofstadter some time ago.[101] Figure 54a shows Fourier-transform spectra at a fixed magnetic field in the regime of gate voltages in which the electron dots become electrically connected. Capacitance measurements show that at voltage $V_g = -2.7$ V, the dots still are electrically connected, although there already are voids in the electron gas so that an electron mesh has formed. At voltage $V_g = -3.1$ V, isolated dots are created. In the regime of coupled dots, the ω_- mode at wave number $\bar{v} \approx 20$ cm^{-1} splits into two. The bulk ω_+ mode centered at about 60 cm^{-1} increases rapidly in its strength as the voltage is increased and an additional resonance becomes apparent as a shoulder on its high-energy side. Figure 54b presents resonance

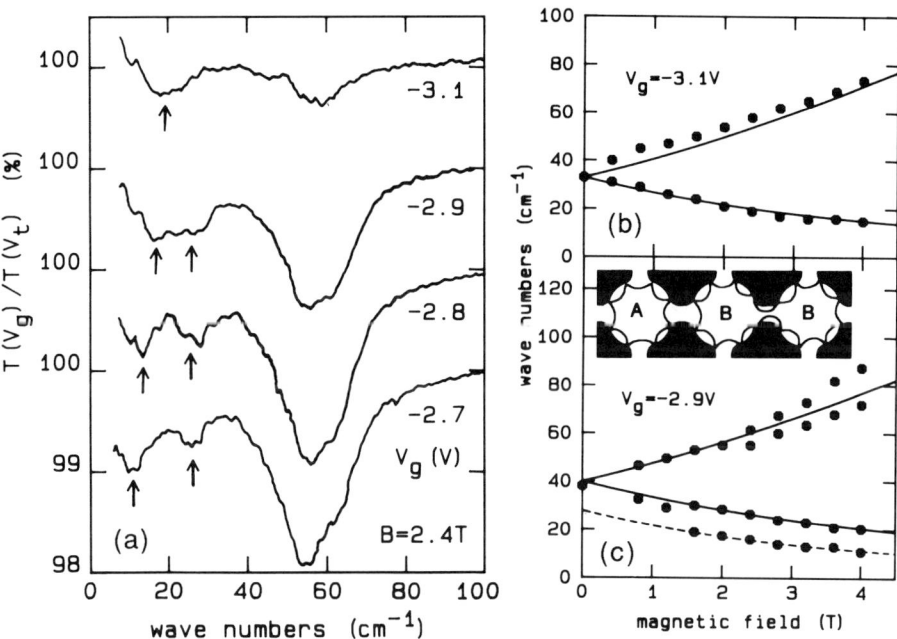

FIG. 54. Transmission spectra and resonance positions for quantum dots on an $Al_xGa_{1-x}As/GaAs$ heterojunction. (a) Spectra recorded at different gate voltages V_g. Above $V_g = -3.1$ V, the system is in the coupled-dot state. Two edge modes appear (arrows) as the bias is increased. The shoulder on the high-energy side of the bulk mode is a magneto-plasmon. (b) Measured resonance positions for isolated and (c) coupled quantum dots. The solid lines are calculated from Eq. (36). The inset illustrates trajectories responsible for the upper- (A) and lower- (B) frequency edge modes. The dashed line indicates the dispersion expected for a type B edge mode. (From Ref. 221.)

positions as a function of the magnetic field strength in the isolated regime and Fig. 54c in the coupled regime. In the regime of coupling, four modes are observed in stronger magnetic fields ($B \geq 2.5$ T). Tuning the electron system from a homogeneous 2D electron gas at zero gate voltage over an electron mesh to an isolated-dot array allows one to identify the various modes.

The highest-lying mode is a magneto-plasmon resonance; the adjacent one is the confined cyclotron resonance. The two lower modes both are edge modes that can be visualized in a classical trajectory picture given in the inset of Fig. 54c. In this picture, the charges move along boundaries and the mode frequency is given by Eq. (34) with the squared depolarization frequency ω_d^2 being inversely proportional to the dot perimeter. The lowest-lying resonance is well-described as the mode in which the charge density wave moves along the boundary of two adjacent dots.

V. Conclusions and Perspectives

In this chapter, we have tried to give an up-to-date summary of electronic properties of laterally periodic nanostructures based on semiconductor devices with a 2D electron system. In an extremely rapidly developing and comparatively young field such as nanostructure device physics, a summary as the one presented here can be only a snapshot, parts of which may be outdated before this manuscript appears in print. Nevertheless, we hope that this chapter can serve as an introduction into this exciting branch of physics for some time to come.

We have focused on the intraband electronic properties of laterally periodic nanostructures. We have tried to show that increased laterally periodic modulation of an originally 2D electron system results in novel transport phenomena specific for lateral superlattices as well as 1D and 0D systems. These reveal the modified electronic density of states as well as the ballistic motion of carriers in such device structures. Also, we have demonstrated that the spectrum of intraband electronic excitations becomes altered strongly by the periodic modulation of the electron system in the lateral dimensions. With increasing lateral confinement, it changes from a spectrum of nearly classical plasma oscillations to a spectrum of excitations between 1D subbands in quantum wires or atomic-like bound states in quantum dots.

We have expanded our discussion on those aspects that we believe to be sufficiently well understood as well as experimentally verified. Thus, some recent work that may prove to be extremely important in the future may be mentioned only briefly or even omitted just because we subjectively felt the results to be not sufficiently well-established yet. This also is the main reason why we have discussed properties of 1D and 0D electron systems in more detail than properties of true lateral superlattices. Only very recently has it become possible to realize artificial lateral superlattices based on 2D electron systems, in which the elastic electronic mean free path well exceeds the superlattice period. We are convinced that the study of electronic phenomena in such lateral superlattices is one essential branch of current and future research in the physics of laterally periodic nanostructures. Thus, experimental realization of the Bloch oscillator[68] and the Hofstadter butterfly[101] are goals that are essential driving forces of present research efforts.

Another challenge that we expect to be a focus point of future research is the electronic transport in strictly one-dimensional quantum wires, in which only the energetically lowest 1D subband is occupied. Here, it remains to be shown whether electron mobilities in true quantum wires[121] exceed those presently achieved in 2D electron systems. Also, the device aspects of such single-mode electron waveguides appear as a fascinating playground for future investigations.

In the *atomic* physics of artificial quantum dots, we see a third essential branch of future studies.[233] Here, one can hope that with control of the number of electrons contained per quantum dot, characteristics of artificial *elements* in a *periodic table* will become visible. One may even speculate that specific signatures of such artificial elements may be employed in data storage and manipulation on the basis of arrays of interacting quantum dots.[171,232,234] All of the preceding provides strong motivation for future research in the field of nanostructure device physics. In such studies, periodic nanostructures as discussed here will continue to play an important part. The relative ease with which the smallest electronic feature sizes well below 100 nm can be realized in periodic nanostructures, as well as the specific averaging built into measurements on such periodic devices, makes them well-suited model systems for nanoelectronic research.

Acknowledgments

We wish to thank all colleagues that have supported us in writing this review through correspondence and discussions, by providing preprints, and by generously granting permission to reproduce their results. For many stimulating discussions and helpful contributions, we particularly want to thank J. Alsmeier, E. Batke, F. Brinkop, A. Chaplik, C. Dahl, D. Heitmann, A. Lorke, T. P. Smith, F. Stern, D. A. Wharam, and R. W. Winkler. We gratefully acknowledge continuous support by the Volkswagenstiftung, the Deustsche Forschungsgemeinschaft, and the Esprit Basic Research Actions.

References

1. L. Esaki and R. Tsu, *IBM J. Res. and Dev.* **14**, 61 (1970).
2. R. Tsu and L. Esaki, *Appl. Phys. Lett.* **19**, 246 (1971).
3. R. Tsu and L. Esaki, *Appl. Phys. Lett.* **22**, 562 (1973).
4. L. Esaki, in *Physics and Applications of Quantum Wells and Superlattices* (E.E. Mendez and K. von Klitzing, eds.), Plenum Press, New York, p. 1 (1987).
5. H. Sakaki, K. Wagatsuma, J. Hamasaki, and S. Saito, *Thin Solid Films* **36**, 497 (1976).
6. R. T. Bate, *Bull. Am. Phys. Soc.* **22**, 407 (1977).
7. P. J. Stiles, *Surf. Sci.* **73**, 252 (1978).
8. J. M. Worlock, ed., "Proceedings of the 7th International Conference on Electronic Properties of Two-Dimensional Systems," *Surf. Sci.* **196** (1988).
9. J. Y. Marzin, Y. Guldner, and J. C. Maan, eds., "Proceedings of the 8th International Conference on Electronic Properties of Two-Dimensional Systems," *Surf. Sci.* **229** (1990).
10. T. Ando, A. B. Fowler, and F. Stern, *Rev. Mod. Phys.* **54**, 437 (1982).
11. R. Dingle, H. L. Störmer, A. C. Gossard, and W. Wiegmann, *Appl. Phys. Lett.* **33**, 665 (1978).
12. A. Sasaki, *Phys. Rev.* **B30**, 7016 (1984).
13. F. Capasso, in *Semiconductors and Semimetals*, (R. Dingle Vol. 24, ed.), Academic Press, p. 319 (1987).
14. G. H. Döhler, in *Festkörperprobleme (Advances in Solid State Physics)* Vol. 23, (P. Grosse, ed.), Vieweg, Braunschweig, Federal Republic of Germany, p. 207 (1983).
15. S. J. Allen, Jr., H. L. Störmer, and J. C. M. Hwang, *Phys. Rev.* **B28**, 4875 (1983).
16. S. J. Allen, F. DeRosa, G. J. Dolan, and C. W. Tu, in "Proceedings of the 17th International Conference on the Physics of Semiconductors (San Francisco)," (J. D. Chadi and W. A. Harrison, eds.), Springer, New York, p. 313 (1985).
17. W. Hansen, J. P. Kotthaus, A. Chaplik, and K. Ploog, in *High Magnetic Fields in Semiconductor Physics* (G. Landwehr, ed.), *Springer Series in Solid-State Sciences*, Vol. 71, Springer, Heidelberg, Federal Republic of Germany, p. 266 (1987).
18. K. Kash, J. M. Worlock, M. D. Sturge, P. Grabbe, J. P. Harbison, A. Scherer, and P. S. D. Lin, *Appl. Phys. Lett.* **53**, 782 (1988).
19. P. M. Petroff, A. C. Gossard, and W. Wiegmann, *Appl. Phys. Lett.* **45**, 620 (1984).
20. M. Tanaka and H. Sakaki, *Jpn. J. Appl. Phys.* **27**, L2025 (1988).
21. M. Tsuchiya, J. M. Gaines, R. H. Yan, R. J. Simes, P. O. Holtz, L. A. Coldren, and P. M. Petroff, *Phys. Rev. Lett.* **62**, 466 (1989).
22. M. Tanaka and H. Sakaki, *Appl. Phys. Lett.* **54**, 1326 (1989).
23. U. Mackens, D. Heitmann, L. Prager, J. P. Kotthaus, and W. Beinvogl, *Phys. Rev. Lett.* **53**, 1485 (1984).
24. J. P. Kotthaus and D. Heitmann, *Surf. Sci.* **113**, 481 (1982).
25. H. I. Smith, *Superlattices and Microstructures* **2**, 129 (1986).
26. T. Cole, A. A. Lakhani, and P. J. Stiles, *Phys. Rev. Lett.* **38**, 722 (1977).
27. L. J. Sham, S. J. Allen, Jr., A. Kamgar, D. C. Tsui, *Phys. Rev. Lett.* **40**, 472 (1978).
28. V. A. Volkov and V. B. Sandormirskii, *Pis'ma Zh. Eksp. Teor. Fiz* **27**, 688 (1978) [*JETP Lett.* **27**, 651 (1978)].
29. V. A. Volkov, V. A. Petrov, and V. B. Sandomirskii, *Usp. Fiz. Nauk* **131**, 423 (1980) [*Sov. Phys. Usp.* **23**, 375 (1980)].
30. D. C. Tsui, M. D. Sturge, A. Kamgar, and S. J. Allen, Jr., *Phys. Rev. Lett.* **40**, 1667 (1978).
31. A. Kamgar, M. D. Sturge, and D. C. Tsui, *Phys. Rev.* **B22**, 841 (1980).
32. T. G. Matheson and R. J. Higgins, *Phys. Rev.* **B25**, 2633 (1982).
33. T. Evelbauer, A. Wixforth, and J. P. Kotthaus, *Z. Phys.* **B64**, 69 (1986).

34. W. Sesselmann and J. P. Kotthaus, *Solid State Commun.* **31**, 193 (1979).
35. L. V. Keldysh, *Fiz. Tverd. Tela (Leningrad)* **4**, 2265 (1962) [*Sov. Phys. Solid State* **4**, 1658 (1962)].
36. "Microcircuit Engineering 88, Proceedings of the International Conference on Microlithography," (F. Paschke, W. Fallmann, and H. Löschner, eds.), North-Holland, Amsterdam (1989).
37. "Proceedings of the 33rd International Symposium on Electron, Ion and Photon Beams," (Monterey 1989), *J. Vac. Sci. Technol.* **B7**, 1373 (1989).
38. M. A. Reed and W. P. Kirk, eds. *Nanostructures Physics and Fabrication*, Academic Press, Boston (1989).
39. S. P. Beaumont and C. M. Sotomayor-Torres, eds., *Science and Engineering of One- and Zero-Dimensional Semiconductors*, Plenum Press, New York (1990).
40. J. Cibert, P. M. Petroff, G. J. Dolan, S. J. Pearton, A. C. Gossard, and J. H. English, *Appl. Phys. Lett.* **49**, 1275 (1986).
41. E. Kapon, D. M. Hwang, and R. Bhat, *Phys. Rev. Lett.* **63**, 430 (1989).
42. M. R. Frei and D. C. Tsui, *Appl. Phys. Lett.* **55**, 2432 (1989).
43. T. N. Theis, *Surf. Sci.* **98**, 515 (1980).
44. D. Heitmann, in *Festkörperprobleme (Advances in Solid State Physics)*, Vol. 25 (P. Grosse, ed.), Vieweg, Braunschweig, Federal Republic of Germany, p. 429 (1985).
45. E. Batke, D. Heitmann, and C. W. Tu, *Phys. Rev.* **B34**, 6951 (1986).
46. E. Batke, D. Heitmann, and E. G. Mohr, Physica **117B** and **118B**, 643 (1983).
47. A. C. Warren, D. A. Antoniadis, H. I. Smith, and J. Melngailis, *IEEE Electron Device Letters* **6**, 294 (1985).
48. T. J. Thornton, M. Pepper, H. Ahmed, D. Andrews, and G. J. Davies, *Phys. Rev. Lett.* **56**, 1198 (1986).
49. G. Bernstein and D. K. Ferry, *Superlattices and Microstructures* **2**, 373 (1986).
50. K. Ismail, W. Chu, D. A. Antoniadis, and H. I. Smith, *J. Vac. Sci. Technol.* **B6**, 1824 (1988).
51. F. Brinkop, W. Hansen, J. P. Kotthaus, and K. Ploog, *Phys. Rev.* **B37**, 6547 (1988).
52. R. W. Winkler, J. P. Kotthaus, and K. Ploog, *Phys. Rev. Lett.* **62**, 1177 (1989).
53. W. Hansen, M. Horst, J. P. Kotthaus, U. Merkt, Ch. Sikorski, and K. Ploog, *Phys. Rev. Lett.* **58**, 2586 (1987).
54. T. P. Smith, III, H. Arnot, J. M. Hong, C. M. Knoedler, S. E. Laux, and H. Schmid, *Phys. Rev. Lett.* **59**, 2802 (1987).
55. K. Y. Lee, T. P. Smith, III, H. Arnot, C. M. Knoedler, J. M. Hong, D. P. Kern, and S. E. Laux, *J. Vac. Sci. Technol.* **B6**, 1856 (1988).
56. H. van Houten, B. J. van Wees, M. G. J. Heijman, and J. P. Andre, *Appl. Phys. Lett.* **49**, 1781 (1986).
57. A. Scherer, M. L. Roukes, H. G. Craighead, R. M. Ruthen, E. D. Beebe, and J. P. Harbison, *Appl. Phys. Lett.* **51**, 2133 (1987).
58. K. K. Choi, D. C. Tsui, and K. Alavi, *Appl. Phys. Lett.* **50**, 110 (1987).
59. T. Demel, D. Heitmann, P. Grambow, and K. Ploog, *Appl. Phys. Lett.* **53**, 2176 (1988).
60. Ch. Sikorski and U. Merkt, *Phys. Rev. Lett.* **62**, 2164 (1989).
61. P. M. Petroff, A. C. Gossard, R. A. Logan, and W. Wiegmann, *Appl. Phys. Lett.* **41**, 635 (1982).
62. H. Temkin, L. R. Harriott, R. A. Hamm, J. Weiner, and M. B. Panish, *Appl. Phys. Lett.* **54**, 1463 (1989).
63. Y. Hirayama, Y. Suzuki, S. Tarucha, and H. Okamoto, *Jpn. J. Appl. Phys.* **24**, L516 (1985).
64. J. Cibert, P. M. Petroff, G. J. Dolan, S. J. Pearton, A. C. Gossard, and J. H. English, *Appl. Phys. Lett.* **49**, 1275 (1986).

65. Y. Hirayama and H. Okamoto, in *Physics and Technology of Submicron Structures*, (H. Heinrich, G. Bauer, and F. Kuchar, eds.), *Springer Series in Solid-State Sciences*, Vol 83, Springer, Berlin, p. 45 (1988).
66. T. Hiramoto, K. Hirakawa, Y. Iye, and T. Ikoma, *Appl. Phys. Lett.* **54**, 2103 (1989).
67. A. D. Wieck and K. Ploog, *Surf. Sci.* **229**, 252 (1990).
68. J. B. Krieger and G. J. Iafrate, *Phys. Rev.* **B33**, 5494 (1986) and references therein.
69. H. Krömer, *Phys. Rev.* **109**, 1856 (1958).
70. L. Esaki, L. L. Chang, W. E. Howard, and V. L. Rideout, in "Proceedings of the 11th International Conference on the Physics of Semiconductors" (ed. by the Polish Academy of Sciences), PWN–Polish Scientific Publishers, Warsaw, Poland, p. 431 (1972); L. Esaki and L. L. Chang, *Phys. Rev. Lett.* **33**, 495 (1974).
71. E. E. Mendez and F. Agulló-Rueda, *J. Lumin.* **44**, 223 (1989).
72. P. A. Lebwohl and R. Tsu, *J. Appl. Phys.* **41**, 2664 (1970).
73. R. F. Kazarinov and R. A. Suris, *Fiz. Tekh. Poluprovodn.* **6**, 148 (1972) [*Sov. Phys. Semicond.* **6**, 120 (1972)].
74. L. Esaki and L. L. Chang, *Crit. Rev. Solid State Sci.* **6**, 195 (1976).
75. H. Kroemer, *Phys. Rev.* **B15**, 880 (1977).
76. R. K. Reich, R. O. Grondin, and D. K. Ferry, *Phys. Rev.* **B27**, 3483 (1983).
77. A. B. Pippard, *Proc. R. Soc. London*, Series **A270**, 1 (1962).
78. M. J. Kelly, *J. Phys.* **C17**, L781 (1984).
79. Y. Tokura and K. Tsubaki, *Appl. Phys. Lett.* **51**, 1807 (1987).
80. K. Ismail, W. Chu, D. A. Antoniadis, and H. I. Smith, *Appl. Phys. Lett.* **52**, 1071 (1988).
81. K. Ismail, W. Chu, A. Yen, D. A. Antoniadis, and H. I. Smith, *Appl. Phys. Lett.* **54**, 460 (1989).
82. M. J. Kelly, *J. Phys.* **C18**, 6341 (1985).
83. U. Wulf, *Phys. Rev.* **B35**, 9754 (1987).
84. G. Bernstein and D. K. Ferry, *Z. Phys.* **B67**, 449 (1987).
85. D. Weiss, K. von Klitzing, K. Ploog, and G. Weimann, in *High Magnetic Fields in Semiconductor Physics II* (G. Landwehr, ed.), *Springer Series in Solid-State Sciences*, Vol. 87, Springer, Berlin, p. 357 (1989).
86. R. R. Gerhardts, D. Weiss, and K. von Klitzing, *Phys. Rev. Lett.* **62**, 1173 (1989).
87. C. W. J. Beenakker, *Phys. Rev. Lett.* **62**, 2020 (1989).
88. M. Büttiker, in *Nanostructures Physics and Fabrication* (M. A. Reed and W. P. Kirk, eds.), Academic Press, Boston, p. 319 (1989).
89. E. F. Schubert, J. Knecht, and K. Ploog, *J. Phys.* **C18**, L215 (1985).
90. T. N. Theis and S. L. Wright, *Appl. Phys. Lett.* **48**, 1374 (1986).
91. K. Tsubaki, H. Sakaki, J. Yoshino, and Y. Sekiguchi, *Appl. Phys. Lett.* **45**, 663 (1984).
92. A. V. Chaplik, *Solid State Commun.* **53**, 539 (1985).
93. C. Zhang and R. R. Gerhardts, *Phys. Rev.* **B41**, 12850 (1990).
94. R. R. Gerhardts and C. Zhang, *Surf. Sci.* **229**, 92 (1990).
95. D. Weiss, C. Zhang, R. R. Gerhardts, K. von Klitzing, and G. Weimann, *Phys. Rev.* **B39**, 13020 (1989).
96. M. L. Roukes and A. Scherer, *Bull. Am. Phys. Soc.* **34**, 633 (1989).
97. E. S. Alves, P. H. Beton, M. Henini, L. Eaves, P. C. Main, O. H. Hughes, G. A. Toombs, S. P. Beaumont, and C. D. W. Wilkinson, *J. Phys.: Condens. Matter* **1**, 8257 (1989).
98. D. Weiss, K. von Klitzing, K. Ploog, and G. Weimann, *Surf. Sci.* **229**, 88 (1990).
99. D. Langbein, *Phys. Rev.* **180**, 633 (1969).
100. A. Rauh, *Phys. Status Solidi.* **B69**, K9 (1975).
101. D. R. Hofstadter, *Phys. Rev.* **B14**, 2239 (1976).
102. G. H. Wannier, *Phys. Status Solidi.* **B88**, 757 (1978).
103. D. J. Thouless, M. Kohmoto, M. P. Nightingale, and M. den Nijs, *Phys. Rev. Lett.* **49**, 405 (1982).

104. A. H. MacDonald, *Phys. Rev.* **B28**, 6713 (1983).
105. J. B. Sokoloff, *Phys. Rep.* **126**, 189 (1985).
106. R. Rammal, in *The Physics and Fabrication of Microstructures and Microdevices* (M. J. Kelly and C. Weisbuch, eds.), *Springer Proceedings in Physics*, Vol. 13, Springer, Berlin, p. 303 (1986).
107. B. Pannetier, J. Chaussy, R. Rammal, and J. C. Villegier, *Phys. Rev. Lett.* **53**, 1845 (1984).
108. S. Tanaka and K. Uchinokura, eds., "Physics and Chemistry of Quasi-One-Dimensional Conductors," *Physica* **B134** (1986).
109. G. Grüner, *Rev. Mod. Phys.* **60**, 1129 (1988).
110. S. Kagoshima, H. Nagasawa, and T. Sambongi, eds., *One-Dimensional Conductors*, Springer Series in Solid-State Sciences, Vol. 72, Springer, Berlin (1988).
111. P. A. Lee and T. V. Ramakrishnan, *Rev. Mod. Phys.* **57**, 287 (1985).
112. W. J. Skocpol, L. D. Jackel, E. L. Hu, R. E. Howard and L. A. Fetter, *Phys. Rev. Lett.* **49**, 951 (1982).
113. K. K. Choi, D. C. Tsui, and S. C. Palmateer, *Phys. Rev.* **B33**, 8216 (1986).
114. H. Z. Zheng, H. P. Wei, D. C. Tsui, and G. Weimann, *Phys. Rev.* **B34**, 5635 (1986).
115. H. van Houten, C. W. J. Beenakker, M. E. I. Broekaart, M. G. H. J. Heijman, B. J. van Wees, H. E. Mooij, and J.-P. Andre, *Acta Electronica* **28**, 27 (1988).
116. V. Chandrasekhar, D. E. Prober, and P. Santhanam, *Phys. Rev. Lett.* **61**, 2253 (1989).
117. S. Washburn and R. Webb, *Adv. Phys.* **35**, 375 (1986).
118. A. D. Stone, in *Physics and Technology of Submicron Structures* (H. Heinrich, G. Bauer, and F. Kuchar, eds.), *Springer Series in Solid-State Sciences*, Vol. 83, Springer, Berlin, p. 108 (1988).
119. C. L. Kane, P. A. Lee, and D. P. DiVincenzo, *Phys. Rev.* **B38**, 2995 (1988).
120. G. Timp, H. U. Baranger, P. deVegvar, J. E. Cunningham, R. E. Howard, R. Behringer, and P. M. Mankiewich, *Phys. Rev. Lett.* **60**, 2081 (1988).
121. H. Sakaki, *Jpn. J. Appl. Phys.* **19**, L735 (1980).
122. J. Lee and M. O. Vassell, *J. Phys.* **C17**, 2525 (1984).
123. D. G. Cantrell and P. N. Butcher, *J. Phys.* **C18**, 5111 (1985).
124. G. Fishman, *Phys. Rev.* **B34**, 2394 (1986).
125. M. J. Kearney and P. N. Butcher, *J. Phys.* **C20**, 47 (1987).
126. S. Das Sarma and X. C. Xie, *Phys. Rev.* **B35**, 9875 (1987).
127. A. Kumar, S. E. Laux, and F. Stern, *Appl. Phys. Lett.* **54**, 1270 (1989).
128. B. J. van Wees, H. van Houten, C. W. J. Beenakker, J. G. Williamson, L. P. Kouwenhoven, D. van der Marel, and C. T. Foxon, *Phys. Rev. Lett.* **60**, 848 (1988).
129. D. A. Wharam, T. J. Thornton, R. Newbury, M. Pepper, H. Ahmed, J. E. F. Frost, D. G. Hasko, D. C. Peacock, D. A. Ritchie, and G. A. C. Jones, *J. Phys.* **C21**, L209 (1988).
130. A. C. Warren, D. A. Antoniadis, and H. I. Smith, *Phys. Rev. Lett.* **56**, 1858 (1986).
131. A. C. Warren, D. A. Antoniadis, and H. I. Smith, *IEEE Electron Device Lett.* **7**, 413 (1986).
132. S. E. Laux and A. C. Warren, "International Electron Devices Meeting, Los Angeles 1986," *IEDM Technical Digest*, 567 (1986).
133. S. E. Laux and F. Stern, *Appl. Phys. Lett.* **49**, 91 (1986).
134. S. E. Laux, D. J. Frank, and F. Stern, *Surf. Sci.* **196**, 101 (1988).
135. K.-F. Berggren, T. J. Thornton, D. J. Newson, and M. Pepper, *Phys. Rev. Lett.* **57**, 1769 (1986).
136. J. H. Davies, *Semicond. Sci. Technol.* **3**, 995 (1988).
137. T. P. Smith, B. B. Goldberg, P. J. Stiles, and M. Heiblum, *Phys. Rev.* **B32**, 2696 (1985).
138. T. P. Smith, III, W. I. Wang, and P. J. Stiles, *Phys. Rev.* **B34**, 2995 (1986).
139. U. Wulf, V. Gudmundsson, and R. R. Gerhardts, *Phys. Rev.* **B38**, 4218 (1988).
140. W. Y. Lai and S. Das Sarma, *Phys. Rev.* **B33**, 8874 (1986).

141. U. Wulf and R. R. Gerhardts, in *Physics and Technology of Submicron Structures* (H. Heinrich, G. Bauer, and F. Kuchar, eds.), *Springer Series in Solid-State Sciences*, Vol. 83, Springer-Verlag, Berlin, p. 162 (1988).
142. W. Hansen, in *Festkörperprobleme (Advances in Solid State Physics)*, Vol. 28, (U. Rössler, ed.), Vieweg, Braunschweig, Federal Republic of Germany, p. 121 (1988).
143. S. B. Kaplan and A. C. Warren, *Phys. Rev.* **B34**, 1346 (1986).
144. M. Wagner, *Superlattices and Microstructures* **6**, 333 (1989).
145. K.-F. Berggren and D. J. Newson, *Semicond. Sci. Technol.* **1**, 327 (1986).
146. D. A. Wharam, U. Ekenberg, M. Pepper, D. G. Hasko, H. Ahmed, J. E. F. Frost, D. A. Ritchie, D. C. Peacock, and G. A. C. Jones, *Phys. Rev.* **B39**, 6283 (1989).
147. K.-F. Berggren, G. Roos, and H. van Houten, *Phys. Rev.* **B37**, 10118 (1988).
148. J. F. Weisz and K.-F. Berggren, *Phys. Rev.* **B40**, 1325 (1989).
149. J. H. Rundquist, *Semicond. Sci. Technol.* **4**, 455 (1989).
150. J. Alsmeier, Ch. Sikorski, and U. Merkt, *Phys. Rev.* **B37**, 4314 (1988).
151. U. Merkt, *Superlattices and Microstructures* **6**, 341 (1989).
152. K. Kern, T. Demel, D. Heitmann, P. Grambow, K. Ploog, and M. Razeghi, *Surf. Sci.* **229**, 256 (1990).
153. J. H. F. Scott-Thomas, S. B. Field, M. A. Kastner, H. I. Smith, and D. A. Antoniadis, *Phys. Rev. Lett.* **62**, 583 (1989).
154. U. Meirav, M. A. Kastner, M. Heiblum, and S. J. Wind, *Phys. Rev.* **B40**, 5871 (1989).
155. H. van Houten and C. W. J. Beenakker, *Phys. Rev. Lett.* **63**, 1893 (1989).
156. T. P. Smith, III, K. Y. Lee, C. M. Knoedler, J. M. Hong, and D. P. Kern, *Phys. Rev.* **B38**, 2172 (1988).
157. W. Hansen, T. P. Smith, III, K. Y. Lee, J. A. Brum, C. M. Knoedler, J. M. Hong, and D. P. Kern, *Phys. Rev. Lett.* **62**, 2168 (1989).
158. U. Sivan and Y. Imry, *Phys. Rev. Lett.* **61**, 1001 (1988).
159. M. Robnik, *J. Phys.* **A19**, 3619 (1986).
160. V. Fock, *Z. Phys.* **47**, 446 (1928).
161. C. G. Darwin, *Proc. Cambridge Philos. Soc.* **27**, 86 (1930).
162. W. Hansen, T. P. Smith, III, J. A. Brum, J. M. Hong, K. Y. Lee, C. M. Knoedler, D. P. Kern, and L. L. Chang, in *Nanostructures Physics and Fabrication* (M. A. Reed and W. P. Kirk, eds.), Academic Press, Boston, p. 97 (1989).
163. G. W. Bryant, *Phys. Rev. Lett.* **59**, 1140 (1987).
164. R. E. Prange and S. M. Girvin, eds., *The Quantum Hall Effect*, Springer, New York (1987).
165. T. Chakraborty and P. Pietiläinen, *The Fractional Quantum Hall Effect*, Springer Series in Solid-State Sciences, Vol. 85, Springer, New York (1988).
166. T. P. Smith, III, K. Y. Lee, J. M. Hong, C. M. Knoedler, H. Arnot, and D. P. Kern, *Phys. Rev.* **B38**, 1558 (1988).
167. W. Hansen, T. P. Smith, III, K. Y. Lee, J. M. Hong, and C. M. Knoedler, *Appl. Phys. Lett.* **56**, 168 (1990).
168. P. A. Maksym and T. Chakraborty, *Phys. Rev. Lett.* **65**, 108 (1990).
169. T. Inoshita, S. Ohnishi, and A. Oshiyama, *Phys. Rev. Lett.* **57**, 2560 (1986).
170. K. J. Vahala, *IEEE J. Quantum Electron.* **24**, 523 (1988).
171. K. Obermayer, G. Mahler, and H. Haken, *Phys. Rev. Lett.* **58**, 1792 (1987).
172. D. C. Tsui, S. J. Allen, Jr., R. A. Logan, A. Kamgar, and S. N. Coppersmith, *Surf. Sci.* **73**, 419 (1978).
173. E. Batke and D. Heitmann, *Infrared Phys.* **24**, 189 (1984).
174. J. F. Koch, *Surf. Sci.* **58**, 104 (1976).
175. M. von Ortenberg, in *Infrared and Millimeter Waves*, Vol. 3 (K. J. Button, ed.), Academic Press, New York, Chap. 6, pp. 275–345 (1980).
176. D. Heitmann, J. P. Kotthaus, and E. G. Mohr, *Solid State Commun.* **44**, 715 (1982).

177. Z. Schlesinger, J. C. M. Hwang, and S. J. Allen, Jr., *Phys. Rev. Lett.* **50**, 2098 (1983).
178. A. D. Wieck, F. Thiele, U. Merkt, K. Ploog, G. Weimann, and W. Schlapp, *Phys. Rev.* **B39**, 3785 (1989).
179. A. Pinczuk and G. Abstreiter, in *Light Scattering in Solids V, Topics in Applied Physics*, Vol. 66 (M. Cardona and G. Güntherodt, eds.), Springer, Berlin, p. 168 (1989).
180. J. S. Weiner, G. Danan, A. Pinczuk, J. Valladares, L. N. Pfeiffer, and K. West, *Phys. Rev. Lett.* **63**, 1641 (1989).
181. J. C. Maan, *Surf. Sci.* **196**, 518 (1988).
182. M. Kohl, D. Heitmann, P. Grambow, and K. Ploog, *Phys. Rev. Lett.* **63**, 2124 (1989).
183. A. D. Wieck, dissertation thesis, Universität Hamburg (1987).
184. S. J. Allen, Jr., D. C. Tsui, and F. DeRosa, *Phys. Rev. Lett.* **35**, 1359 (1975).
185. S. J. Allen, Jr., D. C. Tsui, and R. A. Logan, *Phys. Rev. Lett.* **38**, 980 (1977).
186. T. N. Theis, J. P. Kotthaus, and P. J. Stiles, *Solid State Commun.* **24**, 273 (1977).
187. T. N. Theis, J. P. Kotthaus, and P. J. Stiles, *Solid State Commun.* **26**, 603 (1978).
188. E. Batke, D. Heitmann, A. D. Wieck, and J. P. Kotthaus, *Solid State Commun.* **46**, 269 (1983).
189. E. Batke and D. Heitmann, *Solid State Commun.* **47**, 819 (1983).
190. D. Heitmann, *Surf. Sci.* **170**, 332 (1986).
191. E. Batke, D. Heitmann, J. P. Kotthaus, and K. Ploog, *Phys. Rev. Lett.* **54**, 2367 (1985).
192. D. Heitmann and U. Mackens, *Phys. Rev.* **B33**, 8269 (1986).
193. H. Pohlmann, M. Wassermeier, and J. P. Kotthaus, *Superlattices and Microstructures* **2**, 293 (1986).
194. M. V. Krasheninnikov and A. V. Chaplik, *Fiz. Tekh. Poluprovodn.* **15**, 32 (1981) [(*Sov. Phys. Semicond.* **15**, 19 (1981)].
195. J. P. Kotthaus, W. Hansen, H. Pohlmann, M. Wassermeier, and K. Ploog, *Surf. Sci.* **196**, 600 (1988).
196. C. Dahl, *Phys. Rev.* **B41**, 5763 (1990).
197. W. Que and G. Kirczenow, *Phys. Rev.* **B37**, 7153 (1988).
198. A. V. Chaplik, *Superlattices and Microstructures* **6**, 329 (1989).
199. V. Shikin, T. Demel, and D. Heitmann, *Surf. Sci.* **229**, 276 (1990).
200. D. B. Mast, A. J. Dahm, and A. L. Fetter, *Phys. Rev. Lett.* **54**, 1706 (1985).
201. D. C. Glattli, E. Y. Andrei, G. Deville, J. Poitrenaud, and F. I. B. Williams, *Phys. Rev. Lett.* **54**, 1710 (1985).
202. J. Alsmeier, E. Batke, and J. P. Kotthaus, *Phys. Rev.* **B40**, 12574 (1989).
203. T. Demel, D. Heitmann, P. Grambow, and K. Ploog, *Phys. Rev.* **B38**, 12732 (1988).
204. G. Eliasson, J.-W. Wu, P. Hawrylak, and J. J. Quinn, *Solid State Commun.* **60**, 41 (1986).
205. L. R. Walker, *Phys. Rev.* **105**, 390 (1957).
206. T. Demel, D. Heitmann, P. Grambow, and K. Ploog, *Superlattices and Microstructures* **5**, 287 (1989).
207. J. K. Jain and P. B. Allen, *Phys. Rev. Lett.* **54**, 2437 (1985).
208. G. Fasol, N. Mestres, H. P. Hughes, A. Fischer, and K. Ploog, *Phys. Rev. Lett.* **56**, 2517 (1986).
209. A. Pinczuk, M. G. Lamont, and A. C. Gossard, *Phys. Rev. Lett.* **56**, 2092 (1986).
210. U. Merkt, Ch. Sikorski, and J. P. Kotthaus, *Superlattices and Microstructures* **3**, 679 (1987).
211. U. Merkt, M. Horst, T. Evelbauer, and J. P. Kotthaus, *Phys. Rev.* **B34**, 7234 (1986).
212. U. Merkt, Ch. Sikorski, and J. Alsmeier, in *Spectroscopy of Semiconductor Microstructures* (G. Fasol, A. Fasolino, and P. Lugli, eds.), Plenum, Press, New York, p. 89 (1989).
213. W. Que and G. Kirczenow, *Phys. Rev.* **B39**, 5998 (1989).
214. L. Brey, N. F. Johnson, and B. I. Halperin, *Phys. Rev.* **B40**, 10647 (1989).
215. W. Kohn, *Phys. Rev.* **123**, 1242 (1961).

216. A. L. Fetter, *Phys. Rev.* **B32**, 7676 (1985).
217. S. A. Govorkov, M. I. Reznikov, A. P. Senichkin, and V. I. Talyanskii, *Pis'ma Zh. Eksp. Teor. Fiz.* **44**, 380 (1986) [*JETP Lett.* **44**, 487 (1986)].
218. E. Y. Andrei, D. C. Glattli, F. I. B. Williams, and M. Heiblum, *Surf. Sci.* **196**, 501 (1988).
219. V. I. Talyanskii, M. Wassermeier, A. Wixforth, J. Oshinowo, J. P. Kotthaus, I. E. Batov, H. Nickel, and W. Schlapp, *Surf. Sci.* **229**, 40 (1990).
220. J. Alsmeier, E. Batke, and J. P. Kotthaus, *Phys. Rev.* **B41**, 1699 (1990).
221. A. Lorke, J. P. Kotthaus, and K. Ploog, *Phys. Rev. Lett.* **64**, 2559 (1990).
222. B. A. Wilson, S. J. Allen, Jr., and D. S. Tsui, *Phys. Rev.* **B24**, 5887 (1981).
223. R. P. Leavitt and J. W. Little, *Phys. Rev.* **B34**, 2450 (1986).
224. W. Que and G. Kirczenow, *Phys. Rev.* **B38**, 3614 (1988).
225. R. B. Dingle, *Proc. Roy. Soc. London*, Series A **211**, 500 (1952); A **212**, 38 (1952).
226. Ch. Sikorski and U. Merkt, *Surf. Sci.* **229**, 282 (1990).
227. U. Merkt and Ch. Sikorski, *Semicond. Sci. Technol.* **5**, 182 (1990).
228. Y. Takada, K. Arai, N. Uchimura, and Y. Uemura, *J. Phys. Soc. Jpn.* **49**, 1851 (1980).
229. Y. Takada, *J. Phys. Soc. Jpn.* **50**, 1998 (1981).
230. W. Zawadzki, *J. Phys.* **C16**, 229 (1983).
231. A. V. Chaplik, *Pis'ma Zh. Eksp. Teor. Fiz.* **50**, 38 (1989) [*JETP Lett.* **50**, 44 (1989)].
232. G. W. Bryant, *Phys. Rev.* **B40**, 1620 (1989).
233. U. Merkt, J. Huser, and M. Wagner, *Phys. Rev.* **B43**, 7320 (1991).
234. K. Kempa, D. A. Broido, and P. Bakshi, *Phys. Rev.* **B43**, 9343 (1991).

Index

A

Adiabatic quantum transport, 10, 81–105
  collimation, 47
  quantum Hall effect regime, 81
Aharonov–Bohm effect, 99–105, 167–174, 192, 196, 204, 237
  ballistic ring, 167–169
  inter-edge channel tunneling, 99
  multiply-connected, 99
  periodicity, 100
  phase-difference, 100
  singly connected geometry, 99
  suppression in quantized Hall regime, 169–174, 237
Anomalous quantum Hall effect, 84–94, 248, *see also* Quantum Hall effect
Anomalous Shubnikov–de Haas effect, 94–99
Asymptotic regions, 197

B

Back contact, 314
Background resistance, 16, 80
Backscattering, 227, 240–248, 251
  geometrical, 39
  magnetic suppression, 35, 39–45, 81
  at quantum point contact, 28, 39–45
  selective, 82–84, 91, 95–99
Ballistic device, 310
Ballistic quantum transport, 10, 17–80
  coherent electron focusing, 56–64
  collimation, 45–56
  conductance quantization of a quantum point contact, 19–33
  magnetic depopulation of subbands, 33–39
  non-linear, 64–81
Band-gap, modulation, 280
Bare potential, 370
Bend resistance, *see* Resistance
Bloch oscillator, 296, 299
Bolometer, 335
Boundary scattering
  diffuse, 32
  spectular, 32, 64
Breit–Wigner formula, 98, 267

C

Cap layer, 292, 312
Capacitance measurement, 314, 327
Center coordinate, 317
Channel, *see also* Edge channel; Mode; Subband
  for electrons, 197, 223
  width, 321
Charge neutrality, 196
Chemical potential, 193, 199–202
  elimination, 206
  oscillations, 217
Coherent electron focusing, 56–64
Collimated beam, *see* Collimation; Landauer–Buttiker formalism
Collimation, 27, 45–56
  cone, 48, 55
  deflection by magnetic field, 51, 54

factor, 47
horn and barrier, 56
peak, 55
series resistance, 48
Commensurability problem, 307
Conductance, 193, see also Magnetoconductance
  adiabatic transport, 81–105
  ballistic transport, 17–80
  basic elements of, 195
  channel, 316
  contact, 25, 93
    quantum point, 19–39, see also Quantum point contact
  corrections, 16, 84
  four-terminal
    diagonal, 44, 84
    Hall, 44, 84
    longitudinal, 40, 43, 84
  plateau, 20, 132
  quantization, 11, 13–17, 19–39
    effect of impurity scattering on, 132–135
    theory of, 21–29, 208–212
  quantized, 19–39
    breakdown, 64
    fractions, 43, 48
    in magnetic field, 35
  spreading, 16
  step, 21
  three-terminal, 15, 93, 97, 136–147
  two-terminal, 16, 19–39, 84, 123–135, 227–233
Conductivity
  dynamical, 336, 347
  negative differential, 279, 296
  sheet, 336
Confinement
  classical, 342
  lateral, 282
  parabolic, 33, 358, 368
  square well, 24, 33
  potential, 222
    effective, 312
    electrostatic, 282, 329
    harmonic, 318, 360, 366, 370
    parabolic, see Harmonic confinement model
    self-consistent, 312
  square-well, 329
  strain-induced, 282
  quantum, 342
Confining potential, see Confinement
Constriction, 130–131, 208, 235, see also Quantum point contact
Contact, see also Quantum point contact
  conductance, 25
  to conductor, 192, 208, 212
  equilibration, 97
  with internal reflection, 244
  resistance, 25, 208, 212
Coulomb blockade, 6
Coulomb gap, 326, 332
Current, incident, 201
  conservation, 201, 203
Current–voltage relation
  quantum point contact, 70
  semi-classical point contact, 69
Cyclotron
  energy, 222
    frequency, 222
  orbit, 62
  radius, 300
  resonance, 343
    subband-shifted, 356

D

Density
  modulation, 283, 337
  states, 201, 280
Depletion, 14
  gate, 291
  lateral, 14
  potential, 14
  sidewall, 282
  threshold, 15, 291
  width, 293, 324
Depolarization
  frequency, 342, 363
  mode, 347, 361
Dielectric constant, effective, 337
Dimensional resonance, 212, 348, 269–271
Dimensionality, 308
Dipole
  field, at injector, 79

matrix element, 366
transitions, 367
Disc, 361
Dot, *see* Quantum dot
Drift, *see* Guiding center

E

Edge
  magneto-plasmon, 361, 367
  mode, 361, 371
Edge channel, 34, 61–64, 223, 241
  backscattering of, 82–84, 95–99
  bound in cavity, 102
  detection of, 91–94
  equilibration, 83, 227
  mixer, 97
  selective population, 91
  selective detection, 91
Electrochemical potential, 193, 199
Electron bath, 195
Electron-beam lithography, 3, 14, 114
Electron focusing
  classical, 58, 59
  coherent, 56
  hot, 65, 75–81
Electron system
  one-dimensional, 280, 307
  two-dimensional, 280
  zero-dimensional, 280, 326
Electron waveguide, 114, 117–123
  in a magnetic field, 144–147
Electrostatic saddle-point, 209
Equilibration, at ideal contact, 84, 97
Etching techniques
  deep mesa, 293, 324
  shallow mesa, 292, 323
External reflection, 246

F

Far-infrared spectroscopy, 334–336
Field
  effect device, 290
  electrode, nanostructured, 281
Filling factor, fractional, 332
Focused-ion-beam technique, 295
Fourier transform, spectrometer, 334

Four-terminal, *see* Conductance; Quantum point contact
Fractional quantized Hall effect, *see* Quantized Hall effect

G

Gate, *see also* Split-gate
  dual-stacked, 289
  grating-type, 289
  insulator, 338
  modulated, 283
Grating
  coupler, 336
  metal, 289
  photoresist, 287
Group velocity, 24, 72
Guiding center
  drift, 82
  energy, 82
  ExB, 304

H

Hall effect, *see* Conductance; Quantized Hall effect; Quantum Hall; Quenched Hall effect; Resistance
Hard wall potential, 223
Harmonic onfinement model, 318, 370
High-index surface, 284, 297
  MBE growth, 295
Hofstadter butterfly, 307
Holes, focusing, 78
Holographic Lithography, 287
Hot electron focusing, 65, 75–81

I

Ideal contact, 239
InGaAs/InAlAs heterostructure, 324
Inelastic forward scattering, 213
InSb, 294, 321, 368
Interconnected probes, 240
Interedge state scattering, 256
  absence, 253
Interference
  in coherent electron focusing, 61

mode, 61
  in quantum point contacts, 30–31
Internal reflection, 244, 246, 251
Intersubband
  plasmon, 359
  resonance
    one-dimensional, 357
    two-dimensional, 334
Intraband excitations, 286, 334
Intraedge state scattering, 256
Intrinsic resonances, *see* Dimensional Resonance
Inversion, 289
  subgrating mode, 289, 311, 348
  subgap mode, 290
Irreversibility (due to reservoirs), 199

K

Kohn's theorem, 360

L

Landau
  band, 305
  damping, 342
  gauge, 305, 317
  level, 33, 43, 223, 300
Landauer–Büttiker formalism
  collimated beam experiment, 54
  electron focusing geometry, 93
Landauer formula, 24, 130, 196
Laser, far-infrared, 334
Lift-off, 289
Lithography, 287
  holographic, 287
Localization length, 233
Localized states, 226, 271

M

Magnetic breakdown, 297
Magnetic length, 222, 300
Magnetoconductance, 312, 315, *see also* Magnetoresistance
Magneto-electric subbands, 33, 144, 223, 269
Magnetoresistance, *see also* Resistance
  cavity, 102
  quantum point contacts in series, 52

Mass, effective
  modulation, 281
  negative, 296
  optical, 337
Microreversibility, 199
Microwave transmission, 324
Mode, *see also* One-dimensional
  evanescent, 26
  interference, 12, 61
  longitudinal wavelength, 28
  matching, 26
  selective detection, 91
  selective population, 91
Modulation doping, 1
MOS device, 280
  dual-gate, 284, 298
Multilayered wire, 351
Multiprobe resistance formula, 203

N

Nanostructure, 2, 6
  lateral, 280
Non-equilibrium injection, 245, 253
Non-ideal contact, 244
Non-linear transport
  classical point contact, 69
  hot electron focusing, 75–81
  mechanisms, 64
  quantum point contact, 66, 70
Non-local electron transport, 254
  coefficient, 308
Non-parabolicity, 358, 369

O

One-dimensional, *see also* Subbands
  conductor, 23–25, 306
  density of states, 25, 72, 281
Optical spectroscopy, 334
Oscillator strength, 340, 367

P

Percolating paths, edge states, 229
Phase-coherence, 204, 216, 257
  electron focusing, 46
  length, 308
Phase randomization, 208

Phase-sensitive voltage measurement, 215
Plasmon, 336–344
  in a box, 351
  confined, 348
  magneto-, 343
Plateau, *see* Conductance
Point contact, *see also* Quantum point contact
  ballistic, 17
  diffusive, 17
  Maxwell, 16, 17
  quantized, 19–39
  Sharvin, 17, 23
Polarization
  dynamic, 344
  light, 334, 347
Potential
  harmonic, 317, 366
  landscape
    boundary of electron gas, 92
    at quantum point contact, 23, 92
    of a split gate, 120–122
    of two split gates, 139
  periodic, 279

Q

Quantization energy, 317, 355
Quantized conductance, 19–39, 208, 227, *see also* Resistance, Quantum Point Contact
  four-terminal, 143–182
  in a magnetic field, 164–174
  three-terminal, 136–142
  two-terminal (Sharvin resistance), 117–135
Quantized resistance, 114–115, 123–127
  accuracy of, 125–127
Quantized Hall effect, 192, 238
  fractional, 174–182, 271, 332
  integer, 164–167
Quantum channel, 197, 223
Quantum dot, 3, 5, 261, 280, 326, 363
  coupled, 370
Quantum Hall effect, 2, 192
  adiabatic transport, 81–105
  anomalous, *see* Anomalous quantum Hall effect
  cavity conductance, 102
  conductance, 93
    quantization, 34
  contacts, 84
  disordered, 84
  edge channel picture, 82
  four terminal measurement, 164–167
  fractional, *see* Quantized Hall effect
  guiding center drift, 82
  ideal, 84
  magnetoresistance of annulus, 167–174
  open conductors, 238–240
  resistance overshoot, 99, 232
Quantum interference
  in coherent electron focusing, 61
  at quantum point contact, 30
Quantum limit, 309, 326
Quantum point contact, 4, 9–105, *see also* Resistance
  anomalous quantum Hall effect, 90
  coherent electron focusing, 56–64
  as collector, 54, 58
  collimated beam experiment, 54
  density, 38
  four-terminal resistance, 39, 85, 203, 212
    diagonal, 44, 84
    Hall, 44, 84
    longitudinal, 40, 43, 84
    negative, 40–43
    positive, 43, 88
  impurity near, 31, 132–135
  as injector, 54, 58
  series conductance, 50
  series resistance, 48
Quantum well, 1, 2
Quantum wire, 3, 5, 280, 307
  multiple well, 293, 350
Quasi-accumulation, 350
Quasi-ballistic, 308
Quasi-one-dimensional, 280, 309–326, 345–361
Quasi-two-dimensional, 280
Quasi-zero-dimensional, 280, 326–334, 361–372
Quenched Hall effect, 207, 265, 267
  criterion for, 160–162
  as a function of mode number, 147–164

R

Reciprocity, 155–157, 202–205
  electron focusing, 59
  injector–collector, 59, 61

tests, 204
violation, 100
Reflection probability, 200
Reservoir, 193, 196
Resistance, 193, 202, 203, *see also* Quantized conductance
  background, 16, 80
  bend, 55, 142–164
    observation of negative resistance, 147–164
    transfer resistance, nonlocal, 162–163
  four-terminal, 143–182, 202, 212–221, *see also* Quantum point contact
  Hall, 44, 192, 193
  longitudinal, 40, 43, 193
  Maxwell point contact, 16
  measurement, invasive leads, 141
  negative, 60, 219, 266
    differential, 279, 296, 299
  quantum point contact, 19, 25
  Sharvin point contact, 23, 123
  spreading, 16
  three-terminal, 136–147, 205, 246
  two-terminal, 19–39, 41, 100, 123–135, 202, 221, 246
Resonance, *see* Cyclotron, Dimensional resonance, Intersubband
Resonant Tunneling, 257–269
Resonant tunneling diode, 1

Skipping orbit, 58
  and edge state, 62
Specular boundary scattering, 32, 64
Spin
  degeneracy, 24
  splitting, 35
Split-gate, 13, 118–127, *see also* Resistance
  width, 38
Stark ladder, 296
Subband
  equipartioning, 24, 86, 96
  magneto-electric, 33, 144, 223, 269
  one-dimensional, 23, 72, 309
  separation, 314, 322
  spacing, 129, 358
Substrate bias, 353, 363
Superlattice, 279
  artificial, 298
  band-gap, 282
  lateral, 281
  miniband, 296
  minigap, 284, 296
  mini-zone, 286
  potential, 280
  effect on Shubnikov–Haas oscillations, 300–307
  tight-binding, 284
Suppression of Hall effect, *see* Quenched Hall effect

S

Saddle-shaped constriction, 209, 235
Scanning tunneling microscope, 17
Scattering
  elastic, 309
  impurity, 309
  matrix, 199, 216, 260, 263
    permutation, 265
  phonon, 309
  problem, 197
Schottky gate, 291
  grating-type, 321
Self-consistency, 207
Sharvin, *see* Resistance
Shubnikov–de Haas
  anomalous effect, *see* Anomalous Shubinkov–de Haas effect
  minima, 88
  oscillations, 96, 252, 255, 286, 318

T

Telegraph noise, 135
Thomas Fermi model, 298
Three-terminal, *see* Conductance; Resistance
Threshold
  energy, 209, 269
  voltage, 290
Transconductance oscillations, 298
Transmission
  amplitude, 24
  geometry, 334
    coefficient, 335
  probability, 24, 198, 200
    direct, 49
  and resonance, 202
  resonances
    and Aharonov–Bohm effect, 104
    in quantum point contact, 28

Transport, *see also* Non-linear transport; Non-local electron transport
  magneto-transport, 300, 315
  static
    in lateral superlattices, 295
    in 1D wires, 307
Transverse wave function, 197, 225
Tunneling; *see also* Resonant tunneling
  backscattering, 83
  through bound edge state, 102
  evanescent mode, 28
  in magnetic field, 100
Two-dimensional electron gas (2DEG), 1, 2, 4, 10, 13, 114, 192, 280
Two-probe resistance, *see* Resistance
Two-terminal, *see* Conductance; Quantized conductance; Resistance

U

Universal conductance fluctuations, 2, 18, 308

V

Vacuum admittance, 336
Valley projection model, 284
Vicinal plane, *see* High-index surface
Voltage contact, 192

W

Waveguide, *see* Electron waveguide, Quantum point contact
Waveguide-like mode, 348
Weakly coupled probe, 216

Z

Zeeman splitting, 369
Zener tunneling, 297
Zero-current condition, 197, 203
Zero-dimensional, 326
Zone folding, 341

Contents of Volumes in This Series

Volume 1 Physics of III–V Compounds

C. Hilsum, Some Key Features of III–V Compounds
Franco Bassani, Methods of Band Calculations Applicable to III–V Compounds
E. O. Kane, The $k \cdot p$ Method
V. L. Bonch-Bruevich, Effect of Heavy Doping on the Semiconductor Band Structure
Donald Long, Energy Band Structures of Mixed Crystals of III–V Compounds
Laura M. Roth and Petros N. Argyres, Magnetic Quantum Effects
S. M. Puri and T. H. Geballe, Thermomagnetic Effects in the Quantum Region
W. M. Becker, Band Characteristics near Principal Minima from Magnetoresistance
E. H. Putley, Freeze-Out Effects, Hot Electron Effects, and Submillimeter Photoconductivity in InSb
H. Weiss, Magnetoresistance
Betsy Ancker-Johnson, Plasmas in Semiconductors and Semimetals

Volume 2 Physics of III–V Compounds

M. G. Holland, Thermal Conductivity
S. I. Novkova, Thermal Expansion
U. Piesbergen, Heat Capacity and Debye Temperatures
G. Giesecke, Lattice Constants
J. R. Drabble, Elastic Properties
A. U. Mac Rae and G. W. Gobeli, Low Energy Electron Diffraction Studies
Robert Lee Mieher, Nuclear Magnetic Resonance
Bernard Goldstein, Electron Paramagnetic Resonance
T. S. Moss, Photoconduction in III–V Compounds
E. Antončik and J. Tauc, Quantum Efficiency of the Internal Photoelectric Effect in InSb
G. W. Gobeli and F. G. Allen, Photoelectric Threshold and Work Function
P. S. Pershan, Nonlinear Optics in III–V Compounds
M. Gershenzon, Radiative Recombination in the III–V Compounds
Frank Stern, Stimulated Emission in Semiconductors

Volume 3 Optical of Properties III–V Compounds

Marvin Hass, Lattice Reflection
William G. Spitzer, Multiphonon Lattice Absorption
D. L. Stierwalt and R. F. Potter, Emittance Studies
H. R. Philipp and H. Ehrenreich, Ultraviolet Optical Properties
Manuel Cardona, Optical Absorption above the Fundamental Edge
Earnest J. Johnson, Absorption near the Fundamental Edge
John O. Dimmock, Introduction to the Theory of Exciton States in Semiconductors

B. Lax and J. G. Mavroides, Interband Magnetooptical Effects
H. Y. Fan, Effects of Free Carries on Optical Properties
Edward D. Palik and George B. Wright, Free-Carrier Magnetooptical Effects
Richard H. Bube, Photoelectronic Analysis
B. O. Seraphin and H. E. Bennett, Optical Constants

Volume 4 Physics of III–V Compounds

N. A. Goryunova, A. S. Borschevskii, and D. N. Tretiakov, Hardness
N. N. Sirota, Heats of Formation and Temperatures and Heats of Fusion of Compounds $A^{III}B^{V}$
Don L. Kendall, Diffusion
A. G. Chynoweth, Charge Multiplication Phenomena
Robert W. Keyes, The Effects of Hydrostatic Pressure on the Properties of III–V Semiconductors
L. W. Aukerman, Radiation Effects
N. A. Goryunova, F. P. Kesamanly, and D. N. Nasledov, Phenomena in Solid Solutions
R. T. Bate, Electrical Properties of Nonuniform Crystals

Volume 5 Infrared Detectors

Henry Levinstein, Characterization of Infrared Detectors
Paul W. Kruse, Indium Antimonide Photoconductive and Photoelectromagnetic Detectors
M. B. Prince, Narrowband Self-Filtering Detectors
Ivars Melngailis and T. C. Harman, Single-Crystal Lead-Tin Chalcogenides
Donald Long and Joseph L. Schmit, Mercury-Cadmium Telluride and Closely Related Alloys
E. H. Putley, The Pyroelectric Detector
Norman B. Stevens, Radiation Thermopiles
R. J. Keyes and T. M. Quist, Low Level Coherent and Incoherent Detection in the Infrared
M. C. Teich, Coherent Detection in the Infrared
F. R. Arams, E. W. Sard, B. J. Peyton, and F. P. Pace, Infrared Heterodyne Detection with Gigahertz IF Response
H. S. Sommers, Jr., Macrowave-Based Photoconductive Detector
Robert Sehr and Rainer Zuleeg, Imaging and Display

Volume 6 Injection Phenomena

Murray A. Lampert and Ronald B. Schilling, Current Injection in Solids: The Regional Approximation Method
Richard Williams, Injection by Internal Photoemission
Allen M. Barnett, Current Filament Formation
R. Baron and J. W. Mayer, Double Injection in Semiconductors
W. Ruppel, The Photoconductor-Metal Contact

Volume 7 Application and Devices
PART A

John A. Copeland and Stephen Knight, Applications Utilizing Bulk Negative Resistance
F. A. Padovani, The Voltage-Current Characteristics of Metal-Semiconductor Contacts
P. L. Hower, W. W. Hooper, B. R. Cairns, R. D. Fairman, and D. A. Tremere, The GaAs Field-Effect Transistor
Marvin H. White, MOS Transistors
G. R. Antell, Gallium Arsenide Transistors
T. L. Tansley, Heterojunction Properties

PART B

T. *Misawa,* IMPATT Diodes
H. C. *Okean,* Tunnel Diodes
Robert B. *Campbell and Hung-Chi Chang,* Silicon Carbide Junction Devices
R. E. *Enstrom, H. Kressel, and L. Krassner,* High-Temperature Power Rectifiers of GaAs$_{1-x}$P$_x$

Volume 8 Transport and Optical Phenomena

Richard J. *Stirn,* Band Structure and Galvanomagnetic Effects in III–V Compounds with Indirect Band Gaps
Roland W. *Ure, Jr.,* Thermoelectric Effects in III–V Compounds
Herbert *Piller,* Faraday Rotation
H. Barry *Bebb and E. W. Williams,* Photoluminescence 1: Theory
E. W. *Williams and H. Barry Bebb,* Photoluminescence II: Gallium Arsenide

Volume 9 Modulation Techniques

B. O. *Seraphin,* Electroreflectance
R. L. *Aggarwal,* Modulated Interband Magnetooptics
Daniel F. *Blossey and Paul Handler,* Electroabsorption
Bruno *Batz,* Thermal and Wavelength Modulation Spectroscopy
Ivar *Balslev,* Piezooptical Effects
D. E. *Aspnes and N. Bottka,* Electric-Field Effects on the Dielectric Function of Semiconductors and Insulators

Volume 10 Transport Phenomena

R. L. *Rode,* Low-Field Electron Transport
J. D. *Wiley,* Mobility of Holes in III–V Compounds
C. M. *Wolfe and G. E. Stillman,* Apparent Mobility Enhancement in Inhomogeneous Crystals
Robert L. *Peterson,* The Magnetophonon Effect

Volume 11 Solar Cells

Harold J. *Hovel,* Introduction; Carrier Collection, Spectral Response, and Photocurrent; Solar Cell Electrical Characteristics; Efficiency; Thickness; Other Solar Cell Devices; Radiation Effects; Temperature and Intensity; Solar Cell Technology

Volume 12 Infrared Detectors (II)

W. L. *Eiseman, J. D. Merriam, and R. F. Potter,* Operational Characteristics of Infrared Photodetectors
Peter R. *Bratt,* Impurity Germanium and Silicon Infrared Detectors
E. H. *Putley,* InSb Submillimeter Photoconductive Detectors
G. E. *Stillman, C. M. Wolfe, and J. O. Dimmock,* Far-Infrared Photoconductivity in High Purity GaAs
G. E. *Stillman and C. M. Wolfe,* Avalanche Photodiodes
P. L. *Richards,* The Josephson Junction as a Detector of Microwave and Far-Infrared Radiation
E. H. *Putley,* The Pyroelectric Detector–An Update

Volume 13 Cadmium Telluride

Kenneth *Zanio,* Materials Preparation; Physics; Defects; Applications

CONTENTS OF VOLUMES IN THIS SERIES

Volume 14 Lasers, Junctions, Transport

N. Holonyak, Jr. and M. H. Lee, Photopumped III–V Semiconductor Lasers
Henry Kressel and Jerome K. Butler, Heterojunction Laser Diodes
A. Van der Ziel, Space-Charge-Limited Solid-State Diodes
Peter J. Price, Monte Carlo Calculation of Electron Transport in Solids

Volume 15 Contacts, Junctions, Emitters

B. L. Sharma, Ohmic Contacts to III–V Compound Semiconductors
Allen Nussbaum, The Theory of Semiconducting Junctions
John S. Escher, NEA Semiconductor Photoemitters

Volume 16 Defects, (HgCd)Se, (HgCd)Te

Henry Kressel, The Effect of Crystal Defects on Optoelectronic Devices
C. R. Whitsett, J. G. Broerman, and C. J. Summers, Crystal Growth and Properties of $Hg_{1-x}Cd_xSe$ Alloys
M. H. Weiler, Magnetooptical Properties of $Hg_{1-x}Cd_xTe$ Alloys
Paul W. Kruse and John G. Ready, Nonlinear Optical Effects in $Hg_{1-x}Cd_xTe$

Volume 17 CW Processing of Silicon and Other Semiconductors

James F. Gibbons, Beam Processing of Silicon
Arto Lietoila, Richard B. Gold, James F. Gibbons, and Lee A. Christel, Temperature Distributions and Solid Phase Reaction Rates Produced by Scanning CW Beams
Arto Lietoila and James F. Gibbons, Applications of CW Beam Processing to Ion Implanted Crystalline Silicon
N. M. Johnson, Electronic Defects in CW Transient Thermal Processed Silicon
K. F. Lee, T. J. Stultz, and James F. Gibbons, Beam Recrystallized Polycrystalline Silicon: Properties, Applications, and Techniques
T. Shibata, A. Wakita, T. W. Sigmon, and James F. Gibbons, Metal-Silicon Reactions and Silicide
Yves I. Nissim and James F. Gibbons, CW Beam Processing of Gallium Arsenide

Volume 18 Mercury Cadmium Telluride

Paul W. Kruse, The Emergence of $(Hg_{1-x}Cd_x)Te$ as a Modern Infrared Sensitive Material
H. E. Hirsch, S. C. Liang, and A. G. White, Preparation of High-Purity Cadmium, Mercury, and Tellurium
W. F. H. Micklethwaite, The Crystal Growth of Cadmium Mercury Telluride
Paul E. Petersen, Auger Recombination in Mercury Cadmium Telluride
R. M. Broudy and V. J. Mazurczyck, (HgCd)Te Photoconductive Detectors
M. B. Reine, A. K. Sood, and T. J. Tredwell, Photovoltaic Infrared Detectors
M. A. Kinch, Metal-Insulator-Semiconductor Infrared Detectors

Volume 19 Deep Levels, GaAs, Alloys, Photochemistry

G. F. Neumark and K. Kosai, Deep Levels in Wide Band-Gap III–V Semiconductors
David C. Look, The Electrical and Photoelectronic Properties of Semi-Insulating GaAs
R. F. Brebrick, Ching-Hua Su, and Pok-Kai Liao, Associated Solution Model for Ga-In-Sb and Hg-Cd-Te
Yu. Ya. Gurevich and Yu. V. Pleskov, Photoelectrochemistry of Semiconductors

CONTENTS OF VOLUMES IN THIS SERIES

Volume 20 Semi-Insulating GaAs

R. N. Thomas, H. M. Hobgood, G. W. Eldridge, D. L. Barrett, T. T. Braggins, L. B. Ta, and S. K. Wang, High-Purity LEC Growth and Direct Implantation of GaAs for Monolithic Microwave Circuits
C. A. Stolte, Ion Implantation and Materials for GaAs Integrated Circuits
C. G. Kirkpatrick, R. T. Chen, D. E. Holmes, P. M. Asbeck, K. R. Elliott, R. D. Fairman, and J. R. Oliver, LEC GaAs for Integrated Circuit Applications
J. S. Blakemore and S. Rahimi, Models for Mid-Gap Centers in Gallium Arsenide

Volume 21 Hydrogenated Amorphous Silicon
Part A

Jacques I. Pankove Introduction
Masataka Hirose, Glow Discharge; Chemical Vapor Deposition
Yoshiyuki Uchida, dc Glow Discharge
T. D. Moustakas, Sputtering
Isao Yamada, Ionized-Cluster Beam Deposition
Bruce A. Scott, Homogeneous Chemical Vapor Deposition
Frank J. Kampas, Chemical Reactions in Plasma Deposition
Paul A. Longeway, Plasma Kinetics
Herbert A. Weakliem, Diagnostics of Silane Glow Discharges Using Probes and Mass Spectroscopy
Lester Guttman, Relation between the Atomic and the Electronic Structures
A. Chenevas-Paule, Experiment Determination of Structure
S. Minomura, Pressure Effects on the Local Atomic Structure
David Adler, Defects and Density of Localized States

Part B

Jacques I. Pankove, Introduction
G. D. Cody, The Optical Absorption Edge of a-Si:H
Nabil M. Amer and Warren B. Jackson, Optical Properties of Defect States in a-Si:H
P. J. Zanzucchi, The Vibrational Spectra of a-Si:H
Yoshihiro Hamakawa, Electroreflectance and Electroabsorption
Jeffrey S. Lannin, Raman Scattering of Amorphous Si, Ge, and Their Alloys
R. A. Street, Luminescence in a-Si:H
Richard S. Crandall, Photoconductivity
J. Tauc, Time-Resolved Spectroscopy of Electronic Relaxation Processes
P. E. Vanier, IR-Induced Quenching and Enhancement of Photoconductivity and Photoluminescence
H. Schade, Irradiation-Induced Metastable Effects
L. Ley, Photoelectron Emission Studies

Part C

Jacques I. Pankove, Introduction
J. David Cohen, Density of States from Junction Measurements in Hydrogenated Amorphous Silicon
P. C. Taylor, Magnetic Resonance Measurements in a-Si:H
K. Morigaki, Optically Detected Magnetic Resonance
J. Dresner, Carrier Mobility in a-Si:H

T. Tiedje, Information about Band-Tail States from Time-of-Flight Experiments
Arnold R. Moore, Diffusion Length in Undoped a-Si:H
W. Beyer and J. Overhof, Doping Effects in a-Si:H
C. R. Wronski, The Staebler-Wronski Effect
R. J. Nemanich, Schottky Barriers on a-Si:H
B. Abeles and T. Tiedje, Amorphous Semiconductor Superlattices

Part D

Jacques I. Pankove, Introduction
D. E. Carlson, Solar Cells
G. A. Swartz, Closed-Form Solution of I–V Characteristic for a-Si:H Solar Cells
Isamu Shimizu, Electrophotography
Sachio Ishioka, Image Pickup Tubes
P. G. LeComber and W. E. Spear, The Development of the a-Si:H Field-Effect Transitor and Its Possible Applications
D. G. Ast, a-Si:H FET-Addressed LCD Panel
S. Kaneko, Solid-State Image Sensor
Masakiyo Matsumura, Charge-Coupled Devices
M. A. Bosch, Optical Recording
A. D'Amico and G. Fortunato, Ambient Sensors
Hiroshi Kukimoto, Amorphous Light-Emitting Devices
Robert J. Phelan, Jr., Fast Detectors and Modulators
Jacques I. Pankove, Hybrid Structures
P. G. LeComber, A. E. Owen, W. E. Spear, J. Hajto, and W. K. Choi, Electronic Switching in Amorphous Silicon Junction Devices

Volume 22 Lightwave Communications Technology
Part A

Kazuo Nakajima, The Liquid-Phase Epitaxial Growth of InGaAsP
W. T. Tsang, Molecular Beam Epitaxy for III–V Compound Semiconductors
G. B. Stringfellow, Organometallic Vapor-Phase Epitaxial Growth of III–V Semiconductors
G. Beuchet, Halide and Chloride Transport Vapor-Phase Deposition of InGaAsP and GaAs
Manijeh Razeghi, Low-Pressure Metallo-Organic Chemical Vapor Deposition of $Ga_xIn_{1-x}As_yP_{1-y}$ Alloys
P. M. Petroff, Defects in III–V Compound Semiconductors

Part B

J. P. van der Ziel, Mode Locking of Semiconductor Lasers
Kam Y. Lau and Amnon Yariv, High-Frequency Current Modulation of Semiconductor Injection Lasers
Charles H. Henry, Spectral Properties of Semiconductor Lasers
Yasuharu Suematsu, Katsumi Kishino, Shigehisa Arai, and Fumio Koyama, Dynamic Single-Mode Semiconductor Lasers with a Distributed Reflector
W. T. Tsang, The Cleaved-Coupled-Cavity (C^3) Laser

Part C

R. J. Nelson and N. K. Dutta, Review of InGaAsP/InP Laser Structures and Comparison of Their Performance

N. Chinone and M. Nakamura, Mode-Stabilized Semiconductor Lasers for 0.7–0.8- and 1.1–1.6- μm Regions
Yoshiji Horikoshi, Semiconductor Lasers with Wavelengths Exceeding 2 μm
B. A. Dean and M. Dixon, The Functional Reliabilty of Semiconductor Lasers as Optical Transmitters
R. H. Saul, T. P. Lee, and C. A. Burus, Light-Emitting Device Design
C. L. Zipfel, Light-Emitting Diode Reliability
Tien Pei Lee and Tingye Li, LED-Based Multimode Lightwave Systems
Kinichiro Ogawa, Semiconductor Noise-Mode Partition Noise

Part D

Federico Capasso, The Physics of Avalanche Photodiodes
T. P. Pearsall and M. A. Pollack, Compound Semiconductor Photodiodes
Takao Kaneda, Silicon and Germanium Avalanche Photodiodes
S. R. Forrest, Sensitivity of Avalanche Photodetector Receivers for High-Bit-Rate Long-Wavelength Optical Communication Systems
J. C. Campbell, Phototransistors for Lightwave Communications

Part E

Shyh Wang, Principles and Characteristics of Integratable Active and Passive Optical Devices
Shlomo Margalit and Amnon Yariv, Integrated Electronic and Photonic Devices
Takaaki Mukai, Yoshihisa Yamamoto, and Tatsuya Kimura, Optical Amplification by Semiconductor Lasers

Volume 23 Pulsed Laser Processing of Semiconductors

R. F. Wood, C. W. White, and R. T. Young, Laser Processing of Semiconductors: An Overview
C. W. White, Segregation, Solute Trapping, and Supersaturated Alloys
G. E. Jellison, Jr., Optical and Electrical Properties of Pulsed Laser-Annealed Silicon
R. F. Wood and G. E. Jellison, Jr., Melting Model of Pulsed Laser Processing
R. F. Wood and F. W. Young, Jr., Nonequilibrium Solidification Following Pulsed Laser Melting
D. H. Lowndes and G. E. Jellison, Jr., Time-Resolved Measurements During Pulsed Laser Irradiation of Silicon
D. M. Zehner, Surface Studies of Pulsed Laser Irradiated Semiconductors
D. H. Lowndes, Pulsed Beam Processing of Gallium Arsenide
R. B. James, Pulsed CO_2 Laser Annealing of Semiconductors
R. T. Young and R. F. Wood, Applications of Pulsed Laser Processing

Volume 24 Applications of Multiquantum Wells, Selective Doping, and Superlattices

C. Weisbuch, Fundamental Properties of III–V Semiconductor Two-Dimensional Quantized Structures: The Basis for Optical and Electronic Device Applications
H. Morkoc and H. Unlu, Factors Affecting the Performance of (Al, Ga)As/GaAs and (Al, Ga)As/InGaAs Modulation-Doped Field-Effect Transistors: Microwave and Digital Applications
N. T. Linh, Two-Dimensional Electron Gas FETs: Microwave Applications
M. Abe et al., Ultra-High-Speed HEMT Integrated Circuits
D. S. Chemla, D. A. B. Miller, and P. W. Smith, Nonlinear Optical Properties of Multiple Quantum Well Structures for Optical Signal Processing

F. Capasso, Graded-Gap and Superlattice Devices by Band-gap Engineering
W. T. Tsang, Quantum Confinement Heterostructure Semiconductor Lasers
G. C. Osbourn et al., Principles and Applications of Semiconductor Strained-Layer Superlattices

Volume 25 Diluted Magnetic Semiconductors

W. Giriat and J. K. Furdyna, Crystal Structure, Composition, and Materials Preparation of Diluted Magnetic Semiconductors
W. M. Becker, Band Structure and Optical Properties of Wide-Gap $A^{II}_{1-x}Mn_xB^{VI}$ Alloys at Zero Magnetic Field
Saul Oseroff and Pieter H. Keesom, Magnetic Properties: Macroscopic Studies
T. Giebultowicz and T. M. Holden, Neutron Scattering Studies of the Magnetic Structure and Dynamics of Diluted Magnetic Semiconductors
J. Kossut, Band Structure and Quantum Transport Phenomena in Narrow-Gap Diluted Magnetic Semiconductors
C. Riqaux, Magnetooptics in Narrow Gap Diluted Magnetic Semiconductors
J. A. Gaj, Magnetooptical Properties of Large-Gap Diluted Magnetic Semiconductors
J. Mycielski, Shallow Acceptors in Diluted Magnetic Semiconductors: Splitting, Boil-off, Giant Negative Magnetoresistance
A. K. Ramdas and S. Rodriquez, Raman Scattering in Diluted Magnetic Semiconductors
P. A. Wolff, Theory of Bound Magnetic Polarons in Semimagnetic Semiconductors

Volume 26 III–V Compound Semiconductors and Semiconductor Properties of Superionic Materials

Zou Yuanxi, III–V Compounds
H. V. Winston, A. T. Hunter, H. Kimura, and R. E. Lee, InAs-Alloyed GaAs Substrates for Direct Implantation
P. K. Bhattacharya and S. Dhar, Deep Levels in III–V Compound Semiconductors Grown by MBE
Yu. Ya. Gurevich and A. K. Ivanov-Shits, Semiconductor Properties of Superionic Materials

Volume 27 High Conducting Quasi-One-Dimensional Organic Crystals

E. M. Conwell, Introduction to Highly Conducting Quasi-One-Dimensional Organic Crystals
I. A. Howard, A Reference Guide to the Conducting Quasi-One-Dimensional Organic Molecular Crystals
J. P. Pouget, Structural Instabilities
E. M. Conwell, Transport Properties
C. S. Jacobsen, Optical Properties
J. C. Scott, Magnetic Properties
L. Zuppiroli, Irradiation Effects: Perfect Crystals and Real Crystals

Volume 28 Measurement of High-Speed Signals in Solid State Devices

J. Frey and D. Ioannou, Materials and Devices for High-Speed and Optoelectronic Applications
H. Schumacher and E. Strid, Electronic Wafer Probing Techniques
D. H. Auston, Picosecond Photoconductivity: High-Speed Measurements of Devices and Materials
J. A. Valdmanis, Electro-Optic Measurement Techniques for Picosecond Materials, Devices, and Integrated Circuits
J. M. Wiesenfeld and R. K. Jain, Direct Optical Probing of Integrated Circuits and High-Speed Devices

G. Plows, Electron-Beam Probing
A. M. Weiner and R. B. Marcus, Photoemissive Probing

Volume 29 Very High Speed Integrated Circuits: Gallium Arsenide LSI

M. Kuzuhara and T. Nozaki, Active Layer Formation by Ion Implantation
H. Hashimoto, Focused Ion Beam Implantation Technology
T. Nozaki and A. Higashisaka, Device Fabrication Process Technology
M. Ino and T. Takada, GaAs LSI Circuit Design
M. Hirayama, M. Ohmori, and K. Yamasaki, GaAs LSI Fabrication and Performance

Volume 30 Very High Speed Integrated Circuits: Heterostructure

H. Watanabe, T. Mizutani, and A. Usui, Fundamentals of Epitaxial Growth and Atomic Layer Epitaxy
S. Hiyamizu, Characteristics of Two-Dimensional Electron Gas in III–V Compound Heterostructures Grown by MBE
T. Nakanisi, Metalorganic Vapor Phase Epitaxy for High-Quality Active Layers
T. Mimura, High Electron Mobility Transistor and LSI Applications
T. Sugeta and T. Ishibashi, Hetero-Bipolar Transistor and Its LSI Application
H. Matsueda, T. Tanaka, and M. Nakamura, Optoelectronic Integrated Circuits

Volume 31 Indium Phosphide: Crystal Growth and Characterization

J. P. Farges, Growth of Discoloration-free InP
M. J. McCollum and G. E. Stillman, High Purity InP Grown by Hydride Vapor Phase Epitaxy
T. Inada and T. Fukuda, Direct Synthesis and Growth of Indium Phosphide by the Liquid Phosphorous Encapsulated Czochralski Method
O. Oda, K. Katagiri, K. Shinohara, S. Katsura, Y. Takahashi, K. Kainosho, K. Kohiro, and R. Hirano, InP Crystal Growth, Substrate Preparation and Evaluation
K. Tada, M. Tatsumi, M. Morioka, T. Araki, and T. Kawase, InP Substrates: Production and Quality Control
M. Razeghi, LP-MOCVD Growth, Characterization, and Application of InP Material
T. A. Kennedy and P. J. Lin-Chung, Stoichiometric Defects in InP

Volume 32 Strained-Layer Superlattices: Physics

T. P. Pearsall, Strained-Layer Superlattices
Fred H. Pollack, Effects of Homogeneous Strain on the Electronic and Vibrational Levels in Semiconductors
J. Y. Marzin, J. M. Gerárd, P. Voisin, and J. A. Brum, Optical Studies of Strained III–V Heterolayers
R. People and S. A. Jackson, Structurally Induced States from Strain and Confinement
M. Jaros, Microscopic Phenomena in Ordered Superlattices

Volume 33 Strained-Layer Superlattices: Materials Science and Technology

R. Hull and J. C. Bean, Principles and Concepts of Strained-Layer Epitaxy
William J. Schaff, Paul J. Tasker, Mark C. Foisy, and Lester F. Eastman, Device Applications of Strained-Layer Epitaxy
S. T. Picraux, B. L. Doyle, and J. Y. Tsao, Structure and Characterization of Strained-Layer Superlattices

E. Kasper and F. Schaffler, Group IV Compounds
Dale L. Martin, Molecular Beam Epitaxy of IV–VI Compound Heterojunctions
Robert L. Gunshor, Leslie A. Kolodziejski, Arto V. Nurmikko, and Nobuo Otsuka, Molecular Beam Epitaxy of II–VI Semiconductor Microstructures

Volume 34 Hydrogen in Semiconductors

J. I. Pankove and N. M. Johnson, Introduction to Hydrogen in Semiconductors
C. H. Seager, Hydrogenation Methods
J. I. Pankove, Hydrogenation of Defects in Crystalline Silicon
J. W. Corbett, P. Deák, U. V. Desnica, and S. J. Pearton, Hydrogen Passivation of Damage Centers in Semiconductors
S. J. Pearton, Neutralization of Deep Levels in Silicon
J. I. Pankove, Neutralization of Shallow Acceptors in Silicon
N. M. Johnson, Neutralization of Donor Dopants and Formation of Hydrogen-Induced Defects in n-Type Silicon
M. Stavola and S. J. Pearton, Vibrational Spectroscopy of Hydrogen-Related Defects in Silicon
A. D. Marwick, Hydrogen in Semiconductors: Ion Beam Techniques
C. Herring and N. M. Johnson, Hydrogen Migration and Solubility in Silicon
E. E. Haller, Hydrogen-Related Phenomena in Crystalline Germanium
J. Kakalios, Hydrogen Diffusion in Amorphous Silicon
J. Chevallier, B. Clerjaud, and B. Pajot, Neutralization of Defects and Dopants in III–V Semiconductors
G. G. DeLeo and W. B. Fowler, Computational Studies of Hydrogen-Containing Complexes in Semiconductors
R. F. Keifl and T. L. Estle, Muonium in Semiconductors
C. G. Van de Walle, Theory of Isolated Interstitial Hydrogen and Muonium in Crystalline Semiconductors

Volume 35 Nanostructured Systems

M. Reed, Introduction
H. van Houten, C. W. J. Beenakker, and B. J. van Wees, Quantum Point Contacts
G. Timp, When Does a Wire Become an Electron Waveguide?
M. Büttiker, The Quantum Hall Effect in Open Conductors
W. Hansen, J. P. Kotthaus, and U. Merkt, Electrons in Laterally Periodic Nanostructures